T0335725

METHODS IN MOLECULAR BIOLOGY

Series Editor
John M. Walker
School of Life and Medical Sciences
University of Hertfordshire
Hatfield, Hertfordshire, AL10 9AB, UK

For further volumes:
http://www.springer.com/series/7651

Mouse Models for Drug Discovery

Methods and Protocols

Second Edition

Edited by

Gabriele Proetzel

Takeda Pharmaceutical Company, Cambridge, MA, USA

Michael V. Wiles

The Jackson Laboratory, Bar Harbor, ME, USA

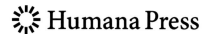 Humana Press

Editors
Gabriele Proetzel
Takeda Pharmaceutical Company
Cambridge, MA, USA

Michael V. Wiles
The Jackson Laboratory
Bar Harbor, ME, USA

ISSN 1064-3745   ISSN 1940-6029   (electronic)
Methods in Molecular Biology
ISBN 978-1-4939-3659-5   ISBN 978-1-4939-3661-8   (eBook)
DOI 10.1007/978-1-4939-3661-8

Library of Congress Control Number: 2016933104

Printed on acid-free paper

This Humana Press imprint is published by Springer Nature
The registered company is Springer Science+Business Media LLC New York

Preface

The cost of drug development continues to climb, reaching in 2014 an estimated $2.6 billion per approved compound to reach market. A major cost contributor in this exorbitant expense is the number of potential drugs that fail during clinical development. Recently, it has been suggested that this high failure rate can perhaps be addressed, at least in part, by improving preclinical research with a focus on more accurate and imaginative use of animal models and a greater understanding of the role of genetics in drug interactions. Linked with this, and as our understanding of disease biology grows deeper, animals models are growing in sophistication and are increasingly more capable of emulating aspects of human disease. Combined, these developments are having a positive impact in reducing or at least slowing the drug development cost spiral.

It is clear that drug interactions with living organisms are a complex interplay of target and response. This interplay is the direct result of, and is defined by, an organism's genetic makeup plus its environment. In this volume, we present the mouse as a principal tool or, perhaps better stated, reagent. Mouse models can be completely genetically defined and as such can provide clear repeatable phenotypes and experimental data. In this regard, Dr. Michael Festing describes here how using good experimental design plus defined animals is critical to the acquisition of meaningful and reproducible data.

In the development of novel model animals, a major break-through has been in genetic engineering. Systems now exist allowing for simple, efficient, and near universally precise genetic manipulation directly in any organism, including the mouse. These genetic editing tools were first based on Zinc Finger Nuclease (ZFN), then Transcription Activator-Like Effector Nucleases (TALEN), and more recently, the development of a simple and highly efficient system based on Clustered Regularly Interspaced Short Palindromic Repeats (CRISPR). Using these genetic editing tools, it is possible to create novel disease models based on, for example, GWAS data within months. Such tools can also be applied to preexisting models facilitating their sequential genetic modification and hence further development and refinement (Chapter 2, Low et al.). These newly created mouse strains require economic archiving and/or safeguarding against diseases or other disasters which can impact animal rooms and, for this, sperm cryopreservation is a simple solution (Chapter 3, Low et al.).

The creation and use of humanized mice, even with the intrinsic limitations of specific organ systems or proteins, allows fast and direct translation to the human condition. Humanization can be loosely divided into two often overlapping approaches: (1) the addition, modification, and/or replacement of mouse genes with human, thus emulating human conditions, or (2) engraftment of human cells into immunocompromised or often genetically "conditioned" recipients facilitating tissue/cell humanization, for example, hematopoietic stem cells (HSC) or human hepatocytes. Such humanized animals have multiple uses in both disease and drug research, and a few examples are outlined here: Chapters by Brehm, Ploss, Roopenian, and Wiles.

Mouse models can be used in a wide field of disease areas. Here we have included models of type 1 and 2 diabetes (Serreze and Baribault), cardiovascular disease (Howles), skin disorders (Sundberg), cancer (Hannun and Li), neurodegenerative diseases (Janus), neuromuscular diseases (Burgess), and retinal disorders (Krebs and Chang). Behavioral models also exist, and we outline examples for depression, anxiety (Kalueff), and autism (Sukoff Rizzo). Lastly, there is the growing awareness that *we* are not alone, and that our bodies are ecosystems each of which includes and is impacted by our bacterial flora, collectively known as the microbiome (Kashyap).

This new volume contains a broad review of the use of mouse models in drug discovery and research and is aimed to equip the reader with the overview of possibilities of mice in drug development.

We would like to thank all the contributors, their discussions, and their patience in making this an important volume on mouse models and drug discovery.

Cambridge, MA, USA   *Gabriele Proetzel*
Bar Harbor, ME, USA   *Michael V. Wiles*

Contents

Contributors

KEN EDWIN ARYEE • *Program in Molecular Medicine, The University of Massachusetts Medical School, Worcester, MA, USA*

HELENE BARIBAULT • *Ardelyx Inc., Fremont, CA, USA*

CARISA L. BERGNER • *Department of Physiology and Biophysics, Georgetown University Medical School, Washington, DC, USA*

YOGESH BHATTARAI • *Department of Gastroenterology and Hepatology, Mayo Clinic, Rochester, MN, USA*

AGNIESZKA B. BIALKOWSKA • *Department of Medicine, Stony Brook University, Stony Brook, NY, USA*

MICHAEL A. BREHM • *Program in Molecular Medicine, The University of Massachusetts Medical School, Worcester, MA, USA*

ROBERT W. BURGESS • *The Jackson Laboratory, Bar Harbor, ME, USA*

BO CHANG • *The Jackson Laboratory, Bar Harbor, ME, USA*

GREGORY J. CHRISTIANSON • *The Jackson Laboratory, Bar Harbor, ME, USA*

GREGORY A. COX • *The Jackson Laboratory, Bar Harbor, ME, USA*

VICTORIA DELELYS • *Department of Neuroscience and CTRND, McKnight Brain Institute, University of Florida, Gainesville, FL, USA*

TERESA P. DILORENZO • *Department of Microbiology and Immunology, Albert Einstein College of Medicine, Bronx, NY, USA; Division of Endocrinology, Department of Medicine, Albert Einstein College of Medicine, Bronx, NY, USA*

BRETT D. DUFOUR • *Department of Animal Sciences, Purdue University, West Lafayette, IN, USA*

RUPERT J. EGAN • *Department of Physiology and Biophysics, Georgetown University Medical School, Washington, DC, USA*

MICHAEL F.W. FESTING • *c/o Medical Research Council Toxicology Unit, University of Leicester, Leicester, UK*

AMR M. GHALEB • *Department of Medicine, Stony Brook University, Stony Brook, NY, USA*

DALE L. GREINER • *Program in Molecular Medicine, The University of Massachusetts Medical School, Worcester, MA, USA*

YUSUF A. HANNUN • *Department of Medicine, Stony Brook University, Stony Brook, NY, USA; Stony Brook Cancer Center, Stony Brook University, Stony Brook, NY, USA*

PETER C. HART • *Department of Physiology and Biophysics, Georgetown University Medical School, Washington, DC, USA*

SUHEYLA HASGUR • *Program in Molecular Medicine, The University of Massachusetts Medical School, Worcester, MA, USA*

CAROLINA HERNANDEZ • *Department of Neuroscience and CTRND, McKnight Brain Institute, University of Florida, Gainesville, FL, USA*

WANDA HICKS • *The Jackson Laboratory, Bar Harbor, ME, USA*

PHILIP N. HOWLES • *Department of Pathology and Laboratory Medicine, Center for Lipid and Arteriosclerosis Studies, University of Cincinnati College of Medicine, Metabolic Diseases Institute, Cincinnati, OH, USA*

GABRIELA HREBIKOVA • *Department of Molecular Biology, Princeton University, Princeton, NJ, USA*

CHRISTOPHER JANUS • *Department of Neuroscience and CTRND, McKnight Brain Institute, University of Florida, Gainesville, FL, USA*

ALLAN V. KALUEFF • *Department of Physiology and Biophysics, Georgetown University Medical School, Washington, DC, USA; Stress Physiology and Research Center (SPaRC), Georgetown University Medical Center, Washington, DC, USA; Department of Pharmacology, Tulane University Medical Center, LA, USA; Department of Physiology and Biophysics, Georgetown University Medical School, Washington, DC, USA; Research Institute for Marine Drugs and Nutrition, College of Food Science and Technology, Zhanjiang, China; Institute of Translational Biomedicine, St. Petersburg State University, St. Petersburg, Russia; ZENEREI Research Center, Slidell, LA, USA; Chemicotechnological Institute, Ural Federal State University, Ekaterinburg, Russia, New Orleans, LA, USA*

PURNA C. KASHYAP • *Department of Gastroenterology and Hepatology, Mayo Clinic, Rochester, MN, USA*

LLOYD E. KING JR. • *Division of Dermatology, Department of Medicine, Vanderbilt Medical Center, Nashville, TN, USA*

MARK P. KREBS • *The Jackson Laboratory, Bar Harbor, ME, USA*

JOHN KULIK • *The Jackson Laboratory, Bar Harbor, ME, USA*

PETER M. KUTNY • *The Jackson Laboratory, Bar Harbor, ME, USA*

JUSTIN L. LAPORTE • *Stress Physiology and Research Center (SPaRC), Department of Physiology and Biophysics, Georgetown University Medical School, Washington, DC, USA*

SHAOGUANG LI • *Division of Hematology/Oncology, Department of Medicine, University of Massachusetts Medical School, Worcester, MA, USA*

BENJAMIN E. LOW • *The Jackson Laboratory, Bar Harbor, ME, USA*

MARIJKE NIENS • *The Jackson Laboratory, Bar Harbor, ME, USA*

PATSY M. NISHINA • *The Jackson Laboratory, Bar Harbor, ME, USA*

LINA M. OBEID • *Northport VA Medical Center, Northport, NY, USA; Department of Medicine, Stony Brook University, Stony Brook, NY, USA*

CONG PENG • *School of Dental Medicine, Tufts University, Boston, MA, USA*

ALEXANDER PLOSS • *Department of Molecular Biology, Princeton University, Princeton, NJ, USA*

C. HERBERT PRATT • *The Jackson Laboratory, Bar Harbor, ME, USA*

GABRIELE PROETZEL • *Takeda Pharmaceutical Company, Cambridge, MA, USA*

STACEY J. SUKOFF RIZZO • *The Jackson Laboratory, Bar Harbor, ME, USA*

HANNO RODER • *TauTaTis, Inc., Jacksonville, FL, USA*

DERRY C. ROOPENIAN • *The Jackson Laboratory, Bar Harbor, ME, USA*

MARKUS VON SCHAEWEN • *Department of Molecular Biology, Princeton University, Princeton, NJ, USA*

KEVIN L. SEBURN • *The Jackson Laboratory, Bar Harbor, ME, USA*

DAVID V. SERREZE • *The Jackson Laboratory, Bar Harbor, ME, USA*

KEITH SHEPPARD • *The Jackson Laboratory, Bar Harbor, ME, USA*

LEONARD D. SHULTZ • *The Jackson Laboratory, Bar Harbor, ME, USA*

KATHLEEN A. SILVA • *The Jackson Laboratory, Bar Harbor, ME, USA*

AMANDA N. SMOLINSKY • *Department of Physiology and Biophysics, Georgetown University Medical School, Washington, DC, USA*

ASHLEY J. SNIDER • *Northport VA Medical Center, Northport, NY, USA; Department of Medicine, Stony Brook University, Stony Brook, NY, USA*

THOMAS J. SPROULE • *The Jackson Laboratory, Bar Harbor, ME, USA*

JOHN P. SUNDBERG • *The Jackson Laboratory, Bar Harbor, ME, USA*

ROB A. TAFT • *The Jackson Laboratory, Bar Harbor, ME, USA*

HANS WELZL • *Division of Neuroanatomy and Behavior, Institute of Anatomy, University of Zürich, Zürich, Switzerland*

MICHAEL V. WILES • *The Jackson Laboratory, Bar Harbor, ME, USA*

MEI XIAO • *The Jackson Laboratory, Bar Harbor, ME, USA*

VINCENT W. YANG • *Department of Medicine, Stony Brook University, Stony Brook, NY, USA; Department of Physiology and Biophysics, Stony Brook University, Stony Brook, NY, USA*

Chapter 1

Genetically Defined Strains in Drug Development and Toxicity Testing

Michael F.W. Festing

Abstract

There is growing concern about the poor quality and lack of repeatability of many pre-clinical experiments involving laboratory animals. According to one estimate as much as $28 billion is wasted annually in the USA alone in such studies. A decade ago the FDA's "Critical path" white paper noted that "The traditional tools used to assess product safety—animal toxicology and outcomes from human studies—have changed little over many decades and have largely not benefited from recent gains in scientific knowledge. The inability to better assess and predict product safety leads to failures during clinical development and, occasionally, after marketing." Repeat-dose 28-days and 90-days toxicity tests in rodents have been widely used as part of a strategy to assess the safety of drugs and chemicals but their repeatability and power to detect adverse effects have not been formally evaluated.

The guidelines (OECD TG 407 and 408) for these tests specify the dose levels and number of animals per dose but do not specify the strain of animals which should be used. In practice, almost all the tests are done using genetically undefined "albino" rats or mice in which the genetic variation, a major cause of inter-individual and strain variability, is unknown and uncontrolled. This chapter suggests that a better strategy would be to use small numbers of animals of several genetically defined strains of mice or rats instead of the undefined animals used at present. Inbred strains are more stable providing more repeatable data than outbred stocks. Importantly their greater phenotypic uniformity should lead to more powerful and repeatable tests. Any observed strain differences would indicate genetic variation in response to the test substance, providing key data. We suggest that the FDA and other regulators and funding organizations should support research to evaluate this alternative.

Key words Toxicity testing, Preclinical development, Inbred strains, Drug development, Factorial experimental designs, Statistics, Signal/noise ratio, Experimental design, Statistical analysis

1   Introduction

Several papers have been published recently expressing concern over the lack of repeatability of preclinical studies involving laboratory animals. In one study, for example, the authors [1] identified 53 "landmark" papers in cancer research and tried to repeat them. They could only do so in six cases. A similar study [2] was only able to repeat the results of 15 of 67 papers. More than 50 papers have been published claiming that a particular drug alleviated the

Gabriele Proetzel and Michael V. Wiles (eds.), *Mouse Models for Drug Discovery: Methods and Protocols*,
Methods in Molecular Biology, vol. 1438, DOI 10.1007/978-1-4939-3661-8_1, © Springer Science+Business Media New York 2016

symptoms of amyotrophic lateral sclerosis (ALS) in transgenic mice carrying 23 copies of the human Superoxide Dismutase 1 ($SOD1^{G93A}$) gene, the standard mouse model of the disease. But none were effective in humans. A detailed study [3] of the papers found that there were a number of confounding factors, such as the copy number of the transgene, which could influence the outcome of the experiments. A new and improved protocol was used to re-screen all 50 plus another 20 candidate drugs. It took 5 years and 18,000 mice. It was found that *none* of the drugs were effective in the mouse model and concluded that "The majority of published effects are most likely measurements of noise in the distribution of survival means as opposed to actual drug effects." According to one estimate, badly designed preclinical experiments are leading to an annual waste of research resources of $28 billion in the United States alone [4].

The exact reason for lack of repeatability of so many experiments is not entirely clear. One contributory factor is probably the lack of training of investigators in experimental design and statistics. Experiments where the subjects are not correctly randomized and the investigators are not blinded can lead to false conclusions. Excessive inter-individual variation, arising from the use of phenotypically variable animals due to infectious disease, suboptimal environmental conditions, poorly genetically defined animals, or excessive variation in body weight can lead to false negative results. Incorrect statistical analysis or interpretation of the results, technical errors, and unreliable reagents such as poorly utilized or incorrect monoclonal antibodies [5] are additional possible causes. It has also been suggested that some of the problems could be due to the use of the wrong animals [6]. While little is known about the reproducibility of repeat dose toxicity tests, this chapter suggests that the use of the wrong animals may reduce statistical power leading to too many false negative results. The candidate drugs are then rejected following expensive clinical trials.

1.1 The Need to Improve Toxicity Screening

There is a need to improve methods of toxicity screening. According to the FDA 2004 "Critical path" white paper "The traditional tools used to assess product safety—animal toxicology and outcomes from human studies—have changed little over many decades and have largely not benefited from recent gains in scientific knowledge. The inability to better assess and predict product safety leads to failures during clinical development and, occasionally, after marketing" [7]. This is expensive and wasteful and needs to be addressed.

The attrition rate of new chemical entities (excluding "me-too" drugs) has been about 96% including 27% rejected for toxicity and 46% rejected due to lack of efficacy [8]. Clinical trials are considerably more expensive than preclinical testing. Therefore, if the number of misleading results could be decreased the cost of developing new drugs would be substantially reduced [9]. The FDA

"Critical path initiative" and the European Union *Innovative Medicines Initiative (IMI)* [10] both aim to improve methods for developing new drugs, but there appears to have been no attempt to improve the repeat dose rodent tests in order to make them more powerful and more repeatable.

More than 50 years ago Russell and Burch, in their classical book on *The Principles of Humane Experimental Technique* [11], suggested that toxicologists should use small numbers of animals of several inbred strains rather than using outbred stocks in toxicological screening. This was based on principles first formulated by RA Fisher who emphasized the need to control *within group* (i.e. inter-individual) variation because this reduces the ability of the experiment to detect differences between treatment groups. In contrast the effects of *between-group* variation, such as the effect of strain and gender differences, on the outcome of an experiment should be *actively explored* in order to increase the generality of the findings. Clearly, if there are important strain differences and an insensitive strain should happen to have been chosen for a toxicity test, there is a danger of a false negative result. Much of this inter-group variation can be explored by using factorial experimental designs which make it possible to include more than one strain at no extra cost. After more than 50 years and numerous papers [12–17] explaining these principles, it is surely time to explore the ways in which these tests can be improved so as to be more predictive of human responses. The choice of strains should be based on evidence rather than intuition or worse, short term convenience [18].

1.2 Brief Summary of In-Vivo Toxicity Testing

Currently, the preclinical testing of new drugs involves a number of formal experiments required by the regulatory authorities [19]. These include 28-day, 90-day, and two-year studies in rodents usually involving four dose levels (including the control) and both sexes, with ten animals per group, or a total of 80 animals for the short-term tests. Many outcomes are measured. A 90-day study in a small numbers of dogs may also be required. Two-year carcinogenesis studies are usually done in rats and/or mice with both sexes, four dose levels, and 50 animals per group, or a total of 400 animals.

Additional data on one other species, often the mouse, may also be required. More recently, novel methods such as the use of transgenic strains like "Big Blue" [20] and Trp53 knockout mice (B6.129S2-Trp53tm1Tyj/J) are being used for short-term carcinogenicity testing [21]. Reproductive toxicity studies and in some cases multi-generation studies may also be required. Many new/additional non-animal tests are in development such as in-vitro tests for ocular toxicity, skin sensitisation, and DNA damage. The tests depend upon on individual circumstances, and will also differ between pharmaceutical and industrial/environmental chemicals. It is an anachronism that the repeat-dose toxicity tests have remained virtually un-changed for more than half a century.

Traditionally, toxicologists have used the rat because larger samples of organs and tissues can be obtained from this species than from the mouse. However, the majority of assays have now been miniaturized for work on even smaller species, such a *C. elegans* and *Drosophila* and therefore the small size of the mouse is no longer a limitation, except in a few special situations such as when delicate surgery is involved. Many protocols are also available for studying mouse biology such as the 270 standard operating procedures for phenotyping the mouse developed by the EUMORPHIA consortium [22]. Many of these can be used directly by toxicologists and pharmacologists when studying the effects of xenobiotics. There is also extensive historical data on the characteristics of many inbred and genetically modified mouse strains (www.informatics.jax.org and http://jaxmice.jax.org/). Mice are also less expensive to maintain and use less of the test agent than rats. Even the use of techniques such as telemetry are now available for the mouse and new methods of whole body imaging such as SPECT and SPECT/CT [23] favor smaller animals; i.e. the balance is changing. According to one anonymous toxicologist in a major drug firm [24] "Although the rat genome is now available, the knowledge about differences between mouse strains, the huge access to mouse knockouts and the lower cost for maintaining mice will make mice preferred as an experimental species. However, there will be situations where it is very difficult to make measurements in mice, and sometimes this can be done in rats instead." Certainly some 28-day repeat-dose toxicity studies are being done using mice rather than rats.

The National Institute of Environmental Health Sciences (NIEHS) has recognized the value of isogenic mouse strains in identifying genes associated with response to xenobiotics. "One aim of the Host Susceptibility Program is to evaluate chemicals identified as toxicants in the research and testing program and evaluate them in multiple genetically diverse isogenic mouse strains to determine which strains are particularly sensitive or insensitive to the chemicals causing toxicity and associated disease." (http://ntp.niehs.nih.gov). "Ultimately, the National Toxicology Program (NTP) expects to learn more about the key genes and pathways involved in the toxic response and the etiology of disease mediated by substances in our environment. Such an understanding of genes and environment interactions will lead to more specific and targeted research with testing strategies for the NTP scientists to use for predicting the potential toxicity of substances in our environment and their presumptive risk to humans and disease susceptibility."

2   Genetic Types of Laboratory Animals

The albino mouse has been domesticated for thousands of years and is known to be a cross between three subspecies *Mus musculus domesticus, M.m. musculus,* and *M castaneous.* The laboratory rat

(*Rattus norvegicus*) is thought to have been domesticated from occasional albino animals trapped for rat baiting in the eighteenth Century.

Table 1 provides a summary of the properties of outbred stocks and inbred strains.

Table 1
Brief summary of differences between inbred strains and outbred stocks

Inbred strains	Outbred stocks
Isogenic All animals genetically identical. This is also true of F1 hybrids between two inbred strains	Heterogenic Each individual genetically different and unique
Homozygous at all gene loci. Parents and offspring are genetically identical. However F1 hybrids will segregate genetically in the next generation	Heterozygous Do not breed true. May carry recessive genes. Parents and offspring genetically different
Phenotypically uniform Phenotypic variation is essentially only due to nongenetic, environmental factors	Phenotypically variable Phenotypic variation is due both to genetic and environmental factors
Identifiable Each individual can be authenticated by genetic markers. Genetic quality control easy	Not identifiable There is no set of markers which can be used to authenticate an outbred stock so genetic quality control of individuals is impossible
Genetically stable Genetic drift slow and due only to new mutations	Genetically unstable Genetic drift in small colonies can be rapid due to changes in gene frequency caused by selection, random drift, and mutation
Consistent data Extensive, reliable, data on genotype and phenotype of common strains which continues to accumulate	Variable data No data on genotype. Reliability of historical data questionable because of lack of strain authentication
Multi-strain experiments common Strains differences show the inheritance of the phenotype. Often the starting point for identifying sensitivity genes	Single stock experiments most common Most investigators only use a single stock so differences between stocks are not commonly seen and the investigator is unaware of genetic variation in response
Distribution universal Internationally distributed in academic and commercial organizations. Investigators in different countries can use genetically identical animals	Distribution limited Source often limited to one or more commercial breeders. Each colony is unique so investigators in different countries have no access to a particular genotype
Consistent naming Genetic nomenclature well established with extensive curated lists of individual strains and their properties	Inconsistent naming Genetic nomenclature rules often ignored. No curated listings of stock characteristics. Designations unreliable

2.1 Outbred Stocks

The term "stock" is used for outbred and "strain" for inbred colonies. Outbred stocks are usually closed colonies (i.e. no new genetic material is introduced) in which each animal is genetically different and unique [25]. These stocks may be designated by such names as "Sprague–Dawley" or "Wistar" rats and "Swiss" or "CD-1" mice, however there can be variation depending upon their source. Colonies are usually named according to their origin, but these origins are usually somewhat obscure and marked genetic changes can occur within each stock over a period of time. Of particular note, when a new breeding colony is established it may go through a "genetic bottleneck" if it is founded by only a small number (say less than 100) of individuals. This can lead to genetic and phenotypic changes in the colony due to alterations in gene frequency and the levels of heterozygosity. A similar situation arises if the colony is "re-derived" in order to eliminate infectious agents. Genetic changes can also occur as a result of genetic contamination or as a result of deliberate (and at times forgotten) mixing of colonies to improve breeding performance. Stocks can also change as a result of selective breeding. In the absence of predation, starvation, exposure, and disease, found in the wild, animals with the genetic ability to produce the largest litters will leave the most offspring. This is also a character wanted by the breeders. So over a period of many generations litter size has increased. Litter size is positively correlated with body size. As a result, the body weight of outbred stocks of mice and rats has also increased. They are now almost always much larger than inbred strains. Selective breeding is only effective in the presence of genetic variation within the strain. This is absent in inbred strains, so that they have remained un-changed by natural or artificial selection.

2.1.1 The Genetic Properties of Outbred Stocks

Two studies of outbred mouse stocks using DNA-based genetic markers have produced some unexpected results. The first [26] was a survey of 66 commercial outbred mouse colonies using, in most cases, 351 genetic markers and samples of about 20–48 mice per colony. Four of the colonies (6%) were almost inbred with heterozygosities of less than 5% and a further five were relatively inbred with heterozygosities of 5–10%. The mean level was 18%, but it ranged up to 34%. Forty five percent of the total variation was due to differences between colonies, the rest being due to variation within the colonies. Six colonies were re-sampled at least a year later. In one case after 4 years the level of heterozygosity fell from 30 to 5% as a result of re-derivation.

Agglomerative clustering of the colonies produced two large clusters and a couple of smaller ones. The largest cluster had colonies with names including CD-1, ICR, and NMRI, but the other large cluster also had colonies of NMRI ancestry. Stocks differ as a result of differences in gene frequency, rather than as a result of private alleles. Perhaps the most surprising finding was that "...

variation between colonies is large. Fixation index (F_{ST}), a measure of variation within and between populations, is 0.454 (in contrast, human population values are typically less than 0.05)." Thus the total variation among these colonies is substantially greater than the variation found in the whole human population.

There is evidence of a history of genetic bottlenecks as well as "genetic contamination". In the past, breeders have probably been quite careless about starting new colonies with small numbers of breeding stock, and in crossing stocks to enhance breeding performance. Names have also been changed, often for commercial (trade mark) reasons.

The over-all conclusion from this study is that

1. Stock names and designations may have little meaning.

2. Colonies differ widely in the amount of heterozygosity which substantially exceeds that found in humans. This may be because laboratory mice are derived from three subspecies.

3. Genetic change is minimal in large colonies but drastic changes can occur as a result of genetic bottlenecks during re-derivation, the establishment of new colonies, and as a result of genetic contamination.

A study of CD-1 mice from three locations from the same commercial breeder [27] found that patterns of heterozygosity and linkage disequilibrium were similar to that found in wild mice, and the difference between two colonies was similar to that found in two closely related human populations, consistent with being due to a founder effect. Each population was genetically distinct even though they had the same name. The study included inbred strains representative of the three subspecies from which it was possible to estimate that the ancestry of CD-1 mice was 75 % *M. m. domesticus*, 19 % *M. m. musculus* and 6 % *M. m. castaneous*. In a behavioral test of freezing-to-tone scores there were statistically highly significant differences in mice from each of the three locations although it was not possible to determine whether this was due to the genetic or the environmental (locational) differences between the colonies.

In summary, stocks can range from genetically variable outbred to near, but undefined inbred. Unless each batch of animals is individually tested using DNA genetic markers (which would be expensive) investigators using outbred stocks can have no idea about the genetic background of the animals which they are using.

Unfortunately, no similar studies have been done on outbred rats. *Rattus norvegicus*, the laboratory rat has a very different ancestry from mice. The species originated in the Middle East and only immigrated to Europe and, subsequently, the USA in the eighteenth Century. At that time large numbers of rats were trapped for rat baiting in which about 80–100 rats were introduced into an arena where terrier would be introduced. Bets would be taken on the time it took for the terrier to kill all the rats. Occasionally albino rats would be found, and it is thought that it was these that were tamed and kept for show or as pets.

The origin of the laboratory rat has been extensively reviewed [28]. The Wistar rat stock was founded at the Wistar Institute in 1906 and these have been distributed throughout the world, although the breeding colony was sold to a commercial company in 1960. Other established stocks include the Osborne Mendel, the black and white hooded Long-Evans and the Sprague–Dawley. The latter is thought to have been established in about 1925 from a cross between a single hooded male of exceptional size and some female Wistar rats.

There clearly is a substantial amount of genetic variation within many rat outbred stocks, as they are now widely used in genome wide association studies (GWAS). But the stocks used in controlled experiments, such as in toxicity testing, have not been genetically characterized in the same was as in the mouse. Like the mouse, they probably range from being almost inbred to being quite outbred, depending on their recent ancestry. However an investigator who claims to have used "Wistar" or "Sprague–Dawley" rats can have no scientific evidence to support such a claim because there is no set of genetic markers defining these stocks.

2.2   Inbred Strains

An inbred strain is like an *immortal clone of genetically identical individuals*. They are produced by at least 20 generations of brother x sister mating with each individual being derived from a single breeding pair in the 20th or a subsequent generation. As a result the strain is "isogenic" (i.e. all animals within the strain are genetically identical). They are also homozygous at practically all genetic loci, although recent mutations will segregate within the strain until they are eliminated or fixed by further inbreeding [29]. Sometimes these mutations are of biomedical interest leading to new inbred variants [30].

Inbred strains are designated by a code such as F344 or LEW rats or BALB/c or C57BL mice. Different sub-strains (i.e. branches of a strain which are, or are presumed to be, slightly different) are indicated by further codes following a forward slash, e.g. C57BL/6J (www.informatics.jax.org). Nomenclature rules for inbred strains are well established and widely used (*see* www.informatics.jax.org/mgihome/nomen/), and although complex are worth being aware of. Investigators using inbred strains of rats should stick to the correct designation such as F344 *not* "Fischer" and "BN", *not* "Brown Norway".

2.2.1   The Properties of Inbred Strains

Inbred strains (or F1s made from them) are usually phenotypically more uniform for most characteristics of toxicological and pharmaceutical interest than outbred stocks. This is because all animals within a strain are genetically identical. This leads to less "noise" in an experiment, [31] allowing either smaller numbers of animals for each experiment, or providing an experiment which will be more powerful, with less chance of false negative results [18].

Inbred strains are genetically much more stable that outbred stocks. Any accidental genetic contamination is likely to be noticed by a sudden increase in litter size and genetic drift is only due to the rare fixation of new mutations not to changes in gene frequency as found in outbred stocks. However, many of the more widely used inbred strains were exchanged between investigators before they were fully inbred, this lead to the creation of major sublines such as C57BL/6 and C57BL/10 which differed initially because they were separated before all loci were made homozygous. Over time, new spontaneous mutations [11] will either be fixed or eliminated by further brother x sister mating. Non geneticists sometimes get the impression that inbred strains are unstable because of the attention attached to genetic drift. But new mutations, hence genetic drift, are of intrinsic interest to geneticists. But once the strain is fully inbred any further change is in actually quite low. Mutation is the only way in which inbred strains can change whereas outbred stocks can change as a result of mutation and much greater changes in gene frequency. Mutational genetic drift can be essentially eliminated by maintaining banks of frozen embryos and restoring the strains from the freezer regularly [32]. Recovery of a single sibling (brother-sister or parent–child) pair of animals is sufficient to recover the whole strain as each mouse has all the alleles present in the strain. Each inbred strain can be identified and differentiated from other strains using DNA-based genetic markers.

There is extensive data on the characteristics of most mouse and rat strains, and many genetic modifications (GM) and mutations are maintained on an inbred background so as to reduce genetic drift and increase the sensitivity when comparing GM versus wild-type animals. In short, inbred strains are the nearest representation of a "standard analytical grade reagent" that is possible when doing experiments with mammals. More details are given on The Jackson Laboratory web site www.informatics.jax.org.

2.3 Mutant and Genetically Modified Strains

Spontaneous and induced mutations and genetic modifications are of increasing importance in drug development including the study of toxic mechanisms. For example, knockout mice in which various genes associated with resistance to cancer are used to test chemicals for carcinogenesis. Genetic editing has advanced particularly rapidly with advances to directly edit the genome in zygotes (*see* Chapter 2 Low etc). Such genetic modifications should always be done and maintained on defined inbred genetic backgrounds as their phenotype is modified by other "background" genes. If such modifications are maintained on an outbred background (such as sometimes seems to be case with rat GMs) their phenotypic characteristics will change as a result of the individual animal's genetic background, local genetic drift and/or natural selection. Investigators should demand an F344, *not* a "Sprague–Dawley" background strain.

3   The Randomized Controlled Repeat-Dose Toxicity Test

The purpose of this type of toxicity test as defined in guidelines TG 407 and 408 (http://www.fda.gov/downloads/AnimalVeterinary/GuidanceComplianceEnforcement/GuidanceforIndustry/ucm052523.pdf) is to estimate the dose at which a potential new drug or chemical is toxic in the species and strain of animals used, and to obtain information about target organs and nature of the toxicity. The outcomes which are studied usually include histology, hematology, biochemistry, organ weights, urine analysis, and possibly some aspects of behavior. Once the level and nature of the toxicity is established, the potentially safe dose in humans is assessed, often by specifying a dose some orders of magnitude less than that which is toxic in the test animals.

It is assumed that the tests will be done in rats or mice but there is no specification of the strain of animals to be used.

Controlled experiments need to be powerful, i.e. they need to have a good chance of detecting a toxic effect at low levels, if they exist. Power is maximized by having uniform subjects, large sample sizes, and a strain of animals which is sensitive to the test substance. In practice, sample size is fixed by guidelines, usually at ten animals per group but no attention has been paid to sensitivity. As this can rarely be predetermined a sensible strategy is to use more than one strain if this can be done without increasing the total number of animals. The animals should be of uniform weight, age and genotype, and free of infectious disease.

3.1   Inbred Strains Versus Outbred Stocks in Repeat-Dose Toxicity Testing?

With a few exceptions, such as the NTP Carcinogenesis bioassay, toxicologists have always used outbred stocks. Innovation in toxicity testing has not been encouraged by the drug regulators, and drug companies have had little incentive to change, and have not done so. Moreover, for the last 30 or more years there seems to have been no serious discussion of whether inbred or outbred animals should be used. A brief review in 1987 noted a couple of committees which had recommended the use of outbred stocks, but these did not include mammalian geneticists and biostatisticians, and their reasons for recommending the use of outbred stocks were based on intuition rather than evidence [16].

The UK Committee on Toxicity stated (incorrectly) that toxicologists use inbred strains [19] and went on to say that

"A potential disadvantage of such tight controls of experimental conditions is that this approach reduces the chance of detecting an adverse effect that occurs only in a subgroup of the experimental animals. The use of larger groups of more outbred animals might increase the chances of detecting such groups, but this could not be guaranteed."

This is a bad suggestion. Testing whether a compound is toxic and searching for genes associated with individual susceptibility require different experimental designs. In the former case a randomized controlled study is needed to obtain a good estimate of the dose–response relationship and the target organs. In the latter case a genome wide association study (GWAS) is needed. This would only be done if the toxicity test gave a favorable result. It involves a *large* number of genetically heterogeneous animals from a carefully chosen outbred stock known to be genetically heterogeneous and suitable for such studies. *All* of the animals are treated with the toxic compound. Individual responses to the test compound are then associated with individual single nucleotide polymorphism (SNPs) genetic markers. Any attempt to combine the two objectives into a single experiment is likely to result in failure to achieve either objective.

As another example, an anonymous toxicologist suggested that "The variability of toxicity obtained in less-well-defined animals is a strength in itself, not a problem, when trying to predict safety margin in the non-isogenic human population." Again, this confuses the two objectives discussed above. In a controlled experiment genetic variation is like any other type of variation. Toxicologists don't use animals which are very variable in body weight on the grounds that humans differ in body weight. They know that using variable animals leads to bad experiments which lack the ability to detect treatment effects.

Technically an experiment involving two or more factors such as "treatment" and "strain" is known as a factorial design. According to RA Fisher [33] by using a factorial design "…. *an experimental investigation, at the same time as it is made more comprehensive, may also be made more efficient if by more efficient we mean that more knowledge and a higher degree of precision are obtainable by the same number of observations.*" The use of more than one inbred strain increases the power of the experiment in two ways. First, by the better control of phenotypic variation found in inbred strains so that there is less "noise"; and second by decreasing the chance that the strain or stock which is chosen is resistant to the test compound.

3.2 Choice of Strains

The NTP carcinogenesis bioassay used B6C3F1 hybrid mice and used F344 inbred rats for over 30 years. The B6C3F1 mice gradually became more obese over time, but this was due to environmental effects (possibly changes in diets or better husbandry) rather than genetic drift as breeding nuclei had been maintained as frozen embryos, preventing any genetic change.

Over the same period the F344 rats became less fecund and developed sporadic seizures and idiopathic chylothorax. Whether this was also due to environmental changes or genetic drift is unknown, but the NTP decided to change to outbred Wistar Han rats which are widely used in toxicity testing [34]. These have proved to be unsatisfactory because of poor breeding performance when transported from the breeder as plugged females [35]. Accordingly, they changed again to the Sprague–Dawley rat from Charles River. Whether this is a good choice remains to be determined as there is a tendency for this stock to become obese and diabetic, [36] and it is known to be resistant to certain steroidal compounds such as diethylstilboestrol [37].

It is difficult to find suitable long lived strains for carcinogenesis studies. Invariably there is some condition which affects old animals in each strain or stock. There is probably no single outbred stock or inbred strain which is ideal. But there is considerable literature on the characteristics of inbred mouse strains, and it would make sense to choose strains with no obvious defects such as a short lifespan due to a particular type of tumor or disease.

3.3 A Note on the Statistical Analysis of 90-Day Toxicity Tests

The statistical analysis of toxicity tests poses problems due to the large number of outcomes which are measured. These include hematology, clinical chemistry, organ weights and often, some types of behavior. A total of 80–100 data points per animal is not unusual.

Each character is usually analyzed separately using an analysis of variance. Post-*hoc* comparisons of control versus treated means using Dunnett's test, which correct for multiple testing across dose levels, may be used. In some cases nonparametric tests are used, such as when the variances are heterogeneous or the residuals are not normally distributed.

Unfortunately, it is not possible to apply corrections for multiple testing of each separate character as these are correlated to an unknown extent. As a result, it is not always clear whether a statistically significant difference between groups is real or whether it is one of the 5 % false positives expected when using a 5 % significance level. A subjective estimate has then to be made of whether a "significant" difference between a treated and control mean is a real effect or whether it is a false positive. Also, effects which just fail to be statistically significant at $p = 0.05$ will usually be ignored even though they might be real and of toxicological importance.

These problems led to the suggestion [38, 39] that toxicity test results should be subjected to an "extended" statistical analysis which emphasizes the magnitude of the response as well as its statistical significance and provides good graphical presentation of the results. The means of each character at each dose level are converted to a standardized effect size (SES) by subtracting the control mean and dividing by the pooled standard deviation. These can be plotted to show the magnitude of the response for each character and the absolute values of the responses can be averaged to give a mean response across all or any chosen subset of characters. So this places the emphasis on the magnitude of the mean response across all characters and whether it is statistically significant. A "bootstrap" test can be used to determine whether differences in responses between dose levels are statistically significant. This "extended" analysis also provides a good graphical representation of the results but still allows access to the usual presentation of means and standard deviations for each dose and for each character.

3.4 Example of a Multi-strain Assay

No comparative study of the use of inbred strains versus outbred stocks in a repeat dose toxicity test has ever been published. There are very few papers describing controlled experiments of any sort in which both inbred and outbred stocks have been used. However, a study of the effects of chloramphenicol succinate administered to mice of four inbred strains and one outbred stock was published some years ago. Six dose levels were used with eight mice per group, and ten hematological parameters were measured. This experiment provides the nearest available example to a repeat dose toxicity test. Full details are given elsewhere [40]. Means and standard deviations for each outcome were converted to standardized effect sizes and mean absolute responses were calculated [38]. Figure 1 shows these responses in standard deviation units averaged across all mice and all hematological parameters in the four inbred strains, the outbred CD-1, and a multi-strain group made up from the records of two animals, taken at random, of each of the four inbred strains.

All of the inbred strains and the multi-strain group were more sensitive than the outbred stock (i.e. responded significantly at lower dose levels). Strains CBA, C3H, and the Multi-strain group showed significant ($p < 0.05$) differences in response between dose levels 1 and all other doses. In BALB/c doses 3–5 differed significantly from dose 1. The least sensitive inbred strains was C57BL/6 in which only doses 4 and 5 differed significantly from dose 1. The outbred stock CD-1 was less sensitive than any of the inbred strains, with only dose 5 being significantly different from dose 1. These results show clearly that the inbred strains were detecting hematological responses to chloramphenicol at lower doses than was the outbred stock.

4   Conclusions

There is increasing evidence that many preclinical experiments are producing results which are not reproducible, leading to a waste of resources. Much greater attention now needs to be given

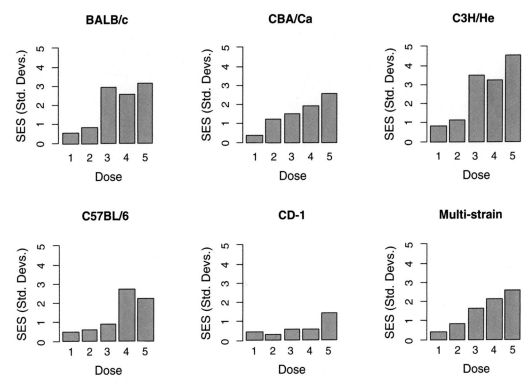

Fig. 1 Mean absolute response (in standard deviation units) to treatment with chloramphenicol succinate at dose levels of 500–2500 mg/kg, averaged across ten hematological parameters. Responses are shown for four inbred strains (BALB/c, CBA/Ca, C3H/He, and C57BL/6), one outbred stock (CD-1), and a multi-strain group consisting of two mice of each of the four inbred strains. Differences between dose level 1 and dose levels 2–5 were statistically significant ($p < 0.05$) using a bootstrap test in strains CBA, C3H, and the Multi-strain group. In BALB/c dose levels 3–5 were significantly different from dose level 1. In C57BL dose levels 4 and 5 differed significantly from dose level 1. In CD-1 only dose level 5 differed significantly from dose level 1

to the design and conduct of preclinical studies if this waste is to be reduced.

Most toxicity studies are based on published guidelines and have hardly changed in several decades. However, that should not make them exempt from further scrutiny, particularly in light of modern genetic developments.

This chapter suggests that repeat dose toxicity experiments are poorly designed because they use the wrong sort of animals. It makes no sense to do careful experiments to GLP standards where the main reagent in the studies, the animal, is completely undefined. Thousands of experiments are done each year using so called "Sprague–Dawley" or "Wistar" rats but if a referee asked the authors of one of these papers to substantiate the claim that they used Sprague–Dawley rats, they would be unable to do so because there is no specification by which a Sprague–Dawley rat can be identified. Too often everyone just depends on the label on the box with no real knowledge of what they are working with.

Genetic studies of outbred stocks of mice show that: (1) Some "outbred" colonies are quite inbred while others are not. (2) Stocks with the same name from the same breeder but different locations can be genetically and phenotypically different, (3) Stocks can change drastically as a result of genetic bottlenecks, such as when a new colony is established or the stock is re-derived to eliminate disease.

Controlled experiments require uniform animals and a carefully controlled environment in order to maximize the statistical power of the experiment. Controlled genetic variation can be included by using several strains in a factorial experimental design without reducing statistical power or increasing the number of animals. However, *uncontrolled* genetic variation, as found in outbred stocks, simply leads to a loss of power and an increased chance of a false negative result.

As far as possible in a controlled experiment animals should be of the same weight and age, have identical environments and diets, be free of disease (i.e. SPF) and be genetically uniform. The anonymous toxicologist who stated that "The variability of toxicity obtained in less-well-defined animals is a strength in itself, not a problem, when trying to predict safety margin in the non-isogenic human population" has clearly failed to grasp these elementary experimental design principles. If the experimental animals are highly variable, then this variation will hide differences between treatments, leading to false negative results.

It clearly *is* sensible to want to study individual variation. But this should be done using GWAS where all animals are given the same dose of the test substance and variation in characters of interest can be associated with the genetic segregation of marker genes. A carefully selected outbred stock, or genetically engineered inbred strains should be used for these studies. But this is expensive and should only be done *after* a, randomized controlled toxicity test (preferably using several inbred strains) has shown that the test substance has an acceptable safety profile.

We now know that many preclinical experiments are giving unreproducible results. Regulatory preclinical toxicity tests have hardly changed in many decades. It is time for the regulators to ensure that the experiments they specify, as part of an application to market a new drug or chemical, are designed to the highest scientific standards if they are not to be complicit in a waste of money and other scientific resources. They should be investigating the advantages and possible disadvantages of using genetically defined animals in these studies.

References

1. Begley CG, Ellis LM (2012) Drug development: raise standards for preclinical cancer research. Nature 483:531–533

2. Prinz F, Schlange T, Asadullah K (2011) Believe it or not: how much can we rely on published data on potential drug targets? Nat Rev Drug Discov 10:712

3. Scott S, Kranz JE, Cole J et al (2008) Design, power, and interpretation of studies in the standard murine model of ALS. Amyotroph Lateral Scler 9:4–15

4. Freedman LP, Cockburn IM, Simcoe TS (2015) The economics of reproducibility in preclinical research. PLoS Biol 13:e1002165

5. Baker M (2015) Reproducibility crisis: blame it on the antibodies. Nature 521:274–276

6. Collins FS, Tabak LA (2014) Policy: NIH plans to enhance reproducibility. Nature 505:612–613

7. Food and Drug Administration (2004) Challenge and opportunity on the critical path to new medical products. [electronic article]. http://www.fda.gov/oc/initiatives/critical-path/whitepaper html.

8. Caldwell GW, Ritchie DM, Masucci JA et al (2001) The new pre-preclinical paradigm: compound optimization in early and late phase drug discovery. Curr Top Med Chem 1:353–366

9. Garner JP (2014) The significance of meaning: why do over 90% of behavioral neuroscience results fail to translate to humans, and what can we do to fix it? ILAR J 55:438–456

10. Innovative Medicines Initiative (2008) The innovative medicines initiative. http://www.imi.europa.eu/

11. Russell WMS, Burch RL (1959) The principles of humane experimental technique. Universities Federation for Animal Welfare, Potters Bar, England, Special Edition

12. Festing MF (2010) Inbred strains should replace outbred stocks in toxicology, safety testing, and drug development. Toxicol Pathol 38:681–690

13. Festing MFW (1975) A case for using inbred strains of laboratory animals in evaluating the safety of drugs. Food Cosmet Toxicol 13:369–375

14. Festing MFW (1980) The choice of animals in toxicological screening: inbred strains and the factorial design of experiment. Acta Zool Pathol Antverp 75:117–131

15. Festing MFW (1986) The case for isogenic strains in toxicological screening. Arch Toxicol Suppl 9:127–137

16. Festing MFW (1987) Genetic factors in toxicology: implications for toxicological screening. CRC Crit Rev Toxicol 18:1–26

17. Festing MFW (1990) Contemporary issues in toxicology: use of genetically heterogeneous rats and mice in toxicological research: a personal perspective. Toxicol Appl Pharmacol 102:197–204

18. Festing MF (2014) Evidence should trump intuition by preferring inbred strains to outbred stocks in preclinical research. ILAR J 55:399–404

19. Committee on Toxicity (2007) Variability and uncertainty in toxicology of chemicals in food, consumer products and the environment. COT, London, Food Standards Agency

20. Ashby J, Tinwell H (1994) Use of transgenic mouse lacI/Z mutation assays in genetic toxicology. Mutagenesis 9(3):179–181

21. Floyd E, Mann P, Long G et al (2002) The Trp53 hemizygous mouse in pharmaceutical development: points to consider for pathologists. Toxicol Pathol 30:147–156

22. Brown SD, Chambon P, de Angelis MH (2005) EMPReSS: standardized phenotype screens for functional annotation of the mouse genome. Nat Genet 37:1155

23. Franc BL, Acton PD, Mari C et al (2008) Small-animal SPECT and SPECT/CT: important tools for preclinical investigation. J Nucl Med 49:1651–1663

24. Petit-Zeman S (2004) Rat genome sequence reignites preclinical model debate. Nat Rev Drug Discov 3:287–288

25. Chia R, Achilli F, Festing MF et al (2005) The origins and uses of mouse outbred stocks. Nat Genet 37:1181–1186

26. Yalcin B, Nicod J, Bhomra A et al (2010) Commercially available outbred mice for genome-wide association studies. PLoS Genet 6:e1001085

27. Aldinger KA, Sokoloff G, Rosenberg DM et al (2009) Genetic variation and population substructure in outbred CD-1 mice: implications for genome-wide association studies. PLoS One 4(3):e4729

28. Lindsey JR (1979) Historical Foundatins. In: Baker HJ, Lindsey JR, Weisbroth SH (eds) The laboratory rat. Academic, New York, pp 1–36

29. Festing MFW (1979) Inbred strains in biomedical research. Macmillan Press, London, Basingstoke

30. Taft RA, Davisson M, Wiles MV (2006) Know thy mouse. Trends Genet 22:649–653

31. Festing MFW (1976) Phenotypic variability of inbred and outbred mice. Nature 263:230–232

32. Stevens JC, Banks GT, Festing MF et al (2007) Quiet mutations in inbred strains of mice. Trends Mol Med 13:512–519

33. Fisher RA (1960) The design of experiments. Hafner Publishing Company, Inc., New York

34. King-Herbert A, Thayer K (2006) NTP workshop: animal models for the NTP rodent cancer bioassay: stocks and strains--should we switch? Toxicol Pathol 34:802–805

35. Pritchet K, Clifford CB, Festing MF (2013) The effects of shipping on early pregnancy in laboratory rats. Birth Defects Res B Dev Reprod Toxicol 98:200–205

36. Keenan KP, Hoe CM, Mixson L et al (2005) Diabesity: a polygenic model of dietary-induced obesity from ad libitum overfeeding of Sprague–Dawley rats and its modulation by moderate and marked dietary restriction. Toxicol Pathol 33:650–674

37. Richter CA, Birnbaum LS, Farabollini F et al (2007) In vivo effects of bisphenol A in laboratory rodent studies. Reprod Toxicol 24:199–224

38. Festing MF (2014) Extending the statistical analysis and graphical presentation of toxicity test results using standardized effect sizes. Toxicol Pathol 42:1238–1249

39. Festing MF (2014) The extended statistical analysis of toxicity tests using standardised effect sizes (SESs): a comparison of nine published papers. PLoS One 9(11):e112955

40. Festing MFW, Diamanti P, Turton JA (2001) Strain differences in haematological response to chloramphenicol succinate in mice: implications for toxicological research. Food Chem Toxicol 39:375–383

Chapter 2

Simple, Efficient CRISPR-Cas9-Mediated Gene Editing in Mice: Strategies and Methods

Benjamin E. Low, Peter M. Kutny, and Michael V. Wiles

Abstract

Genetic modification of almost any species is now possible using approaches based on targeted nucleases. These novel tools now bypass previous limited species windows, allowing precision nucleotide modification of the genome at high efficiency, rapidly and economically. Here we focus on the modification of the mouse genome; the mouse, with its short generation time and comparatively low maintenance/production costs is the perfect mammal with which to probe the genome to understand its functions and complexities. Further, using targeted nucleases combined with homologous recombination, it is now possible to precisely tailor the genome, creating models of human diseases and conditions directly and efficiently in zygotes derived from any mouse strain. Combined these approaches make it possible to sequentially and progressively refine mouse models to better reflect human disease, test and develop therapeutics. Here, we briefly review the strategies involved in designing targeted nucleases (sgRNAs) providing solutions and outlining in detail the practical processes involved in precision targeting and modification of the mouse genome and the establishing of new precision genetically modified mouse lines.

Key words Cas9, CRISPR, Genome editing, RNA-guided endonucleases, Genetic engineering, Mouse model, Microinjection, transgenic, Humanization, Transgenic, sgRNA

1 Introduction

Since the late 1980s mouse Embryonic Stem Cells (ES) have been the primary tool in engineering genetically modified animals [1, 2]. Although powerful in its time, the use of ES cells to create these animals carried a number of limitations including the poor ES cell strain (and species) availability, the high skill sets needed, and the length of time required to obtain a modified strain. With the advent of targeted nuclease technologies, genetic editing directly in the mouse zygote has become universally efficient, rapid, and economic [3–5]. Direct targeted nuclease modification of zygotes was first demonstrated with Zinc Finger Nucleases (ZFN) [6, 7]. This innovative approach was rapidly followed by the development and use of Transcription Activator-Like Effector Nucleases (TALEN) [8, 9]. Both these

Gabriele Proetzel and Michael V. Wiles (eds.), *Mouse Models for Drug Discovery: Methods and Protocols*,
Methods in Molecular Biology, vol. 1438, DOI 10.1007/978-1-4939-3661-8_2, © Springer Science+Business Media New York 2016

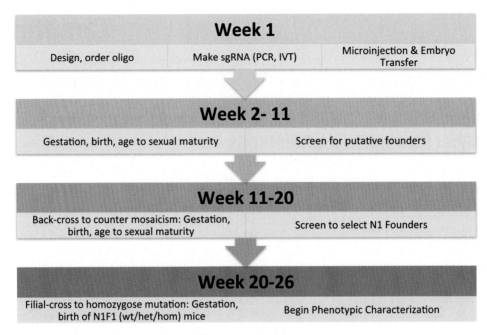

Fig. 1 Time line to construct genetically modified mouse strain

systems are based on custom engineered proteins designed to bind to defined target DNA sequences, leading to a targeted double-stranded DNA cut. As such, their construction is complex, requiring time and a fair degree of skill. In sharp contrast, the CRISPR-Cas9 (Clustered Regularly Interspaced Short Palindromic Repeats, and CRISPR Associated protein 9) [10, 11] system specificity relies on a custom synthetic RNA and its base pairing to the DNA target. These custom RNAs are exceptionally simple to design and to construct. These characteristics, coupled with the observed accuracy and efficiency of genome targeted cutting makes CRISPR-Cas9 the predominant methodology for genetic editing in mouse and other species [4, 5, 12, 13]. Here, we outline the time line (Fig. 1), practical steps and challenges in constructing and using targeted nuclease technologies enabling simple genetic editing directly in mouse zygotes, leading to founder offspring and germline transmission of desired genetic modifications.

1.1 CRISPR-Cas9 and Genome Editing in Zygotes

CRISPR-Cas9 is an RNA-guided endonuclease comprised of two principle parts, a universal DNA endonuclease (Cas9 protein) and a custom-designed synthetic single guide RNA (sgRNA). The sgRNA complexes with and acts as a targeting guide for the Cas9 nuclease. At the time of writing the most commonly used Cas9 nuclease is derived from *Streptococcus pyogenes* (SpCas9). However, novel Cas9 orthologs and modified versions of Cas9 with different properties are being developed, and novel systems discovered [14–17]. The sgRNA sequence is established and construction is simple, being based on a synthetic RNA scaffold plus a 5′ variable 17–20 nucleotide sequence that defines the genomic target by sequence base homology. Once introduced into a cell, the Cas9-sgRNA complexes, enters the nucleus and scans the genome for the sgRNA defined target sequence. This is first achieved by identifying a triplet sequence called the proto-spacer adjacent motif (PAM) [18, 19]. The PAM, is defined by the originating species of Cas9 (e.g. "NGG" for SpCas9) and although part of the targeting recognition sequence, it is *not* represented in the sgRNA guide sequence. Following recognition of a PAM sequence, the sgRNA in association with Cas9 exploits Watson-Crick base pairing of its complementary 5′ targeting region to the putative target genomic sequence. Where the sequences match, the complex causes a blunt-ended double-strand break in the targeted DNA [19] (Fig. 2).

Double-stranded DNA breaks occur naturally often in cells and are rapidly repaired using a Non-Homologous End Joining (NHEJ) repair pathway. Although NHEJ is in general accurate, dsDNA breaks induced by RNA-guided endonucleases often lead to indels; i.e. insertions of one to two base pairs, or more commonly deletions of one to ten's of base pairs, or occasionally hundreds of base pairs [20, 21]. However, in the presence of homologous (donor) DNA template, the cell's internal repair systems can make use of an alternative process, the homology-directed repair (HDR) pathway which can lead to incorporation of the exogenous donor sequence albeit at lower frequencies (Fig. 2) [4, 7, 13, 22]. In recent years these approaches have been used to directly edit the genome of many different species, ranging from bacteria, plants, to mice and humans [4, 23–25]. In the final analysis, the power of CRISPR-Cas9 is that it is simple to design, fast to construct and use, and of crucial importance, is highly efficient and accurate in its action.

1.2 Definitions

Definitions for zygote genome editing outlined in this chapter.

(a). *Simple Knockout.*
   Use of one targeting sgRNA aimed at genomic DNA sequence disruption, for example targeting an initiation codon and causing an indel by NHEJ, without donor DNA, and leading to serendipitous frame shift and disruptive mutations (Fig. 2a) [26].

(b). *Dropout Knockout.*
   Use of multiple targeting sgRNAs simultaneously, flanking a contiguous section of genome with the aim of deleting the intervening sequence. This can be designed to disrupt the gene, or with forethought if desired a hypomorph. The targeted regions can be separated by tens to thousands of bases, without donor DNA (Fig. 2b) [27].

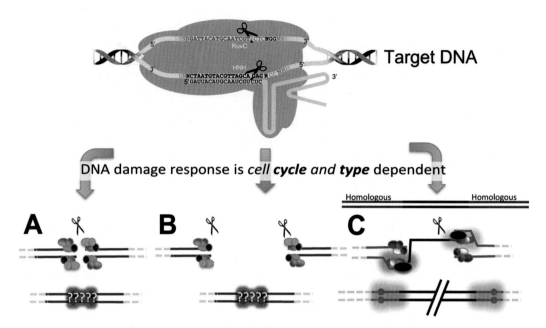

Fig. 2 Cas9/DNA—simplified outline of CRISPR-Cas9 action and cell response. The CRISPR-Cas9 sgRNA complex uses the PAM and then the guide (protospacer) sequence to locate its complementary genomic target DNA sequence. The complex then causes a dsDNA break ~3 bp upstream of the PAM [19]. Upon a genomic dsDNA break being detected in a cell, a rapid repair response is initiated the nature of which is dependent upon the cell type, cell cycle, and if homologous DNA is present. In example (**a**), genomic DNA has been cut with a single Cas9-sgRNA complex leading to a rapid NHEJ repair and often a serendipitous indel of plus/minus a few base pairs, to tens or occasionally hundreds of base pairs. In example (**b**), two sgRNA's have been applied flanking a region of interest, this configuration will often lead to the "drop-out" of the intervening region and a NHEJ repair (in such drop-out KOs, the actual region deleted will often be larger than that defined by the two guides (MVW, BEL observations)). In example (**c**), one, or more Cas9-sgRNA initiated dsDNA break have occurred, however when donor template DNA with sequence homology is present, for example double-stranded DNA (e.g. plasmid) or ssDNA (e.g. oligo), an alternative repair using HDR *can* occur, although often with lower frequency than observed NHEJ. HDR leads to the precise modification of a region, e.g. adding, subtracting, or changing single to many hundreds, or thousands of nucleotides

(c). *Simple KnockIn.*

Use of one or more targeting sgRNA's to disrupt a target sequence followed by Homology Directed Repair (HDR) designed to incorporate small genetic modifications (1–~50 bp) using a single-stranded DNA (ssDNA) oligo (100–200 bases) as donor template; for example introducing a SNP, modifying coding sequence to effect amino acids changes, introducing a tag sequence (e.g. V5), or designed to give a small precise deletion (Fig. 2c) [7, 28].

(d). *Large KnockIn.*

Use of one or more targeting sgRNAs to disrupt a target sequence that is repaired by HDR using a donor DNA template (e.g. plasmid) with extensive sequence homology arms

(2–10 kb) to the target region. This aims at integrating longer, >50 bp, to many kb of novel DNA sequences precisely (e.g. multiple LoxP sites, visualization markers, humanization), or to effect precise larger deletions (Fig. 2c) [29, 30].

1.3 Challenges Using CRISPR-Cas9 to Genetically Modify Mice

It is only since late 2012 that CRISPR-Cas9 has been used to genetically alter mammalian cells hence protocols are in a state of flux and are still being refined. From our own experiences and others it is already apparent that there are a number of common challenges and needs to address when using this system; these include:

1.3.1 Off Target Events

CRISPR-Cas9 interacts with its target sequence via an RNA guide sequence, binding by base pairing to its complementary DNA target strand via Watson-Crick-type interactions. Although such base pairing is conceptually simple, it is known that base pair mismatches between the guide sequence (especially at the 5′ end, distal to the PAM) and the target sequence can be tolerated to varying and at times surprising degrees. Such near matches can lead to off-target dsDNA breaks and consequently unintended genetic modification [31–36]. In an attempt to address this issue a number of partial solutions have been developed:

1. *Guide Design Software.*
   There are numerous web-based software resources where target sequences can be submitted and suggested guide sequences returned, *see* Table 1. These various web routines are generally simple to use. Crucially many will scan the target species genome and provide a listing of near matching, potential off-target sites. Some of these programs attempt to rank guide sequences however, the rules which govern off target matches are still very much under development and hence any ranking should be regarded as advisory and not absolute [36]; for recent reviews and approaches *see* [37–39]. As a rule of thumb and after assessment by a web-based program for near matching sites, we further select target guide sequences with GC content of between 40 and 60%, whilst avoiding runs of four of more nucleotides of the same base, and aim for higher G/C content at the 3′ seed sequence end [19, 32]. In some cases, for example with simple KnockIn (KI)'s the choice of guides sequence may be very limited due to the location of the desired precision modification, which needs to be within the guide recognition sequence (preferably at its 3′, PAM proximal end) in order to prevent recutting post modification. At times, however, due to local sequence constraints the only practical approach is pragmatism coupled with an acceptance that off-target cutting may occur and may require additional screening, and/or back-crossing (relying on allelic segregation) to eliminate any unintended

Table 1
Web-based resources. Non-exhaustive listing of web sites to assist and advise on designing targeting sequence, gene analysis databases, and physical reagent resources, see also http://omictools.com/crispr-cas9-category

Cas9 guide advisory programs	
Benchling CRISPR design	https://benchling.com/crispr
CHOPCHOP	https://chopchop.rc.fas.harvard.edu
CRISPR Design	http://crispr.mit.edu
CRISPR direct	http://crispr.dbcls.jp/doc/
E-CRISP	http://e-crisp.org/ECRISP/
RGEN Tools	http://www.rgenome.net
sgRNA Designer	http://www.broadinstitute.org/rnai/public/analysis-tools/sgrna-design
ZIFIT	http://zifit.partners.org/ZiFiT/Introduction.aspx
DNA/gene sequence analysis and resources	
MGI	http://www.informatics.jax.org
Ensembl	http://www.ensembl.org/index.html
NCBI	http://www.ncbi.nlm.nih.gov/pubmed
DNA resource	
Addgene	https://www.addgene.org
IDT	https://www.idtdna.com/site
Eurofins/Operon	http://www.eurofinsgenomics.com
CRISPR discussions	https://groups.google.com/forum/#!forum/crispr
OXfCRISPR	www.dpag.ox.ac.uk/research/liu-group/liu-group-news/oxfcrispr

effects in a new mouse strain. It should also be noted that the reference genome used by many of these programs (e.g. C57BL/6J) may not truly represent the actual strain/genome you are targeting. For example, each inbred mouse line has a multitude of strain-specific polymorphisms. This challenge becomes more significant (and at times useful) when working with outbred animals. Also, if sequentially modifying a strain, previous modifications (i.e. transgenes) may need to be considered when designing the guide sequence.

2. *Modified versions of Cas9.*
    SpCas9 can be mutated to function as a nickase, cutting only one of the two strands of the targeted dsDNA. In this inception, two sgRNAs are used in close proximity targeting opposite

DNA strands, resulting in a staggered dsDNA break. Data suggest that this dual sgRNA nickase strategy can considerably lower off-target cutting [31, 40]. A further enhancement of this approach is to use a "dead" Cas9 (i.e. Cas9 with no nuclease activity), fused to an obligate heterodimer, FokI nuclease [41, 42]. Here, only when the heterodimer FokI nucleases are in the correct configuration will a (staggered) dsDNA cut occur. However, the use of paired Cas9's with their configuration restrictions (e.g. dual co-located PAMs) and possible Cas9-sgRNA differential activities may lower overall efficiency. Modified Cas9's are readily available from Addgene, *see* Table 1.

3. *Truncated guides.*
    The actual binding interaction of the guide or seed part of sgRNA to target DNA, leading to the cutting of the DNA is not fully understood. However, it has been empirically determined that truncating the target guide sequence from the 20 bases to 18 bases substantially reduces off target cutting [36, 43]. The truncated guide approach is simple to implement and the efficiency of cutting does not seem to be significantly impaired; MVW and BEL's own observation and [43]. However, when designing these truncated guides it should be noted that most guide design programs listed in Table 1, use a default of 20 bases and will need to be truncated manually at the 5′ end by two bases. Also if the T7 promoter is used to make the sgRNA, a "G" is required at the 5' end of the sgRNA (see below).

4. *Breeding Founders.*
    When creating a genetically modified mouse line, the putative founder mice should be backcrossed at least once. This will help reduce any potential off target effects via genetic segregation, unless they are closely (physically) linked to the desired modification. In special cases where it is known that a guide may modify loci which are not the primary objective (e.g. when targeting a gene family motif, or unintentionally modifying pseudogenes etc.), several cycles of back-crossing coupled with active selection by genotyping against identified off-target alleles may be necessary.

1.3.2 Mosaicism in Founder Animals

The introduction of CRISPR-Cas9 into the zygote does not necessarily ensure that a NHEJ or HDR event occurs before S-phase or zygote cleavage. In fact, CRISPR-Cas9 events often appear to occur later than the zygote stage, including at 2-, 4-, or even perhaps 8-cell stages (MVW and BEL, own observations). Founder animals that progress from such independent CRISPR-Cas9 blastomere events will develop as genetic mosaics composed of cells where different nuclease-mediated repair events have occurred. Practically this means that when a tail tip, ear notch, or other *somatic* tissue is collected from founder animals, it is only a *sample* of the potential genetic heterogeneity within the mosaic animal. Further, even where clear results are obtained from such somatic tissue samples; e.g. a perfect biallelic modification, the animal's gametes (sperm or oocytes) may not totally reflect this (Fig. 3). Such somatic mosaicism also jeopardizes, or minimally complicates using founder animals for direct phenotypic analysis. The take-home message here is that somatic cell biopsies of founder animals cannot be regarded as a complete reflection of the whole animal, including importantly the germ-line. As such, any putative founder animal must be back-crossed to the background strain and the resulting offspring (N1) sequence analyzed to determine precisely their specific mutant allele. At times, such germline mosaicism can be viewed as an opportunity, as a single founder may provide multitude alleles, offering more genome editing events from fewer animals. Once N1 animal(s) that contain the desired events are identified they can then be used to establish the new strain.

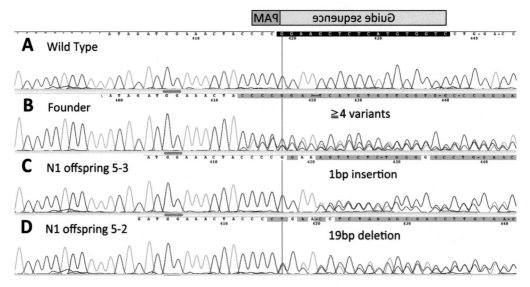

Fig. 3 Sequence chromatographs showing sequence derived from mosaic founder. An alignment of DNA sequence chromatographs obtained from PCRs covering the flanking regions of a CRISPR-Cas9 event using mouse tail derived DNA (inbred mouse line). The sgRNA guide sequence and PAM (inverted and hence CCN) is shown above, with the expected dsDNA break ~3 bp upstream of the PAM. Panel (**a**) is from a wild-type mouse showing the known homozygous target region. Panel (**b**) is derived from a founder animal and shows at least four alleles with varying degrees of signal strength for the targeted region, i.e. the tail is genetically mosaic. Panel (**c**) is derived from an N1 offspring derived from the founder "**b**" and clearly shows a wild-type and a single-base insertion allele at the expected site, i.e. this N1 offspring is heterozygous with a 1-base insertion allele. Panel (**d**) shows another, different example of a founder "**b**" N1 offspring, showing clearly a wild-type and a mutated allele; however, this time the mutation is a 19 base deletion. This demonstrates that although modified founder animals can be identified, it can be difficult to predict what the actual germline contains due to their often mosaic nature. This problem also occurs with KI's where the correction can be masked by a host of overlaying NHEJ events

1.3.3 Founder Breeding

The objective in developing a genetically modified mouse is to generate a resource to either study gene function or develop a disease model - fast. This generally requires germline transmission of the genetic modification, often to homozygosity. In the pursuit of speed it is tempting to pair modified founder animals with the objective of obtaining homozygous modified animals rapidly. The authors *strongly* caution against this approach. With especially serendipitous indel KOs, this course of action will often result in variable compound heterozygous offspring due to founder animal mosaicism, and *will* complicate analysis considerably. In the case of founders with precise HDR events, although often still mosaics, these could be intercrossed resulting in rapid development of offspring homozygous for the desired precise event. However, the resulting frequency of homozygous offspring is dependent upon the *actual* allele frequency in the gametes and not the tail! Further, there is an elevated risk that any off-target events present in founders will become fixed in the nascent homozygous strain, potentially complicating later phenotypic analysis. We suggest that if this latter short cut is practiced, it be done only as a quick but potentially flawed approach to obtain preliminary data. Hence we strongly recommend that founder animals be backcrossed at *least once* before developing a homozygous strain.

1.4 Strategies for Mouse Strain Creation: Clearly Define Your Aims

1.4.1 Background Strain

A key advantage in using targeted nucleases is that they can be applied to zygotes derived from any strain or genetic background. It should be remembered and possibly exploited, genetic background will have profound impacts upon mutation phenotype [44]. For the majority of our work we have used inbred backgrounds including C57BL/6J, FVB/J, DBA/2J, Balb/cJ, NOD/ShiLTJ, MRL/MpJ, and CAST/EiJ. Currently, we are increasingly using strains that have undergone previous (often multiple) genetic modification, including KI's, KO's, and transgenics, e.g. NSG, NRG, Tg32 FcRn (respectively, The Jackson Laboratory mouse strain reference # 5557, 7799, and 14565). The use of previously modified and characterized strains (models) allows rapid and sequential genomic editing of animals, leading to the development of highly customized mouse models. The use of these perhaps more obscure strains, however, may need to be tempered against availability, embryo yield, and survivability post zygote injection/transfer, each of which can be quite unique to a strain. Recently approaches based on improved superovulation and/or collection and cryopreservation of oocytes have been developed which may alleviate these challenges [45, 46].

A number of publications have used F1 or F2 zygotes derived from crossing different inbred strains for CRISPR-Cas9-mediated modifications [26, 29]. Creation of these zygotes, due to hybrid vigor uses fewer resources and the zygotes often exhibit increased resilience to the traumatic processes involved in microinjection and embryo transfer. Such embryos have their uses, for example in optimization of methods and training, but caution that their use will severely compromise coherent genetic analysis due to genetic segregation; i.e. offspring will be genetically unique and highly heterogeneous which will cause variability to their phenotype confounding analysis [47, 48].

1.4.2   Target Sequence

When sgRNA guides are designed, it is crucial that the target sequence is *fully* defined as unintentional mismatches between target and guide may cause a guide to fail, or work very inefficiently (*see* Table 1 for resources to assist). This is of special concern where the precise sequence of a particular strain is not available. Further, if using F2 or outbred animals, the existence of sequence polymorphisms may raise both on-target and off-target complications (at times this can be used to experimental advantage). Where there is any doubt regarding the genomic target sequence, a simple PCR using the desired strain's genomic DNA as template, followed by sequence analysis across the region of interest is recommended. This simple step in advance of the microinjection can save significant time and resources, and has the added benefit that the PCR will provide useful data to assist in the screening of putative founders post-microinjection for the CRISPR/Cas9-mediated events.

1.4.3   Delivery of Nuclease: KO/ KI

The choice of cytoplasmic vs. pronuclear microinjection of the sgRNA and Cas9 is determined by the objective: KO vs. KI. For simple KO's, where no donor DNA template is utilized, delivery of sgRNA and Cas9 by microinjection to the cytoplasm is the most commonly used approach [49]. Most Cas9's contain one or two nuclear localization signals and the Cas9-sgRNA complex appears to be actively conveyed to the nucleus. A novel approach avoiding microinjection, based on electroporation of zygotes with sgRNA and Cas9 is also a possibility [50]. Where KI's are desired which use donor DNA (oligo or plasmid), most groups use pronuclear microinjection to deliver the templates to the nucleus, whilst also "lingering" in the cytoplasm to deposit the RNA's into their appropriate subcellular location. While the skill set needed for pronuclear microinjection is higher, and results in reduced live born compared to cytoplasmic microinjection, this approach has proven to be very efficient for KI's (MVW and BEL own observations) [26, 29, 51].

1.4.4   KO Targeting

When creating KO animals it is essential to define the goal, i.e., if a null mutation is the absolute need or if potential hypomorphs would be of use. To help ensure that the desired event occurs the gene domain structure needs to be understood, facilitating targeting of critical regions; e.g. the initiation (ATG) codon, a transmembrane domain, or active binding site (*see* Table 1 for web resources). In this context our experience when using a single targeting sgRNA will often result in an indel of only a few bases around the targeted region.

Although this may be effective, many genes do have complex, often not fully characterized splice variations and may also possess cryptic initiation sites and/or splice acceptors. As such, a single targeting sgRNA may potentially lead to a hypomorph instead of a null mutation, and even the doubt of failing to produce a null can lead to substantial extra work. A more effective approach to gene disruption is to use two sgRNA flanking an essential element within the target, >100 bp–~10 kb apart; however, note smaller dropouts occur more efficiently, although 50–100 kb dropouts can be found (Fig. 2b). If carefully designed, this two sgRNA strategy will increase the probability of a null mutation or minimally give two opportunities to disrupt the target region. The use of an oligo or a plasmid containing flanking regions homologous to the intended deleted region is also an option as these can mediate much larger and controlled deletions ($\gg$10 kb) [52, 53].

1.4.5   KI Targeting

For subtle precision KI with changes in the 1–50 bp range, e.g. SNP humanization, tags, integrase sites, and precision deletions, the use of a single Cas9-sgRNA mediated dsDNA cut followed by HDR using a predominantly homologous donor ssDNA oligo ($\leq$200 base) has proven to be highly efficient. A key consideration when designing this type of modification is that once the desired HDR event occurs in the genome that the Cas9-sgRNA *will no longer* (re)cut the modified targeted region. To achieve this the guide + PAM target site must be sufficiently compromised post-HDR so that the Cas9-sgRNA complex can no longer target the region. We and others have found that this is best done by introducing target changes post modification which have one or more mismatches in the 3′ "seed" end of the guide sequence proximal to the PAM. The PAM of SpCas9 PAM can be used as the mismatch region, although it should be noted that the NGG can functionally substitute NAG and NGA [32, 54]. Synonymous, or silent substitutions to the target genome could also be used, however we suggest that unnecessary changes to the target region, even so-called silent ones be avoided as they may have unintended consequences. For larger KIs (e.g. >100 bp), at the time of writing and in the experience of the authors approaches to incorporate larger donor sequences are still poorly defined, are locus sensitive, and appear to be less efficient in inbred mice vs. F1 or F2 strains (MVW, BEL own observations).

1.4.6   Targeting Multiple Sites by Design

Theoretically many regions could be targeted provided Cas9 does not become limiting. We and others have used multiple sgRNAs targeting three or more genes (with 2xsgRNA's/gene) simultaneously [29]. This approach decreases the cost/gene, although when multiple modifications occur in a single animal they will, unless linked, segregate independently. In any case, care should be exercised as each guide has a finite probability of off-target cutting. It is also possible to use a single guide to target multiple copies of a sequence (e.g. gene family motif, putative pseudogenes, or other functional sequence motifs); i.e. all contain the same target sequence (including PAM) (MVW own observations). The efficiencies for the various target sites will vary and screening offspring can be challenging as it requires resolving which of the multiple copies of a target sequence have been modified.

1.4.7 An Abundance of Riches: Too Many Alleles

An unexpected challenge we have seen is having too many founders and offspring with a multitude of various (potentially interesting) mutations. Handling these animals with this abundance of alleles can rapidly overwhelm mouse space, analysis, and financial constraints. The authors have found that selecting two or at most, three (where possible, male) founders with the desired mutation has been sufficient to obtain a desired line rapidly. Once offspring with the desired mutations are identified the other lines are terminated, or where additional lines may need to be preserved as potentially interesting, sperm cryopreservation is used (*see* also Chapter 3 for details on sperm cryopreservation) [55, 56].

1.5   Future Directions

Nuclease-mediated genetic editing technologies have only recently come onto the scene and their future potential can only be guessed. However, it is obvious that their future use will literally change the world. Key elements that are under active research include:

1. Approaches to increase HDR efficiency vs. NHEJ; e.g. through the use of NHEJ inhibitors [57, 58].

2. Elimination of mosaicism, potentially allowing biallelic modified founder animals to define phenotype [59, 60].

3. Developing systems for larger scale genomic integrations (i.e. hundreds of kb of DNA) using, for example safe harbor sites (e.g. ROSA26 locus), facilitating controlled transgenesis, synthetic biology approaches to gene expression and novel gene control systems [3, 61–65].

4. Redefinition and control of PAM sequence specificity, allowing universal targeting by engineering the Streptococcus pyogenes Cas9 PAM site [15] and/or the use of orthologous Cas9's [14, 23, 66–68]. For example, Cpf1 is a novel targeting nuclease, with an alternative PAM and perhaps more importantly, it exhibits different physical properties from Cas9, being smaller and causing a staggered dsDNA cut outside the guide sequence [16, 17].

5. The elimination or reduction of off-target concerns providing genetic editing without significant collateral damage; e.g. by the development of proven rules in guide design based on nonbiased off target assessment methods, and a better understanding of RNA-guided endonuclease target interactions [36, 69, 70].

6. Improved control of oocyte production allowing more rapid and sequential modification of modified mouse strains and other species [45, 46].

2 Materials

2.1 Enzymes and Kits

1. Taq DNA Polymerase with supplied PCR Buffer (New England Biolabs Inc., cat. # M0273).

2. High Fidelity PCR; e.g. Phusion® High-Fidelity DNA Polymerase with supplied buffer (New England Biolabs Inc., cat. # M0530).

3. PCR product purification (QIAquick PCR Purification Kit, Qiagen cat. # 28106).

4. PCR product purification kits for sequencing (HighPrep™ PCR MAGBIO, AC-60050).

5. T7 In Vitro Transcription kit for sgRNA (MEGAshortscript™ T7 Transcription Kit, Life Technologies, cat. # AM1354M).

6. SpCas9 mRNA (capped and polyadenylated) home-made or from a commercial source (e.g. Trilink, Cat # L 6125). Store at –80 °C as single-use aliquots (e.g. 5–10 µg aliquots at 1000 ng/µL).

2.2 Reagents

1. RNase-ZAP (Life Technologies, AM9780).

2. Molecular grade water, nuclease-free.

3. Nuclease-free TE pH 8.0, 10 mM Tris pH 8.0, 0.1 mM EDTA (IDT Cat # 11-05-01-09).

4. Phenol:Chloroform:Isoamyl Alcohol 25:24:1, molecular biology grade (Sigma, P3803).

5. Isopropanol (ACS Reagent Grade Plus).

6. Tris Borate EDTA (TBE) electrophoresis buffer, used at 0.5×.

7. Electrophoresis gel UV DNA/RNA stain (EnviroSafe® Helixxtec, cat # HDS001).

8. Microinjection TE buffer, nuclease free TE pH 7.5: 10 mM Tris pH 7.5, 0.1 mM EDTA (IDT Cat # 11-05-01-05).

2.3 DNA Oligos and PCR Primers

1. T7-guide sgDNA overlapping PCR primer 5′-TTAATACGACT CACTATA-(GN17-19)- GTTTAAGAGCTATGCTGGAA-3′, where "N" is the sgRNA guide target sequence (see [71]) and below.

2. Common 80mer ssDNA oligo defining the crRNA stem loop region of the sgRNA 5′-AAAAAAAGCACCGACTCGGTGC CACTTTTTCAAGTTGATAACGGACTAGCCTTAT TTAAACTTGCTATGCTGTTTCCAGCATAGCTC TTAAAC-3′ [71].

3. Various synthetic ssDNA oligos for use as donor templates; e.g. IDT, made up at 1000 ng/μL in 10 mM Tris, 0.1 mM EDTA pH 8.0 aliquoted and stored at –20 °C; e.g. IDT, cat # 11-05-01-09.

2.4 Superovulation and Microinjection Equipment

1. Stereomicroscope (Zeiss Discovery.V8).
2. Inverted microscope (Zeiss AxioObserver.D1).
3. Micromanipuators (Eppendorf NK2).
4. Injectors (Narashige IM-5A, IM-11-2 or Eppendorf Femtojet).
5. Needle puller (Sutter P97, P1000).
6. Microforge (Narashige MF-900).

2.5 Consumables and Reagents for Superovulation and Microinjection

1. Thin wall capillary tubes with filament (World Precision Instruments (WPI) TW100F-4).
2. Thin wall capillary tubes without filament (WPI TW100-4).
3. Glass cover slips (Fisher Scientific 12-545J).
4. Aspirator tube assembly (Sigma A5177; or as an alternate to mouth pipetting, COOK Flexipet).
5. Pregnant Mare Serum Gonadotropin (PMSG) (ProSpec HOR-272).
6. Human Chorionic Gonadotropin (hCG) (ProSpec HOR-250).
7. Embryo Culture Media (COOK K-RCVL).
8. Embryo Handling Media, M2 (e.g. Zenith Biotech ZFM2-100).
9. Silicone fluid (Clearco PSF-20cSt), alternatively mineral oil can be used.
10. Hyaluronidase (Sigma H3506-1G).

2.6 Reagents for Founder Screening

1. NaOH tail lysis buffer: 50 mM NaOH, 0.2 mM EDTA.
2. 1 M Tris–HCl pH 8.3.
3. Proteinase K tail buffer: 50 mM KCl, 20 mM Tris pH 8.3, 1 mM beta-mercaptoethanol, 0.5 % NP40, 0.5 % Tween-20, 1 mM EDTA.
4. Proteinase K.
5. 10 mM Tris pH 8.3.
6. Optional: Qiagen's DNeasy Blood and Tissue Kit (Qiagen 69506).

3 Methods

3.1 Characterization of Target Region

Where the sequence of the region to be modified is not completely known, approximately 1–2 kb centered on the target should be PCR amplified and sequence verified. If the animals/zygotes are heterozygous for the region, e.g. outbred, F1, F2 etc., this may

Table 2
Standard gradient PCR for optimization

Reagent	1× μL	PCR conditions	
Molecular grade water	11.825	1 95 °C	30 s
10× Standard buffer	1.500	2 95 °C	15 s
10 mM dNTPs	0.300	3 48–68 °C	15 s
Primer Mix each @ [10 μM]	0.300	4 68 °C	1 min/KB
Taq Polymerase (e.g. NEB#M0273)	0.075	Repeat 2–4	30×
Template DNA	1.000	5 68 °C	5 min
Final volume	**15.000**	6 4 °C	**Hold**

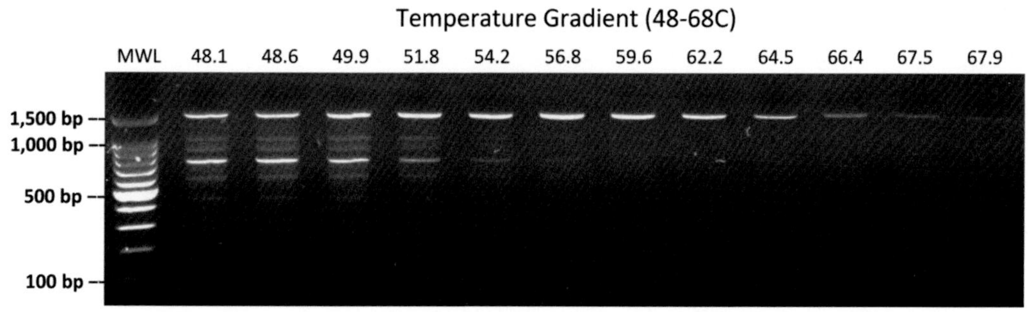

Temperature Gradient (48-68C)

MWL 48.1 48.6 49.9 51.8 54.2 56.8 59.6 62.2 64.5 66.4 67.5 67.9

1,500 bp
1,000 bp
500 bp
100 bp

Fig. 4 Example of optimization PCR. The effect of different PCR annealing conditions. Note that as the annealing temperature increases, specificity and noise to signal ratio improves. In this example, we would select 62 °C as the optimized annealing temperature, as it has a good signal strength combined with high specificity

result in reduced effective HDR. For larger deletions long range PCR (3–10 kb) or other strategies may need to be developed to cover the region (*see* Table 1 for resource listing). These data will also be used for subsequent screening of putative founders.

Using the known sequence and established PCR primer standard design programs, primers are designed and ordered. We routinely test and optimize these in advance using ~1–3 ng of pure genomic or 1 μL (~100–300 ng, *see* below) of crude "tail digest" DNA in a 30 cycle 15 μL PCR, *see* Table 2. To ascertain an optimal annealing temperature with a good specific signal to noise ratio, we use a temperature annealing gradient PCR to optimize the PCR conditions empirically (Fig. 4). Typically we test annealing temperatures in the range of 48 °C–68 °C. If this fails, for example due to polymorphism or repetitive sequences, other primers may need to be designed and tested.

3.2   HDR ssDNA Oligo Donor for Microinjection

For HDR, oligo donor ssDNAs for microinjection can be ordered from various suppliers and are generally supplied lyophilized. When ordering ssDNA as donor DNA, order oligos ~150–200 bp in length and in sufficient quantity (i.e. 100 nM) to allow for PAGE purification. Although we have found that non-PAGE purified ssDNA oligos work well for HDR, PAGE purification will reduce the presence of truncated products that may compete with the full-length donor following CRISPR-induced DSBs. Some publications suggest that the first 2–3 bases 5′ and also the third and second to last 3′ bases are phosphorothioated to improve oligonucleotide stability in vivo [7, 72]. Donor oligos should be dissolved in appropriate volume of microinjection TE buffer (10 mM Tris pH 7.5, 0.1 mM EDTA) made to 1 μg/μL stocks and stored at –20 °C.

3.3 sgDNA Template Preparation

An outline of construction and synthesis of sgRNA is shown in Fig. 5 and is based on [71, 73]. Targeting sgRNAs are synthesized from a dsDNA template containing a T7 promoter and the unique guide sequence (T7-guide) 5′-TTAATACGACTCACTATA-(G_{N17-19})-GTTTAAGAGCTATGCTGGAA-3′, plus a common 80mer defining the crRNA stem loop region of the sgRNA 5′-AAAAAAAGCAC C G A C T C G G T G C C A C T T T T T C A A G T T G ATAACGGACTAGCCTTATTTAAACTTGCTATGCTGTTTC CAGCATAGCTCTTAAAC [43, 71]. The use of an overlapping PCR synthesis to make an in vitro transcription template avoids time-consuming cloning and quality control sequencing, providing speed and complete flexibility in the process of sgRNA design and production. Using separate guide and tracr RNA's made synthetically should also be considered

3.4 Design of Targeting Guide for sgRNA

1. A truncated guide for the sgDNA oligo should ideally be 18 bases in length, although 19 and 20 base guides can be used (*see* **Note 1**) [43].

2. When designing guides we utilize web software to screen for and minimize potential off-target cutting, selecting those guides with the least near matches (e.g. ZIFIT, *see* Table 1).

3. Due to the use of the T7 promoter for in vitro transcription, the guide sequence MUST initiate with at least one guanine ("G"), without this it will fail. If necessary, the guide can begin with a non-recognition "G" (i.e. one that does not exist in the genomic target). This will be the first base of the transcribed sgRNA; for a more complete description, *see* Fu et al 2014 [43]. Once this N17-20 sequence is established it is "inserted" into the T7-guide sequence for oligo DNA synthesis and can be ordered.

3.5 T7-guide PCR Synthesis of sgDNA Template

On ice, assemble reagents for the PCR synthesis of the sgDNA template as outlined in Table 3 (Fig. 5). Once reagents are mixed, gently centrifuge and perform PCR using conditions given in Table 3 (*see* **Notes 2** and **3**).

Table 3
sgDNA Template Preparation (overlapping PCR) for sgDNA template synthesis reaction

Reagent	1× µL	PCR Conditions		
Molecular grade water	76.00	1	98 °C	30 s
5× HF buffer	20.00	2	98 °C	10 s
10 mM dNTPs	2.00	3	60 °C	30 s
Common stem loop 80 bp oligo [100 µM]	0.50	4	72 °C	15 s
Phusion polymerase	1.00	Repeat 2–4		35×
Master mix volume	**99.50**	5	72 °C	10 min
T7-guide (target specific) Primer [100 µM]	0.50	6	4 °C	**Hold**
Final volume	**100.00**			

3.6 sgDNA Template Quality Control

As the next stages involve synthesizing and handling RNA they must be done as clean as possible, avoiding the introduction of RNases that will rapidly degrade sgRNA and Cas9 mRNAs. RNase-ZAP or similar RNase cleaning agents can be used to decontaminate the work area. These stages require to always use *clean* reagents, dust free disposable plastic supplies, and should be executed whilst wearing fresh gloves. RNA should always be kept on ice or stored at −80 °C to reduce RNase activity.

1. Following sgDNA template production, reserve 1–5 µL of PCR product for quality control DNA gel electrophoresis.

2. Purify remaining PCR product using Qiaquick PCR Purification Kit, eluting in 30 µL nuclease-free molecular grade water.

3. Quantify by spectrophotometry. A typical concentration expected is ~75–150 ng/µL; i.e. a total yield of ~2.4–4 µg from a 100 µL PCR reaction.

4. Gel electrophorese using a 2 % agarose made and run in 0.5× TBE, loading 1 µL of PCR (pre and post purification) with appropriate markers to visually verify the sgDNA is present and is a single band at ~120 bp (*see* Fig. 6).

5. Store Purified sgDNA at −20 °C, or proceed immediately to the in vitro transcription using the MEGAshortscript T7 Transcription kit.

3.7 sgRNA In Vitro Transcription

sgRNA synthesis using, e.g. MegaShortscript T7 Kit and the T7 promoter-containing sgDNA template created above (*see* **Note 4**).

1. Follow manufacturers protocol combining reagents at room temperature and in order as outlined in Table 4 (for convenience,

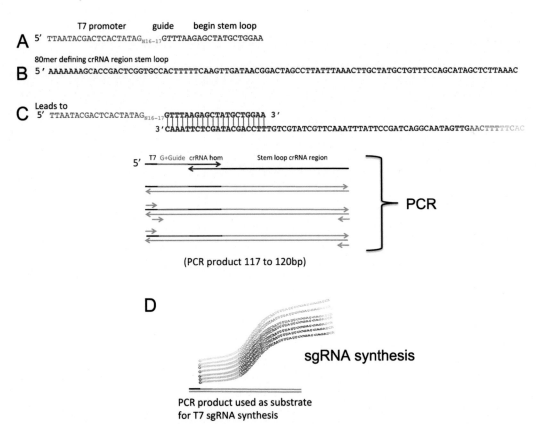

A T7 promoter   guide   begin stem loop
5′ TTAATACGACTCACTATAG$_{N16-17}$GTTTAAGAGCTATGCTGGAA

B 80mer defining crRNA region stem loop
5′ AAAAAAAGCACCGACTCGGTGCCACTTTTTCAAGTTGATAACGGACTAGCCTTATTTAAACTTGCTATGCTGTTTCCAGCATAGCTCTTAAAC

C Leads to
5′ TTAATACGACTCACTATAG$_{N16-17}$GTTTAAGAGCTATGCTGGAA 3′
3′ CAAATTCTCGATACGACCTTTGTCGTATCGTTCAAATTTATTCCGATCAGGCAATAGTTGAACTTTTTCAC

5′ T7 G+Guide crRNA hom   Stem loop crRNA region

PCR

(PCR product 117 to 120bp)

D

sgRNA synthesis

PCR product used as substrate
for T7 sgRNA synthesis

Fig. 5 Outline dsDNA T7 template to sgRNA synthesis. Outline of the process to construct a dsDNA template for sgRNA synthesis using PCR, followed by T7 promoter-driven IVT synthesis of the sgRNA. Section (**a**) is the generalized sequence for the T7-guide PCR primer, which is the main variable and crucial to specificity in sgRNA design. When using truncated guides the G + N16–17 defines the guide protospacer, i.e. the target sequence for the sgRNA. Note, the T7 polymerase promoter has an obligate "G" as it final 3′ base, in order for the T7 to be active it is crucial that the first base of the *guide* sequence is a "G". However, we have found where this cannot be arranged that G + 18 bases, where the "G" is silent and *not* part of the recognition sequence will function well. For guides of 20 bases, G + 20 where the "G" does not match the target also appear to be functional. Section (**b**), shows the common 80 base oligo defining stem loop structure of the sgRNA which also contains 20 bases (3′) which are complementary (overlapping) to the T7-guide PCR primer. Section (**c**), the T7-guide primer and the common 80 base oligo are combined in an overlapping PCR to produce the sgDNA template. Section (**d**), after PCR and purification, the sgDNA template is used in a T7 driven IVT, synthesizing large amounts of sgRNA. The sgRNA construction and synthesis is fast, and the only variable is the T7-guide, which can be synthesized and delivered within 24–72 h. The subsequent processes (PCR, PCR purification, IVT, and sgRNA purification) can be completed within a single day

the nucleotides can be combined and 8 μL of the premix used per reaction).

2. Mix, quick spin, and then incubate at 37 °C for 2–4 h (longer will increased RNA yields slightly).

3. Add 1 μL of Turbo DNase (provided in the kit) and incubate at 37 °C for another 15 min in order to remove the sgDNA template, leaving the sgRNA intact.

QC Gel: sgDNA PCR Preps

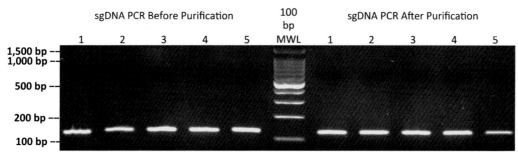

Fig. 6 TBE agarose gel showing sgRNA samples before and after extraction cleanup. Five independent sgDNA's PCR synthesis, before and after clean-up (0.5× TBE, 1.5 % agrose). The expected band is ~120 bp and approximately 100 ng product was loaded in each lane

Table 4
sgRNA In Vitro Transcription (IVT) Preparation

Reagent	1× µL
Purified sgDNA PCR Product [75–150 ng/µL]	8.00
10× T7 Reaction Buffer	2.00
75 mM T7 ATP Solution	2.00
75 mM T7 CTP Solution	2.00
75 mM T7 GTP Solution	2.00
75 mM T7 UTP Solution	2.00
T7 Enzyme Mix	2.00
Final volume	20.00

4. Transfer 20 µL of the in vitro transcription reaction to an Eppendorf tube for phenol chloroform extraction.

5. Reserve the remaining ~1 µL for quality control gel electrophoresis, keep on ice ~4 °C, or better –20 to –80 °C.

6. Add 115 µL molecular grade water to the reaction and 15 µL ammonium acetate (provided in the MegaShortscript T7 kit) giving a total volume of 150 µL (20 + 115 + 15 µL).

7. Add an equal volume (150 µL) of phenol:chloroform:isoamyl alcohol (25:24:1), pH 7.0.

8. Vortex for 30 s, then spin at 20,000 × *g* for 5 min at RT in a bench top Eppendorf centrifuge.

9. Remove ~100 µL of the upper layer containing the sgRNA to a fresh tube (*see* **Note 5**).

10. To precipitate the sgRNA add an equal volume (100 µL) iso-propanol (propan-2-ol), vortex, and quick spin down.

11. Place the tubes into –20 °C, –80 °C, or dry ice and chill for ≥15 min (this can be left overnight if desired).

12. Thaw if needed, and pellet the sgRNA by centrifugation in a refrigerated microcentrifuge; e.g. bench top Eppendorf centrifuge at $20,000 \times g$ for 15 min at 4 °C.

13. Remove supernatant *with care*, leaving a small white pellet of sgRNA at the bottom of the tube.

14. Wash twice in 500 µL 70 % ethanol, spinning at $20,000 \times g$ for 5 min at 4 °C each time, and taking care not to disturb the pellet.

15. Use a P10 tip to remove as much of the ethanol as possible whilst avoiding the pellet.

16. Allow the sgRNA pellet to dry for 5–10 min (*see* **Note 6**).

17. Resuspend the sgRNA pellet in 30 µL in 10 mM Tris, 0.1 mM EDTA, pH 7.5 and keep on ice for immediate use, otherwise, store at –80 °C.

3.8 sgRNA Quality Control

1. Dilute sgRNA sample (e.g. ~1:5, 1–4 µL IDTE) and use to measure the concentration of the sgRNA (keep sgRNA at 4 °C or if for longer periods, –80 °C).

2. Quantify by spectrophotometry (e.g. using a Eppendorf Biospektrometer, Nanodrop), typical concentration is ~300 ng/µL; i.e. ~50 µg yield from 100 µL of the in vitro transcription reaction.

3. Back-calculate to determine the actual sgRNA stock concentration; in this case 5× the measured concentration of the diluted sgRNA.

4. To visualize the sgRNA quality, use some/all of the remaining dilution (~3 µL; aim for ~100–400 ng total sgRNA per lane in order to prevent overloading the gel) and gel-electrophorese on a standard 2.0 % agarose electrophorese 0.5× TBE gel with EnviroSafe DNA Stain to visualize RNA and DNA ladder); *see* **Note 7**. If desired, also run the reserved (unpurified sgRNA from the in vitro transcription step). Figure 7 shows example sgRNA gel images before and after purification.

5. Once sgRNA material is confirmed it should be stored at –80 °C until needed. We have not found it useful to conduct in vitro assay to determine sgRNA activity (*see* **Note 8**). If multiple microinjections are planned using a single sgRNA batch, freezing down multiple aliquots of the sgRNA is recommended to reduce freeze-thaw cycles.

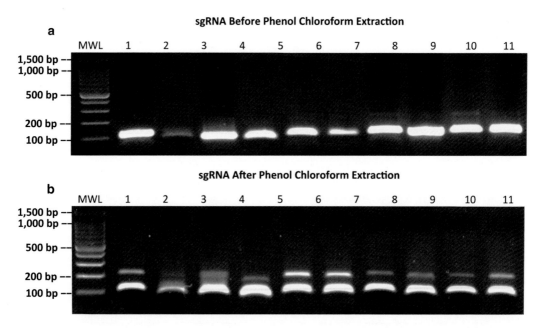

Fig. 7 TBE agarose gel showing 11 sgRNA samples, before (**a**) and after (**b**) clean up via Phenol:Choloroform Extraction. The result of 11 examples of sgRNA IVT using the overlapping PCR to construct the sgDNAs (0.5× TBE, 2.0% agarose). The sgRNA is ~120 bases and runs at about this equivalent size as the dsDNA marker. Note: sample #2 sgRNA had a reduced yield, and the Pre-cleanup gel (**a**) demonstrates that the IVT was poor for this sample, and not sample loss during clean-up. Further, if this were due to RNA degradation it would run as a faint smear (Fig. 8). The secondary structures (*dimers*) seen here in panel (**b**) with purified sgRNA are often seen at this stage

3.9 Assembly of Reagents for Microinjection

Table 5 outlines starting point concentrations of the various CRISPR-Cas9 and donor DNA for zygote microinjection. Once microinjection reagents are mixed they must be kept at 4 °C. It is also *strongly* suggested as a quality control measure, that samples pre *and* post microinjection be kept and gel electrophoresed to ascertain if the preps have been compromised (Fig. 8).

1. Combine RNA reagents: Microinjection TE buffer, sgRNA's, and Cas9 mRNA on ice in a PCR tube. Use a thermocycler to denature RNAs, using a program that steps down from 90 to 4 °C, holding at 90, 80, 70, 60, 50, 40, 30, 20, and 10 °C each for 1 min, and maintaining samples at 4 °C until ready to proceed.

2. Add where required other components (i.e. donor DNA) and RNAsin (*see* **Note 9**).

3. Mix, and transfer the entire preparation to a 1.5 mL microcentrifuge tube and spin at $20,000 \times g$ at 4 °C for 10 min. This helps sediment any microparticulates that will block the injection needle (*see* **Note 10**).

Table 5
Suggested starting point for zygote microinjection reagent concentrations

Simple KO: Cytoplasmic injection + 1 sgRNA	Final concentration
Microinjection TE buffer 10 mM Tris pH7.5, 0.1 mM EDTA	Make to 25 μL
Cas9 mRNA	100 ng/μL
sgRNA	50 ng/μL
RNAsin	0.2 U/μL

Dropout KO: Cytoplasmic injection + 2 sgRNA's	Final concentration
Microinjection TE buffer 10 mM Tris pH7.5, 0.1 mM EDTA	Make to 25 μL
Cas9 mRNA	100 ng/μL
sgRNA #1	25 ng/μL
sgRNA #2	25 ng/μL
RNAsin	0.2 U/μL

Simple KI: Pronuclear injection + oligo	Final concentration
Microinjection TE buffer 10 mM Tris pH7.5, 0.1 mM EDTA	Make to 25 μL
Cas9 mRNA	60 ng/μL
sgRNA	30 ng/μL
Donor Oligo (ssDNA, ~100–200mer)	1–10 ng/μL
RNAsin	0.2 U/μL

Large KI: Pronuclear injection + plasmid	Final concentration
Microinjection TE buffer 10 mM Tris pH7.5, 0.1 mM EDTA	Make to 25 μL
Cas9 mRNA	60 ng/μL
sgRNA	30 ng/μL
Donor Plasmid (dsDNA) Supercoiled	1–20 ng/μL
RNAsin	0.2 U/μL

4. With care, transfer 20 μL of the supernatant into a new 0.2 mL PCR tube, leaving the remaining ~5 μL plus debris in the 1.5 mL tube. Keep both tubes on ice; use the 20 μL supernatant for microinjection. The remaining 5 μL is reserved to confirm that the RNAs were intact prior to microinjection by gel electrophoresis.

5. After microinjection, gel electrophorese the remaining microinjection sample through a 1.5 % agarose 0.5× TBE gel, to check for degradation during the microinjection process (Figs. 8 and 9).

QC Gel: MIJ Prep reagents, Before & After MIJ

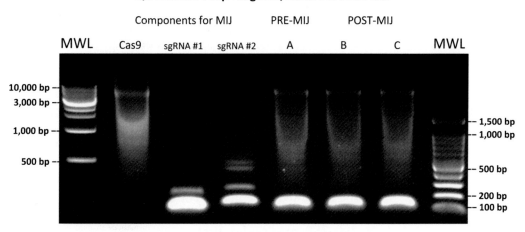

Fig. 8 RNAs pre and post Microinjection. An example of RNA preparations before and after microinjection, gel is 1.5 % agarose in 0.5× TBE (non-denaturing), shown with two size DNA markers, Molecular Weight Ladders (MWL), 100 bp MWL (Promega) and a 1 KB MWL (NEB). Note, the microinjection preparation samples (**a–c**) here have been denatured as outlined in the text and so the sgRNA's move as single band size ~120 bp. In contrast, the individual RNA components of the sample (Cas9, sgRNA #1, and #2) did not receive this treatment and so the secondary structures (e.g. dimers) are evident in those lanes. This gel confirms that the microinjection sample was intact and present before microinjection ("**a**"), and that no significant RNA degradation occurred during the microinjection process ("**b**" and "**c**", two aliquots of the same preparation sample)

Example of RNase Activity Gel: MIJ Prep reagents, before and after MIJ

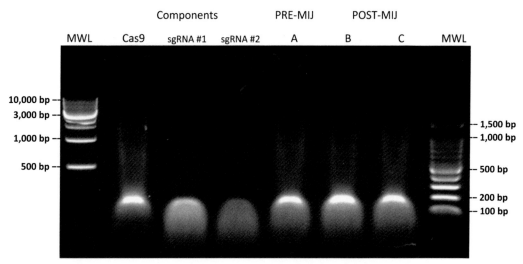

Fig. 9 Visualization of degraded RNA (RNAse Activity). Examples of degraded Cas9 and sgRNA due to RNAse activity. This gel (1.5 % agarose in 0.5× TBE) shows the same samples as used in Fig. 8 but subjected to RNases for a *few seconds* at room temperature and directly loaded into a gel. The result shows RNase degradation leading to a low molecular weight smear of all the samples, which is especially prominent with the Cas9 mRNA. Such degraded RNA would probably fail to yield modified pups post microinjection

3.10 Superovulation: Isolation of Zygotes (C57BL/6 J) for Microinjection

1. The production of large numbers of synchronized zygotes via superovulation greatly increases microinjection throughput, while minimizing the vivarium footprint. Here we will briefly outline their production and isolation from C57BL/6J mice. It must be noted however, that there is no universal protocol for inducing superovulation as genetic background profoundly impacts the procedure and zygote yield. When optimizing superovulation protocols for a specific strain it is important to also consider age and weight, in relation to hormone dose [74], as well as the time of hormone injection in conjunction with the room light cycle [75].

2. Calculate the number of donor females necessary to complete the requested experiment. For example with C57BL/6 J, 10 plugged donor females, 24–28 days old are expected to yield 200–250 good quality zygotes.

3. Three days before microinjection, at ~10.00 AM (mouse room lights 12 h on, 12 h off, 6 AM–6 PM) inject intraperitoneal (IP) 5 international units (IU) of PMSG in C57BL/6J at 24–28 days of age.

4. Forty seven hours post PMSG administration, inject IP 5 IU of hCG and immediately setup donor females at 1:1 with proven C57BL/6J stud males (*see* **Note 11**).

5. The following day at ~7.00 AM check for copulation plug. Those females displaying a plug are segregated for zygote collection (*see* **Note 12**).

6. For zygote collection females are euthanized via cervical dislocation, the oviduct excised and placed in 35 mm Petri dish with 3 mL of M2 media at RT. The ampullae lysed and the oocyte clutches removed. Hyaluronidase is added to the media at ~3 mg/mL to digest the cumulus cells and release zygotes, which are then transferred immediately to fresh M2 media at RT. Zygotes should be washed several times in M2 to remove cumulus cells, debris, and hyaluronidase [75] (*see* **Note 13**).

7. For microinjection, high quality zygotes are selected and placed into 30 μL microdrops of embryo culture media (e.g. COOK K-RCVL) under silicone fluid (*see* **Note 14**).

3.11 Microinjection of CRISPR Reagents

It is important when preparing glasswear, microinjection needles etc. for Cas9-sgRNA zygote injection that extra effort is made to minimize RNase contamination, including wearing gloves. Any reagent degradation will considerably reduce CRISPR efficiency.

1. The holding side pipettes and handling pipettes can be prepared in advance but the microinjection needles should be prepared the day of the experiment (or purchased from a reputable vendor). Table 6 illustrates an effective injection needle pulling

Table 6
Suggested starting point for pulling injection needles on a Sutter P97 using WPI TW100-F thin wall capillary tubes

Filament size	Pressure	Heat	Pull	Velocity	Delay
2.5×4.5 (FB245B)	500	Ramp minus 13–15	75	70	80

program for a Sutter P97 using WPI TW100F-4 thin wall capillary tubes (*see* **Note 15**).

2. At microinjection and immediately before use, CRISPR reagents should be prepared and stored on ice. After loading the needle into the injector, back pressure must be established immediately to prevent dilution of the reagents by the injection medium.

3. After the needle is loaded and mounted, collected zygotes are placed onto a slide containing 150–200 µL of M2 media. The number of zygotes placed on a slide should not exceed the number that can be injected within 20–30 min.

3.11.1 Cytoplasmic injection

1. Move the needle through the zona and into the cytoplasm of the zygote. The needle must pierce the oolemma. Once pierced it is possible to visualize the reagents leaving the needle as a disturbance of the cytoplasm. Exit from the zygote should be swift to minimize zygote lysis. It is estimated that 2–10 pL can be deposited in the cytoplasm without cell lysis (*see* **Note 16**).

2. After zygotes have been injected, remove and wash through three 30 µL drops of equilibrated K-RCVL and place back into culture in a K-MINC-1000 benchtop incubator at 37 °C, 5 % CO_2/5 % O_2/90 % N.

3.11.2 Pronuclear Plus Cytoplasmic Injection

1. Zygotes are loaded onto a slide as outlined above. The needle is prepared also as outlined.

2. The zygote is orientated in order to ensure proper focal plane alignment between the needle tip and the pronucleus. Entry into the pronucleus is confirmed visually by observing swelling of the pronucleus. Upon exit from the pronucleus the injector should consciously linger a few seconds to deposit material in the cytoplasm. After completing injections, remove the zygotes and wash through three 30 µL drops of equilibrated K-RCVL and place into culture in a K-MINC-1000 benchtop incubator at 37 °C, 5 % CO_2/5 % O_2/90 % N_2.

3.12 Transfer Embryos Into Pseudopregnant Recipients

It is essential that high quality pseudos' are available to receive the injected zygotes. We prefer to transfer injected zygotes on the same day (i.e. at one cell stage) as this reduces the time zygotes are in an artificial environment however, transfer at the two cell stage is also commonly used [75] (*see* **Note 17**). Pseudopregnant recipient mice can be readily provided by a good mouse unit [75], or can also be purchased from reputable dealers.

3.13 Screening of Putative Founder Animals for CRISPR Meditated Events

PCR screening for CRISPR meditated events: For crude DNA tail lysate preparation ear punches (~1 mm circle) or tail tips (≤1 mm) are collected at 2–3 weeks of age from putative founder mice and subjected to either NaOH or Proteinase K digestion to make crude DNA available for PCR analysis (*see* **Note 18**).

3.14 Crude NaOH DNA Tail Lysate

NaOH tail lysis buffer is 50 mM NaOH, 0.2 mM EDTA and can be made up in advance and stored at RT.

1. One ear punch or ~1 mm of tail tip tissue is digested in 100 μL NaOH DNA tail lysate at 95 °C for 1 h; e.g. using 200 μL 8-strip tubes and PCR machine.

2. Neutralize the NaOH by the addition of 10 μL 1 M Tris–HCl pH 8.3.

3. Mix samples, then centrifuge at max speed (~6000×g) in a bench top mini-centrifuge at RT to sediment hair, cartilage, and other tissue remnants.

4. For most PCRs 0.1–1 μL (30–300 ng DNA) of this crude supernatant is sufficient in a 15 μL PCR reaction (more may suppress the PCR).

5. NaOH derived DNA tail lysate can be stored at 4 °C if used that day, or at –20 °C for long-term storage.

3.15 Crude Proteinase K DNA Tail Lysate

Proteinase K tail buffer is 50 mM KCl, 20 mM Tris pH 8.3, 1 mM beta-mercaptoethanol, 0.5 % NP40, 0.5 % Tween-20, 1 mM EDTA and can be made up in advance (i.e. without proteinase K) and stored at RT.

1. Immediately before use, add proteinase K (stored at –20 °C) to the buffer to a final concentration of 0.25 mg/mL.

2. One ear punch or ~2 mm of tail tip tissue is digested in 100 μL proteinase K DNA tail lysate with proteinase K using 200 μL 8-strip PCR tubes at 55–60 °C for at 5–18 h.

3. After digestion, the crude proteinase K DNA tail lysate samples are mixed and centrifuged at max speed (~6000×g) in a bench top mini-centrifuge at RT to sediment hair, cartilage, and other debris.

4. Before using Crude Proteinase K DNA tail lysate in the PCR, samples are diluted 1:10 in 10 mM Tris pH 8.3 (e.g. 5 μL/45 μL) and *must* be heated to 94 °C for ~3 min. This *critical* step both denatures the sample DNA and destroys proteinase K activity.

5. For most screening PCRs, 1 µL (~30 ng DNA) of the 1/10 diluted and denatured crude supernatant is sufficient for use in a 15 µL PCR reaction.

6. Crude Proteinase K DNA tail lysate can be stored at 4 °C if used that day or at –20 °C for long-term storage.

3.16 Crude DNA Tail Lysate Purification

Purification of crude DNA lysate is only rarely required and should be avoided except for particularly finicky PCR reactions. Where required, we suggest Qiagen's DNeasy Blood and Tissue Kit be used, following the manufacturer's protocol.

3.17 PCR Screening for Founder Animals

Depending upon the exact strategy, potential founder animals are screened for KO and/or KI by PCR. Where large regions have been potentially deleted or a specific "in-out" PCR strategy has been applied, a positive result PCR is usually quite obvious/predictable. However, we suggest that these PCR products be sequenced to begin defining the exact nature of the modification and help prioritize which animals to breed from.

In Fig. 10 we show a typical gel obtained from a simple KI founder screen. Obvious serendipitous deletions are also apparent. In this example, while only 3/24 PCR bands show an obvious CRISPR-mediated effect (insertion or deletion), subsequent sequence analysis identified 15 additional mutated founders (although mostly mosaic), three of which appeared to contain the desired HDR KI.

3.18 Selection of Optimal Founders to breed: KO and KI, What to Look for

1. The selection criteria for which identified founder animals/s to breed is project specific and often not clear cut due to mosaicism. As a general rule, we recommend selecting ~3 putative founder animals where the desired event can be seen, ideally males due to their ability to sire more offspring rapidly.

2. For KO's, the putative founder(s) animals with the largest deletions are typically the most likely to result in a functional null. However, an understanding of the gene structure will allow appropriate ranking of founders.

3. For simple KI's, when examining founder animal derived sequence it is not uncommon for the desired sequence modification to be obscured by other mutant (NHEJ) alleles often due to mosaicism. As such, this process can take some patience, and good sequence analysis software is invaluable (e.g. CodonCode Aligner). Aligning the sequence chromatograms to the reference sequence will identify where the mutations begin (and end, if sequencing from both sides of the target). This will eliminate some candidates based on the likelihood that the desired location is modified. We find when assessing founder carrier animals that a simple although subjective rating helps; e.g. *A* = clear trace evidence of HDR, *B* = strong trace evidence of HDR, *C* = some trace evidence of HDR and *D* = faintest trace evidence HR.

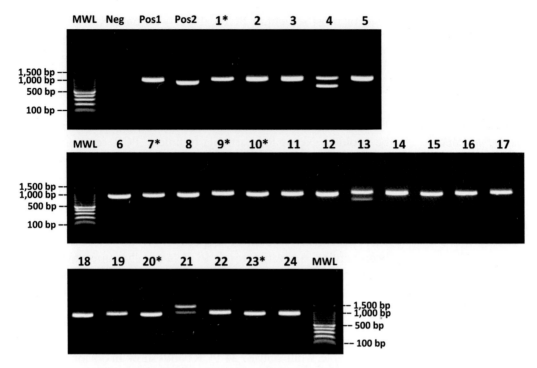

Fig. 10 Examples PCR and electrophoresis gel for screening putative Founders. Electrophoresis gel (1.2 % agarose, in 0.5× TBE) of a genotyping PCR assay of potential Founder animals using tail derived DNA from a project designed to introduce a small modification (simple KI). A total of 24 mice were born and tissue biopsies were subjected to PCR/Sequencing analysis. Clear evidence of a serendipitous deletion is shown in two samples #4 and #13, while an insertion is also obvious in sample #21. However, the majority of the mutants, including the correctly targeted KI Founders, were not identified until after sequence analysis of the PCR product. Sequence analysis identified an additional 15 mutants, 3 with the desired HDR event, and only 6/24 mice were shown to be WT by sequence analysis (indicated in the gel image with *). Gel shows: Neg PCR, i.e. negative control; Pos1, i.e. PCR positive control using crude DNA lysate isolated from wild-type; Pos2, i.e. PCR positive control using purified DNA (the slight differences in running is probably due to salt concentration differences in the PCR due to DNA used and isolation method)

4. For Large KI's, the challenge is to distinguish a successfully targeted KI event from a random transgenic (i.e. an illegitimate, nontargeted genomic integration). We screen founders using two approaches sequentially. First, using a PCR specific to the exogenous sequence we identify any/all animals that contain the donor sequences, regardless of integration position. Typically, this reduces the candidate pool considerably. Next, we screen this smaller candidate pool using a long range "In/Out" PCR strategy, using an appropriate PCR kit (e.g. LongAmp™ Taq PCR Kit, NEB cat # E5200S). This PCR uses a primer designed within the donor DNA region, plus a second primer that lies in the genome outside of the donor HDR region; i.e. beyond the homology arms. Using this In/Out PCR design a correct sized product is produced only when HDR has occurred at the intended locus. For better verification, this PCR should be

performed across both junction points. The region should also be sequenced to verify the completeness and accuracy of the HDR event in the founder, and in N1 offspring.

3.19 Breeding and Maintaining New Lines

Upon selecting the most promising founders animals, these mice should be set up for breeding typically at 6–7 weeks of age, to wild-type (same) background animals. To maximize offspring numbers, male founders are preferred and used in trio, or harem matings, or by rotating females after they become pregnant. In this manner, more N1 offspring can be produced to rapidly assess the germline transmission of the modified allele. With putative female founders a single male is sufficient, although if the females are not pregnant within 4 weeks, this male should be replaced. It is essential that defined criteria be established to select those N1 animals which will be used to establish the strain as soon as possible to contain costs. At this time we recommend sperm cryopreservation be considered. This would use 1–2 male carriers and will safeguard against catastrophic loss [55, 56].

Lastly, if not already established, it is essential that a *simple* and *robust* PCR genotyping assay be developed for the established line; i.e. not based on sequencing. Often this cannot be fully optimized until after the mutant allele has been characterized from sequences of an N1 mouse. Once developed, this PCR assay should allow simple rapid genotyping of the new line without the time and cost of sequencing.

Now it is only necessary to phenotype the new mouse lines.

4 Notes

1. The PAM sequence; e.g. NGG is intrinsic to the CAS9 and must not be part of the targeting sgDNA primer sequence.

2. Lyophilized oligos (T7-guide oligo and the common step loop 80 bp oligo) should be reconstituted at 100 μM in 10 mM Tris, 0.1 mM EDTA pH 8.0. To prevent freeze/thaw degradation, it is recommended that these oligo stocks be aliquotted and stored at –20 °C until use. Also it is recommend to use Phusion® or other high-fidelity DNA polymerase for sgDNA synthesis to ensure fidelity.

3. The common stem loop 80 bp oligo can be replaced with a shorter primer: 5′-AAAAAAAGCACCGACTCGGTGC-3′ and a plasmid at very low concentration (<1 ng/μL) which contains the 80 bp stem loop sequence (e.g. Addgene 51024).

4. RNA is very sensitive to RNases, which are ubiquitous, keeping the sample cold or preferably frozen will reduce any contaminating activity and prevent your sample from "disappearing".

5. To avoid bringing over contaminating material from the interface we suggest leaving behind a considerable amount of the aqueous phase, e.g. in this case, collecting only ~100 μL of the

~150 µL aqueous phase. We have found that the amount of sgRNA isolated is still >50 µg with this approach, which is more than adequate for tens of standard microinjections.

6. It is crucial that there is no remaining wash liquid left in the tube.

7. The 0.5× TBE gels are non-denaturing gel, however they provide a rapid and good indication as to sgRNA quality (i.e. degraded or not). For RNA work we recommend that the gel box and "teeth" are clean, and that only fresh running buffer be used.

8. The CRISPR-Cas9 approach exhibits a very high routine functionality in zygotes without prior testing. Also in the case of mouse, germline testing would require a high level of resources/time; i.e. it is faster and directly relevant to use the sgRNA and assess its activity directly in vivo rather than in cell lines.

9. RNAsin is a protein and would be destroyed by the denaturation protocol if added before this step. Also, dsDNA (plasmid) should not be subjected to the denaturation protocol as resulting nicks/breaks could result in a more difficult sample to microinject.

10. It is unclear where such microparticulate debris can accrue from, however we have found this centrifugation step prevents frustration during microinjection due to blocked microinjection needles.

11. Stud male colonies used for zygote production must be maintained with detailed records on the mating performance (i.e. success of females with copulation plug) and age. This will allow purging of nonproductive males to maintain optimal fecundity.

12. Often checking for the presence of a copulation plug is considered optional. We have found that for certain strains, including C57BL/6J, C57BL/6NJ, females without a copulation plug can yield zygotes. As such, harvesting all paired females is a viable option. For other strains including NOD/ShiLtJ, NSG, NRG (respectively, JAX mouse strain reference # 1976, 5557, and 7799), paired females without visible copulation plugs rarely contain zygotes. When plug rates fall below 50 % it is not worth the effort to harvest these females. For expensive or difficult to obtain mouse strains, females (depending on the starting age) with no copulation plug may be reused after 2–4 weeks (if not pregnant), however with variable limited success.

13. At this stage processed zygotes can be graded for fertility and morphology. Careful grading will allow more efficient use of injection systems. High quality optics are required to identify any zygotes worth injecting which should show polar bodies and the presence of two distinct pronuclei.

14. Culture dishes containing microdrops under silicone fluid need to have been prepared ~24 h in advance to equilibrate to 37 °C, in 5 % CO_2/5 % O_2/90 % N_2; e.g. in a K-MINC-1000 benchtop incubator (or equivalent).

15. CRISPR-Cas9 modification often requires the microinjection of high concentrations of nucleic acids, this may require developing a needle pulling protocol for pronuclear/cytoplasmic injections using a gradual elongated taper from the shoulder to the point. In addition, the temper of the glass should be such to allow for multiple chips while retaining a sharp point.

16. If possible the reagents should be deposited deep within the cytoplasm and the needle should swiftly exit the cell. If deposited close to the oolemma the chances of the cell lysing increase greatly. If a plume of cellular material is seen external to the oolemma (under the zona) after exiting, the likelihood of zygote survival is small.

17. Survival of zygotes to live born is dependent upon zygote quality, mouse background, microinjectionist skill, cleanness, and reagent concentrations of microinjected material. Survival also depends upon the nature of the zygote injection, for example with C57BL/6 J zygotes post cytoplasmic injections, we see 30–40 % survival to live born, whilst with pronuclear injection, routinely we observe 20–30 % survival. Lastly, putative genetically edited founder animals are a precious resource. To help ensure success surgically implanted females should be monitored closely on and around their due date. It is also strongly advised that dedicated foster litters be produced in conjunction with all surgically generated litters. This allows in the case of dystocia or other birthing difficulties, that viable pups can be rescued by performing a Caesarean section and using arranged foster mothers.

18. It can be tempting to believe that using more tissue is better, however this is not the case. These methods are easily overloaded, and the use of too much tissue will inhibit the subsequent PCR reaction. For most robust PCR reactions, 0.1–1 µL of crude NaOH lysate (estimated at ~30–300 ng) will work well, however especially for longer range PCRs (>2 kb) the proteinase K digestion preps may be preferable.

Acknowledgements

This work was done in close cooperation with The Jackson Laboratory microinjection team. We also fully acknowledge the help and excellent assistance of Cindy Avery and Deb Woodworth for genotyping and animal maintenance, plus discussions regarding CRISPR with Drs. Vishnu Hosur, Wenning Qin, and Gabriele Proetzel. Funding for this work was provided by the National Institutes of Health Grant OD011190 (M.V.W.) and The Jackson Laboratory. We also give our thanks to Addgene for supplying many of the plasmids used in our studies.

References

1. Doetschman T, Gregg RG, Maeda N, Hooper ML, Melton DW, Thompson S, Smithies O (1987) Targetted correction of a mutant HPRT gene in mouse embryonic stem cells. Nature 330:576–578

2. Capecchi MR (1989) Altering the genome by homologous recombination. Science 244:1288–1292

3. Gaj T, Gersbach CA, Barbas Iii CF (2013) ZFN, TALEN, and CRISPR/Cas-based methods for genome engineering. Trends Biotechnol 31:397–405

4. Hsu PD, Lander ES, Zhang F (2014) Development and applications of CRISPR-Cas9 for genome engineering. Cell 157:1262–1278

5. Singh P, Schimenti JC, Bolcun-Filas E (2015) A mouse geneticist's practical guide to CRISPR applications. Genetics 199:1–15

6. Carbery ID, Ji D, Harrington A, Brown V, Weinstein EJ, Liaw L, Cui X (2010) Targeted genome modification in mice using zinc-finger nucleases. Genetics 186:451–459, 10.1534/genetics.110.117002

7. Orlando SJ, Santiago Y, DeKelver RC, Freyvert Y, Boydston EA, Moehle EA, Choi VM, Gopalan SM, Lou JF, Li J, Miller JC, Holmes MC, Gregory PD, Urnov FD, Cost GJ (2010) Zinc-finger nuclease-driven targeted integration into mammalian genomes using donors with limited chromosomal homology. Nucleic Acids Res 38:e152

8. Cermak T, Doyle EL, Christian M, Wang L, Zhang Y, Schmidt C, Baller JA, Somia NV, Bogdanove AJ, Voytas DF (2011) Efficient design and assembly of custom TALEN and other TAL effector-based constructs for DNA targeting. Nucleic Acids Res 39:e82

9. Bedell VM, Wang Y, Campbell JM, Poshusta TL, Starker CG, Krug RG 2nd, Tan W, Penheiter SG, Ma AC, Leung AY, Fahrenkrug SC, Carlson DF, Voytas DF, Clark KJ, Essner JJ, Ekker SC (2012) In vivo genome editing using a high-efficiency TALEN system. Nature 491(7422):114–118

10. Ishino Y, Shinagawa H, Makino K, Amemura M, Nakata A (1987) Nucleotide sequence of the iap gene, responsible for alkaline phosphatase isozyme conversion in Escherichia coli, and identification of the gene product. J Bacteriol 169:5429–5433

11. Horvath P, Barrangou R (2010) CRISPR/Cas, the immune system of bacteria and archaea. Science 327:167–170

12. Jinek M, Chylinski K, Fonfara I, Hauer M, Doudna JA, Charpentier E (2012) A programmable dual-RNA-guided DNA endonuclease in adaptive bacterial immunity. Science 337:816–821

13. Doudna JA, Charpentier E (2014) Genome editing. The new frontier of genome engineering with CRISPR-Cas9. Science 346:1258096. doi:10.1126/science.1258096

14. Esvelt KM, Mali P, Braff JL, Moosburner M, Yaung SJ, Church GM (2013) Orthogonal Cas9 proteins for RNA-guided gene regulation and editing. Nat Methods 10:1116–1121

15. Kleinstiver BP, Prew MS, Tsai SQ, Topkar V, Nguyen NT, Zheng Z, Gonzales APW, Li Z, Peterson RT, Yeh JRJ, Aryee MJ, Joung JK (2015) Engineered CRISPR-Cas9 nucleases with altered PAM specificities. Nature 523:481–485

16. Shmakov S, Abudayyeh OO, Makarova KS, Wolf YI, Gootenberg JS, Semenova E, Minakhin L, Joung J, Konermann S, Severinov K, Zhang F, Koonin EV (2015) Discovery and functional characterization of diverse class 2 CRISPR-Cas Systems. Mol Cell 60:1–13

17. Zetsche B, Gootenberg JS, Abudayyeh OO, Slaymaker IM, Makarova KS, Essletzbichler P, Volz SE, Joung J, van der Oost J, Regev A, Koonin EV, Zhang F (2015) Cpf1 is a single RNA-guided endonuclease of a novel Class 2 CRISPRCas system. Cell 163:759–771

18. Nishimasu H, Ran FA, Hsu PD, Konermann S, Shehata SI, Dohmae N, Ishitani R, Zhang F, Nureki O (2014) Crystal Structure of Cas9 in Complex with Guide RNA and Target DNA. Cell 156:935–949

19. Sternberg SH, Redding S, Jinek M, Greene EC, Doudna JA (2014) DNA interrogation by the CRISPR RNA-guided endonuclease Cas9. Nature 507:62–67

20. Jinek M, East A, Cheng A, Lin S, Ma E, Doudna J (2013) RNA-programmed genome editing in human cells. Elife 2:e00471. doi:10.7554/eLife.00471

21. Bétermier M, Bertrand P, Lopez BS (2014) Is non-homologous end-joining really an inherently error-prone process? PLoS Genet 10:e1004086. doi:10.1371/journal.pgen.1004086

22. Rogakou EP, Pilch DR, Orr AH, Ivanova VS, Bonner WM (1998) DNA double-stranded breaks induce histone H2AX phosphorylation on serine 139. J Biol Chem 273:5858–5868

23. Cong L, Ran FA, Cox D, Lin S, Barretto R, Habib N, Hsu PD, Wu X, Jiang W, Marraffini LA, Zhang F (2013) Multiplex genome engineering using CRISPR/Cas systems. Science 339:819–823

24. Jiang W, Bikard D, Cox D, Zhang F, Marraffini LA (2013) RNA-guided editing of bacterial genomes using CRISPR-Cas systems. Nat Biotechnol 31:233–239

25. Liang P, Xu Y, Zhang X, Ding C, Huang R, Zhang Z, Lv J, Xie X, Chen Y, Li Y, Sun Y, Bai Y, Songyang Z, Ma W, Zhou C, Huang J (2015) CRISPR/Cas9-mediated gene editing in human tripronuclear zygotes. Protein Cell 6:363–372

26. Yang H, Wang H, Jaenisch R (2014) Generating genetically modified mice using CRISPR/Cas-mediated genome engineering. Nat Protoc 9:1956–1968

27. Chen X, Xu F, Zhu C, Ji J, Zhou X, Feng X, Guang S (2014) Dual sgRNA-directed gene knockout using CRISPR/Cas9 technology in Caenorhabditis elegans. Sci Rep 4:7581. doi:10.1038/srep07581

28. Dicarlo JE, Norville JE, Mali P, Rios X, Aach J, Church GM (2013) Genome engineering in Saccharomyces cerevisiae using CRISPR-Cas systems. Nucleic Acids Res 41:4336–4343

29. Wang H, Yang H, Shivalila CS, Dawlaty MM, Cheng AW, Zhang F, Jaenisch R (2013) One-step generation of mice carrying mutations in multiple genes by CRISPR/Cas-mediated genome engineering. Cell 153:910–918

30. Yang H, Wang H, Shivalila CS, Cheng AW, Shi L, Jaenisch R (2013) One-step generation of mice carrying reporter and conditional alleles by CRISPR/Cas-mediated genome engineering. Cell 154:1370–1379

31. Mali P, Aach J, Stranges PB, Esvelt KM, Moosburner M, Kosuri S, Yang L, Church GM (2013) CAS9 transcriptional activators for target specificity screening and paired nickases for cooperative genome engineering. Nat Biotechnol 31:833–838

32. Hsu PD, Scott DA, Weinstein JA, Ran FA, Konermann S, Agarwala V, Li Y, Fine EJ, Wu X, Shalem O, Cradick TJ, Marraffini LA, Bao G, Zhang F (2013) DNA targeting specificity of RNA-guided Cas9 nucleases. Nat Biotechnol 31:827–832

33. Fu Y, Foden JA, Khayter C, Maeder ML, Reyon D, Joung JK, Sander JD (2013) High-frequency off-target mutagenesis induced by CRISPR-Cas nucleases in human cells. Nat Biotechnol 31(9):822–826

34. Wu X, Scott DA, Kriz AJ, Chiu AC, Hsu PD, Dadon DB, Cheng AW, Trevino AE, Konermann S, Chen S, Jaenisch R, Zhang F, Sharp PA (2014) Genome-wide binding of the CRISPR endonuclease Cas9 in mammalian cells. Nat Biotechnol 32:670–676

35. Lin Y, Cradick TJ, Brown MT, Deshmukh H, Ranjan P, Sarode N, Wile BM, Vertino PM, Stewart FJ, Bao G (2014) CRISPR/Cas9 systems have off-target activity with insertions or deletions between target DNA and guide RNA sequences. Nucleic Acids Res 42:7473–7485

36. Tsai SQ, Zheng Z, Nguyen NT, Liebers M, Topkar VV, Thapar V, Wyvekens N, Khayter C, Iafrate AJ, Le LP, Aryee MJ, Joung JK (2015) GUIDE-seq enables genome-wide profiling of off-target cleavage by CRISPR-Cas nucleases. Nat Biotechnol 33:187–198

37. Gabriel R, Von Kalle C, Schmidt M (2015) Mapping the precision of genome editing. Nat Biotechnol 33(2):150–152. doi:10.1038/nbt.3142

38. Prykhozhij SV, Rajan V, Gaston D, Berman JN (2015) CRISPR multitargeter: a web tool to find common and unique CRISPR single guide RNA targets in a set of similar sequences. PLoS One 10:e0119372, 10.1371/journal.pone.0119372

39. Wiles MV, Qin W, Cheng A, Wang H (2015) CRISPR-Cas9 mediated genome editing and guide RNA design. Mamm Genome 26:501–510

40. Ran FA, Hsu PD, Lin CY, Gootenberg JS, Konermann S, Trevino AE, Scott DA, Inoue A, Matoba S, Zhang Y, Zhang F (2013) Double nicking by RNA-guided CRISPR Cas9 for enhanced genome editing specificity. Cell 154:1380–1389

41. Guilinger JP, Thompson DB, Liu DR (2014) Fusion of catalytically inactive Cas9 to FokI nuclease improves the specificity of genome modification. Nat Biotechnol 32:577–582

42. Tsai SQ, Wyvekens N, Khayter C, Foden JA, Thapar V, Reyon D, Goodwin MJ, Aryee MJ, Joung JK (2014) Dimeric CRISPR RNA-guided FokI nucleases for highly specific genome editing. Nat Biotechnol 32:569–576

43. Fu Y, Sander JD, Reyon D, Cascio VM, Joung JK (2014) Improving CRISPR-Cas nuclease specificity using truncated guide RNAs. Nat Biotechnol 32:279–284

44. Montagutelli X (2000) Effect of the genetic background on the phenotype of mouse mutations. J Am Soc Nephrol 11:S101–S105

45. Nakagawa Y, Sakuma T, Sakamoto T, Ohmuraya M, Nakagata N, Yamamoto T (2015) Production of knockout mice by DNA microinjection of various CRISPR/Cas9 vectors into freeze-thawed fertilized oocytes. BMC Biotechnol 15:33

46. Tanaka Y, Yamada Y, Ishitsuka Y, Matsuo M, Shiraishi K, Wada K, Uchio Y, Kondo Y, Takeo T, Nakagata N, Higashi T, Motoyama K, Arima H, Mochinaga S, Higaki K, Ohno K, Irie T (2015) Efficacy of 2-hydroxypropyl-β-cyclodextrin in Niemann-Pick disease type C

model mice and its pharmacokinetic analysis in a patient with the disease. Biol Pharm Bull 38:844–851

47. Festing M (1999) Warning: the use of heterogeneous mice may seriously damage your research. Neurobiol Aging 20(2):237–244, discussion 245–236

48. Festing M (2004) The choice of animal model and reduction. Altern Lab Anim 32(Suppl 2):59–64

49. Shen B, Zhang J, Wu H, Wang J, Ma K, Li Z, Zhang X, Zhang P, Huang X (2013) Generation of gene-modified mice via Cas9/RNA-mediated gene targeting. Cell Res 23:720–723

50. Qin W, Dion SL, Kutny PM, Zhang Y, Cheng A, Jillette NL, Malhotra A, Geurts AM, Chen YG, Wang H (2015) Efficient CRISPR/Cas9-mediated genome editing in mice by zygote electroporation of nuclease. Genetics 200: 423–430

51. Li D, Qiu Z, Shao Y, Chen Y, Guan Y, Liu M, Li Y, Gao N, Wang L, Lu X, Zhao Y (2013) Heritable gene targeting in the mouse and rat using a CRISPR-Cas system. Nat Biotechnol 31:681–683

52. Chen F, Pruett-Miller SM, Huang Y, Gjoka M, Duda K, Taunton J, Collingwood TN, Frodin M, Davis GD (2011) High-frequency genome editing using ssDNA oligonucleotides with zinc-finger nucleases. Nat Methods 8:753–755

53. Zhang L, Jia R, Palange NJ, Satheka AC, Togo J, An Y, Humphrey M, Ban L, Ji Y, Jin H, Feng X, Zheng Y (2015) Large genomic fragment deletions and insertions in mouse using CRISPR/Cas9. PLoS One 10:e0120396. doi:10.1371/journal.pone.0120396

54. Zhang Y, Ge X, Yang F, Zhang L, Zheng J, Tan X, Jin ZB, Qu J, Gu F (2014) Comparison of non-canonical PAMs for CRISPR/Cas9-mediated DNA cleavage in human cells. Sci Rep 4:5405. doi:10.1038/srep05405

55. Ostermeier GC, Wiles MV, Farley JS, Taft RA (2008) Conserving, distributing and managing genetically modified mouse lines by sperm cryopreservation. PLoS One 3(7):e2792

56. Wiles MV, Taft RA (2010) The sophisticated mouse: Protecting a precious reagent. Methods Mol Biol 602:23–36

57. Chu VT, Weber T, Wefers B, Wurst W, Sander S, Rajewsky K, Kühn R (2015) Increasing the efficiency of homology-directed repair for CRISPR-Cas9-induced precise gene editing in mammalian cells. Nat Biotechnol 33:543–548

58. Maruyama T, Dougan SK, Truttmann MC, Bilate AM, Ingram JR, Ploegh HL (2015) Increasing the efficiency of precise genome editing with CRISPR-Cas9 by inhibition of nonhomologous end joining. Nat Biotechnol 33:538–542

59. Chapman KM, Medrano GA, Jaichander P, Chaudhary J, Waits AE, Nobrega MA, Hotaling JM, Ober C, Hamra FK (2015) Targeted germline modifications in rats using CRISPR/Cas9 and spermatogonial stem cells. Cell Rep 10:1828–1835

60. Zhou X, Xin J, Fan N, Zou Q, Huang J, Ouyang Z, Zhao Y, Zhao B, Yi X, Guo L, Esteban MA, Zeng Y, Yang H, Lai L (2015) Generation of CRISPR/Cas9-mediated gene-targeted pigs via somatic cell nuclear transfer. Cell Mol Life Sci 72:1175–1184

61. Zambrowicz BP, Imamoto A, Fiering S, Herzenberg LA, Kerr WG, Soriano P (1997) Disruption of overlapping transcripts in the rosa beta-geo 26 gene trap strain leads to widespread expression of beta-galactosidase in mouse embryos and hematopoietic cells. Proc Natl Acad Sci U S A 94:3789–3794

62. Soriano P (1999) Generalized lacZ expression with the ROSA26 Cre reporter strain. Nat Genet 21:70–71

63. Tan W, Carlson DF, Lancto CA, Garbe JR, Webster DA, Hackett PB, Fahrenkrug SC (2013) Efficient nonmeiotic allele introgression in livestock using custom endonucleases. Proc Natl Acad Sci U S A 110:16526–16531

64. Fineran PC, Dy RL (2014) Gene regulation by engineered CRISPR-Cas systems. Curr Opin Microbiol 18:83–89

65. Akeson AL, Wetzel B, Thompson FY, Brooks SK, Paradis H, Gendron RL, Greenberg JM (2000) Embryonic vasculogenesis by endothelial precursor cells derived from lung mesenchyme. Dev Dyn 217:11–23

66. Oehler D, Poehlein A, Leimbach A, Müller N, Daniel R, Gottschalk G, Schink B (2012) Genome-guided analysis of physiological and morphological traits of the fermentative acetate oxidizer Thermacetogenium phaeum. BMC Genomics 13:723

67. Chylinski K, Makarova KS, Charpentier E, Koonin EV (2014) Classification and evolution of type II CRISPR-Cas systems. Nucleic Acids Res 42:6091–6105

68. Hou Z, Zhang Y, Propson NE, Howden SE, Chu LF, Sontheimer EJ, Thomson JA (2013) Efficient genome engineering in human pluripotent stem cells using Cas9 from Neisseria meningitidis. Proc Natl Acad Sci U S A 110:15644–15649

69. Kim D, Bae S, Park J, Kim E, Kim S, Yu HR, Hwang J, Kim JI, Kim JS (2015) Digenome-

seq: genome-wide profiling of CRISPR-Cas9 off-target effects in human cells. Nat Methods 12:237–243

70. Mandal PK, Ferreira LMR, Collins R, Meissner TB, Boutwell CL, Friesen M, Vrbanac V, Garrison BS, Stortchevoi A, Bryder D, Musunuru K, Brand H, Tager AM, Allen TM, Talkowski ME, Rossi DJ, Cowan CA (2014) Efficient ablation of genes in human hematopoietic stem and effector cells using CRISPR/Cas9. Cell Stem Cell 15:643–652

71. Chen B, Gilbert LA, Cimini BA, Schnitzbauer J, Zhang W, Li GW, Park J, Blackburn EH, Weissman JS, Qi LS, Huang B (2013) Dynamic imaging of genomic loci in living human cells by an optimized CRISPR/Cas system. Cell 155(7):1479–1491

72. Lennox KA, Sabel JL, Johnson MJ, Moreira BG, Fletcher CA, Rose SD, Behlke MA, Laikhter AL, Walder JA, Dagle JM (2006) Characterization of modified antisense oligonucleotides in Xenopus laevis embryos. Oligonucleotides 16:26–42

73. Bassett AR, Tibbit C, Ponting CP, Liu JL (2014) Highly efficient targeted mutagenesis of drosophila with the CRISPR/Cas9 system. Cell Rep 6:1178–1179

74. Luo C, Zuñiga J, Edison E, Palla S, Dong W, Parker-Thornburg J (2011) Superovulation strategies for 6 commonly used mouse strains. J Am Assoc Lab Anim Sci 50:471–478

75. Behringer R, Gertsensten M, Vintersten K, Nagy A (2013) Manipulating the Mouse Embryo; A Laboratory manual, 4th edn. Cold Spring Harbor Press, New York

Chapter 3

Mouse Sperm Cryopreservation and Recovery of Genetically Modified Mice

Benjamin E. Low, Rob A. Taft, and Michael V. Wiles

Abstract

Highly definable genetically, the humble mouse is the "reagent" mammal of choice with which to probe and begin to understand the human condition in all its complexities. With the recent advance in direct genome editing via targeted nucleases, e.g., TALEN and CRISPR/Cas9, the possibilities in using these sophisticated tools have increased substantially leading to a massive increase in the variety of strain numbers of genetically modified lines. With this increase comes a greater need to economically and creatively manage their numbers. Further, once characterized, lines may be of limited use but still need to be archived in a format allowing their rapid resurrection. Further, maintaining colonies on "the shelf" is financially draining and carries potential risks including natural disaster loss, disease, and strain contamination. Here we outline a simple and economic protocol to cryopreserve mouse sperm from many different genetic backgrounds, and outline its recovery via in vitro fertilization (IVF). The combined use of sperm cryopreservation and IVF now allows a freedom and versatility in mouse management facilitating rapid line close down with the option to later recover and rapidly expand as needed.

Key words Genetically modified mice, Inbred mice, Sperm cryopreservation, IVF, In vitro fertilization, Genetic drift, Genetic contamination, CRISPR/Cas9, Speed expansion

1 Introduction

1.1 The Reagent-Grade Mouse

Since the time of alchemy, scientific research has continually strived to refine and define its basic experimental reagents. Reagents are tools of the trade for the scientist and their quality defines the science. The resulting progress seen in the last ~200 years has allowed alchemy to become science, which is now driving a revolution in our ever-increasing understanding of life. In the field of experimental biomedical and genetic research, the evolution of the mouse as a reagent has been similar to that of other reagents with a continual refinement in its precise definition. Now, with the advent of genome sequencing, inbred mouse strains can be *completely* defined and we believe that the term reagent grade can justifiably be applied to the inbred mouse [1]. As with any reagent its

Gabriele Proetzel and Michael V. Wiles (eds.), *Mouse Models for Drug Discovery: Methods and Protocols*,
Methods in Molecular Biology, vol. 1438, DOI 10.1007/978-1-4939-3661-8_3, © Springer Science+Business Media New York 2016

basic attributes must be understood if it is to be used successfully, including purity and/or quality, and how to maintain its full functionality. Mice are living creatures, and to say the least are a highly complex reagent.

Two characteristics make the "reagent-grade" mouse invaluable as a model organism. First, it is possible to introduce precise genetic modifications into the mouse genome at will. These modifications include gene ablation, addition, and introduction of precise genetic changes enabling the visualizing of the resulting phenotype in a complete living organism [2–6]. Second, comparisons of the mouse and human genomes reveal that mice and humans shared a common ancestor, diverging about 75 million years ago, and although by no means perfect analogue, this *tractable* model shows that ~99 % share gene function with humans [7, 8]. However, "living reagents" require continual maintenance if they are not to be lost or degraded. It is here that mouse management by cryopreservation can be used, providing versatile archiving and management of these valuable resources.

1.2 Why Cryopreserve Mouse Strains

Inbred mice, defined by >20 generations of brother-sister matings, are the most precise mammal available for experimental manipulation, with individual mice within each inbred strain being essentially clones (>99 % homozygosed at all loci). This allows precise experimental comparisons within strains, between multiple inbred strains, and between genetically modified vs. non-modified mice [9]. Inbred mice are living animals and are maintained as continual breeding colonies. If this is not done correctly and with care, then their genetic background will change. The major sources of genetic change are (1) fast, disastrous genetic contamination (i.e., in one breeding cycle), and (2) insidious genetic drift which occurs over years. Appropriate use of cryopreservation can forestall the cumulative adverse effects of genetic drift, including copy number variation (CNV), and allow rapid restoration of strains if genetic contamination, disease, phenotypic shifts, etc. occur (for a review [1]). Additionally, cryopreservation increases mouse management options, facilitating cost-effective colony management.

In regard to more custom genetically modified strains, although it is tempting to believe that these strains are safe during active experimental work, *all* vivariums carry risks including the possibility of disease, breeding cessation, genetic contamination, and other disasters. For example, in June 2000, the tropical storm Allison caused the flooding of vivariums at the Texas Medical Center killing more than 30,000 mice and rats, causing incalculable losses [10]. In 2012 hurricane Sandy brought storm surges to a number of research laboratories including the NYU Langone Medical Center in New York. Strain backup is therefore a prudent management step and also facilitates dynamic cost management allowing strains to be closed down and upon demand rapidly re-initiated.

The combined use of sperm cryopreservation and IVF allows a freedom and versatility in mouse management, facilitating rapid line close down with the option to later recover and rapidly expand as needed.

1.3 Mouse Archiving: Options

In the management of any resource a key consideration is "return on investment"; that is, in this case, it is of little value to cheaply store/archive mouse strains in a format that makes their recovery prohibitively expensive and/or unpredictable (*see* Table 1 for summary of approaches), while at the same time it is not viable to invest large sums into the archiving of strains if the likelihood of them ever being wanted at a later date is small.

Mouse strains can be archived as embryos (2–8 cell), gametes (sperm, oocytes), or as sources of gametes (spermatogonial stem cells, ovaries); *see* Table 1. When looking at costs to cryopreserve and recover mouse strains, sperm is in general the most logical choice based on the ease of collection and the sheer numbers of sperm available from a single male ($\sim 30 \times 10^6$ sperm/male). However, until recently recovery of live born, especially when derived from C57BL/6 sperm, the most commonly used mouse strain, was less than 5%, making routine recovery expensive and unpredictable.

Females can provide naturally ~6–8 oocytes, or upon superovulation up to 50 oocytes. However, this is highly strain dependent; for example, C57Bl/6J can give high numbers of oocytes, while 129 strains give very few. There is also evidence to suggest that embryo quality falls with the use of superovulation and high oocyte yields [11, 12].

While occasionally necessary (e.g., multiple modifications that would require extensive time/resources to return alleles to homozygosity), cryopreservation of embryos is considerably more expensive than sperm cryopreservation. This is true financially as well as in terms of animals required and up-front time invested. For example, to produce ~250 two-cell embryos by IVF requires >15 females (assuming a C57BL/6 background), IVF with overnight culture, followed by selection of two-cell embryos and transfer into storage tubes. If the strains are never recovered or only recovered once or twice, then this expense remains forever frozen. By contrast, sperm cryopreservation significantly reduces the initial investment, using only a few males, and relatively little labor and materials are required [13]. Animals and labor are then only invested upon recovery using in vitro fertilization (IVF), and this can be scaled up or down, depending on the desired number of offspring required, allowing for a more rapid deployment of the strain.

Archiving strains in this way can save valuable time and resources while still safeguarding against catastrophic loss. As the ability for even the smallest lab to create a virtually unlimited number of unique GE mouse models continues to increase, we put

Table 1
Comparison of methods and cost for cryopreservation and recovery of mouse strains

Method	Cryopreservation		Recovery		Comparative cost/animal
	Pros	Cons	Pros	Cons	Preserve and recover
Sperm cryopreservation	Very simple Needs only 1–3 carrier males ~30x10⁶ sperm/ male Inexpensive Highly reliable	Only half the genome is preserved Needs IVF quality control to be absolutely safe	Moderately simple to recover Highly strain reproducible Highly scalable	Requires IVF to recover strain Requires appropriate oocyte donor strain Some strains and mutations adversely effect IVF success Offspring will be heterozygotes	$
Ovary cryopreservation	Moderately simple Inexpensive	Only half the genome is preserved Needs multiple donor females, i.e., one female has only two ovaries	Represents female linage, i.e., mitochondrial DNA	Due to potential rejection needs appropriate recipient animals Moderate surgical skill to implant Low yield of offspring/ovary Offspring will be heterozygotes Can transmit disease Not scalable	$$
Two-cell embryo cryopreservation (i.e., heterozygote embryos generated via IVF)	Moderately simple Needs only 1–3 carrier males	Only half the genome is preserved Needs IVF to make embryos Needs female (wild type) oocyte donors Strain dependent 5–50 oocytes/female Only half the genome is persevered	Simple recovery into pseudopregnant animals Can be used to achieve strain rederivation	Low yield of offspring Offspring will be heterozygotes Not scalable (based on initial investment)	$$$

Method	Advantages	Disadvantages		Cost
2–8-cell embryo cryopreservation (i.e., flushed homozygous embryos)	Simple Provides complete "homozygous" storage of a strain	Needs a large colony of the target strain to supply necessary numbers of females for embryos	Simple to implant Can be used to achieve rederivation of strain	Expensive resource to restock Not scalable $$$$
ICSI[a]	Simple Inexpensive storage	Only half the genome is preserved Technically demanding	None	Requires a high level of resources/skill to recover Pathogen transmission could occur Offspring will be heterozygotes associated with genetic damage Not scalable $$$$$
ES (iPS) cells[b]	Simple Inexpensive	Time	None	Need to make germline transmitting chimeras Upon germline transmission only half the ES derived offspring will be heterozygotes ES cells carry tissue culture associated genetic damage Not scalable $$$$$

When comparing costs, it should also be considered how often cryopreserved strains would be recovered, i.e., if infrequent, then lower cryopreservation vs. higher recovery costs become more attractive. Lastly, where a strain has multiple genetic modifications, cryopreservation as sperm with recovery on wild-type animals will only produce heterozygous animals, complicating and extending the time to recover the strain as homozygous animals

[a]Intracytoplasmic sperm injection, ICSI, although not strictly necessary a cryopreservation method—this approach has been used as a method to archive and restore mouse strains

[b]Embryonic stem cells, ES, cells are often generated as part of the process for a targeted genetic modification. Although a strain can be recreated from the original targeted ES line it requires the recreation of germline chimeras and their successful germline breeding before the strain is recovered. Further, ES cells are known to mutate in culture. Alternatively, induced pluripotent stem (iPS) cells can be used; depending on how these have been established and cultured, their germline transmission efficiency varies

Note that all the above approaches are also subject to strain effects and to possible deleterious effects of gene modification or addition

forth this very simple and cost-effective system to archive strains that may not currently be the lab's primary focus, but still hold promise for future investigation. Ultimately, these technologies allow a lab to be more nimble, to follow the science without completely discarding other potentially fruitful avenues of research.

A major disadvantage of sperm cryopreservation is that only half the genome of a strain is being stored by this approach (i.e., donor sperm is haploid!). This does not represent a major problem when the genetic modification is on a readily available standard background, as high-quality stock is readily available from *reputable* breeders for most standard inbred mice. Assuming one allele of interest and Mendelian ratios, recovery of sperm via IVF will produce heterozygous mice in the first generation, and ¼ homozygous mice should result in the second generation, following a filial mating of the recovered offspring. However, in the case of multiple alleles of interest and/or the presence of transgenes, the number of generations (i.e., time, labor, money, and mice) required to recreate the genotype of interest may far exceed the cost of simply cryopreserving two-cell embryos at the outset. These factors should be carefully considered before deciding on which method is better suited to archiving a given strain.

1.4 How Safe Is Gamete Cryopreservation

There are papers discussing the integrity of cryopreserved gametes; however, it is accepted that once material passes through the glass transition temperature of water, i.e., –137 °C, all biological activity ceases. Although gamma radiation can still cause accumulative damage, simulation studies have indicated that this is insignificant over ~2,000 years under normal background radiation levels [14, 15].

Of much greater concern with long-term storage is temperature variation, where gametes are exposed to temperature fluctuations above –137 °C. The most likely causes of temperature variation (or increase) are improper handling of the frozen gametes (e.g., while "rummaging" in the liquid nitrogen storage tanks); failure to fill liquid nitrogen tanks, i.e., they run dry; destruction of the storage facility due to fire, etc.; or the physical failure of the tank's vacuum [16]. As such, it is *strongly* recommended that cryopreserved gametes be stored physically in at least two liquid nitrogen storage tanks and additionally that tanks be in two or more separate facilities as one part of a comprehensive approach to repository operation [17].

1.5 Sperm Cryopreservation and Recovery

The sperm cryopreservation approach outlined below is essentially as published [13]; however there have been recent advances in recovery of sperm for IVF. For example using N-acetyl-L-cysteine (NAC) has been shown to increase the rate of fertilization when added to the IVF medium but did not adversely affect embryo development in vitro or in vivo [18].

Lastly although it is perhaps prudent to test cryopreserved sperm by doing a trial IVF, implanting and obtaining live born, we have found that once you have mastered the approach this is not an absolute requirement. Although leaving out this QC step can increase the risk of recovery failure, it will considerably reduce costs. This risk, which may have seemed unacceptable just a few years ago, should now be reconsidered in light of the current technologies available to rapidly recreate simple genetic modifications using targeted nucleases.

2   Materials

2.1   Cryopreservation of Mouse Sperm

2.1.1   Cryoprotective Media

1. Distilled water (Invitrogen, cat # 15230–238).
2. 18 % w/v raffinose (Sigma cat # R7630).
3. 3 % w/v skim milk (BD Diagnostics cat # 232100).
4. MTG: 447 μM Monothioglycerol (Sigma cat # M6145).

2.1.2   Consumables

1. 0.25 mL French straws (IMV cat# AAA201).
2. Cassettes (Zander Medical Supplies, 145 mm 16980/0601).
3. Styrofoam box internal dimensions ~ 35 cm × ~30 cm, a styrofoam float (piece should be approx. ~2–3 cm thick and be cut to cover ~80 % of internal area of box).
4. Monoject Insulin 1 mL syringe.

2.1.3   Mice

Two to three male mice, preferably 10–16 weeks old (*see* **Note 1**).

2.2   In Vitro Fertilization Method

2.2.1   Hormones for Superovulation

1. Pregnant mare serum gonadotropin (PMSG).
2. Human chorionic gonadotropin (hCG).
3. Sterile physiological saline.

2.2.2   Mice for Superovulation

Five to ten female mice, 3 weeks or 6–12 weeks of age, depending on the strain.

2.2.3   In Vitro Fertilization

1. MVF media: COOKS Mouse Vitro Fert Fertilization medium (Cook MVF, Australia).
2. Mixed gas (5 % O_2, 5 % CO_2, balanced with N_2).
3. One large 60 × 100 mm Falcon dish for every three females (Falcon).
4. One small 35 × 10 Falcon dish for each male (Falcon).
5. Embryo-tested mineral oil.
6. PBS: Phosphate-buffered saline.

3 Methods

It is critical that all media are carefully prepared, and have the correct pH and temperature, as well as batch-tested reagents. Further, it is helpful to pay attention to details and efficient laboratory setup, e.g., small incubators, heated stages, or other devices that ensure proper temperature and pH stability.

3.1 Sperm Cryopreservation

3.1.1 Preparation of Cryoprotective Media (CPM)

1. Place ~80 mL of bottled distilled water in a beaker.
2. Heat for ~40 s in microwave (do not boil).
3. Place beaker on heated stir plate, add 18 g of raffinose, and heat and stir till solution clears (*see* **Note 2**).
4. Add 3 g of skim milk to the raffinose mixture and heat and stir until dissolved (*see* **Note 3**).
5. Transfer solution to volumetric flask and bring to 100 mL with bottled distilled water.
6. Add MTG now (cool to < 30 °C) or after thawing (*see* **Note 4**).
7. Mix well and divide the solution into two 50 mL centrifuge tubes.
8. Centrifuge at $13,000 \times g$ for 15 min at room temperature (~22 °C).
9. Filter through a 0.22 μm cellulose filter (a prefilter may help the flow).
10. Verify that the osmolarity is in the range of 470–490 mOsm.
11. Aliquot 10 mL of filtered cryoprotective media into labeled 15 mL conical tubes.
12. Cap and store at –20 °C until ready for use (*see* **Note 5**).

3.1.2 Sperm Cryopreservation Setup

1. Thaw and warm CPM in 37.5 °C water bath (*see* **Note 6**) while the media is warming:
2. Label and mark straws and affix to a 1 mL Monoject syringe.
3. Fill Styrofoam box to a depth of 6–9 cm with liquid nitrogen.
4. Place Styrofoam float into Styrofoam box.
5. Replace Styrofoam box lid to slow the evaporation of liquid nitrogen.
6. Place the lid from Petri dish on the warming tray and lean the bottom of a Petri dish against it so that one side of the Petri dish is elevated. This arrangement forces the CPM to collect on one side, making it easier to fill the straws.
7. If monothioglycerol was not added prior to freezing, add it now, to a final concentration of 477 μM. Add 1 mL of CPM to the dish for each male from which sperm will be collected.

3.1.3 Sperm Collection	1. Euthanize the males (1–3) and remove the cauda epididymis and vas deferens, carefully removing the testicular artery to avoid contaminating the sperm with blood.
	2. Release sperm into the CPM by making several cuts through the epididymis and vas deferens using a beveled hypodermic needle while holding the tissues with a pair of forceps.
	3. Remove tissue from the CPM after 10 min.

3.1.4 Sperm Cryopreservation

1. Aspirate a 4.5 cm column of CPM into a straw followed by a 2 cm column of air.

2. Aspirate a 0.5 cm column of sperm into the straw and then aspirate additional air until the column of CPM without sperm contacts the PVA powder in the cotton plug.

3. Seal the end of the straw with a brief pulse from an instantaneous heat sealer.

4. Repeat this process until the desired number of straws has been filled (we suggest minimum of 20/strain).

5. Place five straws into one cassette. Repeat until four cassettes have been filled.

6. Place the cassettes in the liquid nitrogen-filled box on the float (i.e., in vapor phase) so that they are not touching.

7. Put the lid on the box and leave for 10–30 min.

8. Plunge the cassettes into the liquid nitrogen.

9. After at least 10 min in liquid nitrogen the cassettes can be removed and *rapidly* placed into storage in liquid nitrogen (*see* **Note 7**).

3.2 In Vitro Fertilization with Frozen Sperm

3.2.1 IVF Setup

1. Prepare oocyte collection dish by adding 2 mL of PBS to a 25 mm × 10 mm dish and keep at 37.5 °C in room air.

2. Prepare IVF dish by placing a 250 μL drop of MVF medium (*see* **Note 8**) in the center of a 60 mm Petri dish. Place four additional 150 μL drops of MVF medium around the 250 μL drop.

3. Carefully add sufficient oil to cover the media and place in an incubator or sealed chamber filled with mixed gas (5 % O_2, 5 % CO_2, 90 % N_2) at least 1 h prior to IVF (*see* **Note 9**).

3.2.2 Thawing Sperm

1. Place the straw in a 37 °C water bath.

2. Rapidly swirl the straw in the water until all ice has melted (about 30 s).

3. Dry the straw with a paper towel.

4. Cut off the sealed end of the straw opposite the cotton plug. Using a metal rod, expel the sperm from the straw into the 250 μL IVF drop.

5. Allow sperm to incubate at 37 °C for 1 h prior to adding oocytes.

Fertilization rates vary widely among commonly available inbred strains; introduction of mutations and genetic modifications can have indirect and unanticipated affects on the quantity and quality of sperm produced [13, 19]. The use of NAC should be considered as it has been demonstrated that with some strains it will increase IVF efficiency [18].

1. Inject females with PMSG.

2. Inject females with hCG 13 h prior to oocyte collection (*see* **Notes 10** and **11**).

3. Euthanize 2–5 superovulated females approximately 13 h post-hCG.

4. Remove the ovary, oviduct, and a small portion of the uterine horn and place in the dish containing PBS.

5. Repeat for all females.

6. Working under low magnification, identify the ampulla. Cumulus-enclosed oocytes should be easily visible within the ampulla of the oviduct. Using a beveled hypodermic needle, open the ampulla to release the cumulus-enclosed oocytes.

7. Repeat until all oocytes have been released.

8. Using a 1 mL pipette (or a wide-bore pipette tip) transfer the cumulus-enclosed oocytes to the dish containing MVF medium, transferring as little medium as possible.

9. Add 10 μL sperm using a wide-bore pipette tip to reduce shearing forces on sperm.

10. Incubate at 37 °C for 4 h under mixed gas.

11. Using a finely drawn glass pipette with a diameter slightly larger than an oocyte wash the oocytes through the 150 μL media drops to remove cumulus cells and sperm.

12. Culture overnight at 37 °C under mixed gas (*see* **Note 12**).

13. Count and evaluate embryos the following morning. Embryos can now be cultured, transferred to a pseudopregnant animal, or cryopreserved (*see* **Notes 13** and **14**).

4 Notes

1. Variations in sperm quality among males within a strain are not unusual—hence it is best to use two or three male donors and pool their sperm.

2. This is a nearly saturated solution; heating the solution makes it easier to get the raffinose into solution.

3. The solution will be opaque after the addition of the skim milk; centrifugation at room temperature ($13,000 \times g$ for 15 min) is recommended.

4. Addition of MTG is recommended immediately prior to use. Alternatively, it can be added in advance and the solution stored at –80 °C for up to 3 months. Solutions containing MTG should **not** be stored at 4 °C for more than a few days. MTG is viscous, making careful pipetting essential. Making a diluted stock solution that is added to the media helps reduce the likelihood that too much MTG will be added to the CPM. MTG diluted stock solution should be used on the day it is made and then discarded.

5. This solution can be stored for at least 6 months at –80 °C without MTG or up to 3 months at –80 °C with MTG.

6. Water baths are a common source of bacterial contamination and liquid nitrogen is not sterile. Straws should be carefully wiped to remove any moisture from the outside of the straw prior to cutting the end off or dispensing sperm to reduce the risk of contaminating the IVF.

7. It is essential that during the transfer from liquid nitrogen to long-term storage that the cassettes be handled rapidly to minimize unintentional warming.

8. COOKS Mouse Vitro Fert is similar to Human Tubal Fluid media reported by Quinn and can be made as described there [20].

9. Oil can be washed and filtered. Oil should be stored in a dark cool place.

10. Typical doses are in the range of 2.5–5 i.u. per mouse. Optimal dose varies by strain, age, and weight of the mouse. Extending the oocyte collection window beyond 14 h post-hCG may reduce fertilization rates and compromise embryo quality due to oocyte aging.

11. The response to superovulation is highly strain dependent [19] and some strains are entirely refractory to superovulation.

12. The use of a low O_2 culture environment may not improve fertilization, but may improve embryo quality [21, 22].

13. The laboratory environment can have a significant effect on the outcome of IVF [23, 24]. Materials that release volatile organic compounds (VOC) and cleaning/sanitizing agents such as bleach and floor waxes should be avoided.

14. Prior to embryo transfer, embryos should be washed following the IETS protocol if the sperm or oocytes were collected from animals with an unknown or unacceptable health status [25].

References

1. Taft RA, Davisson M, Wiles MV (2006) Know thy mouse. Trends Genet 22(12):649–653

2. Doetschman T, Gregg RG, Maeda N, Hooper ML, Melton DW, Thompson S, Smithies O (1987) Targetted correction of a mutant HPRT gene in mouse embryonic stem cells. Nature 330(6148):576–578

3. Thomas KR, Folger KR, Capecchi MR (1986) High frequency targeting of genes to specific sites in the mammalian genome. Cell 44(3): 419–428

4. Jinek M, Chylinski K, Fonfara I, Hauer M, Doudna JA, Charpentier E (2012) A programmable dual-RNA-guided DNA endonuclease in adaptive bacterial immunity. Science 337(6096):816–821

5. Singh P, Schimenti JC, Bolcun-Filas E (2015) A mouse geneticist's practical guide to CRISPR applications. Genetics 199(1):1–15

6. Roopenian DC, Low BE, Christianson GJ, Proetzel G, Sproule TJ, Wiles MV (2015) Albumin-deficient mouse models for studying metabolism of human albumin and pharmacokinetics of albumin-based drugs. MAbs 7(2):344–351

7. Waterston R, Lindblad-Toh K, Birney E, Rogers J, Abril J, Agarwal P, Agarwala R et al (2002) Initial sequencing and comparative analysis of the mouse genome. Nature 420(6915): 520–562

8. Takao K, Miyakawa T (2015) Genomic responses in mouse models greatly mimic human inflammatory diseases. Proc Natl Acad Sci U S A 112(4):1167–1172

9. Bogue MA, Grubb SC, Maddatu TP, Bult CJ (2007) Mouse Phenome Database (MPD). Nucleic Acids Res 35(suppl 1):D720–D730

10. Sincell M (2001) Houston flood: research toll is heavy in time and money. Science 293(5530):589

11. Fortier AL, Lopes FL, Darricarrère N, Martel J, Trasler JM (2008) Superovulation alters the expression of imprinted genes in the midgestation mouse placenta. Hum Mol Genet 17(11):1653–1665

12. Wang Y, Ock SA, Chian RC (2006) Effect of gonadotrophin stimulation on mouse oocyte quality and subsequent embryonic development in vitro. Reprod Biomed Online 12(3):304–314

13. Ostermeier GC, Wiles MV, Farley JS, Taft RA (2008) Conserving, distributing and managing genetically modified mouse lines by sperm cryopreservation. PLoS One 3(7):e2792

14. Whittingham DG (1980) Principles of embryo preservation. In: Ashwood-Smith MJ, Farrant J (eds) Low temperature preservation in medicine and biology pitman medical. Tunbridge Wells, England, pp 65–83

15. Lyon MF (1981) Sensitivity of various germ-cell stages to environmental mutagens. Mutat Res 87:323–345

16. Tomlinson M (2008) Risk management in cryopreservation associated with assisted reproduction. Cryo Letters 29(2):165–174

17. International Society for B, Environmental R (2008) Best practices for repositories: collection, storage, distribution and retrieval of biological materials for research. Cell Preserv Technol 6:3–58

18. Takeo T, Horikoshi Y, Nakao S, Sakoh K, Ishizuka Y, Tsutsumi A, Fukumoto K, Kondo T, Haruguchi Y, Takeshita Y, Nakamuta Y, Tsuchiyama S, Nakagata N (2015) Cysteine analogs with a free thiol group promote fertilization by reducing disulfide bonds in the zona pellucida of mice. Biol Reprod 92(4):90. doi:10.1095/biolreprod.114.125443

19. Byers SL, Payson SJ, Taft RA (2006) Performance of ten inbred mouse strains following assisted reproductive technologies (ARTs). Theriogenology 65(9):1716–1726, S0093-691X(05)00405-X doi: 10.1016/j.theriogenology.2005.09.016 [pii]

20. Quinn P (1995) Enhanced results in mouse and human embryo culture using a modified human tubal fluid medium lacking glucose and phosphate. J Assist Reprod Genet 12(2):97–105

21. Adam AA, Takahashi Y, Katagiri S, Nagano M (2004) Effects of oxygen tension in the gas atmosphere during in vitro maturation, in vitro fertilization and in vitro culture on the efficiency of in vitro production of mouse embryos. Jpn J Vet Res 52(2):77–84

22. Dumoulin JC, Vanvuchelen RC, Land JA, Pieters MH, Geraedts JP, Evers JL (1995) Effect of oxygen concentration on in vitro fertilization and embryo culture in the human and the mouse. Fertil Steril 63(1):115–119

23. Cohen J, Gilligan A, Esposito W, Schimmel T, Dale B (1997) Ambient air and its potential effects on conception in vitro. Hum Reprod 12(8):1742–1749

24. Hall J, Gilligan A, Schimmel T, Cecchi M, Cohen J (1998) The origin, effects and control of air pollution in laboratories used for human embryo culture. Hum Reprod 13(Suppl 4):146–155

25. Stringfellow DA, Seidel SM (1998) Manual of the international embryo transfer society: procedural guide and general information for the use of embryo transfer technology, emphasizing sanitary procedures, 3rd edn. International Embryo Transfer Society, Illinois

Chapter 4

Generation of Immunodeficient Mice Bearing Human Immune Systems by the Engraftment of Hematopoietic Stem Cells

Suheyla Hasgur, Ken Edwin Aryee, Leonard D. Shultz, Dale L. Greiner, and Michael A. Brehm

Abstract

Immunodeficient mice are being used as recipients of human hematopoietic stem cells (HSC) for in vivo analyses of human immune system development and function. The development of several stocks of immunodeficient *Prkdcscid* (*scid*), or recombination activating 1 or 2 gene (*Rag1* or *Rag2*) knockout mice bearing a targeted mutation in the gene encoding the IL2 receptor gamma chain (*IL2rγ*), has greatly facilitated the engraftment of human HSC and enhanced the development of functional human immune systems. These "humanized" mice are being used to study human hematopoiesis, human-specific immune therapies, human-specific pathogens, and human immune system homeostasis and function. The establishment of these model systems is technically challenging, and levels of human immune system development reported in the literature are variable between laboratories. The use of standard protocols for optimal engraftment of HSC and for monitoring the development of the human immune systems would enable more direct comparisons between humanized mice generated in different laboratories. Here we describe a standard protocol for the engraftment of human HSC into 21-day-old NOD-*scid IL2rγ* (NSG) mice using an intravenous injection approach. The multiparameter flow cytometry used to monitor human immune system development and the kinetics of development are described.

Key words Hematopoietic stem cells, SCID, Thymus, HSC, Humanized mice, NSG, IL2rγ, Transplantation

1 Introduction

1.1 Immunodeficient Mice Bearing Mutations Within the IL2rγ Gene as Recipients of Human HSC

The generation of immunodeficient mice that support the engraftment of human immune systems has enabled the in vivo study of human immune system development and function [1]. Early efforts to engraft human immune systems into mice utilized the C.B-17 strain bearing the *Prkdcscid* (*scid*, severe combined immune deficiency) mutation [2], which permitted low levels of human immune cell engraftment after injection with peripheral blood mononuclear cells (PBMC) or hematopoietic stem cells (HSC) but

Gabriele Proetzel and Michael V. Wiles (eds.), *Mouse Models for Drug Discovery: Methods and Protocols*,
Methods in Molecular Biology, vol. 1438, DOI 10.1007/978-1-4939-3661-8_4, © Springer Science+Business Media New York 2016

overall immune system function was limited [1]. The development of NOD-*scid* mice [3] improved engraftment of human immune systems but overall function and levels of take were still suboptimal for the study of human immunobiology [4, 5]. The introduction of immunodeficient *Prkdcscid* (*scid*) or recombination activating 1 or 2 gene (*Rag1* or *Rag2*) knockout mice bearing a mutated interleukin-2 receptor gamma chain (*IL2rγnull*) facilitated greatly the in vivo engraftment and function of human immune cells [6–9]. The IL2rγ chain is required for high-affinity signaling of IL-2, IL-4, IL-7, IL-9, IL-15, and IL-21 cytokines, and *IL2rγnull* mice have severe defects in natural killer (NK) cell activity in addition to T and B cell development [10]. These new strains of immunodeficient *IL2rγnull* mice are now being used for studies of human hematopoiesis, innate and adaptive immunity, autoimmunity, infectious diseases, cancer biology, and regenerative medicine [11].

1.2 HSC Engraftment of Immunodeficient NOD-scid IL2rγ^null^ Mice

NOD.Cg-*Prkdcscid Il2rgtm1Wjl*/Sz (NOD-*scid IL2rγnull* or NSG) mice support engraftment of human HSC from a variety of sources, including G-CSF-mobilized peripheral blood, bone marrow aspirates, umbilical cord blood, and fetal liver [12–15]. In vivo human hematopoietic repopulation through transplantation of human CD34+ HSC in NSG recipients allows high levels of human HSC engraftment, differentiation of human T cells in the murine thymus and human B cells, differentiation of human myeloid subsets, and human immune function in vivo [12–14, 16–20]. A critical aspect for generating HSC-engrafted immunodeficient mice is the use of standardized protocols that enable consistent and robust human immune system development. For example, age of the mouse recipient has important implications for development of human immune cell subsets. One study has shown that newborn NSG mice support more efficient human T cell development after HSC injection than adult NSG mice (8–12 weeks) [20].

HSC injection into newborn mice is challenging, as the injection sites (intrahepatic, intracardiac, and facial vein) require technical expertise and in some instances survival is problematic [6]. Here we describe a standard protocol for the engraftment of human HSC into 21-day-old NSG mice using an intravenous injection approach. A description of the multiparameter flow cytometry used to monitor human immune system development is shown. The kinetics of human immune system development in 21-day-old NSG mice were compared to those in HSC-engrafted newborn NSG mice.

2   Materials

2.1 Human Cord Blood Preparation (See Note 1)

1. Citrate phosphate dextrose anticoagulant solution, USP (CPD) blood pack unit (Fenwal, Lake Zurich, IL): UCB specimens were provided by the University of Massachusetts Memorial

Umbilical Cord Blood Donation Program under Institutional Review Board (IRB) approval.

2. Hespan (6% Hetastarch/0.9% NaCl solution) (Braun Medical Inc, Bethlehem, PA): Hespan is used for volume reduction of UCB and preserves recovery of HSC [21].

3. Histopaque-1077 (Sigma-Aldrich, St. Louis, MO, USA): Histopaque is a density medium and is used for separating mononuclear cells from blood [22].

4. Bovine serum albumin (BSA) (Fisher Scientific, Pittsburgh, PA).

5. DNase I recombinant, RNase free (10,000 U/mL) (Roche, Indianapolis, IN).

6. Phosphate-buffered saline, PBS.

7. RPMI 1640 (Gibco, Life Technologies, Grand Island, NY USA).

8. 50 mL Centrifuge tubes (BD Falcon, Franklin Lakes, NJ, USA).

9. Water bath.

2.2 CD3 T Cell Depletion (See Note 2)

1. Human CD3 microbead kit (Miltenyi Biotech, Auburn, CA, USA).

2. MidiMACS Separator (Miltenyi Biotech, Auburn, CA, USA).

3. MACS multistand (Miltenyi Biotech, Auburn, CA, USA).

4. LD columns (Miltenyi Biotech, Auburn, CA, USA).

5. MACS buffer: PBS supplemented with 0.5% fetal bovine serum (FBS, Atlanta Biologicals, Lawrenceville, GA, USA), 2 mM EDTA, and sterilized with a vacuum flask (0.2 μm filter).

2.3 Flow Cytometry Analysis

1. FACS buffer: PBS supplemented with 1% FBS, and 0.1% sodium azide.

2. Flow cytometry analysis of T cell-depleted UCB.

Human CD3, Clone UCHT1 (BD Biosciences, San Jose, CA, USA).

Human CD34, Clone 581 (BD Biosciences, San Jose, CA, USA).

Human CD45, Clone HI30 (BD Biosciences, San Jose, CA, USA).

3. Fluorescent antibodies for screening of HSC-engrafted NSG mice.

Anti-Mouse CD45 (mLy5), Clone 30-F11 (BD Biosciences, San Jose, CA, USA).

Anti-Human CD45, Clone HI30 (BD Biosciences, San Jose, CA, USA).

Anti-Human CD3, Clone UCHT1 (BD Biosciences, San Jose, CA, USA).

Anti-Human CD20, Clone 2H7 (BD Biosciences, San Jose, CA, USA).

Anti-Human CD33, Clone WM53 (BD Biosciences, San Jose, CA, USA).

Anti-Human CD14, Clone HCD14 (BioLegend, San Diego, CA, USA).

2.4 HSC Injection

The materials listed below are necessary for intravenous injection of HSC into mice that are 21–28 days old. The materials used for injecting newborn mice with HSC have been listed in detail previously [6].

1. Immunodeficient mice: NOD-*Prkdcscid IL2rgtm1Wjl* (NSG) (The Jackson Laboratory, Bar Harbor ME, Stock No:005557) mice between 3 and 4 weeks of age (*see* **Note 3**).

2. ^{137}Cs gamma irradiator.

3. Autoclaved, filtered, ventilated device for holding mice during irradiation.

4. 1 mL Syringe with 27-G needle.

5. Heating pad or warming lamp.

6. Tail vein restrainer for mice (Braintree Scientific, Braintree, MA).

3 Methods

The protocols described below involve the manipulation of primary human cells. All work should be done in a standard laminar flow hood and with appropriate personal protective equipment. Waste materials should be disposed of using protocols approved by an Institutional Biosafety Committee. All mouse injections and handling should be done using protocols approved by an Institutional Animal Care and Use Committee.

3.1 Preparation of Umbilical Cord Blood

1. Allow histopaque and RPMI supplemented with 0.5 % BSA to warm to room temperature.

2. Transfer umbilical cord blood (UCB) to 50 mL conical tubes (30 mL per tube).

3. Add hespan to each tube of UCB to a final concentration of 20 % per volume and mix gently.

4. Incubate for 30 min at room temperature. The incubation period is to allow red blood cells (RBC) to sediment.

5. Remove bottom layer of RBC with 10 mL pipette, leaving approximately 2 mL of the RBC volume.

6. Add RPMI supplemented with 0.5 % BSA to each tube to bring to a total volume of 30 mL.

7. Slowly underlay 14 mL of histopaque to each tube containing UCB. Ensure that a clear interface is maintained between the histopaque and the cell-containing medium.

8. Spin for 30 min at $300 \times g$ and 25 °C with centrifuge brake off.

9. Remove the top layer of plasma, leaving 2 mL volume above buffy coat layer.

10. Remove buffy coat layer, and transfer to a new 50 mL conical tube.

11. Add 30 mL of RPMI supplemented with 0.5 % BSA and centrifuge for 5 min $400 \times g$ and 4 °C.

12. Discard supernatant.

13. Pool pellets from each 50 mL conical tube in a total volume of 10 mL of RPMI supplemented with 0.5 % BSA (*see* **Note 4**).

14. Perform viability count.

15. Reserve 100,000 cells for flow cytometric analysis.

16. Centrifuge cells for 5 min at $400 \times g$ and 4 °C and proceed to Subheading 3.2.

3.2 Depletion of CD3+ T Cells

Our laboratory uses reagents from Miltenyi Biotech to deplete human CD3+ T cells from UCB samples. The manufacturer's protocol was followed for depletion of human T cells (*see* **Note 2**).

1. Resuspend CD3-depleted UCB cells in 1 mL RPMI supplemented with 0.5 % BSA and perform viability counts.

2. To validate depletion of CD3+ T cells, stain recovered cells with antibodies specific for human CD3 and human CD45. T cell levels (cells staining double positive for CD3 and human CD45) should be below 0.5 % of total cells. Cells saved from Subheading 3.1, **step 15**, should be used as a positive staining control.

3. To determine the CD34+ HSC levels, stain the recovered cells with antibodies specific for human CD34 and human CD45. HSC are identified as CD45 dim and CD34 positive.

4. Resuspend cells in RPMI supplemented with 0.5 % BSA. Keep cells on ice until injection.

3.3 HSC Injection

The protocol for injecting newborn mice with HSC has been described in detail previously [6]. Here we focus on intravenous injection of HSC into 21–28-day-old mice.

1. Irradiate recipient mice by whole-body gamma irradiation. For young (21–28 days old) NSG mice a conditioning dose of 100 cGy is normally well tolerated and enables efficient engraftment. For engraftment of newborn NSG mice, 100 cGy is also used. HSC injection should be performed between 4 and 24 h after irradiation.

2. Warm recipient mice using warming lamp.

3. Inject CD3-depleted UCB cells to obtain a total of 1×10^5 CD34+ cells per injection into the lateral tail vein. For newborn mice shown here, 5×10^4 CD34+ cells were injected by intra-cardiac route into 24–72-h-old NSG pups.

4. Ensure that bleeding from injection site has stopped and return recipient mouse to cage.

3.4 Evaluation of Human Cell Chimerism Levels by Flow Cytometry

HSC-engrafted NSG mice are normally screened for human cell chimerism levels in peripheral blood between 12 and 15 weeks post-injection. This evaluation is easily done by flow cytometric analysis of peripheral blood cells to validate the human immune system development. For analysis, all cells positive for mouse CD45 are excluded from the gating strategy. Human immune cells are identified as staining positive for human CD45 and subpopulations are defined from those cells. The immune cell subsets that develop will be dependent on the age of the recipient mouse and the specific time point evaluated. The results shown below compare human immune system development between HSC-engrafted newborn NSG mice and HSC-engrafted young (21-day-old) NSG mice.

1. Levels of human CD45+ cells.

   Newborn and 21-day-old NSG mice were irradiated and injected intravenously by intracardiac route (newborn) or via the tail vein (3 weeks old) with T cell-depleted UCB containing 1×10^5 CD34$^+$ cells as described above. Between 6 and 18 weeks after HSC injection, the percentages of human CD45+ cells in the blood were determined by flow cytometry (Fig. 1). Each symbol represents an individual animal and mice injected with the same HSC are identified by symbols of similar shape and color. Higher proportions of human CD45+ cells were detected in the blood of 21-day-old, HSC-engrafted NSG mice at 6, 9, 12, and 18 weeks as compared to newborn mice (Fig. 1). In addition higher levels of human CD45+ cells were detected in the spleen and bone marrow of the 21-day-old NSG mice compared with levels of circulating cells (Fig. 2). These results show that 21-day-old NSG mice and newborn NSG mice support HSC engraftment and that 21-day-old mice show faster kinetics of human immune system development.

2. Human T cell development.

   The kinetics of human CD3+ T cell development were compared between newborn and 21-day-old NSG mice that had been injected with human HSC as described above (Fig. 3) The levels of human T cells detected in the peripheral blood were low at the 6- and 9-week time points for both groups of HSC-engrafted NSG mice and began to increase by week 12. At weeks 15 and 18 the levels of T cells were significantly

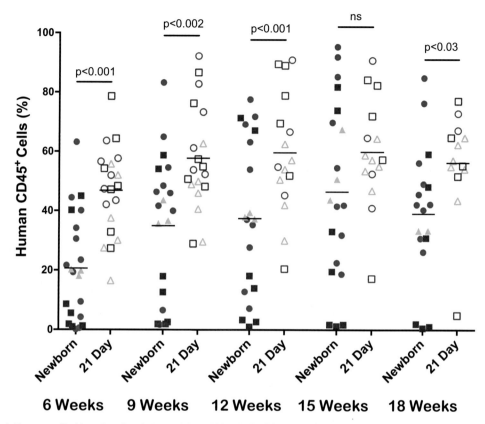

Fig. 1 Human cell chimerism levels in peripheral blood of NSG mice engrafted with human HSC. Newborn and 21-day-old NSG mice were irradiated and injected intravenously via intracardiac route (newborn) or the tail vein (21 days old) with T cell-depleted UCB containing CD34$^+$ cells. At 6–18 weeks after HSC injection, the percentages of human CD45-positive cells were determined by flow cytometry. Each symbol represents an individual animal, and points of the same color and shape are from independent experiments ($n = 3$)

higher in newborn NSG mice as compared to 21-day-old NSG mice. These data suggest that both newborn and 21-day-old NSG mice support human T cell development, but the kinetics are accelerated in mice injected as newborns (*see* **Note 5**).

3. B cell development.

The kinetics of human CD20+ B cell development were compared between newborn and 21-day-old NSG mice that had been injected with human HSC as described above (Fig. 4). Both newborn and 21-day-old injected NSG mice supported B cell development and significantly higher levels were detected in 21-day-old mice at 6, 15, and 18 weeks. Overall a significant proportion of the CD45+ cells in NSG mice were human B cells at all time points.

4. Monocyte/macrophage development.

The kinetics of human CD14+/CD33+ monocyte/macrophage development were compared between newborn and

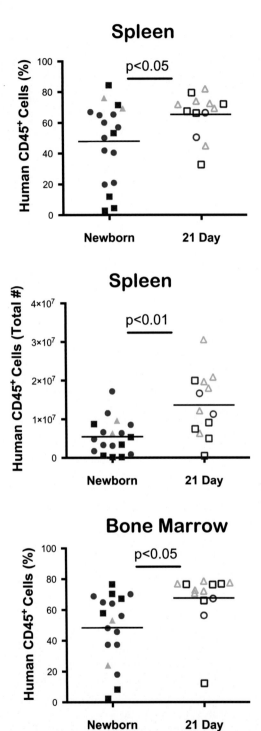

Fig. 2 Human cell chimerism levels in spleen and bone marrow of NSG mice engrafted with human HSC. Newborn and 21-day-old NSG mice were irradiated and injected intravenously via intracardiac route (newborn) or the tail vein (21 days old) with T cell-depleted UCB containing CD34+ cells. Eighteen weeks after HSC injection, the percentages and number of human CD45+ cells in the spleen and percentages in the bone marrow were determined by flow cytometry. Each symbol represents an individual animal, and points of the same color and shape are from independent experiments ($n = 3$)

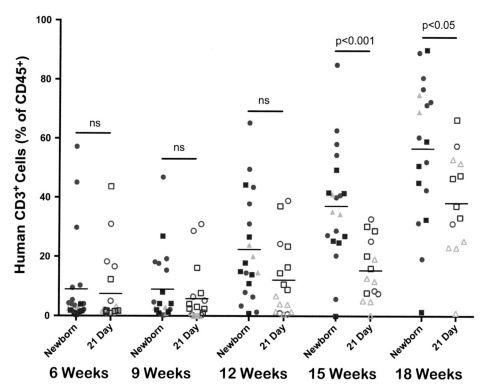

Fig. 3 Human CD3+ T cell development in peripheral blood of NSG mice engrafted with human HSC. Newborn and 21-day-old NSG mice were irradiated and injected intravenously via intracardiac route (newborn) or the tail vein (21 days old) with T cell-depleted UCB containing CD34+ cells. At 6–18 weeks after HSC injection, the percentages of human CD3-positive T cells were determined by flow cytometry. Each symbol represents an individual animal, and points of the same color and shape are from independent experiments ($n = 3$)

21-day-old NSG mice that had been injected with human HSC as described above (Fig. 5). Both newborn and 21-day-old NSG mice supported low levels of human monocyte/macrophage development with slightly higher levels detected in 21-day-old mice at the 15-week time point. At all time points tested CD14+/CD33+ cells were detectable in the peripheral blood.

4 Notes

1. UCB is a reliable source of functional human HSC. Alternative sources include G-CSF-mobilized peripheral blood, bone marrow aspirates, and fetal liver [23]. Each source will have differences in preparation, cell yields, and engraftment capacity and the characteristics of the developed human immune system.

2. The standard protocol for our laboratory is to use CD3-depleted UCB cells for injecting CD34+ HSC into mouse recipients. The depletion of CD3+ T cells is essential to prevent

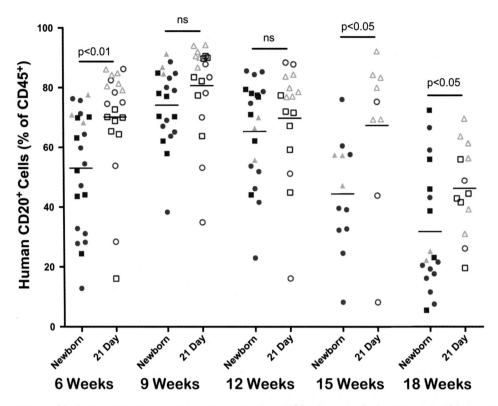

Fig. 4 Human CD20+ B cell development in peripheral blood of NSG mice engrafted with human HSC. Newborn and 21-day-old NSG mice were irradiated and injected intravenously via intracardiac route (newborn) or the tail vein (21 days old) with T cell-depleted UCB containing CD34+ cells. At 6–18 weeks after HSC injection, the percentages of human CD20-positive T cells were determined by flow cytometry. Each symbol represents an individual animal, and points of the same color and shape are from independent experiments ($n = 3$)

development of acute xenogeneic graft-versus-host disease (GVHD) in the mice. Alternatively, purified CD34+ cells can also be injected into the recipient mice [24]. One advantage for the CD3 depletion approach is that accessory cells (CD34 negative) present within the UCB specimen have been shown to enhance engraftment of human HSC in immunodeficient mice [25].

3. Immunodeficient mice bearing mutations within *IL2rγ* are the ideal recipients for the engraftment of human HSC. NSG mice were developed at The Jackson Laboratory by back-crossing a complete null mutation at the *IL2rγ* locus onto the NOD.Cg-*Prkdcscid* (NOD/SCID) strain [12]. Alternative mouse strains bearing mutations within the *IL2rγ* locus have been described previously [7, 23].

4. DNase (15 μL 10,000 U/mL DNase/30 mL) can be added if pellet is viscous after resuspension.

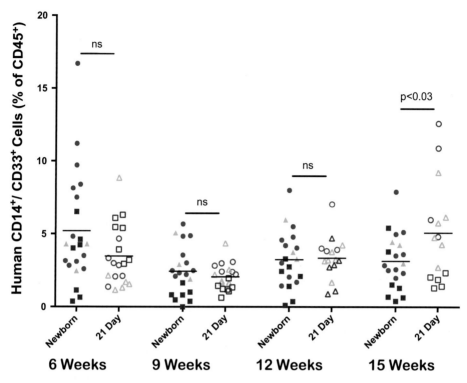

Fig. 5 Human monocyte/macrophage development in peripheral blood of NSG mice engrafted with human HSC. Newborn and 21-day-old NSG mice were irradiated and injected intravenously via intracardiac route (newborn) or the tail vein (21 days old) with T cell-depleted UCB containing CD34+ cells. At 6–15 weeks after HSC injection, the percentages of human CD14/CD33+ positive myeloid cells were determined by flow cytometry. Each symbol represents an individual animal, and points of the same color and shape are from independent experiments ($n = 3$)

5. Immunodeficient mice expressing HLA class I and II have been used to support HLA-restricted human T cell development and are available from The Jackson Laboratory [26].

Acknowledgement

This work was supported by National Institutes of Health Grants AI046629, AI112321, DK104218, CA034196, and OD018259 and grants from the Helmsley Charitable Trust and the Juvenile Diabetes Research Foundation. The authors would like to thank TUBITAK (the Scientific and Technological Research Council of Turkey) for the research support (2214A). MAB and DLR are consultants for The Jackson Laboratory (Bar Harbor, ME). The contents of this publication are solely the responsibility of the authors and do not necessarily represent the official views of the National Institutes of Health.

References

1. Shultz LD et al (2012) Humanized mice for immune system investigation: progress, promise and challenges. Nat Rev Immunol 12(11):786–798

2. Bosma GC, Custer RP, Bosma MJ (1983) A severe combined immunodeficiency mutation in the mouse. Nature 301(5900):527–530

3. Shultz LD et al (1995) Multiple defects in innate and adaptive immunologic function in NOD/LtSz-scid mice. J Immunol 154(1):180–191

4. Greiner DL, Hesselton RA, Shultz LD (1998) SCID mouse models of human stem cell engraftment. Stem Cells 16(3):166–177

5. Hesselton RM et al (1995) High levels of human peripheral blood mononuclear cell engraftment and enhanced susceptibility to human immunodeficiency virus type 1 infection in NOD/LtSz-scid/scid mice. J Infect Dis 172(4):974–982

6. Pearson T, Greiner DL, and Shultz LD (2008) Creation of "humanized" mice to study human immunity. In: Coligan JE (ed) Curr Protoc in Immunol, Chapter 15:Unit 15.21. Wiley, Hoboken, pp 15.21.1-15.21.21

7. Shultz LD, Ishikawa F, Greiner DL (2007) Humanized mice in translational biomedical research. Nat Rev Immunol 7(2):118–130

8. Manz MG, Di Santo JP (2009) Renaissance for mouse models of human hematopoiesis and immunobiology. Nat Immunol 10(10): 1039–1042

9. Sugamura K et al (1996) The interleukin-2 receptor gamma chain: its role in the multiple cytokine receptor complexes and T cell development in XSCID. Annu Rev Immunol 14:179–205

10. Rochman Y, Spolski R, Leonard WJ (2009) New insights into the regulation of T cells by gamma(c) family cytokines. Nat Rev Immunol 9(7):480–490

11. Brehm MA, Shultz LD, Greiner DL (2010) Humanized mouse models to study human diseases. Curr Opin Endocrinol Diabetes Obes 17(2):120–125

12. Shultz LD et al (2005) Human lymphoid and myeloid cell development in NOD/LtSz-scid IL2R gamma null mice engrafted with mobilized human hemopoietic stem cells. J Immunol 174(10):6477–6489

13. Ito M et al (2002) NOD/SCID/gamma(c) (null) mouse: an excellent recipient mouse model for engraftment of human cells. Blood 100(9):3175–3182

14. Ishikawa F et al (2005) Development of functional human blood and immune systems in NOD/SCID/IL2 receptor {gamma} chain(null) mice. Blood 106(5):1565–1573

15. Kalscheuer H et al (2012) A model for personalized in vivo analysis of human immune responsiveness. Sci Transl Med 4(125):125ra30

16. Hiramatsu H et al (2003) Complete reconstitution of human lymphocytes from cord blood CD34+ cells using the NOD/SCID/gammacnull mice model. Blood 102(3):873–880

17. Tanaka S et al (2012) Development of mature and functional human myeloid subsets in hematopoietic stem cell-engrafted NOD/SCID/IL2rgammaKO mice. J Immunol 188(12):6145–6155

18. Traggiai E et al (2004) Development of a human adaptive immune system in cord blood cell-transplanted mice. Science 304(5667): 104–107

19. Legrand N, Weijer K, Spits H (2006) Experimental models to study development and function of the human immune system in vivo. J Immunol 176(4):2053–2058

20. Brehm MA et al (2010) Parameters for establishing humanized mouse models to study human immunity: analysis of human hematopoietic stem cell engraftment in three immunodeficient strains of mice bearing the IL2rgamma(null) mutation. Clin Immunol 135(1):84–98

21. Rubinstein P et al (1995) Processing and cryopreservation of placental/umbilical cord blood for unrelated bone marrow reconstitution. Proc Natl Acad Sci U S A 92(22): 10119–10122

22. Feldman DL, Mogelesky TC (1987) Use of Histopaque for isolating mononuclear cells from rabbit blood. J Immunol Methods 102(2):243–249

23. Rongvaux A et al (2013) Human hematolymphoid system mice: current use and future potential for medicine. Annu Rev Immunol 31:635–674

24. Lan P et al (2006) Reconstitution of a functional human immune system in immunodeficient mice through combined human fetal thymus/liver and CD34+ cell transplantation. Blood 108(2):487–492

25. Covassin L et al (2013) Human immune system development and survival of NOD-scid IL2rgamma (NSG) mice engrafted with human thymus and autologous hematopoietic stem cells. Clin Exp Immunol 174:372–388

26. Brehm MA et al (2013) Overcoming current limitations in humanized mouse research. J Infect Dis 208(Suppl 2):S125–S130

Chapter 5

Generation of Human Liver Chimeric Mice for the Study of Human Hepatotropic Pathogens

Markus von Schaewen, Gabriela Hrebikova, and Alexander Ploss

Abstract

Human liver chimeric mice have become valuable tools for the study of human hepatotropic pathogens and for the investigation of metabolism and pharmacokinetics of novel drugs. The evolution of the underlying mouse models has been rapid in the past years. The diverse fields of applications of those model systems and their technical challenges will be discussed in this chapter.

Key words Humanized mice, Human liver chimeric mice, Viral hepatitis, Hepatitis C virus, Hepatitis B virus, Hepatitis delta virus, Malaria, DMPK

1 Introduction

Pathogens affecting the liver substantially contribute to the global health burden imposed by human infectious diseases. At least 130 million people are chronically infected with hepatitis C virus (HCV) [1] and another 350 million with hepatitis B (HBV) [2]. Either infection, if left untreated, leads to the development of severe liver disease, including liver fibrosis, cirrhosis, and hepatocellular carcinoma. Malaria affects an even larger number of individuals, with over 200 million people infected annually, leading to approximately 600,000 deaths per year. The majority of cases are caused by the parasites *Plasmodium vivax* and *Plasmodium falciparum*, both of which have a liver stage as part of their complex life cycles [3].

The study of hepatotropic viruses and parasites has been challenging as those pathogens that cause disease in humans do not readily infect rodents and only few non-primate mammalian species. To address this technological gap, research efforts have focused on the development of humanized animal models. Mice are the small animal model of choice for humanization as they are readily amenable for genetic manipulation and there are many preexisting tools/reagents to investigate host responses. To overcome species barriers for studying human hepatotropic pathogens, two major

Gabriele Proetzel and Michael V. Wiles (eds.), *Mouse Models for Drug Discovery: Methods and Protocols*, Methods in Molecular Biology, vol. 1438, DOI 10.1007/978-1-4939-3661-8_5, © Springer Science+Business Media New York 2016

strategies have been pursued. One is to genetically humanize mice either by expressing human factors required to render mice susceptible to a given pathogen or by ablating dominant negative restriction factors of the host. In this chapter we focus on humanizing the liver through xenotransplantation.

Genetic humanization requires an in-depth understanding of the pathogen's life cycle to identify potential roadblocks in the murine host. For example for HCV, CD81 [4], scavenger receptor type B class 1 (SRB1) [5], occludin (OCLN) [6, 7], and claudin-1 (CLDN1) [8] are the minimal set of entry factors required to render mouse cell lines susceptible to HCV entry, with at least CD81 and OCLN needing to be of human origin [6]. This knowledge led to the construction of the first inbred mouse model supporting HCV entry into murine hepatocytes [6, 9, 10]. This model utilizes an adenoviral delivery approach to express the human entry factors and has already been used to evaluate the in vivo efficacy of broadly neutralizing anti-HCV antibodies [9, 11, 12]. However, viral replication and production of infectious particles are not supported in this model system. A more refined mouse model in which the human entry factors are expressed transgenically and innate immunity is blunted has overcome this issue. In this model, HCV replication leads to low-level viremia with production of infectious particles observed for up to 100 days [13].

Similarly, the recent discovery of human sodium-taurocholate co-transporting polypeptide (hNTCP) as a common entry factor for HBV and hepatitis D virus (HDV) [14, 15] allowed the construction of a mouse model transgenically expressing hNTCP [14, 16]. This mouse model supports infection with HDV albeit at low levels but not HBV, which is in agreement with previous in vitro observations in which hNTCP expression renders hepatoma cell lines from a variety of species susceptible to HDV. While only human, and not murine cell lines support HBV infection [15, 17].

The focus of the protocols in this chapter is on xenotransplantation for generating humanized mice, in particular human liver chimeric mice, whereby murine livers are engrafted with human hepatocytes (Fig. 1). All mouse strains used for the creation of such human liver chimeric mice have two common attributes: they are immunodeficient to prevent rejection of engrafted human hepatocytes and harbor an endogenous liver injury to provide a growth advantage to the xenograft.

Liver injury in the recipient strains can be inflicted by partial hepatectomy, treatment with chemical agents, e.g., retrorsine or tetrachloride, or more commonly by genetic manipulation. In the earliest genetic approach, murine urokinase plasminogen activator (uPA) was overexpressed transgenically under the control of the albumin promoter (Alb-uPA) [18]. The resulting liver failure was rescued following transplantation of wild-type murine hepatocytes [19]. Subsequent endeavors to backcross this strain to a background harboring a severe

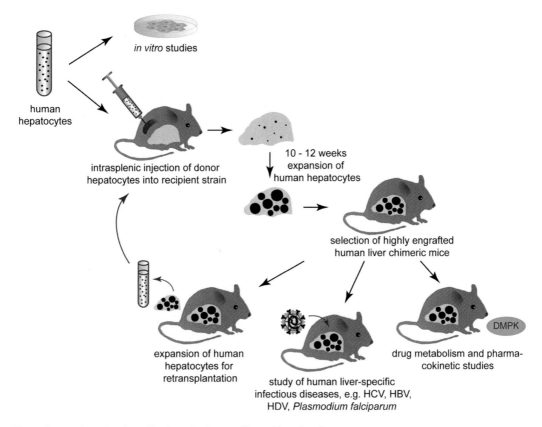

Fig. 1 Generation of and applications for human liver chimeric mice

combined immunodeficiency (SCID) resulted in the creation of a recipient strain with stable engraftment of human hepatocytes and the first small animal model (SCID/Alb-uPA mouse) that supported sustained HCV infection [20]. This model system has been utilized to study aspects of HCV biology, such as viral entry [21] and the role of anti-HCV antibodies [22]. In addition, this SCID/Alb-uPA strain supports HBV monoinfection and HBV/HDV coinfection and has been used to investigate an HBV entry inhibitor [23, 24]. Proof-of-concept has also been established for using this model to study the life cycle of the liver-dependent malaria parasite *Plasmodium falciparum* [25, 26].

As valuable as SCID/Alb-uPA mouse models are, they have several drawbacks. Homozygotes for the transgene cannot be maintained without engraftment of donor cells while heterozygotes yield dissatisfying engraftment levels. Additionally, the mortality of neonates is high due to a significant bleeding tendency and the window for engraftment of donor cells is fairly small. Some of these issues were resolved with the construction of Alb-uPA mice with a highly immunodeficient non-obese diabetic (NOD) SCID interleukin 2 receptor gamma deficient (IL-2Rγ^{null}) background. These uPA-NOG mice enable higher xenoengraftment, the breeding of homozygotes, and successful engraftment later in life [27]. However, no study has yet demonstrated that these animals are equally susceptible to hepatotropic pathogens. In another approach to overcome the limitations of SCID/alb-uPA mice, a model was generated in which uPA expression is driven by the major urinary protein (MUP) promoter, which is activated later in the animal's life. It has been demonstrated that the SCID/MUP-uPA mice can be engrafted with human hepatocytes [28] and are susceptible to infection with HBV and HCV [29].

An additional approach to genetically inflict host liver injury relies on a deficiency in the enzyme fumaryl acetoacetate hydrolase (FAH), which is responsible for hereditary tyrosinemia and liver failure in humans. In this model mice also suffer from liver failure due to the accumulation of hepatotoxic metabolites [30] and can be rescued by treatment with 2(2-nitro-4-trifluoromethylbenzoyl)-1,3 cyclohexanedione (NTBC) [31].

After transfer of human donor cells, withdrawal of NTBC leads to the death of the murine host hepatocytes thereby creating a growth advantage for the donor cells. One method to visualize the engraftment of human hepatocytes is performing an immunohistochemical or -fluorescence stain for FAH. Only human donor hepatocytes that are FAH sufficient will be stained and can be easily distinguished from the recipient hepatocytes (Fig. 2a–c) This model system again requires an immunodeficient background to prevent rejection of donor cells. FAH$^{-/-}$ mice crossed to a Rag2$^{-/-}$ IL-2Rγ^{NULL} background and mice crossed to an Rag1$^{-/-}$ IL2Rg^{NULL} (referred to as either FRG or FNRG mice) have been proven to allow high human hepatocyte engraftment levels [11, 32–35]. FNRG mice offer the advantage that breeding pairs on NTBC are healthy and viable and liver injury can be induced at any age by withdrawal of NTBC.

They have already been shown to be valuable tools for the study of a variety of pathogens. By using FNRG mice, for the first time it was possible to recapitulate the complete *Plasmodium falciparum* liver stage development in a small animal model [36].

Another study added to the understanding of HCV biology by providing proof-of-concept that broadly neutralizing antibodies can abrogate an established HCV infection in FNRG mice [11]. Likewise, proof-of-concept was established for the utility of liver humanized FRG for testing HBV therapeutics [35].

Another transgenic liver injury model relies on the herpes simplex virus thymidine kinase (HSV-TK), called TK-NOG (available from Taconic Biosciences, Hudson, NY, cat. no. 12907), that converts prodrugs like ganciclovir or acyclovir into toxic metabolites. Hasegawa et al. transgenically expressed HSV-TK in the livers of mice with a NOD-SCID IL-2Rγ^{NULL} background. They demonstrated that the gene expression pattern in the liver was representative of a mature human liver and exhibited a human-specific drug metabolism profile. The humanized liver was stably functional for up to 8 months and only one prodrug administration was necessary to induce a liver damage that was sufficient for successful xenoengraftment [37].

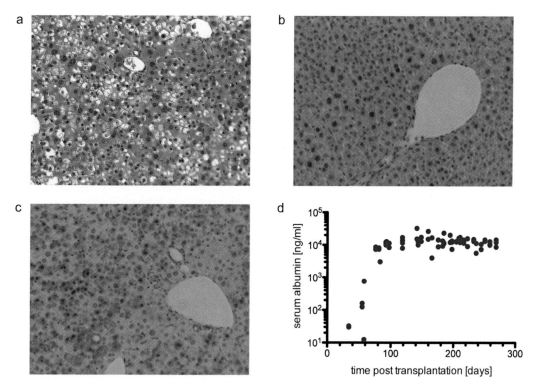

Fig. 2 Characterization of FNRG mice engrafted with human hepatocytes. Hematoxylin and eosin (HE) stain (**a**) and FAH stain (**b** and **c**) of FNRG mice engrafted with human hepatocytes (**a** and **c**) and nonengrafted FNRG mice (**c**). Human, FAH positive hepatocytes appear *brown*. Human albumin concentration in serum of FNRG mice engrafted with human hepatocytes over time (**d**)

A variety of cells, including hepatoma cell lines, primary adult hepatocytes, and induced hepatocyte-like cells (iHeps) derived from induced pluripotent stem cells (iPSCs), have been used to engraft xenorecipients. However, robust engraftment appears to be only possible with human adult hepatocytes. iHeps offer the advantage that they can be easily generated from readily accessible patient material such as skin fibroblasts and iPSCs are readily amenable to genetic manipulation prior to differentiation into HLCs [38, 39]. While engraftment with adult human hepatocytes has been well established, the engraftment of iPSCs has been challenging as they do not seem to respond to proliferation signals in the injured liver. So far only one study describes the successful engraftment of iHeps into a murine xenorecipient that consecutively supported the establishment of an HCV infection [40]. Regardless of the cell type used for the transplantation, the procedure is always the same: Cells are injected intrasplenically and travel via the portal venous system to the liver, traverse the endothelium into the liver parenchyma where they settle in clusters and start to proliferate.

In vivo model systems for studying human infectious diseases provide an important advantage in testing the efficacy and safety of potential drug candidates. Regular mouse models are of limited use as drug metabolism, pharmacokinetics (DMPK), and toxicology are highly species-specific. Additionally, drug-drug interactions resulting from induction or inhibition of metabolic enzymes cannot be predicted in non-humanized mice. To approach this problem, a variety of transgenic mouse models expressing human enzymes and regulators involved in drug metabolism have been engineered and shown able to predict important drug-drug interactions. For example, mouse models expressing the human P450 enzyme CYP3A4, which is involved in the metabolization of approximately 50 % of all clinically used drugs, have been designed [41, 42]. However, an even more complete approach to model human-type metabolic responses can be achieved by utilizing human liver chimeric mice. Studies have shown that expression levels of major P450 enzymes in chimeric livers of SCID/alb-uPA mice are comparable to human hepatocytes [43] and that the enzyme inhibition [44] and induction [45, 46] profiles are similar to those of human livers. Additionally, human liver chimeric mice have been proven to recapitulate the metabolic profile observed in humans for several drugs. For example, excretory metabolites in urine of the nonsteroidal antiphlogistics naproxen and ibuprofen were comparable in human liver chimeric mouse and to that observed in humans [47]. Both drugs are metabolized by P450 and UDP-glucuronyltransferase.

Despite the advantages human liver chimeric mouse models offer, they still have some shortcomings. To avoid graft rejection xenorecipient strains are usually highly immunodeficient and therefore cannot be utilized to study human adaptive immune responses or vaccine candidates. A potential strategy to overcome this is the creation of mouse models dually engrafted with human liver cells and a human immune system. One of the first promising landmark studies to address these issues was conducted by Washburn et al. [48]. This study provided proof-of-concept that the co-engraftment of genetically modified immunodeficient mice (AFC8-huHSC/hep mice) with CD34+ hematopoietic stem cells and human mixed fetal liver cells results in a model susceptible to HCV infection and able to generate human-specific immune responses [48]. More recent reports have demonstrated that extensive double humanization of both the liver and immune system can be achieved with mature hepatocytes and HSCs [49, 50]. Long-term dual reconstitution, without any evidence of hepatocyte rejection by the human immune system, was sustained even when the human cells were mis-matched in their major histocompatibility complex (MHC, [49]). The latter observation is consistent with the limited HLA matching in human liver transplantations, presumably due to the tolerogenic microenvironment of the liver.

The methods part of this chapter will focus on the generation and handling of human liver chimeric mice based on the FNRG model, since it is one of the most advanced and well-established models systems.

2 Materials

2.1 Mice

Mice with a targeted disruption in the fumaryl acetoacetate hydrolase gene ($Fah^{-/-}$) [30] were provided by M. Grompe [Oregon Health & Science University (OHSU)] and crossed for 13 generations onto the non-obese diabetic (NOD) recombinase activating gene 1 null ($Rag1^{-/-}$) interleukin 2 receptor common gamma chain null ($IL2R\gamma^{NULL}$) (NRG) background [33], yielding $Fah^{-/-}$ NOD $Rag1^{-/-}$ $IL2Rg^{NULL}$ (FNRG) mice [11].

FRG mice and FRG-N mice are available from Yecuris Corporation, Tualatin, OR.

2.2 Animal Chow and Drinking Supply

- Modified PicoLab Mouse diet w/ 0.12 % Amoxicillin (LabDiet, St. Louis, Missouri, USA).

- 2(2-nitro-4-trifluoromethylbenzoyl)-1,3 cyclohexanedione (NTBC) (Yecuris, Portland, Oregon, USA).

2.3 Anesthesia and Analgesia

- 100 mg/mL ketamine HCl (Hospira, Lake Forest, Illinois, USA).

- 20 mg/mL xylazine HCl (Lloyd, Shenandoah, Iowa, USA).

- Isoflurane (TW Medical, Lago Vista, Texas, USA).

- Bupivacaine HCl 0.5 % (Hospira, Lake Forest, Illinois, USA).

- Meloxicam for injection 5 mg/mL (Boehringer Ingelheim, Ingelheim, Germany).

- Meloxicam oral suspension 1.5 mg/mL (Boehringer Ingelheim, Ingelheim, Germany).

- Mouse anesthesia system (e.g., compac5, vetequip, Livermore, California, USA).

- 28G × ½ insulin syringes.

2.4 Detection of Human Albumin by Enzyme Linked Immunosorbent Assay (ELISA)

- Human serum albumin (Sigma Aldrich, St. Louis, MO, USA).

- Goat anti-human albumin polyclonal antibody (Bethyl Laboratories, Montgomery, TX, USA).

- Mouse anti-human albumin monoclonal antibody (Abcam, Cambridge, United Kingdom).

- Goat anti-mouse horseradish peroxidase (HRP) antibody (Life Technologies, Carlsbad, CA, USA).

- 96-well ELISA plates (Thermo Fisher Scientific, Waltham, MA, USA).

- 96-well plates, round bottom (Thermo Fisher Scientific, Waltham, MA, USA).

- Coating buffer: 1.59 g of Na_2CO_3, 2.93 g of $NaHCO_3$ to 1 L of distilled water, adjusted to pH of 9.6.

- SuperBlock blocking buffer (Thermo Fisher Scientific, Waltham, MA, USA).

- ELISA wash buffer: 0.05 % Tween 20 (Sigma Aldrich, St. Louis, MO, USA) in phosphate buffered saline (PBS).
- Sample diluent: 10 % SuperBlock and 90 % ELISA wash buffer.
- Tetramethylbenzidine (TMB) substrate (Sigma Aldrich, St. Louis, MO, USA).
- Stop solution: 2 N H_2SO_4.
- ELISA plate reader.

2.5 Hepatocyte Transplantation

- Human adult hepatocytes can be obtained from various commercial vendors, e.g., BioreclamationIVT (Baltimore, MD, USA), Triangle Research Labs (Triangle Research Park, NC, USA), or Life Technologies (Carlsbad, CA, USA).
- Human fetal liver cells can be obtained from Advanced Bioscience Resources (ABR, Inc., Alameda, CA, USA) or the Human Fetal Tissue Repository (Bronx, NY, USA).
- Dulbecco's Modified Eagle Medium (DMEM, Life Technologies, Carlsbad, CA, USA).
- 50 mL conical tubes (Nunc, Roskilde, Denmark).
- Two pairs of Mirco-Dissecting Scissors (Roboz Surgical Instrument, Gaithersburg, MD, USA).
- Two Moloney Forceps (Roboz Surgical Instrument, Gaithersburg, MD, USA).
- Water bath.
- Neubauer Chamber cytometer (Hausser Scientific, Horsham, PA, USA).
- 5 mL serological pipettes (BD Falcon, Franklin Lakes, NJ, USA).
- Ear tags (Kent Scientific, Torrington, CT, USA).
- Ear tag Applicator (Kent Scientific, Torrington, CT, USA).
- Betadine (Thermo Fisher Scientific, Waltham, MA, USA).
- 70 % ethanol spray.
- Stapler (Braintree Scientific, Braintree, MA, USA).
- Staples, autoclaved (Braintree Scientific, Braintree, MA, USA).
- Staple remover (Braintree Scientific, Braintree, MA, USA).
- Needle Feeder (Kent Scientific, Torrington, CT, USA).
- Cauterizer (Bovie Medical, Clearwater, FL, USA).
- Insulin syringe, 28G.
- 4-0 coated VICRYL suture (Ethicon, Somerville, NJ, USA).
- Heat lamp.
- Mouse dissection board.

2.6 Murine Hepatocyte Isolation

- HBM Basal Medium (Lonza Clonetics, Basel, Switzerland).
- HCM Single Quods (Lonza Clonetics, Basel, Switzerland).

- Collagen I coated 24-well plates (Corning, New York, NY, USA).

- CryoStor CS 10 (Bio Life Solutions, Bothell, WA, USA).

- 2 mL cryovial tubes (Corning, New York, NY, USA).

- Perfusor Ismatec (Cole Parmer, Vernon Hills, IL, USA).

- 3-stop silicone tubing 2.79 mm I.D. PK/6 (Cole Parmer, Vernon Hills, IL, USA).

- Intravenous administration set (Exel International Medical Products, Los Angeles, CA, USA).

- Safelet I.V. Catheter 24G×¾ in (Exel International Medical Products, Los Angeles, CA, USA).

- Surgical clamp (Fine Science Tools, Foster City, CA, USA).

- Water bath.

- 100 µm cell strainer nylon mesh (Thermo Fisher Scientific, Waltham, MA, USA).

- 70 µm cell strainer nylon mesh (Thermo Fisher Scientific, Waltham, MA, USA).

- Neubauer chamber cytometer (Hausser Scientific, Horsham, PA, USA).

- 50 mL conical tubes (Nunc, Roskilde, Denmark).

- Two Moloney forceps (Roboz Surgical Instrument, Gaithersburg, MD, USA).

- Two pairs of microdissecting scissors (Roboz Surgical Instrument, Gaithersburg, MD, USA).

- Two Q-tips.

- Insulin syringe, 28G.

- 60 mm petri dish (Thermo Fisher Scientific, Waltham, MA, USA).

- 2, 5, 10, and 25 mL serological pipettes (BD Falcon, Franklin Lakes, NJ, USA).

- 0.4 % Trypan blue (Sigma Aldrich, St. Louis, MO, USA).

- PBS (Life Technologies, Carlsbad, CA, USA).

- 70 % ethanol.

- Mouse dissection board.

2.7 Perfusion Solutions

- Solution A: 5 mL of 1 M 4-(2-hydroxyethyl)-1-piperazineethanesulfonic acid (HEPES), pH 7.3 and 0.5 M ethylene glycol tetraacetic acid (EGTA), pH 8.0 in 500 mL Earle's Balanced Salt Solution (EBSS) without Ca^{2+}, Mg^{2+}, or phenol red.

- Solution B: 5 mL of 1 M HEPES, pH 7.3 in 500 mL of EBSS with Ca^{2+}, Mg^{2+}, and phenol red.

- Solution C: 50 mL Fetal Bovine Serum (FBS), 5 mL of 1 M HEPES, pH 7.3, and 5 mL of Antibiotic/Antimycotic solution (HyClone, Logan, UT) in 500 mL DMEM (Life Technologies, Carlsbad, CA).

- Collagenase Solution: 50 mg of collagenase type 2 (Worthington, Lakewood, NJ) in 10 mL of PBS, store in 8 mL aliquots at –20 °C.

3 Methods

3.1 Introduction

Human adult hepatocytes, as well as human fetal liver cells (HFLCs), can be obtained from various commercial and not-for-profit entities. Cells acquired from these organizations should be tested by the investigator for human pathogens, including but not necessary limited to cytomegalovirus (CMV), Epstein Barr virus (EBV), HIV, and HBV, HCV if these information are not available from the supplier. Human adult hepatocytes are usually isolated from liver tissue obtained during whole liver resections or partial hepatectomies and are provided in a cryopreserved form. The resulting caveat is that the donors of these hepatocytes most likely suffer from a chronic hepatic disease, which could affect the utility of the hepatocytes. This complicates issues already arising from donor variability due to different genetic backgrounds. Cryopreserved adult hepatocytes are generally offered as "cryopreserved platable" or "cryopreserved unplatable." Both qualities are suitable for xenotransplantation, although cryopreserved platable hepatocytes usually result in higher engraftment levels.

In two large studies, factors that influence engraftment success were identified after analyzing SCID/Alb-uPA mice transplanted with hepatocytes from 90 different human donors [51, 52]. Both groups concluded that cells from pediatric donors had a higher viability and a greater potential for engraftment and appropriate liver cell function than cells from adult donors. They also observed that warm ischemia time was a critical determinant of transplantation success and found that minimizing the digestion time of liver sections for hepatocyte isolation optimized cell recovery numbers [51].

Before you start any work, especially with human pathogens in humanized mice, it is imperative that you seek the approval of your Institutional Animal Care and Use Committee (IACUC), Institutional Biosafety Committee (IBC), and Institutional Review Board (IRB) or the corresponding institutions of the country, where you plan to perform your research. You should also consult with your veterinary staff, if your animal facility meets all requirements, especially if you have the infrastructure in place to handle and contain the pathogens you are intending to work with then you can proceed safely.

3.2 Transplantation Protocol for Human Adult Hepatocytes

As described earlier, the only cells that can be robustly engrafted in injury liver models are adult hepatocytes and to a much lesser extent HFLCs, hepatocyte-like cells (HLCs). In this section, we describe the procedure for transplanting adult hepatocytes into FNRG mice. The transplantation of human adult hepatocytes is the best established approach and has proven most successful in terms of engraftment levels for the creation of human liver chimeric mice.

All mouse procedures must be performed in compliance with your local regulatory institutions. Procedures for anesthesia and analgesia, as well as postsurgical monitoring, may vary from the procedures we describe here.

3.2.1 Housing of FNRG Mice

- Keep mice housed under standard, pathogen-free conditions.

- Due to their highly immunocompromised status and therefore increased susceptibility to opportunistic infection mice may have to be maintained on amoxicillin (w/ 0.12 % amoxicillin provided in the rodent chow).

- Keep mice on NTBC-containing drinking water (1.6 mg NTBC/L of drinking water) from the time of birth until subjected to xenogeneic hepatocyte engraftment.

- Provide food and drinking water ad libitum.

- Mark individual mice (e.g., by ear tagging or ear punch) before the start of the experiment.

3.2.2 Preparation of Cryopreserved Adult Hepatocytes for Transplantation

1. Calculate the number of cells you need (number of mice to transplant $\times 1 \times 10^6$ cells). 1×10^6 cells are optimal for successful engraftment (*see* **Note 1**).

2. Warm 2 mL of DMEM per 1×10^6 cells (without additives) in a 50 mL conical tube to 37 °C.

3. Remove the appropriate number of vials containing the human hepatocytes from liquid nitrogen. Only when ready, thaw by immediately placing in water bath at 37 °C. Shake the vials continuously while halfway submerged in the water bath until the cell suspension is thawed.

4. Remove vials from the water bath and spray down the vials with 70 % ethanol.

5. Working in a tissue culture hood, place 1 mL of the prepared DMEM into each tube, using a 5 mL serological pipette. *Do not pipet up and down.*

6. Collect the content of each vial into the 50 mL conical tube, containing DMEM using a 5 mL serological pipette.

7. Place 1 mL of DMEM in to each vial to wash and collect any remaining cells.

8. Centrifuge the 50 mL conical tube at 300 rpm for 4 min at 4 °C and aspirate supernatant.

9. Flick the conical tube softly and slowly resuspend the pellet in 10 mL DMEM, while rotating the conical tube and place on ice.

10. Perform a cell count using a Neubauer Chamber. Count viable and dead cells and calculate the ratio (*see* **Note 1**).

11. Centrifuge the 50 mL conical tube again at 300 rpm for 4 min.

12. Carefully, aspirate the supernatant.

13. Flick the conical tube and resuspend the cell pellet in 50 μL DMEM per 1×10^6 cells. Do not pipette up and down—simply add the DMEM to the flicked cell pellet and transfer the content to a new cryovial. Use a 1000 μL pipette tip to keep the sheer stress on cells as minimal as possible.

14. Place the vial on ice and quickly transfer to the facility where the transplantation procedure will take place.

3.2.3 Anesthesia and Analgesia of Mice for Transplantation and Hepatocyte Isolation from Engrafted Mice

Isoflurane Inhalation Anesthesia

For the hepatocyte transplantation procedure: Operate the anesthesia system according to the manufacturer's instructions and set the oxygen flow rate to 3–5 L/min and the isoflurane concentration to 5 %. Place the mouse into the induction chamber of the anesthesia system for several minutes until the animal is sufficiently anesthetized. Remove the animal from the induction chamber and place it onto a surgical board; continue isoflurane inhalation via a nose cone. Continue isoflurane anesthesia throughout the entire procedure.

Terminal Anesthesia Procedure for Hepatocyte Isolation from Mouse

Perform intraperitoneal injection of ketamine and xylazine for the terminal procedures of hepatocyte isolation from engrafted mice and whole liver explantation. This method is chosen to ensure that mice are effectively anesthetized and sedated while still having sufficient blood circulation to prevent intravascular coagulation that would interfere with liver perfusion procedures.

Mix 90 mg/kg ketamine and 10 mg/kg xylazine in an insulin syringe and administer intraperitoneally. Place the animal into a cage and monitor it until it is anesthetized. Confirm sufficient anesthetic depth by pinching the toe of the mouse. Do not proceed with any invasive procedure until mouse shows no motoric reaction to the toe pinch.

3.2.4 Transplantation Protocol

1. Choose female FNRG mice of at least 6 weeks of age (*see* **Note 2**).

2. Perform isoflurane anesthesia procedure as described in Subheading 3.2.3.

3. Apply ophthalmic ointment to the rodent's eyes to prevent the corneas from drying out.

4. Apply an appropriate systemic analgesic (e.g., 5 mg/kg bodyweight Carprofen or 0.3 mg/kg Meloxicam subcutaneously).

5. Shave a small subcostal left lateral area from the pelvic bone up to the rib cage using an electric shaver.

6. Prepare skin for aseptic surgery (use, e.g., Betadine and 70 % ethanol or sterile water).

7. Place the mouse in a right lateral position on a sterile surgical board.

8. Make a left subcostal skin incision of 5–8 mm using scissors.

9. Make sure you can see the spleen as a dark-red spot through the still closed peritoneum and subsequently using a new pair of scissors, open the peritoneum (usually requires no more than 3–5 mm peritoneal defect).

10. Expose the spleen by grabbing the hilar fat and connective tissue with tweezers and gently pulling it out of the peritoneal cavity, resting it on a 3×10 mm piece of sterile film (*see* **Note 3**).

11. Prepare a syringe with the desired amount of human adult hepatocytes (typically 1×10^6 cells, *see* Subheading 3.2.2) and subsequently fix the spleen in place by holding the hilar connective tissue with tweezers.

12. Use a 28G or 30G needle to inject the cells into the spleen. Upon injection, the spleen should blanch. To prevent rupture, keep the injection volume below 50 μL and apply this volume over a period of ~30 s. If you have to exceed this volume, increase the injection time (*see* **Note 4**).

13. Immediately after removing the needle, use a sterile cotton swab to apply very gentle pressure on the injection site for several seconds until the bleeding stops.

14. Maneuver the spleen back into the peritoneum by using a sterile cotton swab.

15. Apply 20–50 μL of bupivacaine topically at the site of incision and then close the peritoneal defect by using Vicryl 4-0 sutures (typically one or two sutures are needed).

16. Approximate the skin with tweezers and close it up using surgical clips (typically one or two).

17. Monitor postoperatively as described below before placing the animal in a clean cage with food and water.

3.2.5 Postsurgical Monitoring

After completion of the transplantation procedure, continue the anesthesia and place the animal into an empty rodent cage on a heat source. Monitor the animal continuously until it is able to ambulate, access food and water, and is not at risk for aggression by cage mates or bedding inhalation. After the monitoring phase, the animal is transferred to its home cage.

For days 1–5 post-surgery, a systemic analgesic should be provided (e.g., add an oral suspension of meloxicam to drinking water).

For day 1–5 post-surgery, mice must be monitored once daily for signs of pain and/or infection (hunching, piloerection, lethargy, decreased ambulation, weight loss, and/or dehydration). Wound conditions must be inspected for signs of local infection or bleeding. Remove the surgical wound clips 10–14 days post-surgery.

3.3 NTBC Cycling of Transplanted Mice

As described above, FAH deficiency in the FNRG mouse strain used here leads to hypertyrosinemia, which causes liver injury and is lethal in non-transplanted and non-treated animals. NTBC pharmacologically rescues the FAH deficiency in mice [31, 53] and therefore mice must be maintained with drinking water containing NTBC (1.6 mg NTBC/L of drinking water) from birth until the time of xenogeneic hepatocyte engraftment. Directly after the transplantation procedure, take mice off NTBC for 2 weeks. Measure bodyweight of mice once or twice per week and cycle mice off the drug based on weight loss and overall health; i.e., when weight drops by 10–15% restore the drug regime. Expect the mice to be on NTBC approximately every 7–14 days for 3–4 days.

3.4 Human Albumin ELISA

To monitor engraftment with human donor hepatocytes we recommend measuring human serum albumin levels by ELISA, as serum levels correlate well with engraftment levels of human hepatocytes [35, 51]. Serum albumin levels should be measured every 2 weeks and typically reach maximum engraftment (exceeding 1 mg/mL) levels approximately 10–12 weeks post-transplantation (Fig. 2d).

1. Draw 50 μL of whole blood either by submandibular bleeding or retro-orbital bleeding into 1.5 mL microcentrifuge tubes.

2. Let the blood sit for 20 min at room temperature to allow for clotting.

3. Spin blood samples for 10 min at ca 850×g at room temperature.

4. Transfer serum into a new 1.5 mL microcentrifuge tube using a p20 pipette. Store samples at –20 °C until ready to perform the ELISA assay.

5. Coat 96-well ELISA plates with 50 μL per well of goat anti-human albumin diluted 1:500 in coating buffer and incubate for 1 h at 37 °C or 4 °C overnight.

6. Wash plate four times with 300 μL ELISA wash buffer per well.

7. Add 200 μL SuperBlock per well and incubate for 1 h at 37 °C or at 4 °C overnight.

8. Wash plate two times with 300 μL ELISA wash buffer per well.

9. Prepare the standard curve and dilution series of samples in a regular 96-well flat bottom plate (master plate):

   - To generate the standard curve, prepare an albumin stock solution at 140a concentration of 1000 ng/mL albumin in sample diluent.

   - Fill each well of the master plate with 135 μL sample diluent.

   - Create the standard curve by pipetting 135 μL of the albumin stock solution into the first column of the plate and pipetting 135 μL from column 1 into column 2 and so on, thereby creating a 1:2 dilution series.

- Take the samples from **step 4** and pipette 15 μL into one well and then 15 μL into the next well thereby creating a 1:10 dilution series of the sample to ensure you will have a dilution of your sample within the range of the standard curve.

- For the standard curve, as well as for the serum samples, work in duplicates (Fig. 2).

10. Using a multichannel pipette, transfer 50 μL of the prepared samples from **step 9** to the ELISA plate prepared as described in **steps 5–8**.

11. Incubate for 1 h at 37 or 4 °C overnight.

12. Aspirate samples from plate and wash three times with 300 μL/well of ELISA wash buffer.

13. Add 50 μL of mouse anti-human albumin antibody diluted 1:10,000 in sample diluent.

14. Incubate for 2 h at 37 or 4 °C overnight.

15. Dump sample and wash four times with 300 μL/well of ELISA wash buffer.

16. Add 50 μL/well of goat anti-mouse HRP antibody diluted 1:2000 in sample diluent.

17. Incubate for 1 h at 37 or 4 °C overnight.

18. Aspirate sample from plate and wash six times with ELISA wash buffer.

19. Quickly add 100 μL of TMB substrate.

20. Depending on the albumin concentration, wells will start to turn blue. Wait until the well of the standard curve that has a concentration of 1 ng/mL begins to turn blue, then add 12.5 μL of the stop solution to all wells of the plate. This will stop the reaction and cause the color to turn yellow.

21. Within 5–10 min after stopping the reaction using an ELISA plate reader, measure the absorbance at 450 nm.

3.5 Hepatocyte Isolation from Engrafted Mice

Primary human hepatocytes can only be maintained in tissue culture as hepatocytes for only several days before they begin to dedifferentiate and no longer proliferate. One application for human liver chimeric mice is to use such animals as a platform to expand human hepatocytes of particular interest (e.g., specific genotype) in vivo [32]. One mouse injected with 1×10^6 cells can yield as much as $4–8 \times 10^7$ hepatocytes. Here we describe the procedure for isolating hepatocytes from engrafted mice to then use in downstream applications. Isolated hepatocytes will have a high viability that allows for re-transplantation into other recipient mice, tissue culture applications, or cryopreservation for later use of cells. Previous work showed that serial transfer can lead to better engraftment in secondary recipients and may be repeated several times [32]. However, it has to be kept in mind that during each engraftment only a subfraction of injected cells reconstitutes the organ and conceivably a subpopulation with distinct, possibly undesirable/distorted physiology may be selected during this procedure.

3.5.1 Preparations for Hepatocyte Isolation

1. Prewarm water bath to 40 °C.

2. For each mouse that will be perfused, dispense 50 mL of Perfusion Solution A and 32 mL of Perfusion Solution B into a 50 mL conical tube and prewarm the solutions in the water bath (*see* **Notes 5** and **6**).

3. Immediately before starting the perfusion add 8 mL of Collagenase Solution to 32 mL of Solution B, the final concentration of Collagenase Type 2 in Solution B is now 1 mg/mL.

4. Connect the 3-stop silicone tubing to the perfusor.

5. Calibrate the perfusor to deliver 3.0 mL/min.

6. Rinse pump tubing with 30 mL of 70 % ethanol and continue pumping until all of the ethanol is out of the tubing.

7. Rinse pump tubing with 30 mL of PBS and continue pumping until all of the solution is out of the tubing.

8. Attach intravenous administration set and fill all tubing and bubble trap with Perfusion Solution A. Invert the bubble trap as Perfusion Solution A enters until it is above the fill hole then return the trap to the upright position (*see* **Note 7**).

3.5.2 Perfusion of Human Liver Chimeric Mice

1. Anesthetize the mouse as described in Subheading "Terminal Anesthetize Procedure for Hepatocyte Isolation from Mouse" with ketamine and xylazine.

2. Lay the anesthetized mouse onto a blue adsorbent pad on top of the mouse dissection board and double check for the lack of any motoric reaction to toe pinching.

3. Immobilize all four limbs of the mouse with tape or pins.

4. Spray down the mouse with ethanol spray until the mouse's fur is completely soaked.

5. Using the first set of scissors and forceps, perform a median skin incision up to the sternum without injuring the peritoneum. Then perform lateral incisions down to the flank, starting at the top and the bottom of the median incision line. Using forceps and scissors, separate the skin from the peritoneum and flip the skin aside.

6. Use the second set of scissors and forceps to open the abdominal cavity by making a median incision into the peritoneum up to the sternum, followed by lateral incisions down to the flank starting at the top and the bottom of the median incision line.

7. Using a sterile cotton applicator, push intestines to the left side of the animal, thereby exposing the vessels leading into the liver. Identify the inferior vena cava (IVC) and portal vein (PV).

8. Dispense 5 mL of *Perfusion Solution B* (*containing the collagenase type 2*) into a 60 mm dish.

9. Use two sterile cotton applicators, applying weight on the stick, to keep the organs out of the way and apply some tension on the PV as this will provide optimal conditions for catheterization.

10. Cannulate the PV with the Safelet I.V. Catheter. The best proof of a successful catheterization is backflow of blood into the catheter.

11. Secure the catheter with a surgical clip.

12. If there is not enough blood backflow into the catheter to fill the whole catheter, manually fill the catheter with *Perfusion Solution A* using a 28 G insulin syringe.

13. Start the peristaltic pump first and then connect the perfusion tubing to the catheter. The liver should instantly start blanching.

14. Immediately after you start the perfusion, cut the IVC to allow perfusion solutions to be released (*see* **Note 8**).

15. Immediately switch to Solution B. The process of in situ digestion by Solution B will take 10–12 min. At the latest, stop the peristaltic pump when all of Solution B has been exhausted. To check if the liver is thoroughly digested, you can carefully pinch the liver surface with forceps. A digested liver should lose its elasticity and not bounce back. You should be able to see an indentation upon pinching with the forceps.

16. After the liver is digested, remove the surgical clamp and pull the catheter out of the vessel.

17. Using the second set of scissors and forceps, carefully cut out the liver without damaging the liver capsule and transfer the entire liver from the abdominal cavity to the petri dish containing Solution B.

18. Disrupt the liver capsule with forceps and scissors. If the digest was successful, the hepatocytes should form a homogenous suspension and the solution should quickly cloud up.

19. Using a 25 mL serological pipette, transfer the hepatocyte suspension into a 50 mL conical tube. Wash the petri dish with 5 mL of Solution C and also transfer it to the conical tube.

20. Proceed with the following steps under a tissue culture hood.

3.5.3 Processing Isolated Hepatocytes

1. Place a 100 µm cell strainer on top of a 50 mL conical tube and a 70 µm cell strainer on top of a second tube.

2. Using a sterile 25 mL pipet, pass the liver slurry through the 100 µm filter on top of the 50 mL tube. Aid the cell suspension through the filter by gently stirring the tip of the 25 mL pipet on the surface of the filter.

3. Wash the conical tube that initially contained the hepatocyte suspension with 10 mL of Solution C. Pass the solution

through the same 100 μm filter on the same 50 mL conical tube. Discard the filter.

4. Using a sterile 25 mL pipette, adjust the volume in the 50 mL collection tube to 45 mL using Solution C.

5. Using the same pipette, transfer the cell suspension to the 70 μm filter on top of the second 50 mL tube. Discard the filter.

6. Spin the tube containing the 45 mL of cell suspension at $140 \times g$ for 5 min at 4 °C.

7. Carefully aspirate the supernatant from the tube.

8. Carefully resuspend the cell pellet in 45 mL of Solution C. Flicking the pellet and resuspending in a smaller volume before filling the tube up to 45 mL will help to properly resuspend the cell pellet.

9. Spin the tube at $140 \times g$ for 5 min at 4 °C.

10. Repeat **steps 7–9** one more time.

11. Carefully aspirate the supernatant from the tube.

12. Resuspend the cell pellet in 10–20 mL of Solution C, depending on the size of the cell pellet.

13. Manually count the cells using the Neubauer Chamber cytometer using a trypan blue exclusion method. Determine the number of viable and dead cells and calculate the percentage of viable cells. Ideally the percentage of viable cells should be at least 80–90 % (*see* **Note 9**).

14. Follow either the protocol for plating freshly isolated hepatocytes (Subheading 3.5.4) or the protocol for cryopreservation of hepatocytes (Subheading 3.5.5).

3.5.4 Plating of Freshly Isolated Hepatocytes

We found that for most applications and as described here, seeding hepatocytes on to 24-well collagen-coated plates is useful. Generally speaking, hepatocytes should be seeded at a density of 5×10^4 cells/cm². All numbers refer to viable cells.

1. Continue working under a sterile tissue culture hood.

2. Add the HCM single aliquots of growth factors to the HBM Basal Medium (henceforth referred to as HCM medium) and prewarm it to 37 °C in a water bath.

3. Directly proceed from **step 13** of Subheading 3.5.3 and centrifuge the hepatocytes at $140 \times g$ at 4 °C for 5 min.

4. Carefully aspirate the supernatant.

5. Resuspend the cells so that there are 2×10^5 cells/mL in the HCM medium.

6. Pipet 0.5 mL of the cell/media suspension into each well of a 24-well collagen-coated plate.

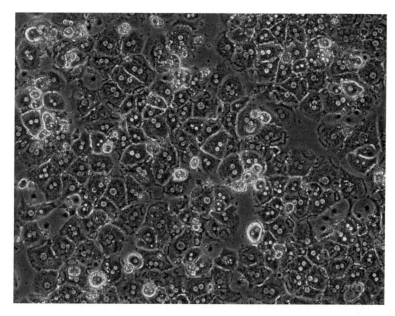

Fig. 3 Primary hepatocyte exhibit characteristic cobblestone pattern after plating

7. Gently swirl the 24-well plate several times to insure even distribution of the cells.

8. Incubate at 37 °C, 5 % CO_2 for 6 h or overnight.

9. Aspirate the media and replace with fresh 0.5 mL of HCM medium.

10. Incubate overnight at 37 °C, 5 % CO_2.

11. Hepatocytes should now have formed a characteristic "cobblestone pattern" and can be used for downstream applications (Fig. 3).

12. Perform medium change every 24–48 h.

3.5.5 Cryopreservation of Freshly Isolated Hepatocytes

1. Put CryoStor 10 on ice and prepare and label cryovials.

2. Directly proceed from **step 13** of Subheading 3.5.3 and centrifuge the hepatocytes at $140 \times g$ at 4 °C for 5 min.

3. Carefully aspirate the supernatant, making sure to remove as much as possible.

4. Resuspend the cell pellet in ice-cold CryoStor 10 at a concentration of 1×10^7 viable cells/mL.

5. Aliquot 1 mL into each of the premarked cryovials.

6. Place the cryovials into a polystyrene container and then into a −80 °C freezer overnight.

7. Place vials in a liquid nitrogen tank the next day.

3.5.6 Thawing of Cryopreserved Hepatocytes

1. Prewarm HCM medium to 37 °C in a water bath.

2. Remove cryovials from the liquid nitrogen tank and immediately thaw by swirling the tubes in a 37 °C water bath.

3. Spray down the tubes with 70 % ethanol and continue work under a tissue culture hood.

4. Using a 2 mL serological pipette, transfer the thawed cell suspension to a 50 mL conical tube.

5. Add dropwise slowly 10 mL of HCM medium to the cell suspension while gently swirling the conical tube.

6. Manually count the cells using the Neubauer Chamber cytometer and trypan blue method. Determine the number of viable and dead cells and calculate the percentage of viable cells (*see* **Note 1**).

7. Proceed with **step 3** of Subheading 3.5.4.

4   Notes

1. For performing the cell count after thawing the cryopreserved hepatocytes, it is not recommended to use an automated cell count device. Our experience is that due to the tendency of the cells to clump, the automated counting is not reliable and even the manual count is challenging, with counting results varying depending on the experimenter.

2. Work in a team of two so while one person is preparing the cryopreserved hepatocytes the other is preparing the mice. The goal is to keep the time from thawing the cells to cell injections as short as possible (ideally not longer than 20 min).

3. Upon transplantation, it is critical to not hold the spleen directly with the tweezers, since the spleen capsule is very thin and vulnerable—only hold it at the hilar connective tissue.

4. Make sure to inject the spleen once, and do not inject the cell suspension too fast as this can rupture the spleen capsule and kill the animal.

5. Only process one mouse at a time to minimize the time needed for all working steps since this will increase the viability of the hepatocytes.

6. Ensure that all solutions are at the temperatures described in the protocol. Especially for the collagenase perfusion it is crucial that the collagenase solution is prewarmed since the optimal enzymatic activity is at 37 °C. For this purpose it can be helpful to put the mouse on a heating pad during the perfusion.

7. Avoid air bubbles in the tubing at all costs. Even small air bubbles can abrogate a successful perfusion.

8. Make sure that you cut the IVC immediately after the start of the liver perfusion. Failing to do so will dramatically decrease the viability of the hepatocytes due to compression damage.

9. If you observe a lot of cell debris and other unwanted particles like platelets during the cell count of the hepatocytes, another low speed centrifugation step at $140 \times g$ can be added.

Acknowledgements

The authors thank Jenna Gaska for edits and critical discussion of the manuscript. Work in the laboratory is in part supported by grants from the National Institutes of Health (2 R01 AI079031-05A1, 1 R01 AI107301-01, 1 R56 AI106005-01, 1R21AI117213-01), a Scholar grant from the American Cancer Society (RSG-15-048-01-MPC), and the Grand Challenge Program of Princeton University. M.v.S. is a recipient of a fellowship from the German Research Foundation (Deutsche Forschungsgemeinschaft).

References

1. Perz JF, Armstrong GL, Farrington LA, Hutin YJ, Bell BP (2006) The contributions of hepatitis B virus and hepatitis C virus infections to cirrhosis and primary liver cancer worldwide. J Hepatol 45:529–538

2. Seeger C, Mason WS (2000) Hepatitis B virus biology. Microbiol Mol Biol Rev 64:51–68

3. Nilsson SK, Childs LM, Buckee C, Marti M (2015) Targeting human transmission biology for malaria elimination. PLoS Pathog 11: e1004871

4. Pileri P, Uematsu Y, Campagnoli S, Galli G, Falugi F, Petracca R, Weiner AJ, Houghton M, Rosa D, Grandi G, Abrignani S (1998) Binding of hepatitis C virus to CD81. Science 282:938–941

5. Scarselli E, Ansuini H, Cerino R, Roccasecca RM, Acali S, Filocamo G, Traboni C, Nicosia A, Cortese R, Vitelli A (2002) The human scavenger receptor class B type I is a novel candidate receptor for the hepatitis C virus. EMBO J 21:5017–5025

6. Ploss A, Evans MJ, Gaysinskaya VA, Panis M, You H, de Jong YP, Rice CM (2009) Human occludin is a hepatitis C virus entry factor required for infection of mouse cells. Nature 457:882–886

7. Liu S, Yang W, Shen L, Turner JR, Coyne CB, Wang T (2009) Tight junction proteins claudin-1 and occludin control hepatitis C virus entry and are downregulated during infection to prevent superinfection. J Virol 83:2011–2014

8. Evans MJ, von Hahn T, Tscherne DM, Syder AJ, Panis M, Wolk B, Hatziioannou T, McKeating JA, Bieniasz PD, Rice CM (2007) Claudin-1 is a hepatitis C virus co-receptor required for a late step in entry. Nature 446:801–805

9. Dorner M, Horwitz JA, Robbins JB, Barry WT, Feng Q, Mu K, Jones CT, Schoggins JW, Catanese MT, Burton DR et al (2011) A genetically humanized mouse model for hepatitis C virus infection. Nature 474:208–211

10. Dorner M, Rice CM, Ploss A (2013) Study of hepatitis C virus entry in genetically humanized mice. Methods 59:249–257

11. de Jong YP, Dorner M, Mommersteeg MC, Xiao JW, Balazs AB, Robbins JB, Winer BY, Gerges S, Vega K, Labitt RN et al (2014) Broadly neutralizing antibodies abrogate established hepatitis C virus infection. Sci Transl Med 6:254ra129

12. Giang E, Dorner M, Prentoe JC, Dreux M, Evans MJ, Bukh J, Rice CM, Ploss A, Burton DR, Law M (2012) Human broadly neutralizing antibodies to the envelope glycoprotein complex of hepatitis C virus. Proc Natl Acad Sci U S A 109:6205–6210

13. Dorner M, Horwitz JA, Donovan BM, Labitt RN, Budell WC, Friling T, Vogt A, Catanese MT, Satoh T, Kawai T et al (2013) Completion of the entire hepatitis C virus life cycle in genetically humanized mice. Nature 501:237–241

14. Yan H, Zhong G, Xu G, He W, Jing Z, Gao Z, Huang Y, Qi Y, Peng B, Wang H et al (2012) Sodium taurocholate cotransporting polypep-

tide is a functional receptor for human hepatitis B and D virus. Elife 1:e00049

15. Ni Y, Lempp FA, Mehrle S, Nkongolo S, Kaufman C, Falth M, Stindt J, Koniger C, Nassal M, Kubitz R et al (2014) Hepatitis B and D viruses exploit sodium taurocholate cotransporting polypeptide for species-specific entry into hepatocytes. Gastroenterology 146:1070–1083

16. He W, Ren B, Mao F, Jing Z, Li Y, Liu Y, Peng B, Yan H, Qi Y, Sun Y et al (2015) Hepatitis D virus infection of mice expressing human sodium taurocholate co-transporting polypeptide. PLoS Pathog 11:e1004840

17. Li H, Zhuang Q, Wang Y, Zhang T, Zhao J, Zhang Y, Zhang J, Lin Y, Yuan Q, Xia N, Han J (2014) HBV life cycle is restricted in mouse hepatocytes expressing human NTCP. Cell Mol Immunol 11:175–183

18. Sandgren EP, Palmiter RD, Heckel JL, Daugherty CC, Brinster RL, Degen JL (1991) Complete hepatic regeneration after somatic deletion of an albumin-plasminogen activator transgene. Cell 66:245–256

19. Rhim JA, Sandgren EP, Degen JL, Palmiter RD, Brinster RL (1994) Replacement of diseased mouse liver by hepatic cell transplantation. Science 263:1149–1152

20. Mercer DF, Schiller DE, Elliott JF, Douglas DN, Hao C, Rinfret A, Addison WR, Fischer KP, Churchill TA, Lakey JR et al (2001) Hepatitis C virus replication in mice with chimeric human livers. Nat Med 7:927–933

21. Law M, Maruyama T, Lewis J, Giang E, Tarr AW, Stamataki Z, Gastaminza P, Chisari FV, Jones IM, Fox RI et al (2008) Broadly neutralizing antibodies protect against hepatitis C virus quasispecies challenge. Nat Med 14:25–27

22. Vanwolleghem T, Bukh J, Meuleman P, Desombere I, Meunier JC, Alter H, Purcell RH, Leroux-Roels G (2008) Polyclonal immunoglobulins from a chronic hepatitis C virus patient protect human liver-chimeric mice from infection with a homologous hepatitis C virus strain. Hepatology 47:1846–1855

23. Lutgehetmann M, Mancke LV, Volz T, Helbig M, Allweiss L, Bornscheuer T, Pollok JM, Lohse AW, Petersen J, Urban S, Dandri M (2012) Humanized chimeric uPA mouse model for the study of hepatitis B and D virus interactions and preclinical drug evaluation. Hepatology 55:685–694

24. Volz T, Allweiss L, Ben MM, Warlich M, Lohse AW, Pollok JM, Alexandrov A, Urban S, Petersen J, Lutgehetmann M, Dandri M (2013) The entry inhibitor Myrcludex-B efficiently blocks intrahepatic virus spreading in humanized mice previously infected with hepatitis B virus. J Hepatol 58:861–867

25. Sacci JB Jr, Alam U, Douglas D, Lewis J, Tyrrell DL, Azad AF, Kneteman NM (2006) Plasmodium falciparum infection and exoerythrocytic development in mice with chimeric human livers. Int J Parasitol 36:353–360

26. VanBuskirk KM, O'Neill MT, De La Vega P, Maier AG, Krzych U, Williams J, Dowler MG, Sacci JB Jr, Kangwanrangsan N, Tsuboi T et al (2009) Preerythrocytic, live-attenuated Plasmodium falciparum vaccine candidates by design. Proc Natl Acad Sci U S A 106:13004–13009

27. Suemizu H, Hasegawa M, Kawai K, Taniguchi K, Monnai M, Wakui M, Suematsu M, Ito M, Peltz G, Nakamura M (2008) Establishment of a humanized model of liver using NOD/Shi-scid IL2Rgnull mice. Biochem Biophys Res Commun 377:248–252

28. Heo J, Factor VM, Uren T, Takahama Y, Lee JS, Major M, Feinstone SM, Thorgeirsson SS (2006) Hepatic precursors derived from murine embryonic stem cells contribute to regeneration of injured liver. Hepatology 44:1478–1486

29. Tesfaye A, Stift J, Maric D, Cui Q, Dienes HP, Feinstone SM (2013) Chimeric mouse model for the infection of hepatitis B and C viruses. PLoS One 8:e77298

30. Grompe M, al-Dhalimy M, Finegold M, Ou CN, Burlingame T, Kennaway NG, Soriano P (1993) Loss of fumarylacetoacetate hydrolase is responsible for the neonatal hepatic dysfunction phenotype of lethal albino mice. Genes Dev 7:2298–2307

31. Grompe M, Lindstedt S, al-Dhalimy M, Kennaway NG, Papaconstantinou J, Torres-Ramos CA, Ou CN, Finegold M (1995) Pharmacological correction of neonatal lethal hepatic dysfunction in a murine model of hereditary tyrosinaemia type I. Nat Genet 10:453–460

32. Azuma H, Paulk N, Ranade A, Dorrell C, Al-Dhalimy M, Ellis E, Strom S, Kay MA, Finegold M, Grompe M (2007) Robust expansion of human hepatocytes in Fah−/−/Rag2−/−/Il2rg−/− mice. Nat Biotechnol 25:903–910

33. Brehm MA, Cuthbert A, Yang C, Miller DM, DiIorio P, Laning J, Burzenski L, Gott B, Foreman O, Kavirayani A et al (2010) Parameters for establishing humanized mouse models to study human immunity: analysis of human hematopoietic stem cell engraftment in three immunodeficient strains of mice bearing the IL2rgamma(null) mutation. Clin Immunol 135:84–98

34. Bissig KD, Le TT, Woods NB, Verma IM (2007) Repopulation of adult and neonatal mice with human hepatocytes: a chimeric animal model. Proc Natl Acad Sci U S A 104:20507–20511

35. Bissig KD, Wieland SF, Tran P, Isogawa M, Le TT, Chisari FV, Verma IM (2010) Human liver chimeric mice provide a model for hepatitis B and C virus infection and treatment. J Clin Invest 120:924–930

36. Vaughan AM, Mikolajczak SA, Wilson EM, Grompe M, Kaushansky A, Camargo N, Bial J, Ploss A, Kappe SH (2012) Complete Plasmodium falciparum liver-stage development in liver-chimeric mice. J Clin Invest 122:3618–3628

37. Hasegawa M, Kawai K, Mitsui T, Taniguchi K, Monnai M, Wakui M, Ito M, Suematsu M, Peltz G, Nakamura M, Suemizu H (2011) The reconstituted 'humanized liver' in TK-NOG mice is mature and functional. Biochem Biophys Res Commun 405:405–410

38. Takahashi K, Yamanaka S (2006) Induction of pluripotent stem cells from mouse embryonic and adult fibroblast cultures by defined factors. Cell 126:663–676

39. Takahashi K, Tanabe K, Ohnuki M, Narita M, Ichisaka T, Tomoda K, Yamanaka S (2007) Induction of pluripotent stem cells from adult human fibroblasts by defined factors. Cell 131:861–872

40. Carpentier A, Tesfaye A, Chu V, Nimgaonkar I, Zhang F, Lee SB, Thorgeirsson SS, Feinstone SM, Liang TJ (2014) Engrafted human stem cell-derived hepatocytes establish an infectious HCV murine model. J Clin Invest 124:4953–4964

41. Yu AM, Fukamachi K, Krausz KW, Cheung C, Gonzalez FJ (2005) Potential role for human cytochrome P450 3A4 in estradiol homeostasis. Endocrinology 146:2911–2919

42. van Herwaarden AE, Wagenaar E, van der Kruijssen CM, van Waterschoot RA, Smit JW, Song JY, van der Valk MA, van Tellingen O, van der Hoorn JW, Rosing H et al (2007) Knockout of cytochrome P450 3A yields new mouse models for understanding xenobiotic metabolism. J Clin Invest 117:3583–3592

43. Katoh M, Matsui T, Nakajima M, Tateno C, Kataoka M, Soeno Y, Horie T, Iwasaki K, Yoshizato K, Yokoi T (2004) Expression of human cytochromes P450 in chimeric mice with humanized liver. Drug Metab Dispos 32:1402–1410

44. Katoh M, Sawada T, Soeno Y, Nakajima M, Tateno C, Yoshizato K, Yokoi T (2007) In vivo drug metabolism model for human cytochrome P450 enzyme using chimeric mice with humanized liver. J Pharm Sci 96:428–437

45. Emoto C, Yamato Y, Sato Y, Ohshita H, Katoh M, Tateno C, Yokoi T, Yoshizato K, Iwasaki K (2008) Non-invasive method to detect induction of CYP3A4 in chimeric mice with a humanized liver. Xenobiotica 38:239–248

46. Katoh M, Watanabe M, Tabata T, Sato Y, Nakajima M, Nishimura M, Naito S, Tateno C, Iwasaki K, Yoshizato K, Yokoi T (2005) In vivo induction of human cytochrome P450 3A4 by rifabutin in chimeric mice with humanized liver. Xenobiotica 35:863–875

47. Sanoh S, Horiguchi A, Sugihara K, Kotake Y, Tayama Y, Uramaru N, Ohshita H, Tateno C, Horie T, Kitamura S, Ohta S (2012) Predictability of metabolism of ibuprofen and naproxen using chimeric mice with human hepatocytes. Drug Metab Dispos 40:2267–2272

48. Washburn ML, Bility MT, Zhang L, Kovalev GI, Buntzman A, Frelinger JA, Barry W, Ploss A, Rice CM, Su L (2011) A humanized mouse model to study hepatitis C virus infection, immune response, and liver disease. Gastroenterology 140:1334–1344

49. Gutti TL, Knibbe JS, Makarov E, Zhang J, Yannam GR, Gorantla S, Sun Y, Mercer DF, Suemizu H, Wisecarver JL et al (2014) Human hepatocytes and hematolymphoid dual reconstitution in treosulfan-conditioned uPA-NOG mice. Am J Pathol 184:101–109

50. Wilson EM, Bial J, Tarlow B, Bial G, Jensen B, Greiner DL, Brehm MA, Grompe M (2014) Extensive double humanization of both liver and hematopoiesis in FRGN mice. Stem Cell Res 13:404–412

51. Kawahara T, Toso C, Douglas DN, Nourbakhsh M, Lewis JT, Tyrrell DL, Lund GA, Churchill TA, Kneteman NM (2010) Factors affecting hepatocyte isolation, engraftment, and replication in an in vivo model. Liver Transpl 16:974–982

52. Vanwolleghem T, Libbrecht L, Hansen BE, Desombere I, Roskams T, Meuleman P, Leroux-Roels G (2010) Factors determining successful engraftment of hepatocytes and susceptibility to hepatitis B and C virus infection in uPA-SCID mice. J Hepatol 53:468–476

53. Kelsey G, Ruppert S, Beermann F, Grund C, Tanguay RM, Schutz G (1993) Rescue of mice homozygous for lethal albino deletions: implications for an animal model for the human liver disease tyrosinemia type 1. Genes Dev 7:2285–2297

Chapter 6

Human FcRn Transgenic Mice for Pharmacokinetic Evaluation of Therapeutic Antibodies

Derry C. Roopenian, Gregory J. Christianson, Gabriele Proetzel, and Thomas J. Sproule

Abstract

Therapeutic monoclonal antibodies are widely recognized to be a most promising means to treat an increasing number of human diseases, including cancers and autoimmunity. To a large extent, the efficacy of monoclonal antibody treatment is because IgG antibodies have greatly extended persistence in vivo. However, conventional rodent models do not mirror human antibody pharmacokinetics. The key molecule responsible for the extended persistence antibodies is the major histocompatibility complex class I family Fc receptor, FcRn. We describe human FcRn transgenic mouse models and how they can be exploited productively for the preclinical pharmacokinetic evaluation of therapeutic antibodies.

Key words Monoclonal antibodies, Fc fusion proteins, FcRn, Transgenic mice, Pharmacokinetics, Serum half-life

1 Introduction

1.1 Antibodies

As realized by Frank Macfarlane Burnet [1], antibodies are considered nature's magic bullets. The antigen binding sites permit the antibody to bind its ligand with extraordinary specificity. Sequences in the Fc fragment then mediate events, such as the activation of complement and Fc receptors, to the antigenic moiety (Fig. 1). This coupling of specificity to effector function is the basis for humoral immunity. Of the various antibody classes (IgM, IgA, IgE, and IgG), IgG has features that make it the preferred antibody for therapeutic development. IgG is a highly stable tetrameric molecule, comprised of two light chains and two heavy chains. It is the dominant antibody class in circulation because it has greatly extended persistence. For example, the half-life of IgG1 in human circulation is 10–20 days, ten times the half-life for antibodies of other isotypes. IgG is also distinguished from IgM, IgA, and IgE in that it readily passes from the circulation to extravascular tissues. Like the effector functions associated with

Gabriele Proetzel and Michael V. Wiles (eds.), *Mouse Models for Drug Discovery: Methods and Protocols*,
Methods in Molecular Biology, vol. 1438, DOI 10.1007/978-1-4939-3661-8_6, © Springer Science+Business Media New York 2016

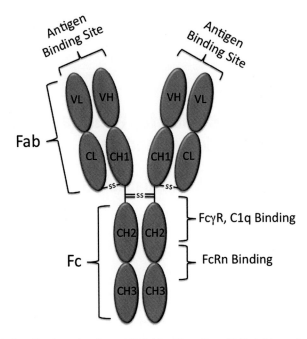

Fig. 1 Rendition of an IgG antibody molecule and its interaction sites. (V) Variable regions; (L) Light chain; (VH1 and CH 1–4) constant domains of the heavy chain. Sites of C1q, FcgR, and FcRn binding are indicated

IgG, it is the Fc fragment of IgG that is responsible for the increased serum persistence and extracellular access.

1.2 FcRn Protects IgG Antibodies from Catabolic Destruction

FcRn is quite distinct from conventional Fc receptors and is an evolutionary offshoot of major histocompatibility complex (MHC) class I proteins [2, 3]. Like other MHC class I family members, FcRn forms an obligate heterodimer with the β_2 microglobulin (β_2m) light chain. FcRn binds the CH2 CH3 hinge region of the Fc fragment in an acidic environment (pH ~6–6.5) with nanomolar affinity, but demonstrates a precipitous affinity drop at neutral pH ~7.2 [4–6]. The steady state localization is primarily in the early endosomes. In cells, such as vascular endothelial cells, extracellular proteins (including IgG) are taken up by fluid phase endocytosis. IgG/FcRn complexes are then recycled to the plasma membrane leading to the dissociation of IgG at neutral pH and release into the extracellular environment (Fig. 2) (reviewed in Refs. [7–9]). In this way, FcRn is thought to selectively "protect" IgG from catabolic elimination by recycling the IgG cargo either apically (back to the bloodstream) or basolaterally (across endothelial or epithelial barriers). As such, FcRn both increases the serum persistence and the extravascular bioavailability of IgG antibodies.

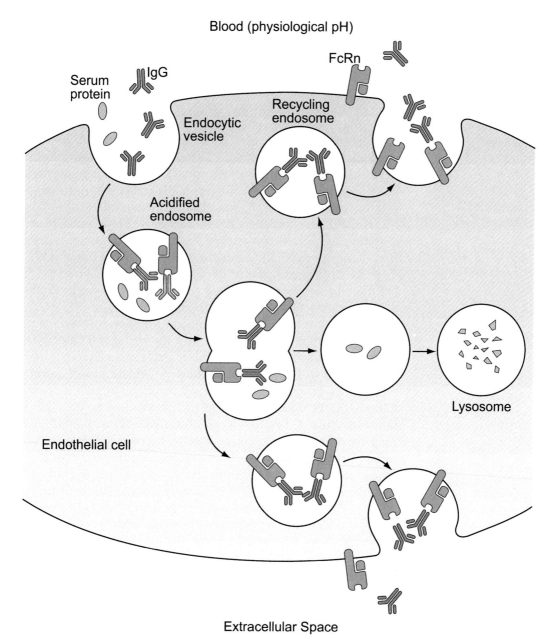

Fig. 2 Model for how FcRn rescues IgG from catabolism by recycling and transcytosis. IgG and many other soluble proteins are present in extracellular fluids. Vascular endothelial cells are active in fluid phase endocytosis of blood proteins. Material taken up by these cells enters the endosomes where FcRn is found as an integral membrane protein. The IgG then binds FcRn in this acidic environment. This binding results in transport of the IgG to the apical plasma membrane for recycling into the circulation, or to the basolateral membrane for transcytosis into the extracellular space. Exposure to a neutral pH in both locations then results in the release of IgG. The remaining soluble proteins are channeled to the lysosomal degradation pathway

In contrast, proteins that to not bind FcRn in this acid pH-dependent manner are not rescued by FcRn and thus channeled into the lysosomal compartment where they are catabolized. The failure to be rescued by FcRn thus explains the abbreviated serum half-life of most proteins in the circulation compared to IgG.

1.3 Therapeutic Antibodies

As of 2015, 45 therapeutic monoclonal antibodies (mAbs) mAbs have been approved for clinical use in the US, and with over 400 antibodies being in preclinical and clinical development further increase of antibody therapies is assured [10, 11]. As a general rule, the Fc fragment is a key component of therapeutic mAb design because it extends their pharmacokinetics. Inclusion of the Fc from IgG is also a key component of other bioactive proteins where prolongation of pharmacokinetics is desired (e.g., the tumor necrosis factor receptor (TNFR) fusion protein etanercept (Enbrel®)) [12]. Thus for both therapeutic antibodies and Fc fusion proteins, the FcRn interaction is a generalized way to exploit FcRn protection achieve the benefits of extended persistence in vivo.

There are substantial constraints on the species source of the Fc fragment that is incorporated into the therapeutic mAb. The first therapeutic mAb, OKT3, exhibited very rapid clearance when administered to humans due to being a standard mouse mAb. A primary cause for OKT3's abbreviated serum persistence is because mouse IgG Fc does not bind in the presence of human IgG to human FcRn and therefore is not protected by human FcRn [13]. To achieve the serum persistence it has become standard to "chimerize" therapeutic mAbs by replacing the mouse Fc with the human IgG Fc counterpart (most commonly of the IgG1 subclass) with the additional benefit of reducing immunogenicity. A further modification to reduce immunogenicity is to more fully "humanize" the mAb by the replacement of certain amino acids within the variable region to match the human scaffold. There are also continuing adaptations to the above "first generation" mAb therapeutics to create more effective designer mAbs. These include, (1) amino acid substitutions in the Fc region to maximize the interaction with human FcRn and prolong the pharmacokinetics, (2) substitutions that eliminate FcRn interaction and minimize half life, and (3) modifications in the Fc region that alter binding to conventional Fc receptors or complement, with the goal of enhancing or eliminating effector functions without interfering with FcRn binding (reviewed in Refs. [7, 14, 15]).

1.4 Conventional In Vivo Model Systems for Therapeutic mAb Evaluation

Rodent model systems have proven problematic as they are not a reliably model for the pharmacokinetics of humanized mAb and Fc-fusion proteins. In contrast to the failure of mouse mAbs to be protected by human FcRn, humanized mAbs have an abnormally high affinity for mouse and rat FcRn, resulting in an artificially prolonged serum persistence [13, 16]. This fact alone has greatly diminished the preclinical utility of standard mice for therapeutic mAb development and testing. The alternative cynamolgus monkey model has proven to be reliable, but it is hampered by considerable expense and ethical concerns that limit its routine use.

2 Materials

2.1 hFCRN Mouse Models

1. B6.Cg-Fcgrt<tm1Dcr>Tg(FCGRT)276DcrJ (The Jackson Laboratory, Bar Harbor, stock number 004919); abbreviated B6.mFcRn$^{-/-}$ hFCRN Tg276.

2. B6.Cg-Fcgrt<tm1Dcr>Tg(FCGRT)32DcrJ (The Jackson Laboratory, stock number 014565); abbreviated B6.mFcRn$^{-/-}$ hFCRN Tg32.

3. B6.129X1-Fcgrt<tm1Dcr>/DcrJ (The Jackson Laboratory, stock number 003982); abbreviated FcRn$^{-/-}$ mice.

2.2 Monitoring Mouse Genotypes and FCRN Expression

1. Wild-type murine genotypic FcRn forward primer (FcRn-F): GGGATGCCACTGCCCTG; reverse primer (FcRn-R): CGA GCCTGAGATTGTCAAGTGTATT.

2. Targeted murine genotypic FcRn forward primer (T-FcRn-F): GGAATTCCCAGTGAAGGGC; reverse primer (T-FcRn-R): CGAGCCTGAGATTGTCAAGTGTATT.

3. Human FCRN forward genotypic primer (hFcRn-F): AGCC AAGTCCTCCGTGCTC; reverse primer (hFcRn-R): CTCAG AGATGCCAGTGTTCC.

4. Antibodies against human FCRN: ADM31 and ADM32 (Dr. Roopenian, The Jackson Laboratory).

5. AlexaFluor 647 label (Invitrogen, Carlsbad, USA, #A20006).

6. FACSCalibur Flow Cytometer (Becton Dickinson).

2.3 Antibodies and Materials for Monitoring Serum Kinetics and Half-Lives

1. Purified human IgG antibody (GammaGard, Baxter Laboratories).

2. Humanized mAb Herceptin® (trastuzumab; Genentech, Inc.).

2.4 Blood Collection

1. Capillary tubes, 75 μl volume, heparinized (Globe Scientific, 51608).

2. Capillary tubes, 25 μl volume (Drummond Scientific, 251-000-0250).

3. Heparin, sodium salt (Sigma, H0777) at 10,000 U/ml in 0.85 % NaCl solution.

4. Microcentrifuge tubes, 1.5 ml (USA Scientific #1615-5500).

2.5 ELISA Reagents

1. 96 well high binding ELISA plates (Greiner Bio-One, 655061).

2. Mouse anti-human IgG-Fc (Southern Biotech, Birmingham, Alabama; 9040-01).

3. Dulbecco's phosphate buffered saline (DPBS), pH 7.2 (Hyclone; SH30013).

4. Manifold for plate washing (Bel-Art, Pequannock, NJ; Vaccu-Pette/96).

5. Bovine serum albumin (BSA, Fitzgerald Industries International, Concord, MA; 30-AB75).

6. Mouse anti-human kappa-AP (Southern Biotech; 9220-04).

7. *P*-nitrophenyl phosphate (Sigma; N2765).

8. Substrate Buffer: 20 mM sodium bicarbonate, 24 mM sodium carbonate, 7 mM magnesium chloride hexahydrate, pH 8.6.

3 Methods

3.1 The hFCRN Transgenic Model

To overcome the limitations in both monkey and conventional rodent models, we generated human (h) FcRn transgenic mice with the idea that they would prove to be an effective surrogate for preclinical evaluation of therapeutic mAbs destined for human use [16–18]. The key elements of this model are: (1) the lack of mouse (m) FcRn accomplished by gene inactivation; (2) the transgenic expression of hFCRN; and (3) a genetically homogenous genetic background C57BL/6J (B6) that reduces biological noise caused by genetic variation in mixed backgrounds.

To produce these mice, we first inactivated the FcRn gene by conventional gene targeting techniques [17]. The resulting 129X1-derived null (–) allele of FcRn (formally designated *Fcgrttm1dcr*) was backcrossed for 11 generations onto C57BL6/J (abbreviated B6) resulting in the congenic strain B6.129X1-Fcgrt<tm1Dcr>/Dcr (B6.mFcRn$^{-/-}$). We also produced a number of hFCRN transgenic (Tg) lines of mice on the B6 background, including B6.hFCRN Tg276. This strain carries the hFCRN cDNA transgene from a B6-derived cDNA construct driven by a tissue ubiquitous CMV promotor/chicken β-actin enhancer. Another strain, B6.hFCRN Tg32, carries a genomic hFCRN cosmid fragment, including ~11 kb encoding the transcription product along with ~10 kb of 5′ and 3′ noncoding human sequence [16, 17, 19]. Both of these transgenic founder lines were produced by microinjection of the respective hFCRN plasmids into B6 zygotes. To produce mice lacking mouse FcRn but expressing the human FcRn transgene, we crossed hFcRn Tgs to mFcRn$^{-/-}$ mice and established lines of B6.mFcRn$^{-/-}$ hFCRN Tg mice (*see* **Note 1**).

3.2 Tracking the hFCRN Transgenes and Monitoring Their Expression

FcRn$^{-/-}$ mice can be genotyped by PCR using primers designed to distinguish the targeted from the wild-type FcRn alleles. The primers for FcRn-F and FcRn-R yield a 248-bp wild-type product, while the primers for T-FcRn-F and T-FcRn-R yield a 378-bp targeted allele product. Tracking the hFCRN transgenes can be carried out by PCR-based genotyping. hFcRn-F and hFcRn-R primers yield an amplified hFCRN product of ~300 bp for the cDNA Tg276 and 740 bp for the genomic Tg32.

Immunofluorescence monitoring of the protein expression of hFCRN can be carried out using mAbs ADM31 or ADM32, both of which are highly specific for hFCRN (*see* **Note 2**). Expression of hFCRN on leukocytes from peripheral blood, spleen, or lymph nodes can be evaluated using conventional flow cytometric procedures. To assess cell surface expression levels, single cell suspensions at approximately $10^6/50$ µl PBS with 1% BSA and 0.05% NaN$_3$ can be stained with 200 ng of ADM31-AlexaFluor 647 (labeled with Alexa Fluor 647 carboxylic acid, succinimidyl ester) and data acquired on a FACSCalibur Flow Cytometer. Expression in solid tissues can be evaluated using conventional immunofluorescence imaging on frozen sections.

3.3 Use of hFCRN Transgenic Models to Evaluate Antibody Kinetics In Vivo

Measurement of the serum concentrations of administered antibodies is a general tool to evaluate their persistence in circulation. This is usually performed by introducing a sufficient amount of the test antibody either by the intravenous or intraperitoneal routes (*see* **Note 3**), in a quantity that can be easily detected and quantified in serum samples, even after a two log reduction in concentration. The antibody "tracer" can be labeled with radioisotope which permits direct quantification in serum samples. To minimize radioactive isotope use, we use an antibody tracer that is unmodified or labeled with biotin or other derivation chemistries, and then determine its serum concentrations by ELISA techniques. We commonly inject 100 µg of the test antibody in a 200 µl volume of phosphate buffered saline into each mouse intraperitoneally (i.p.) (*see* **Note 4**). This amount can vary depending the goals of the experiment and the sensitivity of the detection method. A minimum of five inbred mice, sex-matched and age-matched, at 8–16 weeks of age are recommended for each antibody to be tested.

A typical kinetic profile is divided into alpha and beta phases. The alpha phase occurs rapidly (in the first 24 h) and is considered to be the period in which the administered antibody reaches equilibrium. The beta phase then continues for several days in a direct log-linear relationship of serum concentration to time. It is this beta phase that is considered to be the most reliable indicator of antibody half-life in circulation. To establish beta phase kinetics, we take blood samples (25–75 μl depending on the frequency of sampling) at 1–4 days following administration, followed by several additional samples over a period of days to weeks, depending on the persistence of the test antibody (*see* **Note 3**). A least five blood samples per mouse is recommended.

To illustrate the use of the hFcRn-humanized models to monitor the serum clearance of human IgG, we show an experiment in which we administered 100 μg of purified human IgG on day 0 to B6.mFcRn$^{-/-}$ hFCRN Tg276 and to B6.mFcRn$^{-/-}$ hFCRN Tg32 mice (carrying one or two copies of the respective transgenes). We used five mice in each group and collected 75 μl of blood on days 4, 11, 18, 25, and 32 (Fig. 3a). For comparison, in Fig. 3b, we show results of an experiment in which 100 μg human IgG (hIgG) was administered to five mice lacking FcRn (B6.mFcRn$^{-/-}$ mice). On day zero 25 μl blood was collected, and then on days 3, 5, 6, 7, 9, and 11. The beta phase clearance results are plotted as the percent of tracer signal remaining compared to the start time point.

From these data it is apparent that the serum persistence and the computed serum half-life of hIgG administered to the mice is dependent on the presence of hFCRN, and is influenced by the number of copies of the transgenes, i.e., it is shorter in hFCRN hemizygous mice and extended in homozygous mice. It is also apparent that hIgG has a reduced half-live in B6.mFcRn$^{-/-}$ hFCRN Tg276 mice as compared with B6.mFcRn$^{-/-}$ hFCRN Tg32 mice

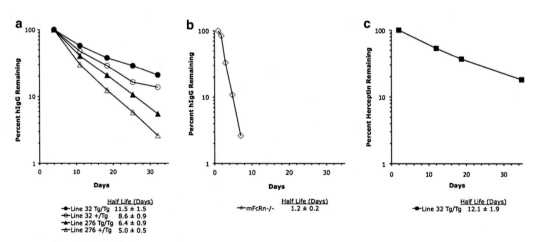

Fig. 3 Beta phase clearance of human IgG in hFcRn transgenic mice. 100 μg of purified human IgG (**a** and **b**) or Herceptin (**c**) was administered through the intraperitoneal route. Blood samples were taken from the retro-orbital sinus at the time points indicated by the symbols. The concentrations of hIgG and Herceptin were determined by ELISA techniques from plasma prepared from these samples, as described in Subheading 3. (**a**) Clearance of human IgG in transgenic mice (*open triangles*): B6.mFcRn$^{-/-}$ hFCRN Tg276 heterozygotes; (*closed triangles*) B6.mFcRn$^{-/-}$ hFCRN Tg276 homozygotes; (*open circles*) B6.mFcRn$^{-/-}$ hFCRN Tg32 heterozygotes; (*closed circles*). B6.mFcRn$^{-/-}$ hFCRN Tg32 homozygotes; B6.mFcRn$^{-/-}$ hFCRN Tg276 homozygotes. (**b**) Clearance of hIgG in FcRn$^{-/-}$ mice. (**c**) Clearance of Herceptin in B6.mFcRn$^{-/-}$ hFCRN Tg32 homozygous mice. Data are presented as the average % of IgG or Herceptin remaining normalized to the first bleed

(see **Note 5**). In general, we find that B6.mFcRn$^{-/-}$ hFCRN Tg276 hemizygous mice are best suited to detect subtle differences in mAb persistence and that B6.mFcRn$^{-/-}$ hFCRN Tg32 homozygous mice are best suited to achieve prolonged serum persistence. Thus one can select the model that is best suited for a specific application.

To illustrate the use of the transgenic mice for determination of the pharmacokinetics of therapeutic mAb, we administered the humanized therapeutic Herceptin mAb into eight B6.mFcRn$^{-/-}$ hFCRN Tg32 homozygous mice. As shown in Fig. 3c, two copies of transgene hFCRN Tg32 resulted in prolonged serum persistence of Herceptin, similar to that observed for purified human IgG. Further illustration of the use of the hFCRN transgenic mice to analyze serum persistence of therapeutic mAbs can be found in Petkova et al. [16]. These and other results show that the hFCRN transgenic model is well suited to evaluate the pharmacokinetics of human therapeutic antibodies [20–22].

3.4 Determination of Serum Concentrations and Half-Life of hIgG and Herceptin

1. hIgG specific ELISA of plasma samples are performed using 96 well high binding ELISA plates coated with mouse anti-human IgG-Fc diluted to 0.5 µg/100 µl/well in Dulbecco's phosphate buffered saline pH 7.2, overnight at 4 °C.

2. Plates are hand washed using a manifold with 96 nozzles connected to a 30 ml syringe, or multichannel pipette, two times with 300 µl per well of DPBS with 0.05 % Tween 20 and 0.05 % sodium azide (ELISA Wash).

3. The ELISA Wash is removed by flicking plates into a sink and blotting the plates on paper towels.

4. Plates are blocked with 300 µl/well of ELISA Block (ELISA Wash with 1 % bovine serum albumin (BSA)) for a minimum of 1 h at room temperature.

5. Plasma or serum samples are diluted 1/200 with ELISA Block.

6. HIgG are diluted in a twofold series from 1000 ng/ml down to 1 ng/ml to act as standards.

7. ELISA Block is removed by flicking and blotting as before (*see* **step 3**).

8. 100 µl of diluted standards and plasma as well as diluent only (to act as blanks) are transferred into the coated and blocked ELISA plates.

9. The plates are incubated at 37 °C for 1 h.

10. The plates are washed three times with 300 µl/well with ELISA Wash (*see* **step 4**).

11. 100 µl/well of mouse anti-human kappa-AP diluted 1/1000 in diluent is then added and incubated at 37 °C for 1 h.

12. Plates are washed as in **step 4** three times with 300 µl/well.

13. 100 μl/well of 1 mg/ml *p*-nitrophenyl phosphate in Substrate Buffer is added.

14. The plates are read $A_{405\ nm}$ at 30 min, or when the maximum optical density exceeds 1.

15. Data are plotted as the percent tracer remaining compared to the first tracer concentration determination. Half-life is calculated using the formula:

$$t_{1/2} = \frac{\log 0.5}{\log A_e / A_o} \times t$$

where $t_{1/2}$ is the half life of the tracer, A_e is the amount of tracer remaining, and A_o is the original amount of tracer at day 0, and t is the elapsed time.

3.5 Analysis of Albumin Pharmacokinetics

While this article focuses on issues related to FcRn and its control over the pharmacokinetics of therapeutic mAbs, there is also strong evidence that FcRn operates in a similar manner to extend the serum persistence of albumin. Indeed, serum albumin has a prolonged half life because it binds FcRn in an acid-dependent manner in a distinct docking site from IgG and is recycled similarly to IgG (reviewed in Ref. [23]). Albumin-conjugated therapeutic proteins are being developed that take advantage of this route to prolong the pharmacokinetics [24]. The human FCRN transgenic models described above have also been used to address the serum persistence of human albumin [18] and should prove also to be an effective general model for the preclinical evaluation of albumin based therapeutics. Recently we have developed a model inactivating mouse albumin to enable to study human albumin without the complication of endogenous mouse albumin levels ([25], and *see* Chapter 7).

3.6 Analysis of Therapeutic mAbs that Are Immunogenic in Mice

The beta phase clearance kinetics of human IgG and therapeutic antibodies is typically monophasic and abides to a log linear relationship of serum concentration with time after administration. However, in rare cases we have found that certain administered mAbs are immunogenic to the mouse and elicit mouse anti-human antibodies. In this case, there is a biphasic kinetic, typically appearing at days 5–6 after antibody administration and resulting in a precipitous antibody loss. This complication can be overcome by the use of immunodeficient hFcRn transgenic mice, described in Subheading 3.7.

3.7 Use of the Model for Efficacy Testing of Therapeutic mAbs

Use of the humanized FCRN model is proving to be an effective preclinical surrogate to evaluate the serum persistence of therapeutic mAbs. With modification, this approach may also assist in the testing of the efficacy. For example, there is need to evaluate the efficacy of therapeutic mAbs directed against neoplastic tissues. Typically, the rodent models used are conventional SCID, nude, and

Rag1-deficient mice. However, as we described in Subheading 3.1, the presence of mouse FcRn makes the mouse unsuitable for reliable modeling of the pharmacokinetics of human therapeutic mAbs. We have intercrossed the SCID and Rag1-null mutations into B6.mFcRn$^{-/-}$ hFCRN Tg276 and B6.mFcRn$^{-/-}$ hFCRN Tg32 mice, making it possible to transplant normal or neoplastic tissue into these mice and then to evaluate both the pharmacokinetics and efficacy of therapeutic mAb treatment [26].

4 Notes

1. The standard B6.mFcRn$^{-/-}$ stock, and FCRN-humanized B6.mFcRn$^{-/-}$ stocks show no evidence of failure to thrive when maintained in conventional, specific pathogen free conditions.

2. Specific antibodies against the human FCRN are used in regular intervals for quality control and for ascertaining the level of hFCRN expression in the transgenic strains. We recommend checking hFCRN protein expression levels in the transgenic colony at least once a year in case of loss of expression.

3. All animal experimental procedures (including handling, housing, husbandry, blood collection, and drug treatment) must be conducted in accordance with national and institutional guidelines for the care and use of laboratory animals.

4. We have not observed a difference in β phase clearance kinetics when comparing tail vein i.v. with i.p. injections.

5. As a general rule, the hFCRN copy number in our strains correlates with the hFCRN expression levels, with B6.mFcRn$^{-/-}$ hFCRN Tg32 homozygous mice having the highest hFCRN protein expression.

References

1. Burnet FM, Freeman M, Jackson AV, Lush D (1941) The production of antibodies. Macmillan and Company Limited, Melbourne

2. Simister NE, Mostov KE (1989) An Fc receptor structurally related to MHC class I antigens. Nature 337:184–187

3. Ahouse JJ, Hagerman CL, Mittal P, Gilbert DJ, Copeland NG, Jenkins NA et al (1993) Mouse MHC class I-like Fc receptor encoded outside the MHC. J Immunol 151:6076–6088

4. Raghavan M, Gastinel LN, Bjorkman PJ (1993) The class I major histocompatibility complex related Fc receptor shows pH-dependent stability differences correlating with immunoglobulin binding and release. Biochemistry 32: 8654–8660

5. Burmeister WP, Gastinel LN, Simister NE, Blum ML, Bjorkman PJ (1994) Crystal structure at 2.2 A resolution of the MHC-related neonatal Fc receptor. Nature 372:336–343

6. Burmeister WP, Huber AH, Bjorkman PJ (1994) Crystal structure of the complex of rat neonatal Fc receptor with Fc. Nature 372:379–383

7. Roopenian DC, Akilesh S (2007) FcRn: the neonatal Fc receptor comes of age. Nat Rev Immunol 7:715–725

8. Ghetie V, Ward ES (2002) Transcytosis and catabolism of antibody. Immunol Res 25:97–113

9. Ghetie V, Ward ES (1997) FcRn: the MHC class I-related receptor that is more than an IgG transporter. Immunol Today 18: 592–598

10. http://antibodysociety.org/news/approved_mabs.php

11. Deckert PM (2009) Current constructs and targets in clinical development for antibody-based cancer therapy. Curr Drug Targets 10: 158–175

12. Jarvis B, Faulds D (1999) Etanercept: a review of its use in rheumatoid arthritis. Drugs 57:945–966

13. Ober RJ, Radu CG, Ghetie V, Ward ES (2001) Differences in promiscuity for antibody-FcRn interactions across species: implications for therapeutic antibodies. Int Immunol 13:1551–1559

14. Presta LG (2008) Molecular engineering and design of therapeutic antibodies. Curr Opin Immunol 20:460–470

15. Liu XY, Pop LM, Vitetta ES (2008) Engineering therapeutic monoclonal antibodies. Immunol Rev 222:9–27

16. Petkova SB, Akilesh S, Sproule TJ, Christianson GJ, Al Khabbaz H, Brown AC et al (2006) Enhanced half-life of genetically engineered human IgG1 antibodies in a humanized FcRn mouse model: potential application in humorally mediated autoimmune disease. Int Immunol 18:1759–1769

17. Roopenian DC, Christianson GJ, Sproule TJ, Brown AC, Akilesh S, Jung N et al (2003) The MHC class I-like IgG receptor controls perinatal IgG transport, IgG homeostasis, and fate of IgG-Fc-coupled drugs. J Immunol 170: 3528–3533

18. Chaudhury C, Mehnaz S, Robinson JM, Hayton WL, Pearl DK, Roopenian DC et al (2003) The major histocompatibility complex-related Fc receptor for IgG (FcRn) binds albumin and prolongs its lifespan. J Exp Med 197:315–322

19. Stein C, Kling L, Proetzel G, Roopenian DC, de Angelis MH, Wolf E, Rathkolb B (2012) Clinical chemistry of human FcRn transgenic mice. Mamm Genome 23:259–269

20. Tam SH, McCarthy SG, Brosnan K, Goldberg KM, Scallon BJ (2013) Correlations between pharmacokinetics of IgG antibodies in primates vs. FcRn-transgenic mice reveal a rodent model with predictive capabilities. MAbs 5:397–405

21. Proetzel G, Roopenian DC (2014) Humanized FcRn mouse models for evaluating pharmacokinetics of human IgG antibodies. Methods 65:148–153

22. Proetzel G, Wiles MV, Roopenian DC (2014) Genetically engineered humanized mouse models for preclinical antibody studies. BioDrugs 28:171–180

23. Anderson CL, Chaudhury C, Kim J, Bronson CL, Wani MA, Mohanty S (2006) Perspective—FcRn transports albumin: relevance to immunology and medicine. Trends Immunol 27: 343–348

24. Osborn BL, Olsen HS, Nardelli B, Murray JH, Zhou JX, Garcia A et al (2002) Pharmacokinetic and pharmacodynamic studies of a human serum albumin-interferon-alpha fusion protein in cynomolgus monkeys. J Pharmacol Exp Ther 303:540–548

25. Roopenian DC, Low BE, Christianson GJ, Proetzel G, Sproule TJ, Wiles MV (2015) A humanized mouse model to study human albumin metabolism and pharmacokinetics of albumin-based drugs. MAbs 7:344–351

26. Zalevsky J, Chamberlain AK, Horton HM, Karki S, Leung IW, Sproule TJ, Lazar GA, Roopenian DC, Desjarlais JR (2010) Enhanced antibody half-life improves in vivo activity. Nat Biotechnol 28:157–159

Chapter 7

A Humanized Mouse Model to Study Human Albumin and Albumin Conjugates Pharmacokinetics

Benjamin E. Low and Michael V. Wiles

Abstract

Albumin is a large, highly abundant protein circulating in the blood stream which is regulated and actively recycled via the neonatal Fc receptor (FcRn). In humans this results in serum albumin having an exceptional long half-life of ~21 days. Some time ago it was realized that these intrinsic properties could be harnessed and albumin could be used as a privileged drug delivery vehicle. However, active development of albumin based therapeutics has been hampered by the lack of economic, relevant experimental models which can accurately recapitulate human albumin metabolism and pharmacokinetics. In mice for example, introduced human albumin is not recycled and is catabolized rapidly. This is mainly due to the failure of mouse FcRn to bind human albumin consequently, human albumin has a half-life of only 2–3 days in mice. To overcome this we developed and characterized a humanized mouse model which is null for mouse FcRn and mouse albumin, but is transgenic for, and expressing functional human FcRn. Published data clearly demonstrate that upon injection of human albumin into this model animal that it accurately recapitulates human albumin FcRn dependent serum recycling, with human albumin now having a half-life ~24 days, closely mimicking that observed in humans. In this practical review we briefly review this model and outline its use for pharmacokinetic studies of human albumin.

Key words Albumin conjugates, Pharmacokinetics, Neonatal Fc receptor, HSA, Analbuminemia, Hypoalbuminemia, Albumin, TALEN, Mouse model

Abbreviations

B6	C57BL/6J
FcRn	Neonatal Fc receptor
HSA	Human serum albumin
kDa	Kilodalton
MSA	Mouse serum albumin
TALEN	Transcription activator-like effector nuclease
Tg32	B6.Cg-*Fcgrttm1Dcr* Tg(FCGRT)32Dcr/DcrJ
Tg32-Alb$^{-/-}$	B6.Cg-*Albem12MvwFcgrttm1Dcr* Tg(FCGRT)32Dcr/MvwJ

Gabriele Proetzel and Michael V. Wiles (eds.), *Mouse Models for Drug Discovery: Methods and Protocols*,
Methods in Molecular Biology, vol. 1438, DOI 10.1007/978-1-4939-3661-8_7, © Springer Science+Business Media New York 2016

1   Introduction

Human serum albumin (HSA) is a 69 kDa highly stable and abundant (35–50 mg/mL in serum) protein synthesized in the liver, with an exceptional serum half-life of ~21 days. HSA is involved in the transport of hormones, fatty acids and is regarded as the primary controller of colloid osmotic pressure. It also serves as a depot and courier of endogenous compounds, including steroids, fatty acids, etc., and of key interest here, exogenously added small molecule drugs and their metabolites. Its abundance and long half-life are the result of rate of synthesis, elimination and crucially, an active recycling pathway using the neonatal Fc receptor (FcRn) [1–6]. FcRn is ubiquitously and functionally expressed in many cell types and organs. Collectively these features have been recognized as a perfect combination for its use as a delivery vehicle of therapeutic drugs, allowing half-life extension combined plus reduced patient side effects and hence, improved compliance [5, 7–9].

Approaches using albumin for compound delivery fall into three broad classes: (1) the coupling of low-molecular weight drugs directly to albumin [7]; (2) direct conjugation or fusion to albumin of bioactive peptides, proteins or siRNAs [5, 8]; (3) encapsulation of drugs into albumin nanoparticles [7, 10]. The development of such compounds represents a new array of effective therapeutics with extended half-life, stability, versatility, increased safety, and manufacture simplicity [9]. Unfortunately, until recently the further rapid development of potential albumin based therapeutics has been hindered by the lack of animal models which can faithfully reproduce human albumin FcRn recycling, metabolism, and half-life [11, 12]. For example, when HSA is introduced into wild type C57Bl/6J mice (abbreviated here to B6), it exhibits a half-life of only ~2.6 days, which is in marked contrast to the ~21 days observed in humans. To address this, a mouse model null for FcRn and transgenic for human FcRn was developed (B6.Cg-*Fcgrttm1Dcr* Tg(FCGRT)32Dcr/DcrJ, abbreviated here to Tg32) ([4, 13] and *see* Chapter 6). Although of use for pharmacokinetic evaluation of therapeutic (human) antibodies, when used for HSA pharmacokinetic studies the mouse's own endogenous serum albumin outcompetes HSA binding to the human FcRn, effectively blocking human albumin recycling. This results in HSA having a half-life only slightly extended in the Tg32 (5.8 days) mouse model compared with that observed wild type mice (2.6 days) (Fig. 1) [12]. To circumvent this we further modified the Tg32 line, using TALENs to disrupt the mouse albumin gene. This line created, B6.Cg-*Albem12MvwFcgrttm1Dcr* Tg(FCGRT)32Dcr/MvwJ, abbreviated here to Tg32-*Alb−/−*, is null for albumin by ELISA, whilst remaining surprisingly healthy. Crucially, upon intravenous injection (IV) injection of human

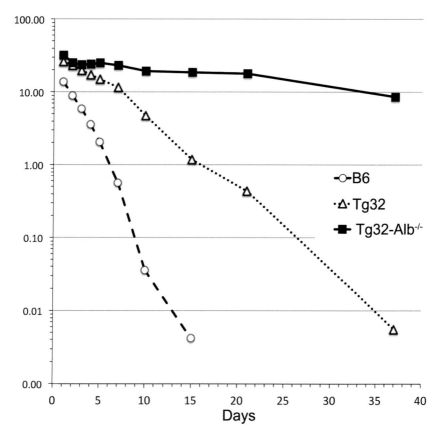

Fig. 1 Data based on seven male mice aged 10–13 weeks from each strain (B6, Tg32, and Tg32-Alb$^{-/-}$) received a single dose via tail-vein injection of HSA. Samples for analysis were collected at 10 min, 12 h, and then 24, 48, 72, 96, and 120 h, then at 7, 10, 15, 21, and 37 days

albumin into Tg32-Alb$^{-/-}$ mice, HSA now demonstrates a half-life of ~24 days.

In this chapter we outline the use of the Tg32-Alb$^{-/-}$ model in determining the half-life of human albumin post IV injection, demonstrating that these animals are of use in the comparative pharmacokinetics testing of albumin and related compounds/ drugs. All mouse lines used are listed in Table 1 and are available from The Jackson Laboratory.

2   Materials

2.1   Mice

Mice age and sex matched, Table 1 and (*see* **Note 1**).

C57Bl/6J abbreviated to B6 (The Jackson Laboratory, Bar Harbor, ME, USA: cat no. 000664).

B6.Cg-*Fcgrttm1Dcr*Tg(FCGRT)32Dcr/DcrJ   abbreviated   to Tg32 (The Jackson Laboratory, Bar Harbor, ME, USA: cat no. 014565).

Tg32-*Alb*$^{-/-}$ B6.Cg-*Albem12MvwFcgrttm1Dcr*Tg(FCGRT)32Dcr/ MvwJ, abbreviated here to Tg32-*Alb*$^{-/-}$ (The Jackson Laboratory, Bar Harbor, ME, USA: cat no. 25201).

Table 1
Mouse strains used in this study and HSA half-life

Short name	Full strain name	JAX stock Ref #	Genotype			$t_{1/2}$ HSA [days] ($n=7$)	SEM (days)
			m_{Alb}	m_{Fcgrt}	h_{FCGRT}		
B6	C57BL/6J	000664	wt	wt	Null	2.6	0.1
Tg32	B6.Cg-*Fcgrttm1Dcr* Tg(FCGRT)32Dcr/DcrJ	014565	wt	–/–	Tg/Tg	5.8	0.5
Tg32-Alb$^{-/-}$	B6.Cg-*Albem12MvwFcgrttm1Dcr* Tg(FCGRT)32Dcr/MvwJ	025201	–/–	–/–	Tg/Tg	24.1	2.8

2.2 Serum Collection

Heparin.

Storage buffer: 50 µL glycerol/40 µL PBS.

HSA (Plasbumin, Talecris, 13533-684-16).

Serum separator tubes (BD Microtainer tube, Cat # 365967).

2.3 HSA Assay System Using Either

Option 1:
    Human Albumin Elisa Kit ELISA Immunoperoxidase Assay for Determination of Albumin in Human Samples (GenWay Biotech, cat. no GWB-64EC6D).
    Option 2:

- 96-well plates (Greiner Bio-One, cat. 655061).

- HSA (Talecris, cat. 13533-684-16).

- Rabbit anti-HSA antibody (US Biological, cat. A1327-46).

- 5 % bovine serum albumin (Sigma-Aldrich, cat. A7284).

- Alkaline phosphatase conjugated goat anti-HSA antibody cross-adsorbed against bovine, mouse, and pig albumins (Bethyl Laboratories Inc., cat. A80-229AP).

- *p*-nitrophenyl phosphate substrate (Amresco Inc., cat. 0405), used at 1 mg/mL dissolved in 10.1 mM $NaHCO_3$, 11.8 mM Na_2CO_3, 2.2 mM $MgCl_2$, pH 8.6.

- Coating Buffer: 45.3 mM $NaHCO_3$, 18.2 mM Na_2CO_3, pH 9.6.

- Storage Buffer: 50 % glycerol in PBS.

- ELISA Wash Buffer: PBS with 0.05 % Tween 20, 0.5 g/L NaN_3.

- ELISA Blocking Buffer: 1 % BSA in PBS with 0.05 % Tween 20, 0.5 g/L NaN_3.

- 3 M NaOH.

3   Methods

Here we describe the protocol for tracking intravenously administered HSA levels in B6, Tg32, and Tg32-Alb$^{-/-}$ (*see* **Note 1**; Fig. 1).

3.1 Pre-bleed of Mice

Control and test animals need to be healthy and of a good weight to tolerate multiple bleedings. Choose three to ten mice per group, and age- and sex-match the mice. Weigh the mice before setting up the groups (*see* **Note 2**).

For each strain to be tested, in this case, B6, TG32, and Tg32-alb$^{-/-}$, select age and gender-matched mice of sufficient numbers for a statistically valid data set. We recommend using 6–8 week old mice. Use a reliable method (e.g., ear notches and ear tags) to identify each individual mouse to enable that the data can be tracked on a per mouse base.

To establish each animal's baseline, collect blood 2 weeks prior to injection date.

This also allows the animals to recover completely before the time point bleeds after dosing. Whole blood samples should be obtained from either the retro-orbital sinus or by cheek bleed (sub-mandibular). Mix blood (~25 μL) and heparin (1 μL) by gently pipetting up and down three times, leaving the blood on ice until all mice have been bled. Spin all preps at $10,062 \times g$ for 5 min, at 4 °C. Carefully remove 10 μL of plasma (straw colored supernatant) to 90 μL of storage buffer and store at –20 °C until ready to perform ELISA. Alternatively, blood can be collected into serum separator tubes, and plasma isolated following the manufacturer's protocol.

3.2 Compound Injection and Serum Collection

Successful completion of these tests require that injection and first bleed times be orchestrated in advance, especially for the first bleed. We have found that this is best done using a team of two, an injector and animal sample collector (bleeding). The analysis method (ELISA) should be optimized beforehand, so that the smallest amount of blood as possible is needed to allow to collect all necessary time points from each mouse.

Upon injection of test compounds a series of plasma samples may need to be collected to ascertain the alpha phase where there is the rapid decrease in compound plasma concentration due to dilution in the circulatory system. This is followed by a quasi-equilibrium beta phase where the compound concentration decrease is principally attributed to metabolism and excretion. It is this latter, beta phase, which is used for measuring compound half-life (Fig. 1). The exact time points may need to be determined depending on the compound to be tested.

One day before injection, weigh each animal and calculate the desired volume to inject (10 mg/kg injected at 1 mg/mL). For example, a 27 g mouse should receive 0.27 mL of HSA at 1 mg/mL.

On the day of injection, prepare an adequate volume of HSA in sterile PBS to a final concentration of 1 mg/mL.

Record time of injection (via tail-vein) and deliver appropriate volumes to each individual mouse (*see* **Note 3**).

Collect ~10–20 μL blood 10 min after the injection is completed, then again at 12 h (*see* **Note 4**). Continue to collect blood 1 day post-injection (24 h), as well as at days 2, 3, 4, 5, 7, 10, 15, 21, and 37 (*see* **Notes 5** and **6**).

3.3 ELISA

In the example shown (Fig. 1), in both the B6 and Tg32 strains HSA is rapidly metabolized, with the alpha and beta phases occurring over a short period of time. With Tg32-Alb$^{-/-}$ animals there is a similarly short alpha phase however, as this strain can successfully recycle HSA there is an elongated beta phase. As such, when testing new compounds a rough calibration series of tests may be required.

For measuring HSA we recommend to use one of the two options described below:

1. Use a commercially available HSA ELISA kit to be measure the human albumin concentrations, and follow manufacturer's directions.

2. Use a "homemade" HSA kit: Antigen-based ELISA.

   - Coat 96-well plate with 5 μg/mL rabbit anti-HSA antibody in Coating Buffer overnight at 4 °C (alternatively, 1 h at 37 °C), using 100 μL per well. Incubations should be done in high humidity (wrap covered plates with wet paper towels, inside plastic bag).

   - Wash plate 2× with 300 μL per well ELISA Wash Buffer.

   - Block for 1 h at RT (or overnight at 4 °C) using 200 μL per well of 1% bovine serum albumin in ELISA Wash Buffer.

   - Wash plate 3× with 300 μL per well ELISA Wash Buffer.

   - Incubate for 1 h at RT (or overnight at 4 °C) using 100 μL per well appropriately diluted plasma samples (1:200 final dilution), as well as standards and blank wells. Plasma is initially diluted 1:10 in Storage Buffer, then dilute an aliquot 1:20 in ELISA Blocking Buffer.

   - Wash plate 3× with 300 μL per well ELISA Wash Buffer.

   - Detect signals using 100 μL per well alkaline phosphatase conjugated goat anti-HSA antibody (0.5 μg/mL prepared in 1% BSA, PBS with 0.05% Tween 20, 0.5 g/L NaN$_3$). Incubate for 1 h at 37 °C, or overnight at 4 °C.

   - Wash plate 3× with 300 μL per well ELISA Wash Buffer.

- Measure at 405 nm after development with colorimetric *p*-nitrophenyl phosphate substrate, used at 1 mg/mL (100 µL per well).

- The reaction can be terminated by the addition of 25 µL 3 M NaOH to each well.

- For quantification, a standard curve was prepared using serial dilutions prepared from HSA. Concentrations were calculated using the Five Parameter Logistics Regression (e.g., SigmaPlot.com, Readerfit.com).

- HSA half-life is calculated from time points within the beta-phase, or linear portion of the log-linear plot of concentration versus time for each individual mouse, and then averaged across the strain [13]. *See* also Chapter 6.

4   Notes

1. This test shows the effect of endogenous mouse albumin on hFcRn mediated HSA recycling. If doing comparative compound testing, tests can be limited to the Tg32-Alb$^{-/-}$ strain alone.

2. Mice are small and it is crucial that they are not over, or too often bled. Local animal welfare oversight needs should be taken into consideration when designing these tests. If a protocol requires more frequent sampling, multiple animals can be used, sampling them in succession. The use of inbred, age and importantly weight matched animals will help in maintaining the integrity of the data obtained.

3. The amount of liquid injected is small, to maintain good replicates is it essential that this be kept accurate and that the injection is cleanly IV. Deviations for the amount and delivery target will impact replicates.

4. The 10 min time point is required to establish the maximal level of the test compound in the circulation. This needs to be orchestrated in advance. We have found that this is best done using a team of two, an injector and animal sample collector (bleeding).

5. All mice need to be healthy and of a good weight to tolerate multiple bleedings. Five to ten weight/sex matched mice/compound are recommended.

6. Consult your local animal welfare compliance officer to ensure the experimental design (especially the blood collection schedule) is within the institutional guidelines.

Acknowledgements

This work was done with help and excellent assistance of Cindy Avery and Deb Woodworth for genotyping and animal maintenance, plus discussions with Gregory J Christianson, Thomas J Sproule, plus Dr's Gabriele Proetzel, Derry C Roopenian, and Vishnu Hosur. This work was supported by National Institutes of Health Grant OD011190 (M.V.W.) and The Jackson Laboratory.

References

1. Ghetie V, Ward ES (2000) Multiple roles for the major histocompatibility complex class I-related receptor FcRn. Annu Rev Immunol 18:739–766

2. Chaudhury C, Brooks CL, Carter DC, Robinson JM, Anderson CL (2006) Albumin binding to FcRn: distinct from the FcRn-IgG interaction. Biochemistry 45:4983–4990

3. Anderson CL, Chaudhury C, Kim J, Bronson CL, Wani MA, Mohanty S (2006) Perspective—FcRn transports albumin: relevance to immunology and medicine. Trends Immunol 27: 343–348

4. Roopenian DC, Akilesh S (2007) FcRn: the neonatal Fc receptor comes of age. Nat Rev Immunol 7:715–725

5. Andersen JT, Dalhus B, Viuff D, Ravn BT, Gunnarsen KS, Plumridge A et al (2014) Extending serum half-life of albumin by engineering neonatal Fc receptor (FcRn) binding. J Biol Chem 289:13492–13502

6. Kuo TT, Baker K, Yoshida M, Qiao SW, Aveson VG, Lencer WI, Blumberg RS (2010) Neonatal Fc receptor: from immunity to therapeutics. J Clin Immunol 30:777–789

7. Kratz F (2008) Albumin as a drug carrier: design of prodrugs, drug conjugates and nanoparticles. J Control Release 132:171–183

8. Nicolì E, Syga MI, Bosetti M, Shastri VP (2015) Enhanced gene silencing through human serum albumin-mediated delivery of polyethylenimine-siRNA polyplexes. PLoS One 10(4):e0122581. doi:10.1371/journal.pone.0122581

9. Sleep D (2015) Albumin and its application in drug delivery. Expert Opin Drug Deliv 12: 793–812

10. Lytton-Jean AKR, Kauffman KJ, Kaczmarek JC, Langer R (2015) Cancer nanotherapeutics in clinical trials. Cancer Treat Res 166: 293–322

11. Knudsen Sand KM, Bern M, Nilsen J, Noordzij HT, Sandlie I, Andersen JT (2015) Unraveling the interaction between FcRn and albumin: opportunities for design of albumin-based therapeutics. Front Immunol 5:682. doi:10.3389/fimmu.2014.00682

12. Roopenian DC, Low BE, Christianson GJ, Proetzel G, Sproule TJ, Wiles MV (2015) Albumin-deficient mouse models for studying metabolism of human albumin and pharmacokinetics of albumin-based drugs. MAbs 7: 344–351

13. Petkova SB, Akilesh S, Sproule TJ, Christianson GJ, Al Khabbaz H, Brown AC, Presta LG, Meng YG, Roopenian DC (2006) Enhanced half-life of genetically engineered human IgG1 antibodies in a humanized FcRn mouse model: potential application in humorally mediated autoimmune disease. Int Immunol 18:1759–1769

Chapter 8

Germ-Free Mice Model for Studying Host–Microbial Interactions

Yogesh Bhattarai and Purna C. Kashyap

Abstract

Germ-free (GF) mice are a relevant model system to study host–microbial interactions in health and disease. In this chapter, we underscore the importance of using GF mice model to study host–microbial interactions in obesity, immune development and gastrointestinal physiology by reviewing current literature. Furthermore, we also provide a brief protocol on how to setup a gnotobiotic facility in order to properly maintain and assess GF status in mice colonies.

Key words Germ-free, Gnotobiotic, Gut microbiome

1 Introduction

The use of modern techniques such as genomics and metabolomics has recently started to unravel the enormous genetic diversity and the metabolic complexity of the microbiota in the gastrointestinal (GI) tract. It is now known that the human GI tract consist of more than 100 trillion commensal bacteria and a large number of other microorganisms including viruses, fungi, protozoa and archaea [1–3]. Among these microorganisms, commensal bacteria form the predominant part of the gut microbiota and interacts intricately with the host to regulate the development and function of the GI tract. Study of microbial communities in humans poses a challenge because targeted microbial manipulation in humans is challenging, and it is difficult to uncouple the effect of microbes from host genetics and environment. Therefore, to better understand how microbiota affect GI function, gnotobiotic animal models are often used. The term *gnotobiotic* is derived from the Greek word "*gnotos*" and "*biota*" meaning known flora and fauna [4]. Gnotobiotic animal models are essential to translational medicine as they help deconstruct complex interactions and allow the study of the effects of specific bacteria on the host in a highly

Gabriele Proetzel and Michael V. Wiles (eds.), *Mouse Models for Drug Discovery: Methods and Protocols*,
Methods in Molecular Biology, vol. 1438, DOI 10.1007/978-1-4939-3661-8_8, © Springer Science+Business Media New York 2016

controlled experimental environment, as well as provide a "frame-of-reference" to understand the role of microbes in regulating host function.

1.1 Historical Perspective: Establishment of Gnotobiotics as a Tool for Studying Host–Microbial Interactions

Gnotobiotic experimentation for studying host–microbial interactions began with Louis Pasteur in 1855, when he postulated that life was dependent on microbial colonization [5]. To test whether germ-free (GF) life of an animal host is possible, Nuttall and Thierfelder raised guinea pigs under GF conditions for the first time at the University of Berlin [6]. The cesarean delivered GF guinea pigs appeared to be healthy but few striking differences such as decreased body weight and enlarged cecum (five to ten times the cecal volume relative to the conventional counterpart) were observed in GF guinea pigs [6]. The caecum in these animals was filled with "a brown liquid which contained cheese like coagula" [6]. Further study of the cecal contents in GF animals showed that it contains bioactive substances that are toxic to the animal and causes altered smooth muscle contractility [7]. These early observations led researchers to hypothesize that although GF life is possible, bacteria are essential for regulating host physiology including "normal" digestion in order to inactivate or reduce toxic substances produced by the host [6]. Subsequent studies suggest that bacteria not only help eradicate/neutralize toxic substances but also interact closely with the host to affect the development and function of the GI tract.

Rearing and maintenance of healthy GF animals was a challenging task due to technological constraints until mid-1900s. Studies with GF animals started systematically when Reyniers and his colleagues established academic organizations in mid-1940s and early 1950s devoted to understand host–microbial interactions [8]. By late 1950s, researchers were successfully rearing GF guinea pigs, mice, and chickens [9]. The availability of gnotobiotic animal model gave researchers not only the ability to compare GF animals with conventionally raised animals with a microbiota but also the ability to introduce one or few bacterial species at a time to understand host–microbial interactions in a simplified environment [10]. Utilizing the advantages these gnotobiotic animal models had to offer, scientists have examined various microbial species in GF mice to understand microbe–microbe interaction, gene–microbe interaction, diet–microbe interactions and factors affecting microbial colonization of the GI tract.

The development of tools such as specific culture media for different bacteria and incubation techniques were enormously helpful to enumerate the bacterial species present in the gut. However, given the fact that the vast majority of gut bacteria are unculturable it was insufficient to study the microbial ecology of the GI tract in detail [11]. Now with recent advancement in next generation sequencing technology, it is possible to investigate the gut microbial ecology in much greater detail. 16s rRNA based microbial community sequencing [12] and metagenomic sequencing [13–15], has now revealed that there are about ~1000 different species of bacteria in the human and the mouse intestine [16] among which those belonging to the phylum Bacteroidetes and the Firmicutes are the most abundant [1, 16]. With the advent of next generation sequencing and advances in microbial ecology, the use of gnotobiotic models that started as an underutilized tool in the last century is now an invaluable resource to understand host–microbial interactions in health and disease.

1.2 GF Mice Model as a Tool for Studying Host–Microbial Interactions in Human Health and Disease

GF mice are gnotobiotic mice that are free from all forms of microbial life including bacteria, viruses, protozoa, archaea, and parasites. Different strains of laboratory mice serve as an important genetic tool to study host–microbial interactions for three main reasons. First, the genotypes of laboratory mice have been well characterized [17]. Second, mice and human genomes are 99 % similar in gene function [18]. Third, transgenic mice unlike any other lab mammalian species, allows us to introduce precise genetic modifications (genetic addition, ablation and modulation) and examine of the effects of these modifications in a living organism [19]. The genotypic characterization and similarity with humans enables us to use mice as a model organism to understand how gene interacts within an organism, and help extrapolate these data to human biology [17], while ability to perform genetic modification facilitates the identification of genes participating in normal and disease pathways in order to help understand host–microbial interaction in health and disease [20]. The ability to rederive these genetically modified mice as GF further allows us to understand the effect of environment and the host genes on the microbial community, as well as the effects of microbial community on host gene expression, epigenetic changes and host physiology.

1.3 GF Mice as a Tool to Understand the Role of Gut Microbiome in Obesity

16s rRNA based microbial community profiling and metagenomic sequencing from humans and genetic mice models have revealed that the phyla Firmicutes and Bacteroidetes are the predominant part of gut microbiota [21, 22]. Interestingly, obesity has been associated with a change in Firmicutes to Bacteroidetes ratio [21]. In particular, obese mice have been shown to have 50 % reduction in the abundance of Bacteroidetes and a significant concomitant increase in Firmicutes compared to lean mice [21, 23]. However, the studies that correlate disease phenotype with gut microbiota composition cannot determine if gut microbiota is driving the phenotype or an innocent bystander that changes in response to a disease. Gnotobiotic mice provide an ideal tool to address hypotheses generated from such studies to help investigate the role of gut microbiota in driving a disease phenotype.

As an example, Turnbaugh et al. colonized GF mice with microbiota from obese mice and found the microbiota transfer from obese mice led to greater adiposity in gnotobiotic mice implicating gut microbiota as one of the factors driving the obese phenotype [22]. Turnbaugh et al. subsequently used a humanized mouse model (ex-GF mice colonized with human fecal microbial communities) and showed that humanized mice when fed a high-fat western diet shifts the structure of gut microbiota with increased representation of Firmicutes. Furthermore, transplantation of microbiome from these high-fat fed humanized mice to GF mice leads to increased adiposity in the recipient mice [24]. Together these studies suggest that the differences in gut microbial ecology in lean and obese individuals may in part be responsible for the metabolic disturbance. An area of interest in this regard is the metabolic potential (capacity to harvest energy from diet) of different microbial communities. Although the physiological contributions of increased Firmicutes to the intestinal ecosystem and to fuel partitioning are unclear in obesity, few studies have reported that certain members in Firmicutes could affect the metabolic potential of the host because they are highly enriched for glycoside hydrolases and polysaccharide lysases and help in efficient extraction of calories from otherwise indigestible common polysaccharides in the diet [22, 25]. Thus, the increased Firmicutes–Bacteroidetes ratio potentially creates a microbial mix that is highly efficient in extracting energy from diets and could potentially promote adiposity [22, 25]. In follow-up studies, Riduara et al. showed that gnotobiotic mice colonized with microbiota from an obese co-twin gain weight as compared to those colonized with microbiota from the lean co-twin [26]. Interestingly, microbes from mice associated with lean microbiome can invade the mice colonized with obese microbiome and prevent weight gain as long as mice were fed a healthy diet [26].

Bariatric surgery is one of the most effective therapies for medically complicated obesity [27]. Tremaroli et al. compared gut microbiota of patients who underwent Roux-en-Y gastric bypass to obese subjects who did not undergo surgery [28]. They found that the surgical procedures causes long term durable changes on the gut microbiota including a decrease in Firmicutes compared to control obese subjects [28]. The variation in microbial ecology corresponds to lower respiratory quotient and decreased utilization of carbohydrates in subjects who underwent Roux-en-Y bypass [28]. While this finding suggested a potential role of gut microbiota in mediating beneficial effects of bariatric surgery on the metabolic phenotype, the authors used a gnotobiotic mouse model to further investigate the relevance of this finding. GF mice were colonized with gut microbiota from patients that either underwent surgery or had no surgical intervention and interestingly microbiota from patients following bariatric surgery led to reduced fat deposition in recipient mice compared to microbiota from control obese subjects [28]. This suggests that the gut microbiome plays an important role in influencing metabolism and adiposity after a bariatric surgery and highlights the utility of using GF mice to investigate the role of gut microbiome in driving obesity.

1.4   GF Mice as a Tool to Understand the Effect of Microbiota on the Development and Function of the Immune System

The role of microbiota in development and regulation of the immune system has been extensively studied. Recent studies suggest that exposure to microbes early in life is essential for the proper development and function of the immune system [29–31]. This is especially true in the GI tract and gnotobiotic mouse model has been extensively utilized to show that commensal gut microbiota interacts with the host to enhance host immunity and defend against enteric pathogens.

Multiple studies have demonstrated that surface antigen and metabolic-end products of gut microbiota modulate immune system activation and production of cytokines [32–37]. Franchi et al. using a GF and conventionally raised mouse model showed that commensal bacteria modulate immune system and cytokine production by priming intestinal macrophages for pathogenic infection via upregulation in pro-IL-1β activity [33, 38]. This in turn leads to an increase in "mature" enzymatically active IL-1β production and ultimately causes neutrophil recruitment for pathogen eradication [33]. This observation is supported by the fact that neutrophil count and macrophage function such as phagocytosis and microbicidal activities including phagocytic superoxide anion production is lower in GF mice [39, 40].

Autoimmune disorders such as inflammatory bowel disease (IBD) and rheumatoid arthritis have been associated with alteration in gut microbial ecology [41]. In IBD, a reduction in Firmicutes and Bacteroidetes and a concomitant overgrowth of Proteobacteria has been observed [42, 43]. Although it is not well understood how the alteration in the gut microbiota composition (dysbiosis) results in inflammation in IBD, it is hypothesized that systemic $CD4^+$ T cell might play a role. In this regard, Mazmanian et al. showed that colonization of GF mice with *Bacteroides fragilis* directs the cellular and physical maturation of the developing immune system via a bacterial polysaccharide (PSA) mediated pathway and corrects systemic $CD4^+$ T cell deficiencies by restoring T helper 1 (T_H1; crucial for the host defense against microbial infection) and T helper 2 (T_H2; crucial for eliminating parasitic infections) balance as the GF immune response is biased towards T_H2 response [32]. Microbial fermentation end products such as short chain fatty acids have also been shown to regulate colonic regulatory T cells and protect against colitis in gnotobiotic mice [44].

Besides impacting T cell lineages, changes in microbiota also differentially impact microbial recognition by affecting pattern-recognition receptors (PRRs) such as Toll-like receptors (TLRs) and nucleotide-binding oligomerization domain (NOD) receptors to cause disease phenotypes [33, 45]. Since, genetic knockdown of TLR receptors and NOD2 gene can contribute to development of inflammatory disorders such as IBD [46], and absence of gut microbiota impairs the development and production of TLR receptors [46, 47], it is possible that commensal microbiota play an important role in protecting against inflammation by impacting the development of PRRs. To this end, GF mice have been used to show that recognition of commensal microbiota by TLR is required for maintenance of intestinal homeostasis [48], while recognition of NOD by intestinal microbiota is responsible for regulation of innate immunity [45, 47]. Overall, these studies highlight the utility of GF mice as a model to understand how microbiota can regulate development and function of different aspects of the immune system.

1.5 GF Mice as a Tool to Understand the Effects of Microbes on GI Physiology Including Motility and Secretion

The GI tract harbors a more diverse microbial ecosystem than any other part of the body. These bacteria can modulate the development of enteric nerves to influence colonic motility and secretion [49, 50].

1.5.1 GI Motility

GF mouse model has played an important role in evaluating the effect of gut microbiota on GI motility. Abrahams and Bishop in 1967, using a non-absorbable radioactive tracer found that GI transit time is significantly faster in conventionally raised mice as compared to GF mice (conventional mice passed >90% of radioactivity in feces within 16 h while it was less than 30% in GF) [51]. Subsequently, several investigators have shown introduction of mouse-derived or human gut-derived bacteria into GF mice alters GI motility and transit time [52, 53]. Recent work in this area has increased our understanding of the effect of gut microbes on the neuromuscular apparatus. Anitha et al. showed that gut microbiome interacts with enteric neurons via LPS mediated activation of TLR-4 in order to increase neuronal survivability and intestinal motility using gnotobiotic mouse model [50]. Furthermore, human or mouse-derived complex microbial communities introduced in GF mice have been shown to accelerate GI transit by increasing serotonin (5-HT) biosynthesis and release in the gut, an effect which can in part be blocked by systemic 5-HT antagonism [52, 54]. These studies highlight the utility of GF mice as a model to elucidate host–microbial interaction and how microbes modulate GI motility.

1.5.2 GI Secretion

Epithelial ionic and water secretion is an important physiological function of the GI tract. Gut microbiota regulates GI secretion possibly via 5-HT production [54, 55]. Since, 5-HT is a neurotransmitter that stimulates ion (bicarbonates and chloride) secretion in the colon to balance luminal fluidity [56–58], imbalance in 5-HT secretion caused by alterations in gut microbiome can potentially disrupt luminal fluidity and lead to dehydration of the feces and disruption of bowel movements, a hallmark of GI motility disorders.

Lomansey et al. reported similar response of colonic mucosa–submucosa preparations from both GF and conventionally raised mice, to neural, epithelial and bacterial stimulation using Ussing chamber [49]. However, GF mice exhibit a heightened response to forskolin/cAMP response, suggesting that commensal gut microbes may influence colonic ion transport, via cAMP-mediated responses. Similarly bacterial toxins from pathogenic bacteria such as Cholera toxin, have been shown to irreversibly activate adenylate cyclase producing copious amounts of cAMP which ultimately results in continuous salt and water secretion [59]. This is yet another example of how GF mice are important to advance our understanding of effect of microbes on host physiology.

1.6 Limitations

GF animal model is very useful tool to study microbe–microbe and microbe–host interaction in health and disease; however, a few limitations have been noted. Previous studies show that GF mice have biochemical and physiological abnormalities which causes altered immune systems [32], mild chronic diarrhea [60], disrupted metabolism [61], and reduced reproductive abilities [62]. Although these observations raise some concerns, these changes likely represent normal physiology needed to survive in the GF state. The introduction of complex microbial communities in GF mice leads to changes in physiological parameters such that they resemble conventionally raised mice. This ability of GF mice to respond to introduction of bacteria suggests they are indeed a good model system to study the effects of microbes on host development and function. In fact, we know that babies start in a GF state and acquire a microbiota right before or at birth from the environment. This primary succession is somewhat similar to introducing microbial communities in GF mice. Thus, even though there are a few concerns as with any animal model, gnotobiotic mice serve as an important tool to understand host–microbe interactions in health and disease as well as a preclinical model to test microbiota directed therapies.

2 Materials

2.1 Setting Up a Gnotobiotic Facility

Flexible film isolators with isolator port, transfer sleeves, and HEPA filtered air inlets.

Autoclavable diets.

Sterilized tap water.

Shelves.

Autoclavable cotton bags for feed and bedding.

Sterilizing cylinder for autoclaving supplies.

Clidox®.

Tweezers.

2.2 Monitoring of GF Status

Fecal pellets.

Brain heart infusion (BHI) broth, Sabouraud dextrose, and nutrient broth (All prepared using manufacturers' recommendation).

1.5 ml sterile eppendorf tube.

Culture tubes.

Anaerobic chamber.

37 °C incubator.

Bunsen burner.

Sterile microscope slide.

Sterile swabs.

Reagents: Crystal violet, Gram's iodine solution, acetone–ethanol (50:50 v/v; decolorizing agent), 0.1% basic fuchsin solution.

2.3 PCR

Fecal pellets.

Phenol–chloroform DNA isolation kit.

Go Taq.

dNTPs.

Taq Buffer.

PCR cleanup kit.

Nanopure water.

PCR machine.

Primers.

Universal bacterial 16S rRNA primers:

8F-5′-AGAGTTTGATCCTGGCTCAG-3′.

1391R-5′-GACGGGCGGTGWGTRCA-3′.

3   Methods

3.1 Setting Up a Gnotobiotic Facility

The advancement and standardization of gnotobiotic methods has led to rapid expansion of gnotobiotic facilities across the country given the relative ease of setting up a new facility. The facility however requires dedicated infrastructure in terms of space and personnel. A gnotobiotic facility can be set up in small spaces as long as it fulfills certain criteria such as but not limited to, restricted personnel access, adequate sound barrier, emergency power back up and central alarms, HEPA filtered air inlets and option for a positive pressure space. Within the facility mice are housed in flexible film isolators wherein factors such as temperature, humidity, pressure, air flow must be precisely controlled [63]. Food, water and other supplies must be appropriately sterilized using autoclave, irradiation or treatment with disinfectants such as chlorine dioxide based sterilant Clidox®. Supplies can be introduced into isolators from autoclavable cylinders carrying dry or wet loads using transfer sleeves, which have to sterilized with a disinfectant prior to the transfer. Alternately non-autoclavable supplies can be placed in the entry port and disinfected prior to transfer inside the isolator. Similarly supplies or samples can be removed from the isolator using a similar protocol. In order to optimize functionality of a facility, mice are bred in larger isolators with shelving system to accommodate 18–20 cages whereas for experimental purposes, mice are transferred to smaller isolators, which can typically accommodate 4–5 cages. This reduces risk of contamination in a large colony due to specific experimental procedures such as special order diets, specialized equipment such as exercise wheels or introduction of specific bacteria (*see* **Note 1**). The facility should have a dedicated technologist who is responsible the day-to-day activities of the facility as well as periodic maintenance and appropriate tests on the mice.

3.2 Monitoring of GF Status

To access GF status three screening assays are typically used. These include anaerobic/aerobic liquid culture, Gram stain, and PCR using universal and specific 16S rRNA bacterial primers. Recent studies however suggests that in practice bacterial culture and Gram stain are adequate for screening GF status as they both offer high sensitivity and specificity as opposed to PCR which although offers high specificity but has lower sensitivity [64].

3.2.1 Liquid Culture

Liquid culture is a routinely used method to detect GF conditions. However, since cultures can easily be contaminated, precautions must be taken to avoid potential loss of time and animals [64].

Pellets are collected from GF mouse in a sterile eppendorf tubes and transferred into the culture tubes containing nutrient broth, Sabourad dextrose, or BHI broth. One set of tubes is stored in anaerobic chamber while the other set is incubated in an aerobic incubator maintained at 37 °C to detect anaerobic and aerotolerant bacteria respectively. The culture tubes are checked every day for 7 days. Clear culture tube is indication of GF conditions while cloudy tube is an indication of bacterial contamination. Although bacterial culture is a sensitive measure, it might still miss some unculturable and or species that show poor growth.

3.2.2 Gram Stain

Besides culture, Gram stain is also routinely used as an inexpensive tool to detect contamination in GF mice. However, unlike culture, Gram stain can used to screen unculturable bacterial species contamination, provided they are present in large quantities in the intestine [64].

For Gram stain, fecal material is thinly spread over a surface of a sterile glass slide, air-dried, and heat fixed. Subsequently crystal violet stain is added over the fixed slide and allowed to stand for 30 s.

The stain is then rinsed with a stream of sterile water, followed by iodine solution, decolorizer, and basic fuchsin solution. Basic fuchsin is finally washed with water and the slides are later air dried and examined under a microscope. Presence of purple or pink staining is an indication of gram-positive and gram-negative bacteria respectively.

One of the major concerns at present is that positive Gram stain could be due to dead bacteria present in autoclaved/sterilized diet. A recent study however suggests that only very few to no dead bacteria are detected in feces of GF mice [64] (*see* **Note 2**).

3.2.3 PCR

GF status can also be verified by checking the presence of 16S rRNA bacterial gene in mice fecal pellet using universal bacterial primers.

A typical PCR cycle used is:

Initial denaturation at 94 °C for 2:00 min.

35 cycles of denaturation, annealing, and extension at

94 °C for 1:00 min,

55 °C for 0:45 min,

72 °C for 2:00 min, respectively.

Final extension at 72 °C for 20:00 min.

3.3 Rederivation

GF mice can be rederived either via embryo transfer or via hysterectomy and fostering to an axenic mother [65]. To perform embryo transfer, embryos are collected from ovulated females and washed to prevent pathogen contamination. These embryos are then transferred surgically into the uterus of surrogate axenic mother kept in a GF isolator. To perform hysterectomy, uterus from donor strain is removed by a sterile surgical technique and passed through a tank containing germicide. The fetuses are then removed from the uterus in GF isolators and placed on heating pads. These pups are adopted by foster mother whose pups have recently been removed.

4 Notes

1. In order to prevent loss of a colony of valuable mouse strains, which were rederived as GF, it is helpful to keep mice in two separate isolators so that the mouse strain can be expanded again in an event of a contamination.

2. In some instances where few bacteria or "bacteria-like particles" were present they were below the detection limit and did not interfere with the specificity of the assay [64].

References

1. Qin J, Li R, Raes J, Arumugam M, Burgdorf KS, Manichanh C, Nielsen T, Pons N, Levenez F, Yamada T et al (2010) A human gut microbial gene catalogue established by metagenomic sequencing. Nature 464:59–65

2. Ashida H, Ogawa M, Kim M, Mimuro H, Sasakawa C (2012) Bacteria and host interactions in the gut epithelial barrier. Nat Chem Biol 8:36–45

3. Moschen AR, Wieser V, Tilg H (2012) Dietary factors: major regulators of the gut's microbiota. Gut Liver 6:411–416

4. Gordon HA, Pesti L (1971) The gnotobiotic animal as a tool in the study of host microbial relationships. Bacteriol Rev 35:390–429

5. Pasteur L (1885) Observations relatives à la note de M. Duclaux. Compte Rendu Académie des Sciences 100:68–69

6. Nuttall G, Thierfelder H (1895) Tierisches Leben ohne Bakterien im Verdauungskanal. Hoppe Seyler's Zeitschrift Physiol Chem 21:109–112

7. Gordon HA (1965) A bioactive substance in the caecum of germ-free animals: demonstration of a bioactive substance in caecal contents of germ-free animals. Nature 205:571–572

8. Bruckner G (1997) How it started—and what is MAS? In: Heidt PJ, Volker R, van der Waaij D (eds) Old Herborn University Seminar, Monograph 9. Herborn Litterae, Herborn-Dill, Germany, pp 24–34

9. Reyniers JA (1959) The pure-culture concept and gnotobiotics. Ann N Y Acad Sci 78:3–16

10. Williams SC (2014) Gnotobiotics. Proc Natl Acad Sci U S A 111:1661

11. Schaedler RW, Dubos R, Costello R (1965) The development of the bacterial flora in the gastrointestinal tract of mice. J Exp Med 122:59–66

12. Zoetendal EG, Akkermans AD, De Vos WM (1998) Temperature gradient gel electrophoresis analysis of 16S rRNA from human fecal samples reveals stable and host-specific communities of active bacteria. Appl Environ Microbiol 64:3854–3859

13. Riesenfeld CS, Schloss PD, Handelsman J (2004) Metagenomics: genomic analysis of microbial communities. Annu Rev Genet 38:525–552

14. Tringe SG, Rubin EM (2005) Metagenomics: DNA sequencing of environmental samples. Nat Rev Genet 6:805–814

15. Kurokawa K, Itoh T, Kuwahara T, Oshima K, Toh H, Toyoda A, Takami H, Morita H, Sharma VK, Srivastava TP et al (2007) Comparative metagenomics revealed commonly enriched gene sets in human gut microbiomes. DNA Res 14:169–181

16. Eckburg PB, Bik EM, Bernstein CN, Purdom E, Dethlefsen L, Sargent M, Gill SR, Nelson KE, Relman DA (2005) Diversity of the human intestinal microbial flora. Science 308:1635–1638

17. Wiles MV, Taft RA (2010) The sophisticated mouse: protecting a precious reagent. Methods Mol Biol 602:23–36

18. Waterston RH, Lindblad-Toh K, Birney E, Rogers J, Abril JF, Agarwal P, Agarwala R, Ainscough R, Alexandersson M, An P et al (2002) Initial sequencing and comparative analysis of the mouse genome. Nature 420:520–562

19. Doetschman T, Gregg RG, Maeda N, Hooper ML, Melton DW, Thompson S, Smithies O (1987) Targeted correction of a mutant HPRT gene in mouse embryonic stem cells. Nature 330:576–578

20. Bogue MA, Grubb SC, Maddatu TP, Bult CJ (2007) Mouse Phenome Database (MPD). Nucleic Acids Res 35:D643–D649

21. Ley RE, Bäckhed F, Turnbaugh P, Lozupone CA, Knight RD, Gordon JI (2005) Obesity alters gut microbial ecology. Proc Natl Acad Sci U S A 102:11070–11075

22. Turnbaugh PJ, Ley RE, Mahowald MA, Magrini V, Mardis ER, Gordon JI (2006) An obesity-associated gut microbiome with increased capacity for energy harvest. Nature 444:1027–1031

23. Turnbaugh PJ, Hamady M, Yatsunenko T, Cantarel BL, Duncan A, Ley RE, Sogin ML, Jones WJ, Roe BA, Affourtit JP et al (2009) A core gut microbiome in obese and lean twins. Nature 457:480–484

24. Turnbaugh PJ, Ridaura VK, Faith JJ, Rey FE, Knight R, Gordon JI (2009) The effect of diet on the human gut microbiome: a metagenomic analysis in humanized gnotobiotic mice. Sci Transl Med 1:6ra14

25. Wexler HM (2007) Bacteroides: the good, the bad, and the nitty-gritty. Clin Microbiol Rev 20:593–621

26. Ridaura VK, Faith JJ, Rey FE, Cheng J, Duncan AE, Kau AL, Griffin NW, Lombard V, Henrissat B, Bain JR et al (2013) Gut microbiota from twins discordant for obesity modulate metabolism in mice. Science 341:1241214

27. Sjöström L, Narbro K, Sjöström CD, Karason K, Larsson B, Wedel H, Lystig T, Sullivan M, Bouchard C, Carlsson B et al (2007) Effects of bariatric surgery on mortality in Swedish obese subjects. N Engl J Med 357:741–752

28. Tremaroli V, Karlsson F, Werling M, Ståhlman M, Kovatcheva-Datchary P, Olbers T, Fändriks L, le Roux CW, Nielsen J, Bäckhed F (2015) Roux-en-Y gastric bypass and vertical banded gastroplasty induce long-term changes on the human gut microbiome contributing to fat mass regulation. Cell Metab 22:228–238

29. Kaplan JL, Shi HN, Walker WA (2011) The role of microbes in developmental immunologic programming. Pediatr Res 69:465–472

30. Douwes J, Cheng S, Travier N, Cohet C, Niesink A, McKenzie J, Cunningham C, Le Gros G, von Mutius E, Pearce N (2008) Farm exposure in utero may protect against asthma, hay fever and eczema. Eur Respir J 32:603–611

31. Blümer N, Herz U, Wegmann M, Renz H (2005) Prenatal lipopolysaccharide-exposure prevents allergic sensitization and airway inflammation, but not airway responsiveness in a murine model of experimental asthma. Clin Exp Allergy 35:397–402

32. Mazmanian SK, Liu CH, Tzianabos AO, Kasper DL (2005) An immunomodulatory molecule of symbiotic bacteria directs maturation of the host immune system. Cell 122:107–118

33. Kamada N, Seo SU, Chen GY, Núñez G (2013) Role of the gut microbiota in immunity and inflammatory disease. Nat Rev Immunol 13:321–335

34. Ivanov II, Atarashi K, Manel N, Brodie EL, Shima T, Karaoz U, Wei D, Goldfarb KC, Santee CA, Lynch SV et al (2009) Induction of intestinal Th17 cells by segmented filamentous bacteria. Cell 139:485–498

35. Pabst O, Herbrand H, Friedrichsen M, Velaga S, Dorsch M, Berhardt G, Worbs T, Macpherson AJ, Förster R (2006) Adaptation of solitary intestinal lymphoid tissue in response to microbiota and chemokine receptor CCR7 signaling. J Immunol 177:6824–6832

36. Hamada H, Hiroi T, Nishiyama Y, Takahashi H, Masunaga Y, Hachimura S, Kaminogawa S, Takahashi-Iwanaga H, Iwanaga T, Kiyono H et al (2002) Identification of multiple isolated lymphoid follicles on the antimesenteric wall of the mouse small intestine. J Immunol 168:57–64

37. Cario E (2013) Microbiota and innate immunity in intestinal inflammation and neoplasia. Curr Opin Gastroenterol 29:85–91

38. Franchi L, Kamada N, Nakamura Y, Burberry A, Kuffa P, Suzuki S, Shaw MH, Kim YG, Núñez G (2012) NLRC4-driven production of IL-1β discriminates between pathogenic and commensal bacteria and promotes host intestinal defense. Nat Immunol 13:449–456

39. Ohkubo T, Tsuda M, Tamura M, Yamamura M (1990) Impaired superoxide production in peripheral blood neutrophils of germ-free rats. Scand J Immunol 32:727–729

40. Mørland B, Midtvedt T (1984) Phagocytosis, peritoneal influx, and enzyme activities in peritoneal macrophages from germfree, conventional, and ex-germfree mice. Infect Immun 44:750–752

41. Sokol H, Seksik P, Rigottier-Gois L, Lay C, Lepage P, Podglajen I, Marteau P, Doré J (2006) Specificities of the fecal microbiota in inflammatory bowel disease. Inflamm Bowel Dis 12:106–111

42. Frank DN, St Amand AL, Feldman RA, Boedeker EC, Harpaz N, Pace NR (2007) Molecular-phylogenetic characterization of microbial community imbalances in human inflammatory bowel diseases. Proc Natl Acad Sci U S A 104:13780–13785

43. Wu HJ, Wu E (2012) The role of gut microbiota in immune homeostasis and autoimmunity. Gut Microbes 3:4–14

44. Smith PM, Howitt MR, Panikov N, Michaud M, Gallini CA, Bohlooly-Y M, Glickman JN, Garrett WS (2013) The microbial metabolites, short-chain fatty acids, regulate colonic Treg cell homeostasis. Science 341:569–573

45. Sanderson IR, Walker WA (2007) TLRs in the Gut I. The role of TLRs/Nods in intestinal development and homeostasis. Am J Physiol Gastrointest Liver Physiol 292:G6–G10

46. Cario E, Gerken G, Podolsky DK (2007) Toll-like receptor 2 controls mucosal inflammation by regulating epithelial barrier function. Gastroenterology 132:1359–1374

47. Clarke TB, Davis KM, Lysenko ES, Zhou AY, Yu Y, Weiser JN (2010) Recognition of peptidoglycan from the microbiota by Nod1 enhances systemic innate immunity. Nat Med 16:228–231

48. Rakoff-Nahoum S, Paglino J, Eslami-Varzaneh F, Edberg S, Medzhitov R (2004) Recognition of commensal microflora by toll-like receptors is required for intestinal homeostasis. Cell 118:229–241

49. Lomasney KW, Houston A, Shanahan F, Dinan TG, Cryan JF, Hyland NP (2014) Selective influence of host microbiota on cAMP-mediated ion transport in mouse colon. Neurogastroenterol Motil 26:887–890

50. Anitha M, Vijay-Kumar M, Sitaraman SV, Gewirtz AT, Srinivasan S (2012) Gut microbial products regulate murine gastrointestinal motility via Toll-like receptor 4 signaling. Gastroenterology 143:1006–1016.e1004

51. Abrams GD, Bishop JE (1967) Effect of the normal microbial flora on gastrointestinal motility. Proc Soc Exp Biol Med 126:301–304

52. Kashyap PC, Marcobal A, Ursell LK, Larauche M, Duboc H, Earle KA, Sonnenburg ED, Ferreyra JA, Higginbottom SK, Million M et al (2013) Complex interactions among diet, gastrointestinal transit, and gut microbiota in humanized mice. Gastroenterology 144:967–977

53. Husebye E, Hellström PM, Sundler F, Chen J, Midtvedt T (2001) Influence of microbial species on small intestinal myoelectric activity and transit in germ-free rats. Am J Physiol Gastrointest Liver Physiol 280:G368–G380

54. Reigstad CS, Salmonson CE, Rainey JF, Szurszewski JH, Linden DR, Sonnenburg JL, Farrugia G, Kashyap PC (2015) Gut microbes promote colonic serotonin production through an effect of short-chain fatty acids on enterochromaffin cells. FASEB J 29:1395–1403

55. Yano JM, Yu K, Donaldson GP, Shastri GG, Ann P, Ma L, Nagler CR, Ismagilov RF, Mazmanian SK, Hsiao EY (2015) Indigenous bacteria from the gut microbiota regulate host serotonin biosynthesis. Cell 161:264–276

56. Kaji I, Akiba Y, Said H, Narimatsu K, Kaunitz JD (2015) Luminal 5-HT stimulates colonic bicarbonate secretion in rats. Br J Pharmacol 172:4655–4670

57. Stoner MC, Scherr AM, Lee JA, Wolfe LG, Kellum JM (2000) Nitric oxide is a neurotransmitter in the chloride secretory response to serotonin in rat colon. Surgery 128:240–245

58. Kadowaki M, Gershon MD, Kuwahara A (1996) Is nitric oxide involved in 5-HT-induced fluid secretion in the gut? Behav Brain Res 73:293–296

59. Martínez-Augustin O, Romero-Calvo I, Suárez MD, Zarzuelo A, de Medina FS (2009) Molecular bases of impaired water and ion movements in inflammatory bowel diseases. Inflamm Bowel Dis 15:114–127

60. Gordon HA, Wostmann BS (2012) Chronic mild diarrhea in germ free rodents: a model portraying host-flora synergism. In: Heneghan J (ed) Germfree research: biological effect of gnotobiotic environments. Elsevier, Amsterdam

61. Coates ME, Hewitt D, Salter DN (2012) Protein metabolism in germ free and conventional chick. In: Heneghan J (ed) Germfree research: biological effect of gnotobiotic environments. Elsevier, Amsterdam

62. Shimizu K, Muranaka Y, Fujimura R, Ishida H, Tazume S, Shimamura T (1998) Normalization of reproductive function in germfree mice following bacterial contamination. Exp Anim 47:151–158

63. Arvidsson C, Hallén A, Bäckhed F (2012) Generating and analyzing germ-free mice. Curr Protoc Mouse Biol 2:307–316

64. Fontaine CA, Skorupski AM, Vowles CJ, Anderson NE, Poe SA, Eaton KA (2015) How free of germs is germ-free? Detection of bacterial contamination in a germ free mouse unit. Gut Microbes 6:225–233

65. Fridland GH (2010) Science AAfLA: gnotobiotics. In: LATG: laboratory animal technologist training manual. Drumwright & Co, Germantown, TN, pp 117–121

Chapter 9

Bridging Mice to Men: Using HLA Transgenic Mice to Enhance the Future Prediction and Prevention of Autoimmune Type 1 Diabetes in Humans

David V. Serreze, Marijke Niens, John Kulik, and Teresa P. DiLorenzo

Abstract

Similar to the vast majority of cases in humans, the development of type 1 diabetes (T1D) in the NOD mouse model is due to T-cell mediated autoimmune destruction of insulin producing pancreatic β cells. Particular major histocompatibility complex (MHC) haplotypes (designated *HLA* in humans; and *H2* in mice) provide the primary genetic risk factor for T1D development. It has long been appreciated that within the MHC, particular unusual class II genes contribute to the development of T1D in both humans and NOD mice by allowing for the development and functional activation of β cell autoreactive CD4 T cells. However, studies in NOD mice have revealed that through interactions with other background susceptibility genes, the quite common class I variants (K^d, D^b) characterizing this strain's *H2g7* MHC haplotype aberrantly acquire an ability to support the development of β cell autoreactive CD8 T cell responses also essential to T1D development. Similarly, recent studies indicate that in the proper genetic context some quite common HLA class I variants also aberrantly contribute to T1D development in humans. This review focuses on how "humanized" HLA transgenic NOD mice can be created and used to identify class I dependent β cell autoreactive CD8 T cell populations of clinical relevance to T1D development. There is also discussion on how HLA transgenic NOD mice can be used to develop protocols that may ultimately be useful for the prevention of T1D in humans by attenuating autoreactive CD8 T cell responses against pancreatic β cells.

Key words Type 1 diabetes (T1D), Autoimmunity, "Humanized" NOD mice, T cells, HLA transgenics

1 Introduction

The development of type 1 diabetes (T1D) in both humans and the NOD mouse model is due to the aberrant autoimmune destruction of insulin producing pancreatic β cells by T lymphocytes (reviewed in [1–3]). In both genera, multiple susceptibility genes (designated *Idd* in mice; *IDDM* in humans) interactively contribute to T1D development. However, while polygenically controlled, particular major histocompatibility complex (MHC) haplotypes provide the primary risk factor for T1D development.

Gabriele Proetzel and Michael V. Wiles (eds.), *Mouse Models for Drug Discovery: Methods and Protocols*,
Methods in Molecular Biology, vol. 1438, DOI 10.1007/978-1-4939-3661-8_9, © Springer Science+Business Media New York 2016

The MHC (designated *H2* in mice; *HLA* in humans) encodes two primary types of gene products termed class I and class II molecules (reviewed in [4]). Class I molecules present peptides primarily derived from intracellular proteins to CD8 T cells that usually exert cytotoxic functions. Virtually all cells express MHC class I molecules. In contrast, MHC class II expression is largely limited to a specialized subset of hematopoietically derived antigen presenting cells (APC) that include B cells, macrophages, and dendritic cells (DC). MHC class II molecules expressed by APC display peptides largely derived from internalized and processed extracellular proteins to CD4 T cells, which produce cytokine molecules that amplify other components of the immune response, including cytotoxic CD8 T cells. Within the MHC, specific combinations of HLA-DQ and DR class II variants provide a large component of T1D susceptibility in humans by mediating the selection and functional activation of β cell autoreactive CD4 T cells [5, 6]. Similarly, transgenic analyses [7–11] have demonstrated that T1D development in NOD mice requires that APC homozygously express the unusual H2-A^{g7} MHC class II gene product (homolog of human DQ8) in the absence of H2-E MHC class II molecules (homolog of human DR). However, while MHC class II restricted CD4 T cell responses clearly contribute to T1D development in both humans and NOD mice, there is compelling evidence that MHC class I restricted CD8 T cells also play an essential pathogenic role.

Because they lack expression of MHC class II molecules, pancreatic β cells cannot be directly recognized by autoreactive CD4 T cells contributing to T1D development. Furthermore, studies in NOD mice have shown that purified CD4 T cells from young prediabetic female donors (which would ultimately demonstrate a disease frequency of ~90%) cannot independently transfer T1D to T and B cell deficient NOD-*scid* recipients [12]. This ruled out the possibility that autoreactive MHC class II restricted CD4 T cells could independently initiate T1D development by destroying pancreatic β cells through the release of some soluble cytotoxic factor(s). However, like most other cell types, pancreatic β cells do express MHC class I molecules. In order to bind antigenic peptides and then be transported to, and expressed in a stable fashion on the cell surface, MHC class I molecules must dimerize with β2-microglobulin (β2m) [4]. As a result, mice carrying a *β2m* allele inactivated by homologous recombination (*β2mnull*) fail to express detectable levels of cell surface MHC class I molecules, and hence cannot generate CD8 T cells [13, 14]. NOD.*β2mnull* mice lacking MHC class I expression and CD8 T cells are completely resistant to T1D [15–17]. These results proved that while the K^d and/or D^b class I gene products encoded by the *H2g7* MHC haplotype are common variants also characterizing many strains lacking autoimmune proclivity, when expressed in NOD mice they aberrantly mediate CD8 T cell responses essential to T1D development.

It was recently found that due to strong interactive effects provided by a polymorphic gene in the *Idd7* locus on Chromosome 7, that when the common class I variants characterizing the *H2g7* MHC haplotype are expressed in NOD mice but not a C57BL/6 (B6) background strain, they aberrantly lose the ability to mediate the deletion of autoreactive diabetogenic CD8 T cells during their development in the thymus [18]. Similarly, a recent quite definitive study [19], coupled with earlier suggestive reports [20–29], revealed that when expressed in some genetic contexts, including the presence of particular MHC class II molecules, certain quite common class I variants can also contribute to T1D development in humans.

It is difficult to directly determine the mechanistic role played by MHC class I genes in the initiation and amplification of diabetogenic T cell responses in humans. However, insights to the types of immune responses that are controlled by various human MHC class I alleles have been gained through their transgenic expression in mice. Indeed, the antigenic peptides presented to murine CD8 T cells by a previously described HLA-A2.1 class I transgene product overlap those presented to human CD8 T cells by endogenously encoded HLA-A2.1 molecules [30, 31]. Interestingly, HLA-A2.1 has been implicated as one common human MHC class I variant (present in ~40 % of Caucasians) that when expressed in the right genetic context can aberrantly exert diabetogenic functions in some individuals [19]. Thus, this chapter focuses on how possible diabetogenic roles played by HLA-A2.1, and perhaps other human class I variants, can be assessed through their transgenic expression in NOD mice. There is also a discussion of how such "humanized" NOD transgenic mouse strains might be used to identify approaches to block the development or function of HLA class I mediated diabetogenic T cell responses.

2 Materials

2.1 Transgenic Production

1. Pregnant mare serum gonadotropin (PMSG).
2. Human chorionic gonadotropin (hCG).
3. Freund's adjuvant.

2.2 Mouse Strains

1. NOD/ShiLtJ (The Jackson Laboratory, stock 001976).
2. NOD.CB17-Prkdc^{scid}/J (The Jackson Laboratory, stock 001803).
3. NOD.*β2mnull* mice: NOD.129P2(B6)-B2m^{tm1Unc}/J (The Jackson Laboratory stock 002309).
4. NOD.*β2mnull.HHD* mice: NOD.129P2(B6)-B2m^{tm1Unc} Tg(HLA-A/H2-D/B2M)1Dvs/DvsJ (The Jackson Laboratory, stock 006611).

3 Methods

3.1 Direct Introduction of Transgenes into NOD Zygotes: Advantages and Keys to Success

A wide variety of transgenes have been expressed in NOD mice, either by repeated crossing of non-NOD transgenic strains to NOD mice or by direct introduction of transgenes into NOD zygotes. In the former case, because at least 20 susceptibility loci contribute to T1D development in NOD mice [2], extensive backcrossing is required to insure that the transgenic strain is fixed to homozygosity for each of the NOD-derived *Idd* loci. This process can be facilitated somewhat using a "speed congenic" approach [32], but it remains nonetheless expensive and time-consuming. Furthermore, difficulties can arise if a transgene has integrated within or near an *Idd* gene. For example, when a knockout allele for interferon-γ alpha chain receptor was transferred by breeding from 129 to NOD mice, an original study concluded the engineered mutation contributed to T1D resistance [33]. However, more extensive backcrossing and further analysis revealed that the T1D resistance phenotype was actually mediated by a 129 derived gene(s) flanking the knockout allele [34]. A similar scenario could also arise in the case of a transgene. For these reasons, direct introduction of transgenes into NOD zygotes is the method of choice to assess their possible effects on T1D pathogenesis. Differences in the reproductive biology of NOD mice compared to strains more commonly used for the production of transgenic mice must be taken into account in order for this procedure to be successful. Some of these differences have been noted previously [35].

The Jackson Laboratory routinely creates transgenic mice on the NOD background. Transgenic models have been made on NOD/ShiLtJ, NOD/ShiLtDVS, and NOD.CB17-*Prkdcscid*/J. NOD strains can be used successfully for transgenic mouse production following many of the same basic procedures as those used for more common strains such as FVB and B6. There are however, some important considerations when using NOD mice. We have found that using mature females as embryo donors is a key deviation from standard transgenic mouse production practices. The mature females typically produce fewer oocytes compared to prepubescent females, but the oocytes that are released tend to yield more developmentally synchronized populations of zygotes as well as zygotes of similar quality and "injectibility."

1. Females 8–12 weeks of age are superovulated by intraperitoneal (IP) injection of 5 I.U. each of PMSG and hCG delivered 47–48 h apart (*see* **Note 1**).

2. Superovulated females are mated to sexually mature stud males immediately after hCG administration. The females are examined the following morning for the presence of copulation plugs and are separated into two groups, plugged and non-plugged.

All females are used for zygote collection regardless of plug status. We have found that many females without obvious copulation plugs yield fertilized oocytes.

3. Mature NOD stud male mice are singly housed and mated no more than once each week. Injection of Freund's adjuvant may be given to delay onset of the diabetic phenotype (*see* **Note 2**).

4. No special considerations are taken regarding the collection, culture, microinjection, and surgical transfer of NOD embryos beyond those associated with standard strains used for transgenic mouse creation (for details *see* ref. 36). One can expect the release of 15–20 oocytes per superovulated female. As many as 30 oocytes per female can be obtained from certain NOD strains. Fertility averages 50 % for the NOD strains that we have used to make transgenic models. The timing of hormone administration and matings to yield zygotes with pronuclei large enough to microinject will vary depending on factors such as the light/dark cycle of the vivarium and the desired time of DNA microinjection (morning or afternoon). The Jackson Laboratory most often utilizes a 14/10 h light/dark cycle with lights on at 06:00 and lights off at 20:00. PMSG is administered to NOD embryo donors at 10:00, hCG and mating occurs at 09:00. Plugs are checked the following morning by 08:00 and microinjection occurs between 09:00 and 14:00. DNA-microinjected NOD zygotes are transferred to the oviducts of 0.5 days post coitum pseudopregnant recipient females immediately after microinjection.

5. Typically, 20–25 microinjected zygotes are transferred unilaterally to each recipient female. Litter sizes of five to seven pups can be expected. Rates of NOD transgenesis will vary due to various factors surrounding the construction and preparation of the transgene, but production rates of 5–15 % are not uncommon.

3.2 Human HLA-A2.1 Class I Molecules Transgenically Expressed in NOD Mice Mediate Diabetogenic CD8 T Cell Responses

The genetic association studies described earlier implicated HLA-A2.1 as one common human MHC class I variant that may aberrantly exert diabetogenic functions in some individuals. Thus, we assessed whether an HLA-A2.1 transgene product expressed in NOD mice could mediate autoreactive CD8 T cell responses against pancreatic β cells. This was indeed found to be the case, and when added to the responses mediated by endogenous murine H2^{g7} MHC class I molecules, the result was a significantly accelerated rate of T1D development in our NOD.*HLA-A2.1* transgenic strain [37]. The accelerated rate of T1D development in NOD.*HLA-A2.1* mice was not a generic effect of transgenically expressing any human class I variant in addition to the endogenous murine molecules, since introduction of the *B27* allele actually suppressed disease onset [37]. Similarly, the B27 class I variant was also recently found to exert a T1D protective effect in humans [19].

These studies provided the first functional demonstration that specific human MHC class I molecules can mediate diabetogenic immune responses in addition to those previously known to be elicited by particular class II variants. Furthermore, the finding that HLA-A2.1 facilitates T1D development in both humans and NOD mice, while HLA-B27 is protective in both, further supports the usefulness of HLA-transgenic NOD strains as models for the human disease.

We subsequently introduced, directly into NOD mice, a chimeric monochain transgene construct, designated *HHD*, that encodes *h*uman β2m covalently linked to the α1 and α2 domains of *H*LA-A*0201, and the α3, transmembrane, and cytoplasmic domains of murine H-2*D*^{b}. The *HHD* transgene was then transferred to the NOD.*β2mnull* strain. Due to its covalent linkage, the human β2m encoded by the *HHD* construct cannot stabilize the expression of murine class I molecules in a trans-acting fashion. Hence, NOD.*β2mnull.HHD* mice express human HLA-A2.1, but no murine class I molecules [38]. Most importantly, while albeit present at lower levels than in standard NOD mice, the NOD.*β2mnull. HHD* strain generates a sufficient array of β cell autoreactive CD8 T cells to induce T1D [38]. Due to its ability to preferentially interact with murine CD8, the inclusion of a murine class I MHC α3 domain in the *HHD* construct facilitates T cell development and antigen recognition [39]. It should be noted that any human HLA class I coding sequence can be incorporated into the *HHD* transgene vector. For this reason it should be possible to use the *HHD* transgene system to generate NOD background mice that could be used to evaluate the potential diabetogenic function of any chosen human HLA class I variant.

3.3 NOD.β2mnull. HHD Mice Can Be Used to Predict HLA-A2.1 Restricted β Cell Antigens Targeted by CD8 T Cells in Human T1D Patients

Since all of their CD8 T cells must be HLA-A2.1 restricted, we reasoned the NOD.*β2mnull.HHD* strain might provide an excellent model for identifying β cell autoantigens that are presented by this class I variant to pathogenic CD8 T cells in human T1D patients. Indeed, CD8 T cells could be isolated from NOD.*β2mnull. HHD* mice that were specifically cytotoxic to human HLA-A2.1 positive islet cells [38]. This demonstrated murine and human HLA-A2.1-positive islets present one or more peptides in common. Previous studies employing a tetramer-based technology approach found that peptides derived from islet-specific glucose-6-phosphatase catalytic subunit-related protein (IGRP 206–214 epitope), insulin (B chain 15–23 epitope), and dystrophia myotonica kinase (DMK 138–146 epitope) represented the targets of up to ~60 % of H2^{g7} class I restricted diabetogenic CD8 T cells in standard NOD mice [40]. While the proportion of islet derived CD8 T cells that recognized these antigenic epitopes varied greatly in individual NOD mice, usually the most frequent target was IGRP 206–214, followed by insulin B 15–23 and DMK 138–146 [40].

Thus, we have been carrying out studies to determine if HLA-A2.1 restricted T cell lines propagated from the pancreatic islets of NOD.*β2mnull.HHD* mice can also recognize IGRP, insulin, or DMK derived peptides.

To date a series of islet derived CD8 T cell lines from 12- to 13-week-old female NOD.*β2mnull.HHD* mice have been used to screen peptide libraries consisting of all possible 8- to 11-mer sequences that can be derived from IGRP or preproinsulin (both preproinsulin 1 and 2). Three peptide sets from each of these β cell proteins were found to stimulate at least a subset of islet derived CD8 T cell lines from NOD.*β2mnull.HHD* mice, as assessed by ELISPOT analysis of IFNγ production. The individual peptides comprising each positive set were then separately screened for auto-antigenic activity. These analyses revealed that the three individual IGRP derived peptides recognized by islet derived CD8 T cells from NOD.*β2mnull.HHD* mice consisted of amino acid residues 228–236, 265–273, and 337–345 ([38] and Table 1). The 228–236 peptide appeared to represent the immunodominant epitope recognized by IGRP autoreactive CD8 T cells in NOD.*β2mnull. HHD* mice ([38] and Table 1). The IGRP 265–273 epitope was also frequently targeted, while IGRP 337–345 represented a less frequent target ([38] and Table 1). For insulin epitopes, we found amino acid residues 3–11 from the leader sequence and 5–14 from the B chain of Ins1, as well as an A chain 2–10 epitope common to both Ins1 and Ins2 were HLA-A2.1 restricted targets of islet derived CD8 T cell lines from NOD.*β2mnull.HHD* mice ([41] and Table 1). The A chain two to ten peptide was found to be the most frequently targeted epitope of insulin autoreactive CD8 T cells in

Table 1
Murine and human IGRP and insulin peptides recognized in a HLA-A2.1 restricted fashion by islet derived CD8 T cells from NOD.*β2mnull.HHD* mice

A2 restricted murine β cell peptide	A2 restricted murine autoantigenic peptide	#NOD.*β2mnull.HHD* CD8 T cell lines recognizing murine peptide	Homologous human peptide[a]	Human peptide recognized by NOD.*β2mnull. HHD* CD8 T cells?
IGRP$_{228-236}$	FGIDLLWSV	14/16	**LN**IDLLWSV	YES
IGRP$_{265-273}$	VLFGLGFAI	8/16	VLFGLGFAI	YES
IGRP$_{337-345}$	ALIPYCVHM	4/16	A**F**IPY**S**VHM	NO
INS1 L$_{3-11}$	LLVHFLPLL	9/22	L**WMRL**LPLL	n.d
INS1 B$_{5-14}$	HLCGPHLVEA	17/22	HLCG**SH**LVEA	YES
INS1/2 A$_{2-10}$	IVDQCCTSI	20/22	IV**E**QCCTSI	YES

[a]*Bold* and *underlined* letters indicate amino acid differences distinguishing human from murine IGRP or insulin sequences
n.d. not determined

NOD.*β2mnull.HHD* mice ([41] and Table 1). Ins1 B 5–14 was also a frequent target ([41] and Table 1). The Ins1 L3–11 epitope was the least frequent target ([41] and Table 1). It should be added that most islet derived CD8 T cell lines from NOD.*β2mnull.HHD* mice included specificities that recognized at least one peptide each of IGRP or insulin origin.

The studies described above identified the murine variants of IGRP or insulin peptides that are recognized in an HLA-A2.1 restricted fashion by islet derived CD8 T cells from NOD.*β2mnull. HHD* mice. However, it remained possible the human counterparts of these peptides might not bind to HLA-A2.1 class I molecules and be presented to pathogenic CD8 T cells in T1D patients. This turned out not to be a concern for the human (h) IGRP 265–273 peptide since its sequence is identical to the murine homologue ([38] and Table 1). Conversely, the human and murine variants of IGRP 228–236 and IGRP 337–345 differed by two amino acids each ([38] and Table 1). The hIGRP 337–345 variant was found to bind very poorly to HLA-A2.1 molecules, which probably accounted for its inability to stimulate islet derived CD8 T cells from NOD.*β2mnull.HHD* mice [38]. In contrast, the hIGRP 228–236 variant demonstrated strong HLA-A2.1 binding and was recognized by islet derived CD8 T cells from NOD.*β2mnull. HHD* mice. The murine Ins1 B 5–14 and Ins1/2 A 2–10 peptides differ from their human counterparts by a single amino acid ([41] and Table 1). However, the human homologues of the Ins1 B 5–14 and Ins1/2 A 2–10 peptides are also recognized by islet derived CD8 T cells from NOD.*β2mnull.HHD* mice ([41] and Table 1). These results indicated the human Ins A 2–10, Ins B 5–14, IGRP 228–236, and/or IGRP 265–273 peptides may represent relevant autoantigenic targets of pathogenic CD8 T cells in HLA-A2.1 expressing T1D patients.

Partly based on our analyses of NOD.*β2mnull.HHD* mice, there has been a recent acceleration in efforts by other investigators to identify populations of autoreactive MHC class I restricted CD8 T cells potentially contributing to T1D development in humans [42–47]. One such study [42], that was of particular interest from our perspective, reported T1D patients can generate HLA-A2.1 restricted CD8 T cells against the IGRP 228–236 or IGRP 265–273 peptides which we had originally identified to be pathologically relevant targets in NOD.*β2mnull.HHD* mice. Our work describing insulin peptides recognized by islet derived CD8 T cells from NOD.*β2mnull.HHD* mice was published more recently than that focused on IGRP responses. However, we hope our study describing insulin epitopes recognized by islet derived CD8 T cells from NOD.*β2mnull.HHD* mice also prompts clinical investigators to screen HLA-A2.1 expressing human T1D patients for the presence of similar pathogenic effectors.

**3.4   Use
of NOD.β2mnull.HHD
Mice for Identifying
Means to Block
HLA-A2.1 Restricted
Diabetogenic T Cell
Responses**

"Humanized" NOD.*β2mnull.HHD* mice provide a model system for developing strategies to inhibit the generation or function of HLA-A2.1 restricted T cells contributing to T1D development that might ultimately be translatable to individuals expressing this class I variant and deemed to be at high future disease risk. Several different approaches could be envisioned for attenuating diabetogenic HLA-A2.1 restricted CD8 T cell responses in NOD.*β2mnull. HHD* mice that might ultimately be suitable for clinical use. Previous studies have demonstrated that in standard NOD mice, IGRP autoreactive CD8 T cells can be induced to undergo deletion by appropriately dosed injections of a soluble antigenic peptide, and this is sufficient to inhibit T1D development [48]. On the other hand, it has also been reported that in standard NOD mice, a CD8 T cell response against (pro)insulin epitopes must initially be established to allow for the subsequent appearance of IGRP specific effectors [49]. An ability to block such "epitope spreading" could at least partly explain why protocols that induce T cell tolerance to (pro)insulin can also efficiently inhibit T1D development in standard NOD mice [49–51]. However, at this point it is unknown whether in the presence of human HLA-A2.1, rather than murine H2^{g7} class I molecules expressed by standard NOD mice, T1D development still requires the initial generation of autoreactive CD8 T cell responses against (pro)insulin as a prerequisite to trigger a subsequent appearance of IGRP specific effectors. The NOD.*β2mnull.HHD* mouse strain should provide a resource making it possible to determine the relative T1D protective effect that may be achieved by ablating autoreactive HLA-A2.1 restricted CD8 T cells that recognize (pro)insulin and/or IGRP epitopes through a tolerogenic antigenic peptide administration approach. Tolerogenic delivery of peptide epitopes to steady-state DCs using an antibody to the DC endocytic receptor DEC-205 has also shown promise in inducing CD8 T cell deletion in NOD mice and should also be considered for this purpose [52].

Through the use of mixed hematopoietic chimera systems, multiple investigators have found that the development of diabetogenic T cells from precursors in the bone marrow (BM) of standard NOD mice is blocked when they are forced to mature in the presence of APC expressing MHC molecules other than those encoded by the *H2g7* haplotype [53–62]. Thus, provided a relatively benign preconditioning protocol is ultimately developed, hematopoietic chimerization by APCs expressing dominantly protective MHC molecules could conceivably provide a means for blocking progression to overt T1D in humans deemed to be a high future disease risk. An obviously important consideration is what array of MHC molecules APC must express to most efficiently block the development and/or functional activation of diabetogenic T cells. We have found that APC expressing MHC variants that elicit a strong cross-reactive allogeneic response by a mature autoreactive CD8 cell clonotype (AI4) contributing to T1D in standard NOD mice will also have the ability to mediate the negative selection of these pathogenic effectors during thymic development [63]. Thus, a similar analysis of allogeneic cross reactivity could identify MHC haplotypes that might block the development of HLA-A2.1 restricted diabetogenic T cells which are normally generated in NOD.*β2mnull.HHD* mice. As an initial "proof of principal" we have identified murine MHC haplotypes that provide strongly cross-activating allogeneic ligands for CD8 T cells from NOD.*β2mnull.HHD* mice which have previously responded to priming with a cocktail of the HLA-A2.1 restricted peptides Ins1/2 A 2–10, Ins1 B 5–14, IGRP 228–236 and IGRP 265–273 (unpublished). Using a partial BM chimerization approach [61], it will then be determined if APC expressing the potentially appropriate allogeneic MHC molecules have the capacity to block development of any of these HLA-A2.1 restricted (pro)insulin and/or IGRP autoreactive CD8 T cell populations that are normally generated in NOD.*β2mnull.HHD* mice.

The NOD.*β2mnull.HHD* mouse strain is also undergoing further modifications so it can be used in a system to assess the capacity of human APC expressing various HLA molecules to block the development of HLA-A2.1 restricted diabetogenic CD8 T cell responses. Other investigators have found that human hematopoietic stem cells (HSC) engrafted in a NOD-*scid* strain also carrying a functionally inactivated IL-2 receptor gamma chain gene (NOD-*scid.IL2rγnull*) give rise to functional populations of all tested lineages of lymphoid and myeloid cells [64]. NOD-*scid.IL2rγnull* mice can also be engrafted with human peripheral blood mononuclear cells [65], further increasing the utility of models based on this strain. The *HHD* transgene and the *β2mnull* mutation have recently been transferred to the NOD-*scid.IL2rγnull* strain (unpublished). Development of functional human immune cell populations occurs most readily in NOD-*scid.IL2rγnull* mice engrafted at 1–3 days of age with $3–10 \times 10^4$ of the test HSC (marked by CD34 expression) after receiving a low 100R preconditioning dose of irradiation [66]. This protocol could be utilized to separately or co-engraft NOD-*scid.IL2rγnull.β2mnull.HHD* mice with the human HSC of choice plus NOD.*β2mnull.HHD* BM. This will make it possible to determine if human APC expressing particular allogeneic HLA molecules have the capacity to temper the development of HLA-A2.1 restricted diabetogenic CD8 T cells that would normally differentiate from precursors in NOD.*β2mnull. HHD* BM. Such information would be invaluable in determining what donor type cells would be most likely to confer strong T1D protection in any hematopoietic chimerization protocol ultimately approved for clinical use.

3.5   Conclusions and Future Directions

Studies to date have demonstrated the value of using "humanized" HLA transgenic NOD mice to identify pancreatic β cell antigenic peptides that are targets of pathogenic autoreactive T cells in T1D patients. As depicted in Fig. 1, NOD HLA transgenic mice could additionally provide a model system for initially developing clinically translatable protocols for blocking the development or function of autoreactive T cell populations also contributing to T1D in humans. The efficacy of various potential T1D intervention protocols may be enhanced when initiated at the earliest possible stage of disease development. Thus, there is also a need to more clearly identify individuals at future T1D risk at earlier prodromal stages of disease development than now possible. Additional markers that might ultimately be useful in assessing an individual's future risk for T1D development may be circulating levels of various autoantigen specific T cell populations first identified in humanized HLA transgenic NOD mice. To date, the currently available NOD.*β2mnull.HHD* mouse strain has made it possible to identify T cell populations that contribute to T1D by recognizing β cell autoantigens presented by the human HLA-A2.1 class I variant. However, there is now strong evidence that when expressed in the proper genetic context, other human HLA class I molecules, such as the A24 and B39 variants, can also mediate diabetogenic CD8 T cell responses [19]. For this reason there is also a need to utilize the *HHD* transgenic approach to generate NOD background mice only expressing these additional human HLA class I variants, or any others subsequently found to potentially contribute to T1D development. This broadening array of HLA transgenic NOD mouse strains should not only expand identification of T cell

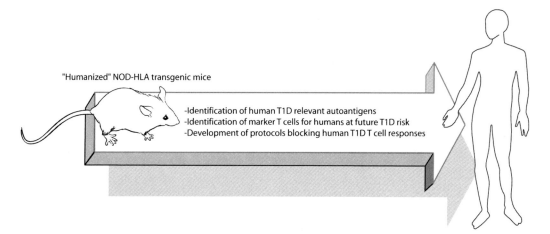

Fig. 1 Potential approaches for using HLA humanized NOD mice to initially develop clinically translatable approaches to block the development or activity of autoreactive T-cell populations contributing to T1D development in humans

populations contributing to, and also potentially providing markers for future risk of T1D in humans, but may also allow for the initial development of intervention protocols that may prevent disease onset in otherwise susceptible individuals.

---

4   Notes

1. Females at 6–7 weeks of age have also been used successfully, but zygote quality is not as consistent as from older females. Some of the sexually mature females may be in proestrus or estrus at the time of gonadotropin administration and injection of exogenous gonadotropins may not successfully induce super-ovulation. In our experience however, the quality of zygotes produced by superovulating mature females (compared to traditional use of prepubescent females) outweighs the loss of overall zygote production due to failure to superovulate.

2. If stud males are not routinely used for mating to superovu-lated females, we find it advantageous to "practice" mate the stud males to any strain of female on-hand 1–2 weeks prior to the next transgenic NOD experiment. This helps ensure the highest plug and fertility rates when it comes time to mate to superovulated NOD females.

References

1. Anderson MS, Bluestone JA (2005) The NOD mouse: a model of immune dysregulation. Annu Rev Immunol 23:447–485
2. Serreze DV, Leiter EH (2001) Genes and path-ways underlying autoimmune diabetes in NOD mice. In: von Herrath MG (ed) Molecular pathology of insulin dependent diabetes melli-tus. Karger Press, New York, pp 31–67
3. Onengut-Gumuscu S, Concannon P (2006) Recent advances in the immunogenetics of human type 1 diabetes. Curr Opin Immunol 18:634–638
4. Mellman I (2007) Private lives: reflections and challenges in understanding the cell biology of the immune system. Science 317:625–627
5. Aly TA, Ide A, Jahromi MM, Barker JM, Fernando MS et al (2006) Extreme genetic risk for type 1A diabetes. Proc Natl Acad Sci U S A 103:14074–14079
6. Erlich H, Valdes AM, Noble J, Carlson JA, Varney M et al (2008) HLA DR-DQ haplo-types and genotypes and type 1 diabetes risk: analysis of the type 1 diabetes genetics consor-tium families. Diabetes 57:1084–1092
7. Hanson MS, Cetkovic-Cvrlje M, Ramiya VK, Atkinson MA, MacLaren NK et al (1996) Quantitative thresholds of MHC class II I-E expressed on hematopoietically derived APC in transgenic NOD/Lt mice determine level of diabetes resistance and indicate mechanism of protection. J Immunol 157:1279–1287
8. Lund T, O'Reilly L, Hutchings P, Kanagawa O, Simpson E et al (1990) Prevention of insulin-dependent diabetes mellitus in non-obese diabetic mice by transgenes encoding modified I-A β-chain or normal I-E α-chain. Nature 345:727–729
9. Miyazaki T, Uno M, Uehira M, Kikutani H, Kishimoto T et al (1990) Direct evidence for the contribution of the unique I-A^{nod} to the development of insulitis in non-obese diabetic mice. Nature 345:722–724
10. Singer SM, Tisch R, Yang X-D, McDevitt HO (1993) An Ab^d transgene prevents diabetes in nonobese diabetic mice by inducing regulatory T cells. Proc Natl Acad Sci U S A 90:9566–9570
11. Slattery RM, Kjer-Nielsen L, Allison J, Charlton B, Mandel T et al (1990) Prevention of diabetes in non-obese diabetic I-A^k trans-genic mice. Nature 345:724–726
12. Christianson SW, Shultz LD, Leiter EH (1993) Adoptive transfer of diabetes into immunodefi-

cient NOD-*scid/scid* mice: relative contributions of CD4[+] and CD8[+] T lymphocytes from diabetic versus prediabetic NOD.NON-*Thy 1a* donors. Diabetes 42:44–55

13. Zijlstra M, Bix M, Simister NE, Loring JM, Raulet DH et al (1990) β2-microglobulin deficient mice lack CD4[-]8[+] cytolytic T cells. Nature 344:742–746

14. Koller BH, Marrack P, Kappler JW, Smithes O (1990) Normal development of mice deficient in b2m, MHC class I proteins, and CD8+ T cells. Science 248:1227–1230

15. Serreze DV, Leiter EH, Christianson GJ, Greiner D, Roopenian DC (1994) MHC class I deficient NOD-*B2mnull* mice are diabetes and insulitis resistant. Diabetes 43:505–509

16. Wicker LS, Leiter EH, Todd JA, Renjilian RJ, Peterson E et al (1994) β2-microglobulin-deficient NOD mice do not develop insulitis or diabetes. Diabetes 43:500–504

17. Sumida T, Furukawa M, Sakamoto A, Namekawa T, Maeda T et al (1994) Prevention of insulitis and diabetes in beta(2)-microglobulin-deficient non-obese diabetic mice. Int Immunol 6:1445–1449

18. Serreze DV, Choisy-Rossi C-M, Grier A, Holl TM, Chapman HD et al (2008) Through regulation of TCR expression levels, an *Idd7* region gene(s) interactively contributes to the impaired thymic deletion of autoreactive diabetogenic CD8 T-cells in NOD mice. J Immunol 180:3250–3259

19. Nejentsev S, Howson JMM, Walker NM, Szeszko J, Field SF et al (2007) Localization of type 1 diabetes susceptibility to the MHC class I genes *HLA-B* and *HLA-A*. Nature 450:887–892

20. Demaine AG, Hibberd ML, Mangles D, Millward BA (1995) A new marker in the HLA class I region is associated with the age at onset of IDDM. Diabetologia 38:622–628

21. Fennessy M, Metcalfe K, Hitman GA, Niven M, Biro PA et al (1994) A gene in the HLA class I region contributes to susceptibility to IDDM in the Finnish population. Diabetologia 37:937–944

22. Honeyman MC, Harrison LC, Drummond B, Colman PG, Tait BD (1995) Analysis of families at risk for insulin-dependent diabetes mellitus reveals that HLA antigens influence progression to clinical disease. Mol Med 1:576–582

23. Mizota M, Uchigata Y, Moriyama S, Tokunaga K, Matsuura N et al (1996) Age dependent association of HLA-A24 in Japanese IDDM patients. Diabetologia 39:371–373

24. Nakanishi K, Kobayashi T, Murase T, Naruse T, Nose Y et al (1999) Human leukocyte antigen-

A24 and-DQA1*0301 in Japanese insulin-dependent diabetes mellitus: independent contributions to susceptibility to the disease and additive contributions to acceleration of beta-cell destruction. J Clin Endocrinol Metab 84:3721–3725

25. Nejentsev S, Reijonen H, Adojaan B, Kovalchuk L, Sochnevs A et al (1997) The effect of HLA-B allele on the IDDM risk defined by DRB1*04 subtypes and DQB1*0302. Diabetes 46:1888–1892

26. Nejentsev S, Gombos Z, Laine A-P, Veijola R, Knip M et al (2000) Non-class II HLA gene associated with type 1 diabetes maps to the 240-kb region near *HLA-B*. Diabetes 49: 2217–2221

27. Robles DT, Eisenbarth GS, Wang T, Erlich HA, Bugawan TL et al (2002) Identification of children with early onset and high incidence of anti-islet autoantibodies. Clin Immunol 102:217–224

28. Tait BD, Colman PG, Morahan G, Marchinovska L, Dore E et al (2003) HLA genes associated with autoimmunity and progression to disease in type 1 diabetes. Tissue Antigens 61:146–153

29. Undlien DE, Lie BA, Thorsby E (2001) HLA complex genes in type 1 diabetes and other autoimmune diseases. Which genes are involved? Trends Genet 17:93–100

30. Geluk A, van Meijgaarden KE, Fraken KLMC, Drijfhout JW, D'Souza S et al (2000) Identification of major epitopes of *Mycobacterium tuberculosis* AG85B that are recognized by HLA-A0201 restricted CD8[+] T cells in HLA-transgenic mice and humans. J Immunol 165:6463–6471

31. Shirai M, Arichi T, Nishioka M, Nomura T, Ikeda K et al (1995) CTL responses of HLA-A2.1-transgenic mice specific for Hepatitis C viral peptides predict epitopes for CTL of humans carrying HLA-2.1. J Immunol 154:2733–2742

32. Serreze DV, Chapman HD, Varnum DS, Hanson MS, Reifsnyder PC et al (1996) B lymphocytes are essential for the initiation of T cell mediated autoimmune diabetes: analysis of a new "speed congenic" stock of NOD.*Igμnull* mice. J Exp Med 184:2049–2053

33. Wang B, Andre I, Gonzalez A, Katz JD, Aguet M et al (1997) Interferon-γ impacts at multiple points during progression of autoimmune diabetes. Proc Natl Acad Sci U S A 94:13844–13849

34. Kanagawa O, Xu G, Tevaarwerk A, Vaupel BA (2000) Protection of nonobese diabetic mice from diabetes by gene(s) closely linked to IFN-γ receptor loci. J Immunol 164: 3919–3923

35. Leiter EH (1997) The NOD mouse: a model for insulin-dependent diabetes mellitus. Curr Protoc Immunol 24:15.19.11–15.19.23

36. Nagy A, Gertsenstein M, Vintersten K, Behringer R (2003) Manipulating the mouse embryo: a laboratory manual, 3rd edn. Cold Spring Harbor Laboratory Press, Cold Spring Harbor, NY

37. Marron MP, Graser RT, Chapman HD, Serreze DV (2002) Functional evidence for the mediation of diabetogenic T cell responses by human HLA-A2.1 MHC class I molecules through transgenic expression in NOD mice. Proc Natl Acad Sci U S A 99:13753–13758

38. Takaki T, Marron MP, Mathews CE, Guttman ST, Bottino R et al (2006) HLA-A*0201-restricted T cells from humanized NOD mice recognize autoantigens of potential clinical relevance to type 1 diabetes. J Immunol 176:3257–3265

39. Irwin MJ, Heath WR, Sherman LA (1989) Species-restricted interactions between CD8 and the alpha 3 domain of class I influence the magnitude of the xenogeneic response. J Exp Med 170:1091–1101

40. Lieberman SM, Takaki T, Han B, Santamaria P, Serreze DV et al (2004) Individual nonobese diabetic mice exhibit unique patterns of CD8 T cell reactivity to three islet antigens including the newly identified widely expressed dystrophia myotonica kinase. J Immunol 173:6727–6734

41. Jarchum I, Baker JC, Yamada T, Takaki T, Marron MP et al (2007) In vivo cytotoxicity of insulin specific CD8+ T cells in HLA-A*0201 transgenic NOD mice. Diabetes 56:2551–2560

42. Mallone R, Martinuzzi E, Blancou P, Novelli G, Afonzo G et al (2007) CD8+ T cell responses identify β-cell autoimmunity in human type 1 diabetes. Diabetes 56:613–621

43. Ouyang Q, Standifer NE, Qin H, Gottlieb PA, Verchere CB et al (2006) Recognition of HLA-class I restricted β-cell epitopes in type 1 diabetes. Diabetes 55:3068–3074

44. Panagiotopoulos C, Qin H, Tan R, Verchere CB (2003) Identification of a β cell specific HLA class I restricted epitope in type 1 diabetes. Diabetes 52:2647–2651

45. Pinske GGM, Tysma OHM, Bergen CAM, Kester MGD, Ossendorp F et al (2005) Autoreactive CD8 T cells associated with β cell destruction in type 1 diabetes. Proc Natl Acad Sci U S A 102:18425–18430

46. Standifer NE, Ouyang Q, Panagiotopoulos C, Verchere CB, Tan R et al (2006) Identification of novel HLA-A*0201-restricted epitopes in recent onset type 1 diabetic subjects and antibody positive relatives. Diabetes 55:3061–3067

47. Toma A, Haddouk S, Briand JP, Camoin L, Gahery H et al (2005) Recognition of a subregion of human proinsulin by class I restricted T cells in type 1 diabetic patients. Proc Natl Acad Sci U S A 102:10581–10586

48. Han B, Serra P, Amrani A, Yamanouchi J, Maree AF et al (2005) Prevention of diabetes by manipulation of anti-IGRP autoimmunity: high efficiency of a low affinity peptide. Nat Med 11:645–652

49. Krishnamurthy B, Dudek N, McKenzie MD, Purcell AW, Brooks AG et al (2006) Responses against islet antigens in NOD mice are prevented by tolerance to proinsulin but not IGRP. J Clin Invest 116:3258–3265

50. Jaeckel E, Lipes MA, von Boehmer H (2004) Recessive tolerance to preproinsulin 2 reduces but does not abolish type 1 diabetes. Nat Immunol 5:1028–1035

51. Nakayama T, Abiru N, Moriyama H, Babaya N, Liu E et al (2005) Prime role for an insulin epitope in the development of type 1 diabetes in NOD mice. Nature 435:220–223

52. Mukhopadhaya A, Hanafusa T, Jarchum I, Chen YG, Iwai Y et al (2008) Selective delivery of beta cell antigen to dendritic cells in vivo leads to deletion and tolerance of autoreactive CD8+ T cells in NOD mice. Proc Natl Acad Sci U S A 105:6374–6379

53. Chen Y-G, Silveira P, Osborne MA, Chapman HD, Serreze DV (2007) Cellular expression requirements for inhibition of type 1 diabetes by a dominantly protective major histocompatibility complex haplotype. Diabetes 56:424–430

54. Ikehara S, Ohtsuki H, Good RA, Asamoto H, Nakamura T et al (1985) Prevention of type 1 diabetes in nonobese diabetic mice by allogeneic bone marrow transplantation. Proc Natl Acad Sci U S A 82:7743–7747

55. Ildstad ST, Chilton PM, Xu H, Domenick MA, Ray MB (2005) Preconditioning of NOD mice with anti-CD8 mAb and co-stimulatory blockade enhances chimerism and tolerance and prevents diabetes while depletion of αβ TCR+ and CD4 T cells negates the effect. Blood 105:2577–2584

56. LaFace DW, Peck AB (1989) Reciprocal allogeneic bone marrow transplantation between NOD mice and diabetes-nonsusceptible mice associated with transfer and prevention of autoimmune diabetes. Diabetes 38:894–901

57. Li H, Kaufman CL, Boggs SS, Johnson PC, Patrene KD et al (1996) Mixed allogeneic

chimerism induced by a sublethal approach prevents autoimmune diabetes and reverses insulitis in nonobese diabetic (NOD) mice. J Immunol 156:380–388

58. Mathieu C, Castells K, Bouillon R, Waer M (1997) Protection against autoimmune diabetes in mixed bone marrow chimeras. J Immunol 158:1453–1457

59. Nikolic B, Takeuchi Y, Leykin I, Fudaba Y, Smith RN et al (2004) Mixed hematopoietic chimerism allows cure of autoimmune diabetes through allogeneic tolerance and reversal of autoimmunity. Diabetes 53:376–383

60. Serreze DV, Leiter EH (1991) Development of diabetogenic T cells from NOD/Lt marrow is blocked when an allo-H-2 haplotype is expressed on cells of hematopoietic origin, but not on thymic epithelium. J Immunol 147:1222–1229

61. Serreze DV, Osborne MA, Chen Y-G, Chapman HD, Pearson T et al (2006) Partial versus full allogeneic hemopoietic chimerization is a preferential means to inhibit type 1 diabetes as the latter induces generalized immunosuppression. J Immunol 177:6675–6684

62. Seung E, Iwakoshi N, Woda BA, Markees TG, Mordes JP et al (2000) Allogeneic hematopoietic chimerism in mice treated with sublethal myeloablation and anti-CD154 antibody: absence of graft-versus-host disease, induction of skin allograft tolerance, and prevention of recurrent autoimmunity in islet-allografted NOD/Lt mice. Blood 95:2175–2182

63. Serreze DV, Holl TM, Marron MP, Graser RT, Johnson EA et al (2004) MHC class II molecules play a role in the selection of autoreactive class I restricted CD8 T cells that are essential contributors to type 1 diabetes development in NOD mice. J Immunol 172:871–879

64. Shultz LD, Ishikawa F, Greiner DL (2007) Humanized mice in translational biomedical research. Nat Rev Immunol 7:118–130

65. King M, Pearson T, Shultz LD, Leif J, Bottino R et al (2008) A new Hu-PBL model for the study of human islet alloreactivity based on NOD-scid mice bearing a targeted mutation in the IL-2 receptor gamma chain gene. Clin Immunol 126:303–314

66. Ishikawa F, Yasukawa M, Lyons B, Yoshida S, Miyamoto T et al (2005) Development of functional human blood and immune systems in NOD/SCID/IL2 γ chain[null] mice. Blood 106:1565–1573

Chapter 10

Mouse Models of Type 2 Diabetes Mellitus in Drug Discovery

Helene Baribault

Abstract

Type 2 diabetes is a fast-growing epidemic in industrialized countries, associated with obesity, lack of physical exercise, aging, family history, and ethnic background. Diagnostic criteria are elevated fasting or postprandial blood glucose levels, a consequence of insulin resistance. Early intervention can help patients to revert the progression of the disease together with lifestyle changes or monotherapy. Systemic glucose toxicity can have devastating effects leading to pancreatic beta cell failure, blindness, nephropathy, and neuropathy, progressing to limb ulceration or even amputation. Existing treatments have numerous side effects and demonstrate variability in individual patient responsiveness. However, several emerging areas of discovery research are showing promises with the development of novel classes of antidiabetic drugs.

The mouse has proven to be a reliable model for discovering and validating new treatments for type 2 diabetes mellitus. We review here commonly used methods to measure endpoints relevant to glucose metabolism which show good translatability to the diagnostic of type 2 diabetes in humans: baseline fasting glucose and insulin, glucose tolerance test, insulin sensitivity index, and body type composition. Improvements on these clinical values are essential for the progression of a novel potential therapeutic molecule through a preclinical and clinical pipeline.

Key words Type 2 diabetes mellitus, Drug discovery, Glucose tolerance test, Insulin tolerance test, Insulin secretion, Insulin sensitivity, Diet-induced obesity, Leptin, Insulin, NEFA

Abbreviations

DEXA	Dual energy X-ray absorptiometry
DIO	Diet-induced obesity
D-PBS	Dulbecco's Phosphate Buffered Saline
ED_{50}	Dose providing 50 % efficacy
GSIS	Glucose-stimulated insulin secretion
GTT	Glucose tolerance test
i.p.	Intraperitoneal
i.v.	Intravenous
ITT	Insulin tolerance test
MRI	Magnetic Resonance Imaging
NEFA	Nonesterified fatty acid

Gabriele Proetzel and Michael V. Wiles (eds.), *Mouse Models for Drug Discovery: Methods and Protocols*,
Methods in Molecular Biology, vol. 1438, DOI 10.1007/978-1-4939-3661-8_10, © Springer Science+Business Media New York 2016

p.o.   per oral gavage
PD   Pharmacodynamics
PK   Pharmacokinetics
s.c.   Subcutaneous
STZ   Streptozotocin
T2DM   Type 2 diabetes mellitus

1   Introduction

The name Diabetes Mellitus, meaning "honey passing through," describes graphically the increased elimination of blood sugar in the urine of diabetic patients or glycosuria. The most common form of diabetes is Type 2 Diabetes Mellitus (T2DM), accounting for 90–95 % of all cases in the United States. It is characterized by insulin resistance, i.e., the inability of insulin to stimulate glucose uptake in target organs such as liver and fat (for review, [1]).

The prevalence of T2DM has steadily increased along with what is thought to be a key causative agent, obesity, in the past few decades. Other risk factors such as aging, lack of physical exercise, and genetic predisposition (family history and/or ethnicity) are also associated with the diseases development. The Center for Disease Control and Prevention estimates that in the United States of America, T2DM affects more than 10 % of the population over 45 years old, and perhaps of greater concern 3–4 % of the younger population [2].

1.1   Diagnostic

T2DM is diagnosed clinically using either of these two criteria: a fasting plasma glucose higher than 126 mg/dL or a 2 h postload glucose level exceeding 200 mg/dL in a glucose tolerance test (GTT). Because these clinical markers are susceptible to daily and other variations, continued monitoring of diabetic patients is performed using the glycated hemoglobin or the HbA_{1c} value to monitor the average plasma glucose concentration over prolonged periods of time [3]. However, HbA1c values are usually not reliable for patients who have undergone recent diet or treatments changes (within 6 weeks). For this reason, fasting glucose and GTT are most commonly used in preclinical animal models.

1.2   Disease Progression

In the early stages of the disease, the pancreas of T2DM patients increases insulin production to compensate for an elevated glycemic index. However, as the disease progresses, these increased plasma insulin levels become insufficient to restore a normal glycemia. Left untreated, this chronically elevated blood glucose becomes cytotoxic, contributing to pancreatic beta cell failure, causing patients to become insulin dependent. Hyperglycemia causes damage in capillaries and other internal tissues. Patients are at serious risks of suffering from retinopathy potentially leading to blindness, nephropathy, and neuropathy causing limb ulceration and even amputation in some cases.

1.3   Treatments

Lifestyle changes, a healthy diet, and an increase in physical activity can improve the symptoms of T2DM and sometimes revert serum glucose values to a normal level and even revert the signs onset of T2DM. Unfortunately, in the vast majority of cases, although it is most often insufficient pharmaceutical intervention is necessary to normalize the patient's glycemic index. Metformin (Glucophage), an agent known to reduce hepatic glucose output, is currently the first-line drug therapy recommended by the American Diabetes Association (ADA) and several other medical associations [4, 5]. As the disease progresses, management of the glycemic index with monotherapy becomes ineffective with T2DM patients requiring a second agent. Second-line drugs are grouped into several classes, each targeting a distinct mechanism of action, yet with respective associated adverse events. Insulin sensitizers such as thiazolidino-diones (TZD) act through PPAR-γ in adipocytes and have been associated with body weight gain. Despite some controversy over the safety of TZD, the FDA has approved to keep this drug on the market [6]. Another class of active drugs are the insulin secreta-gogues, such as sulfonylureas, act through potassium channels in pancreatic beta cells. They do, however, present some risks of hypoglycemia and may also cause body weight gain [7]. Alpha-glucosidase inhibitors, e.g., acarbose, prevent the digestion of complex carbohydrates, but can also cause several gastrointestinal side effects such as flatulence and bloating. Peptide analogs include injectable incretin mimetics, e.g., exenatide, a GLP-1 analog, and dipeptidyl peptidase-4 (DPP-4) inhibitors, can be used to prevent the degradation of GLP-1 [8]. While GLP-1 analogs induce weight loss, DPP-4 inhibitors are weight neutral. Besides stimulating insulin secretion, long-acting GLP-1 act through decreasing gastric emptying which may cause commonly observed nausea side effects. Amylin analogs also slow gastric emptying and suppress glucagon. Like incretins, the most frequent side effect of amylin analog is nausea. Glifozins, a class of renal glucose reabsorption inhibitors, or glycosurics, acting through SGLT2, constitute the newest class of antidiabetic drugs approved by the FDA [9]. They present little risk of hypoglycemia and can even induce mild weight loss. However, urinary tract infection is a common side effect of this latest class.

Despite a wide range of treatment options, a large proportion of diabetic patients eventually become insulin deficient and progress to the insulin-dependent stage of the disease, along with life-threatening diabetic complications.

Bariatric surgery such as sleeve gastrectomy or Roux-en-Y gastric bypass (RYGB) is highly successful in improving the metabolic profile of T2DM patients [10]. While part of its success may be due to the mechanical restriction of food intake, several lines of evidence points to the role of yet unknown additional factors to explain the rapid and long-lasting restoration of normoglycemia even prior to any major body weight loss. The invasive nature of

RYGB hinders its potential for large-scale use. However, it highlights the possible existence of novel potential targets that could be identified with continued discovery research.

1.4 Emerging Discovery Research

Genetic and genomic approaches have led to the identification of a large number of novel potential targets for the treatment of metabolic disorders. For example, genome-wide association studies (GWAS) in T2DM patients have pointed to approximately 75 susceptibility loci influencing the disease [11]. Genome wide expression analysis after bariatric surgery, fasting, or feeding with a high caloric diet has pointed to additional candidate genes for drug discovery [12]. Secretome screening, using large-scale in vivo overexpression in mice has led to the identification of additional potential targets [13].

Fibroblast Growth Factor 21 (FGF21), a secreted protein produced by the liver in response to PPAR-γ activation or fasting, has emerged as an important biomarker of T2DM. Long-acting FGF21 [14] or mimetic antibodies [15] show efficacy in improving the metabolic profile of mice and nonhuman primates. The potential of the FGF21-related molecules to treat T2DM continues to be explored with several ongoing clinical trials [16].

Emerging trends in discovery research over the past decade are highlighting the importance of the gut, the microbiota, muscle, brown fat, and the brain. Variants of FGF19 [17, 18], a secreted protein from the ileum, show promises in the treatment of diabetes and many liver diseases in preclinical models. With regard to the microbiota, fecal transplants into mice from obese and lean twins have demonstrated convincingly the role of gut bacteria in the control of the host energy balance [19]. Efforts have also been made to identify factors secreted by muscles affecting insulin resistance, with the hope of generating "exercise in a pill" form [20]. Finally, our understanding of the activation of brown fat, the so-called good fat, which increases energy expenditure by combustion of fatty acids into heat, has deepened greatly in recent years [21]. Previous attempts at modulating hypothalamic hormones in the brain to suppress appetite have resulted in little success, either due to lack of efficacy or adverse effects. Lorcaserin, a selective serotonin 5-HT-2c receptor agonist, produces modest body weight loss in obese patients and remains the only brain-mediated anorectic drug approved by the FDA to treat obesity, an underlying factor of T2DM [22]. New findings about the role of the reward system in food intake point to the importance of the ventral tegmental area (VTA), in addition to the hypothalamus, as a potential site of intervention [23]. Altogether, the exploration of these pathways offers ample promises for novel classes of therapeutics to treat diabetes.

1.5 Preclinical Models

Many genetic and diet-induced rodent preclinical models of T2DM exist, meeting three fundamental criteria for validation. First, they recapitulate the hallmarks of the disease in humans: elevated fasting glucose and glucose intolerance. Second, all existing T2DM treatments in humans are equally effective at reversing diabetes symptoms in model rodents. Third, they have been mostly predictive of the translation potential of novel therapeutic molecules into clinical trial results. Although large mammals such as pigs and monkeys also develop obesity and insulin resistance as a result of aging and diet, their postprandial glycemic index tends to remain quite low as plasma insulin levels increase. Therefore, fasting glucose levels are used more commonly than GTT in those species. Mouse also has further key advantages over these species because of its small size, high fecundity, the availability of genetic tools to manipulate its genome, and its short generation time.

While symptoms used for a diagnostic of T2DM in humans are well reproduced in mice, differences in the progression of the disease exist. For example, amyloid formation resulting from the aggregation of islet amyloid polypeptide (IAPP), or amylin, has been linked to T2DM in humans. In contrast, IAPP in rodents is not amyloidogenic. In spite of this difference, several mouse models recapitulate the progression of beta-cell failure and pancreatic islets degeneration. Similarly, several models manifest severe symptoms of diabetic nephropathy, yet, none recapitulate all renal histological features seen in humans, e.g., a 50 % decline in glomeruli filtration rate and greater than 10-fold increase in albuminuria.

The methods presented in this chapter cover most commonly used in vivo pharmacology protocols and focus on the glycemic index used to diagnose T2DM per se and other early stage symptoms, rather than subsequent complications of the disease (*see* **Note 1**).

Preclinical models for diabetes can roughly be divided in two types: genetic models and diet-induced obesity (DIO) models.

1.5.1 Genetic Models

Some of the most widely used genetic models of type 2 diabetes are B6.Cg-*Lepob*/J and BKS.Cg-*Dock7m* +/+ *Leprdb*/J [24]. These strains carry single gene spontaneous mutations in either the leptin (*Lepob*) or the leptin receptor (*Leprdb*) genes in an inbred C57BL/6J and a C57BLKS/J background, respectively.

Mice homozygous for the *Lepob* mutation are hyperphagic, gain weight becoming rapidly obese, and become hyperglycemic at a young age. In a C57BL/6J background, hyperglycemia is transient. Fasting glucose level decreases and insulin increases steadily. Mice become normoglycemic, yet hyperinsulinemic by 14 to 16 weeks of age, with the disease progressing similarly in males and females (*see* **Note 2**).

The BKS.Cg-*Dock7m* +/+ *Leprdb*/J strain models are more severe and show advanced stages of the disease. Mice become severely diabetic by 6 weeks of age, suffering pancreatic islet degeneration and renal complications, resulting in lethality sometimes seen as early as 16 to 20 weeks of age. Further, BKS.Cg-*Dock7m* +/+ *Leprdb*/J is considered one of the most robust model of diabetic nephropathy.

While these models are widely used, a concern with such genetic models is that mutations in the leptin gene and its receptor are rare occurrences in humans. Moreover, leptin administration in humans is ineffective in treating T2DM, except for a rare population of patients with mutations in their leptin or leptin receptor genes [25]. In fact, diabetic patients develop hyperleptinemia and leptin resistance. For those reasons, these mutant mouse strains, while convenient, may have shortcomings in modeling all physiological aspects of T2DM in humans.

1.5.2 Diet-Induced Obesity

Obesity is strongly associated with type 2 diabetes, as a high fat diet is a major cause of insulin resistance both in humans and several mouse strains. AKR/J, DBA/2J, and BTBR T+ tf/J strains are very responsive to high fat ("Western") diets, C57BL/6J can be considered a strain of intermediate susceptibility, while A/J and Balb/cJ mice are diet-induced "diabetes resistant" [26–28]. C57BL/6J males develop more insulin resistance in response to the DIO regimen than females, for reasons yet unclear [29].

C57BL/6J is one of the most commonly used mouse strains for diabetes studies. DIO in this strain causes hyperglycemia, hyperinsulinemia, and the development of a fatty liver. One shortcoming of this model, however, is that diet alone is insufficient for the symptoms to progress to a later stage disease, such as beta-cell degeneration and diabetic nephropathy. To mimic late stage T2DM symptoms observed in humans, DIO-C57BL/6J mice can be treated with low doses of streptozotocin (STZ). While mice treated with high doses of STZ are considered a model of type 1 diabetes, when the drug is administered at low doses to DIO mice, it mimics the partial loss of islet cells in the advanced stages of T2DM [30].

B6D2F1/J, a F1 hybrid between C57BL/6 and DBA/2J, has been increasingly used for a DIO model of T2DM [31]. While they have not been as thoroughly characterized as C57BL/6J mice, they develop insulin resistance as quickly. They are also easy to handle being in general gentle, and little less prone to fighting, allowing group housing over extended periods.

2 Materials

2.1 Mice

1. C57BL/6J males, 4 to 6 weeks old (The Jackson Laboratory, stock 000664).

2. Inventoried (DIO) C57BL/6J males, fed for 12 weeks with D12492i, 60 kcal% fat diet (Research Diets, Inc.) (Jackson Laboratory, stock 000664) (*see* **Note 3**).

3. BKS.Cg-*Dock7m* +/+ *Leprdb*/J males, 3 to 4 weeks old (The Jackson Laboratory, stock 000642) (*see* **Note 4**).

4. B6.Cg-Lep^{ob}/J, 3 to 4 weeks old males (The Jackson Laboratory stock 000632) (*see* **Note 5**).

2.2 Feeding and Dosing

1. 10 kcal% fat diet (standard diet) (Research Diets, D12450B).

2. 45 kcal% fat diet (Research Diets, D12451i) (*see* **Note 6**).

3. 60 kcal% fat diet (Research Diets, D12492i) (*see* **Note 6**).

4. 60 kcal% fat diet supplemented with rosiglitazone (Avandia®; GlaxoSmithKline), custom-order (Research Diets) (*see* **Note 7**).

5. Vehicles

   (a) Dulbecco's Phosphate-Buffered Saline (D-PBS), 1X without calcium and magnesium (Mediatech, 21-031-CM).

   (b) Sodium Chloride Injection Solution, Saline: NaCl 0.9 %, 250 mL, Baxter IV solutions (VWR, 68000-342).

   (c) 1 % hydroxypropyl methylcellulose (Alfa Caesar, 9004-65-3). Mix overnight using a magnetic stirrer.

6. Rosiglitazone (Avandia®; GlaxoSmithKline). To prepare a 1 mg/kg solution, put 50 mg rosiglitazone in a mortar and add 500 μL Tween-80 (Alfa Caesar, 9005-65-6). Mix using the pestle. Add 50 mL 1 % methylcellulose and mix with a 10 mL pipette. The solution may remain cloudy. The solution can be stored at 4 °C for the duration of the experiment. Mix well before dosing. Dose per oral gavage (p.o) at 5 mg/kg (100 μL for a mouse of 20 g).

7. Metformin, 1,1-Dimethylbiguanide hydrochloride (Sigma, D5035). Prepare a 200 mg/10 mL metformin solution in D-PBS. Dose via intraperitoneal (i.p.) injection at 100 mg/kg (100 μL for a mouse of 20 g).

8. Exendin-4 (California Peptide Research, 507-77). Prepare a 1 mg/mL solution in D-PBS. Dilute in D-PBS 100-fold for dosing via i.p. injection at 10 μg/kg (200 μL for a mouse of 20 g).

9. Disposable Sterile Animal Feeding Needles, Popper & Sons, 20G × 1½ inch (VWR, 20068-666).

10. Tuberculin syringe with 27 G × ½ inch needle (Becton-Dickinson, 309623).

2.3 Glucose Tolerance, Insulin Tolerance, and Glucose-Stimulated Insulin Secretion Tests

1. Micro-hematocrit centrifuge: Autocrit™ Ultra-3 (Becton Dickinson, 420575).

2. Clean cages with water bottles.

3. Gibco® Glucose, 20 % Solution (Thermo Fisher Scientific, cat# 19002013).

4. Human insulin: NovolinR, 100U/mL, 10 mL (NovoNordisk).

5. Tuberculin syringe with 27 G × ½ inch needle, Becton-Dickinson, Cat no. 309623.

6. Scalpel.

7. Mineral oil, Sigma, cat nr. M3516.

8. Paper towels.

9. Accu-chek Aviva glucometer, Roche Diagnostics (Battery: 3-volt lithium type CR 2032).

10. Accu-chek Aviva test strips (Roche Diagnostics).

11. Batteries, Eveready button cells, ECR2032, 3 V Lithium Cell (Sigma, B0653).

12. SurePrep™ capillary tubes, 75 mm, self-sealing, Becton Dickinson, cat nr. 420315 (*see* **Note 8**).

13. MicroWell 96-well polystyrene plates, round bottom (nontreated), sterile (Nunc, Sigma, P4241).

14. Small Wire Clipper.

15. 20–200 μL pipette with tips.

16. Alumina-Seal (Diversified Biotech, ALUM-1000).

2.4 Insulin ELISA

1. Insulin (Mouse) Ultrasensitive (Alpco Diagnostics, EIA 80-INSMSU-E10).

2. Titer Plate Shaker (Lab-line instruments, 5246).

3. EL406™ Combination Washer Dispenser (BioTek Instruments).

4. Microplate Spectrophotometer, SpectraMax Plus 384 (Molecular Devices).

2.5 NEFA

1. HR Series NEFA-HR(2) kit (Wako Chemicals GmbH, 994-75409), which include:

   (a). Color Reagent A (999-346691).

   (b). Solvent A (995-34791).

   (c). Color Reagent B (991-34891).

   (d). Solvent B (993-35191).

   (e). NEFA standard (oleic acid) solution (276-76491).

   (f). Control sera (410-00101 and 416-00202).

2. Clear flat bottom 96-well plates.

3. Polyester films for ELISA and Incubation (VWR, 60941-120).

2.6 Body Composition

1. EchoMRI-100™ (Echo Medical System, Houston, TX) (*see* **Note 9**)

2.7 Terminal Blood Collection and Tissue Collection for Histology

1. One cc syringes with 25G×5/8 needle, Becton-Dickinson, cat nr. 305122.

2. Microtainer serum separator tubes (BD Diagnostics, 365956).

3. Nunc MicroWell 96-well polystyrene plates, Sigma, P4241. round bottom (nontreated), sterile.

4. Formalin solution, neutral buffered, 10%, Sigma, HT501128.

5. 50 mL conical tubes, VWR, Cat. Nr. 21008-178.

6. Standard dissection tools: scissors and forceps.

3 Methods

Circulating glucose and insulin levels are the key values for a diagnosis of type 2 diabetes. Obesity and elevated levels of nonesterified fatty acids (NEFA) are known to cause insulin resistance and diabetes. Comorbidity of T2DM and dyslipidemia are common in animal models and in clinical populations and therefore, cholesterol, triglycerides, inflammation markers, and blood pressure are often measured within the same experiments. However, for the purpose of this chapter, we will focus on values directly linked to T2DM.

Although the protocols for the genetic and DIO models are largely overlapping, they have a number of important differences, e.g., a high fat diet is required for C57BL/6J mice to develop symptoms of T2DM, while genetic models develop the disease spontaneously. When differences apply, they will be detailed below. Otherwise, the following protocols apply to both types of models.

3.1 Experimental Design

1. Choose a positive control. These need to be chosen depending on the mechanism of action being investigated: for example,

   (a) Insulin secretagogues: exendin-4 is a long-acting homolog of GLP-1, which induces insulin secretion and thereby lowers glucose.

   (b) Insulin resistance: rosiglitazone is an insulin sensitizer that targets PPAR-γ.

   (c) Hepatic glucose output: metformin is an orphan drug which lowers blood glucose by interfering with hepatic glucose output. Its mechanism of action is poorly understood.

2. Choose a neutral vehicle for the test therapeutic molecule(s) (*see* **Note 10**).

3. Choose a negative control. This will be the vehicle used to deliver the positive control and test compounds: e.g., exendin-4 is soluble in D-PBS; therefore, one group of mice injected with D-PBS alone will be included as a negative control.

4. Choose a range of doses to be injected to determine a pharmacokinetic/pharmacodynamic (PK/PD) relationship. Six different concentrations per compound are sufficient to establish an ED_{50} (dose providing 50 % efficacy) value with an acceptable confidence interval. For protein therapeutics, this may range from 0.1 ng/kg to 10 μg/kg. For small molecules this can range from 1 ng/kg to 100 mg/kg.

5. Use 10 to 15 mice per group to obtain statistically significant results. For genetic models, where 80 % of the mice will be

used (*see* preselection in Subheading 3.3), adjust the number of mice accordingly. For example, 100 B6.Cg-*Lepob*/J mice will be sufficient to test one compound at six doses that include a positive and a negative control, using 10 mice per group.

6. Choose the injection schedule. Some compounds, such as rosiglitazone are effective at lowering glucose levels after multiple injections over several days, but not in an acute, one-day injection setting. Other compounds such as Exendin-4 are active in an acute setting. When studying novel compounds, both types of study—acute and chronic dosing—should be performed to determine the properties of the molecule tested. The schedule of injections will depend on the pharmacokinetic properties of the test compounds.

7. Choose a mode of administration. Therapeutic compounds can be administered intraperitoneally (i.p.), orally (p.o.), subcutaneously (s.c.), or intravenously (i.v). Prolonged methods of administration also include osmotic pumps (e.g., Alzet mini-pumps). Alternatively, transgenic delivery of nucleic acids is useful for proof of concept studies: e.g., germ-line transgenic mice, viral expression, or systemic delivery of naked DNA via tail vein injection. The optimal mode of administration for a given compound will depend on its solubility and on whether the therapeutic is a small molecule, a protein, siRNA, or DNA. For example, rosiglitazone is not water soluble and its best modes of administration are via oral gavage or mixed in the diet (*see* **Note 11**).

8. Choose a method and a schedule for blood collection. Common survival blood collection methods for metabolic profiling are tail nick, tail snip, saphenous vein, submandibular (cheek), and retroorbital bleeding. With the tail nick or tail snip blood collection methods, 75 µL samples of blood can be collected up to four times in a one-day experiment, not exceeding a total of 250 µL, and are used as default blood collection procedures. This usually provides up to 30 µL serum samples and is sufficient to measure glucose, insulin, and other metabolic markers. In some cases, larger volumes are needed (up to 250 µL). Then, consider alternatives such as saphenous or submandibular vein collection or retroorbital bleeding (*see* **Note 12**). For terminal blood collection, exsanguination via cardiac puncture or via the vena cava and decapitation are commonly used. It is worth noting that CO_2 asphyxiation and many anesthesia methods can interfere significantly with baseline glucose and insulin measurements.

3.2 Animal Housing

1. Keep mice housed under standard specified pathogen-free conditions with food and water ad libitum and a 12:12 dark:light cycle.

2. Keep mice group housed if possible (*see* **Note 13**).

3. For C57BL/6J, feed mice with 60% fat diet. For genetic models, B6.Cg-Lep^ob/J and BKS.Cg-*Dock7m* +/+ *Leprdb*/J, feed mice with standard diet, D12450B.

4. Perform daily assessment of the bedding conditions. While this is a standard procedure in most vivaria, some strains of diabetic mice such as BKS.Cg-*Dock7m* +/+ *Leprdb*/J mice need additional attention because of related kidney problems, resulting in polyuria. Even with a HEPA-filtered-ventilated caging system, mice caging may need to be changed two to three times weekly as the diabetic condition worsens.

5. Identify individual mice with ear notches several days before performing an experiment as the stress from the procedure may have a transient effect on glucose levels.

3.3 Randomization and Preselection of Mouse Cohorts

Preselection of mouse cohorts based on their baseline glucose and body weight is particularly important for genetic models as baseline glucose can span a range of several hundred units even for mice of the same gender and age: e.g., 200 to 600 mg/dL (*see* **Note 14** and Fig. 1).

1. Fast mice for four hours and measure mouse body weight.

2. Measure baseline fasting glucose: Insert glucometer strip into glucometer. Using gloves, pick up mouse by the tail. Wrap its body in a paper towel restraining it loosely with one hand, leaving the tail exposed. With the other hand, cut approximately one mm off the tail tip with a scalpel or perform a nick of the tail vein on the side of the tail close to the tip. Put one drop of blood onto a glucometer strip and take reading.

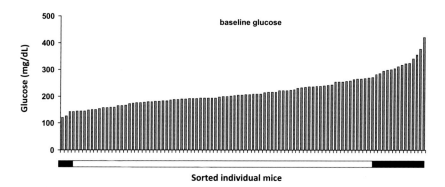

Fig. 1 Selection of mice based on baseline glucose value. Fasting glucose was measured on 100 mice and sorted in increasing order. Each bar represents the glucose value for one mouse. Values ranged from 115 to 430 mg/dL. Mice with glucose values higher than 275 and lower than 140 were excluded (mice over the *black rectangles*). Selection of 80 mice in the middle range was done in a semiquantitative manner (mice at either end of the glucose value range that seem to deviate from the mean the most substantially were discarded) and used for randomization (mice over the *open/white rectangle*)

3. Sort mice by body weight values. Exclude mice with outlier body weight values.

4. Sort glucose values in increasing order.

5. Select mice with mid-range values.

6. Distribute mice of similar baseline glucose values evenly among experimental groups. For example, for three groups, assign the group to mice in sequential order: e.g., A, B, C, C, B , A, A, B, C, C, B, A, etc.

3.4 Tail Snip Blood Collection for Multiple Measurements: Baseline Glucose, Insulin, and NEFA

1. Insert glucometer strip into glucometer.

2. Using gloves, pick up mouse by the tail and cut approximately one mm or less off the tail tip or perform a nick of the tail vein on the side of the tail close to the tip as before. If the tail tip was previously cut, remove the scab off the tail tip with the scalpel.

3. Put one drop of blood on glucometer strip and take reading.

4. Optional: Put a minute amount of mineral oil on index finger and thumb.

5. Rub the sides of the tail on each side with index finger and thumb from the base toward the tail tip. Too much pressure may damage the blood veins, the lighter pressure, the better. Collect the blood into a capillary tube. Repeat this step until the tube is full.

6. Spin the capillary tubes in a micro-hematocrit centrifuge at 10,000 rpm for 8 min.

7. Cut the capillary tubes at the separation between serum and red blood cells with the sharp edge of a file.

8. Use a pipette with a 200 µL tip to gently blow the serum out of the capillary tube into a 96-well plate.

3.5 Insulin ELISA

1. Reagent preparation: Dilute the enzyme conjugate concentrate (11×) with 10 parts of enzyme conjugate buffer. Dilute insulin controls with 0.6 mL of distilled water. Dilute wash buffer concentrate with 20 parts distilled water.

2. Designate wells for standards, controls, and unknown samples. Pipette 5 µL of each in their respective wells.

3. Add 75 µL of enzyme conjugate. Seal the plate with the polyester film provided. Place the plate on an orbital microplate shaker, shaking at 700–900 RPM for 2 h at room temperature.

4. Remove sealing film and wash the plate six times with wash buffer with a microplate washer.

5. Add 100 µL of substrate to each well. Reseal the plate with a polyester film and incubate for 30 min on an orbital microplate shaker at room temperature.

6. Remove sealing film and add 100 µL of stop solution to each well. Gently mix to stop the reaction, remove bubbles before reading with the microplate reader.

7. Program the location of the standards, controls, and unknown samples into an absorption plate reader, with the absorbance at 450 nm with a reference wavelength of 650 nm. Read the plate within 30 min following the addition of the stop solution (*see* **Note 15**).

3.6 NEFA (Nonesterified Free Fatty Acids)

Free fatty acids are elevated in the plasma of obese patients and are known to cause muscle and liver insulin resistance. The Wako HR series NEFA-HR(2) is an in vitro enzymatic colorimetric method assay for the quantitative determination of nonesterified fatty acids (NEFA) in serum. Perform the assay on serum collected from mice fasted for a period greater than 4 h but less than 16 h. Perform the test on samples immediately after collection, without freezing. Also note that hemolysis in the serum samples may interfere with the assay.

1. Prepare reagent solutions. Add 10 mL Solvent A into one vial of Color reagent A and mix gently. Add 20 mL Solvent B into one vial of color reagent B and mix gently. Solvent A and B solutions are stable at 2–10 °C for 5 days.

2. Prepare standard solutions. Stock solution is 1 mEq/L. The test is linear from 0.01 to 4.00 mEq/L. Carry out a serial dilution 1:1 of the standard with water to obtain concentrations of 0.5, 0.25, 0.125 mEq/L.

3. Add samples to 96-well plates in duplicates. Add 10 µL of standard stock solution in two of the wells, to obtain a reading of 2 mEq/L. Then add 5 µL of the 1, 0.5, 0.25, 0.125 mEq/L of the standard solutions in subsequent wells, respectively. Add 5 µL of water in two wells to serve as blanks. Add 5 µL of two control samples. Add 5 µL of unknown samples. If values greater than 2 mEq/L are expected, dilute the samples in PBS.

4. Add 50 µL of reagent A solution to each well. Add polyester film to seal the plate. Mix and incubate at 37 °C for 10 min.

5. Add 100 µL of reagent B solution to each well. Mix and incubate at 37 °C for 10 min. Seal the plate with a polyester film. Mix and incubate at 37 °C for 10 min.

6. Program the location of the standards, controls, and unknown samples into an absorption plate reader. Set the wavelength to 550 nm. Read the plate.

3.7 Glucose Tolerance Test (GTT)

GTTs can be conducted on mice fed ad libitum or following a fasting period. Because food intake in mice occurs mainly during the dark period of the dark–light cycle, we prefer to fast mice during daytime at the beginning of the light period, from 6:00 AM to 10:00

AM. This more closely mimics an overnight fast in humans. Fasting mice also reduces the range of baseline glucose readings and can reveal significant differences between experimental groups that would not reach significance in nonfasted animals.

Glucose can be administered by intraperitoneal injection (i.p.) or by oral gavage (p.o.). Injection is usually faster than gavage; however, it bypasses a potential effect from stimulating incretins in enteroendocrine cells.

1. To initiate a fasting period, transfer mice to clean cages without food but with water ad libitum.

2. Weigh mice. If mice are group housed, draw bars on tails with a black marker pen for quick identification during the test: e.g., zero, one, two, three, or four bars, respectively. Marks on the tail can be read faster than ear notches. These marks will fade after a few days.

3. Use a 20 % (w/v) glucose solution in distilled water. Glucose is dosed at 2 g/kg (glucose/body weight) in a 10 mL/kg volume: e.g., 250 μL for a mouse of 25 g (*see* **Note 16**).

4. Preload one milliliter syringes with the glucose solution. Ensure that all air bubbles have been removed by tapping on the side of the syringes and expressing the air.

5. Shortly before the 4-h fasting is complete, take a measurement of the baseline glucose level. Insert a glucometer strip into a glucometer. Wrap the mouse in a paper towel and restrain loosely with one hand. With the other hand, cut approximately one mm off the tail tip with a scalpel or perform a nick of the tail vein on the side of the tail close to the tip. Put one drop of blood on glucometer strip and take reading. Record glucose value. The range of readable glucose values is 20–600. If the glucometer indicates the value "Hi," i.e., off scale, use 800 as a value (*see* **Note 17**).

6. Upon completion of the 4 h fast, inject mice with glucose, intraperitoneally, or by oral gavage.

7. Repeat blood glucose measurements at 20, 40, 60, and 90 min. Three persons can usually handle up to 120 mice—one dosed mice at 20 s intervals, and a second and third person taking blood measurements starting and 20 and 40 min after dosing, respectively.

8. Add food to mouse cages.

3.8 Glucose-Stimulated Insulin Secretion (GSIS)

Glucose-stimulated insulin secretion (GSIS) is central to normal control of metabolic fuel homeostasis, and its impairment is a key factor in beta-cell failure in T2DM. Some targets may show a phenotype only with GSIS and not in a GTT [32].

GSIS can be performed along with a GTT. The peak of insulin secretion occurs approximately at 7 to 8 min after glucose injection. Blood is typically collected at 7.5, 15, and 30 min after glucose administration. GSIS procedure is more time consuming than a standard GTT and fewer mice can be handled/person. For example, five to eight mice can be handled per person for a GSIS experiment (1 mouse per 1–2 min) compared to 40 mice per person in a standard GTT experiment (1 mouse per 20–25 s).

1. Weigh mice and transfer to clean cages.

2. Rub gently the sides of the tail on each side with index finger and thumb from the base toward the tail tip. Too much pressure may damage the blood veins, the lighter pressure, the better. Collect the blood into a capillary tube. Repeat this step until the tube is half-full (~40 μL). Optional: Use a drop of mineral oil on finger tips.

3. Administer glucose to mice (1 mg/kg in 10 mL/kg volume), $t = 0$.

4. At 7.5, 15, 30, and 60 min after glucose injection, repeat the blood collection procedure.

5. Spin the capillary tubes into a micro-hematocrit centrifuge at 10,000 rpm for 8 min.

6. Cut the capillary tubes at the separation between serum and red blood cells with the sharp edge of a file.

7. Use a pipette with a 200 μL tip to gently blow the serum out of the capillary tube into a 96-well plate.

8. Samples may be frozen at –20 °C for later measurements. Cover the plates with Alumi-Seal.

9. Measure levels of insulin according to the manufacturer's protocol.

3.9 Insulin Tolerance Test (ITT)

Insulin sensitivity can be measured directly by injecting insulin in mice and measuring its effect on circulating glucose levels. The dose of insulin to be administered ranges from 1 to 5 U/kg and is adjusted depending on the model system used. For example, lean C57BL/6J mice will have their baseline glucose reduced substantially at 1 U/kg, while mice fed on a high fat diet will often need to be treated with 1.5 or 2.0 U/kg, and ob/ob mice will sustain up to 5U/kg.

1. Provide access to food and water ad libitum (*see* **Note 18**).

2. Weigh mice.

3. Calculate injection volumes using a volume of 10 mL/kg.

4. Dilute insulin in saline to the desired concentration: 1 to 5 U/10 mL. Keep on ice during preparation. Mix gently and preload syringes.

5. Measure baseline glucose by tail tip or tail snip blood collection, according to procedure described above.

6. Inject mice i.p. with insulin.

7. Measure glucose 15, 30, 60, 90, and 120 min after insulin injection. If the glucometer indicates "Lo," mice should be injected immediately with glucose to prevent loss of consciousness due to hypoglycemia.

3.10 Insulin Sensitivity

Measuring glucose alone is not sufficient to evaluate the state of sugar metabolism. For example, as ob/ob mice age, they become normoglycemic yet they are severely insulin resistant. Further, their pancreatic islets become hypertrophic, and the levels of circulating insulin exceed 10 ng/mL, thereby compensating for the insulin resistance.

The "gold standard" for calculating an insulin sensitivity index is the hyperinsulinemic–euglycemic clamp. It measures the amount of glucose necessary to compensate for an increased insulin level without causing hypoglycemia. This method involves the catheterization of carotid arteries and jugular veins in rodents. It is particularly challenging in mice because of their small size and is not amenable to large-scale studies. Consequently, alternatives that correlate closely to clamp results are often used in determining insulin sensitivity indices. The direct measurement of insulin tolerance, insulin tolerance test (ITT), is often used. Alternatively, simple surrogate indexes for insulin sensitivity/resistance are available (e.g., QUICKI, HOMA, 1/insulin, Matusda index) that are derived from blood insulin and glucose concentrations under fasting conditions (steady state) or after an oral glucose load (dynamic).

The homeostasis model assessment-insulin resistance (HOMA-IR) can be calculated with the following formula:

HOMA-IR index = fasting glucose (mmol/L) × fasting insulin (mU/L)/22.5 [33]

3.11 Body Composition Measurements

Chemical carcass analysis is considered the "gold standard" for accurate whole body composition analysis [34]. It is, however, terminal and time consuming. The adiposity index can also be measured by dissecting and weighing of fat depots in individual animals [35]. This method is also terminal and less accurate. The collection of visceral fat required can be particularly challenging as it is often spread throughout internal organs.

Two imaging systems, Dual Energy X-ray Absorptiometry scanning (DEXA) and Magnetic Resonance Imaging (MRI), allow for longitudinal studies of whole body composition. DEXA measures bone mineral density and content, fat content, and lean content in anesthetized mice. Echo MRI from Echo Medical System, Houston, TX, is used to measure whole body composition parameters such as total body fat, lean mass, body fluids, and total body water in live mice without the need for anesthesia or sedation [36]. The MRI technology is more rapid, less than a minute to scan one mouse, than DEXA which takes about 5 min per mouse. In recent years, GE Medical Systems has discontinued the production and service support for the PIXImus which allowed for DEXA scanning in mice.

3.11.1 Magnetic Resonance Imaging

1. Prior to each run, calibrate the system using a calibrated standard provided by Echo Medical System.

2. Record mouse body weights.

3. Place each mouse into an appropriate size tube and place into the MRI machine.

4. EchoMRI software records spectra on each mouse.

5. The output information is expressed as lean tissue mass, fat mass, and free body fluids in grams.

3.12 Terminal Blood and Tissue Collection

Terminal blood collection is often performed upon completion of an experiment, as it is useful to perform measurements of additional metabolic markers on cohorts that showed compound efficacy. Immediately following blood collection tissue collection can also be performed at the same time from the same animal.

The cytoarchitecture of fat, liver, pancreas, and kidney is seriously affected by T2DM. As mentioned in the introduction, the pancreatic islets of Langerhans become hypertrophic in the early stages of the disease and subsequently undergo degeneration. Adipocytes in fat tissues become hypertrophic. Fatty livers can also be observed by gross morphology and the hepatocytes show accumulation of lipids at the histological level. Finally, in the later stages of the disease, glucotoxicity induces irreversible damage to kidney tubules. Consequently, those are the four main tissues to be examined in histology.

1. Prepare the syringes: Break the vacuum seal of 1 mL syringes by pulling the plunger slightly and push back. Fit the syringes with 25 G × 5/8 inch needles.

2. Before beginning the terminal collection procedures have all serum separation tubes for blood collection labeled and all tubes for tissue collection labeled and filled with 5 to 10 mL 10 % formalin (3.7 % formaldehyde). Organs from the same mouse can be pooled in the same tube as they can simply be embedded in a single block for histological analysis.

3. Euthanize the animal by CO_2 asphyxiation.

4. Put the animal on a paper towel on its back and palpate the sternum with a finger of one hand.

5. With the other hand, insert a needle slightly below and to the left of the sternum, with a mild inclination, aiming into the left cardiac ventricle (*see* **Note 19**).

6. Draw blood slowly so as not to collapse the ventricle. Usually 0.3 to 1.0 mL can be obtained in 10 to 30 s. Once the blood starts flowing in, the depth and the angle of the needle may need to be slightly adjusted to maintain the flow of blood into the needle.

7. To avoid hemolysis remove the needle from the syringe and push the blood into a serum separation tube. Do not exceed the volume indicator on the tube (frosted area).

8. Centrifuge for 10 min at 10,000 rpm in an Eppendorf centrifuge at 4 °C.

9. Transfer the supernatant (serum) to a clean tube or 96-well plate (*see* **Note 20**).

10. Make an incision through the skin and peritoneal membrane to open the abdomen.

11. Cut a lobe of the liver and put in formalin.

12. Collect examples of visceral fat; e.g., the adipose tissue surrounding the intestine.

13. Pull out the spleen and the pancreas which is loosely attached to the spleen and cut out both. Put both organs on a paper towel. Separate the pancreas and put in formalin.

14. Collect one kidney and add to the formalin tube.

15. Proceed using standard histology procedures for paraffin embedding and sectioning (*see* **Note 21**).

16. Stain sections with hematoxylin/eosin (*see* **Note 22**).

4   Notes

1. A large number of additional methods are commonly used in diabetes research to understand mechanisms of action of given drugs or pathways. A few examples are the use of a Comprehensive Laboratory Animal Monitoring System (C.L.A.M.S.) to measure oxygen consumption and thermogenesis, perfusion of pancreatic islets for in vitro studies, gastric emptying measurement protocols, and gastric bypass surgery. Many resources exist elsewhere to cover these methods.

2. In a C57BLKS/J background, the *Lepob* mutation express a phenotype closely related to that observed in the BKS.Cg-*Dock7m* +/+ *Leprdb*/J strain, highlighting the importance of the genetic background in these models.

3. Reported duration of feeding with a high fat diet varies greatly among published studies from a few to up to 20 weeks. Insulin resistance increases steadily during this period. Therefore, the period of high fat diet feeding is a compromise between time efficiency and the window necessary to see an effect.

4. Glucose increases rapidly in these mice and reaches 600 mg/dL by 6 to 8 weeks of age. Therefore, the youngest mice are best. If the availability of BKS.Cg-*Dock7m* +/+ *Leprdb*/J males is limited, females of the same age can be used.

5. B6.Cg-Lep^{ob}/J mice become normoglycemic as they age because of compensatory hyperinsulinemia. We have used these mice successfully up to 12 weeks for pharmacological studies.

6. The most commonly used diets are 45 kcal% fat diet (Research Diets, D12451i) and 60 kcal% fat diet (Research Diets, D12492i). The "i" in "D124xxi" refers to the irradiated form of the diet. Many vivaria use the irradiated form for sterility purposes, but this is not required. 60 % fat diet induces a more severe insulin resistance. However, the high fat pellets crumble more easily, falling in the bedding and make it more difficult to perform accurate food intake measurements. Also, under these regimes more skin lesions can be observed, presumably because of increased subcutaneous fat, greasy fur from the diet, and fighting. Use of high fructose/sucrose diets is not commonly used in mice. While a high carbohydrate diet is efficient at inducing insulin resistance in rats, effects vary strongly between mouse strains. It has little effect on C57BL/6J mice (cf. Research Diets website).

7. Food intake can vary substantially by strain and diet within a range of 2 to 6 g per mouse daily. Therefore, the concentration of rosiglitazone in the food may need to be adjusted. Formulation is custom made by the vendor.

8. Mouse serum collection tubes or the Microvette® CB300 collection system (Kent Scientific) are suitable alternatives to the capillary collection method presented here.

9. The PIXImus™ densitometer from GE Medical Systems was removed from this edition, as this system has now been discontinued. Two systems for MRI in mice are commonly used, EchoMRI™ and the Minispec from Bruker. Both provide reliable body composition measurements. The loading of the mice in the Echo system allows the mice to remain horizontal during measurement regardless of their size—as opposed to vertical—and may cause less stress on the animal.

10. Most peptides and proteins are water soluble, and therefore D-PBS or saline can be used as a vehicle. Often however, limited information is available about the solubility properties of novel small molecules, and the choice of a nontoxic vehicle is more difficult. For example, 200 μL of a 5 % ethanol solution is equivalent to one beer in humans and may affect behavior. A solution of 20 % cyclodextrin has no known side effects in vivo, but in rare cases, some compounds are trapped in the solution and therefore mice have no exposure to the compound. Some

vehicles used for in vitro studies can be toxic in live mice. Some vehicles such as methylcellulose have no side effect when given p.o., but are toxic if administered i.v. Access to information about the pharmacokinetics properties of a test compound can help in the choice of a vehicle. Also, many institutions have internal guidelines determining the use of acceptable vehicles in pharmacological studies. If so, verify that the vehicle chosen is approved in the said guidelines.

11. When mixing a compound to the diet, pilot experiments should be conducted to verify that food intake is not affected by the change of palatability in the diet. This is counterintuitive, as a compound reducing food intake might be considered a benefit for a metabolic disease. However, unspecific effects on metabolism due to diet unpatability should be tested and excluded early on.

12. Retroorbital bleeding can cause blindness in animals and is increasingly discouraged by animal welfare committees. However, alternatives such as saphenous vein blood collection are cumbersome. Submandibular vein blood collection is rapid but causes glucose values to increase (unpublished data.)

13. Mice can be either group housed or single housed. Mice are often singly housed in DIO studies to avoid skin lesions resulting from cage mate aggression. Single housing also allows for individual food intake measurements. On the other hand, mice are social animals and being single housed increases their level of stress and decreases food intake. Additionally, housing space can be a limited resource in many vivariums, in which case group housing may be preferable. Although less commonly used, B6D2F1/J mice are responsive to DIO feeding, can be easily randomized and regrouped with noncage mates into adulthood without increased aggression behavior. In either case, singly- or group-housed, all mice from a given experimental cohort should be housed in the same manner.

14. Given the rapid progression of hyperinsulinemia, the range of body weight and baseline serum glucose value varies greatly. In addition, the date of birth used for shipment is usually a "bin" of the stated date of birth, and if often uses animals born 3 to 4 days before or after the stated date. On a 3-week-old mouse this can lead to significant weight differences. Limits are chosen somewhat arbitrarily to eliminate mice that deviate the most from the average values, while keeping enough mice for cohort size providing statistically significant values.

15. Values may differ slightly depending on the settings chosen in the template used for reading. The test is designed to use a "Blank" value rather than using "standard values = 0" in the template for the wells corresponding to the "zero standard."

The test is also designed to provide a linear correlation of absorbance and concentration when using log–log scales.

16. Lower glucose doses can be used (e.g., 1 g/kg) when mice are hyperglycemic and insulin resistant, and that many of the glucose values collected during a GTT will exceed the detection limit of the glucometer, 800 mg/dL.

17. Alternative glucose measurement methods are available. However, their relevance to glucose metabolism is of limited value considering that when glucose levels exceed 600 mg/dL, glucose is eliminated through urine rather than by glucose uptake in peripheral tissues.

18. GTT and ITT can be conducted either in fasted or nonfasted conditions. This will affect the baseline glucose values and the standard deviation but has little effect on glucose values later in the tests in comparison to the effects of glucose and insulin. Fasting causes an increase in appetite signals and other molecules involved in metabolism. Depending on the test molecule tested, this might interfere with the drug target and study results.

19. Right and left are defined from the mouse perspective, i.e., "left" refers to the mouse's left side.

20. The serum should be yellowish. A reddish color is indicative of erythrocytes lysis (hemolysis) which may interfere with clinical chemistry assays based on colorimetric values. If this happens, you may need to adjust the speed at which the blood is collected and processed or other steps that may cause sheer and red blood cell lysis.

21. All samples can be embedded in a single block for histological analysis as a multitissue block. Histology services are offered by many commercial providers, e.g., IDEXX, a veterinary service with locations worldwide.

22. Hematoxylin–eosin is sufficient to reveal pancreatic islet hyperplasia or degeneration, pancreatitis, liver steatosis, adipocyte hyperplasia, and diabetic nephropathies. Additional staining can be requested, such as immunostaining with anti-insulin or antiglucagon antibodies if additional information is needed about drugs mechanisms of action.

Acknowledgements

I am grateful to Jonitha Gardner, Laura Hoffman, Cheryl Loughery, Drs. Jiangwen Majeti, Alykhan Motani, and Wen-Chen Yeh for scientific discussions and critical review of the manuscript.

References

1. Saltiel AR (2001) New perspectives into the molecular pathogenesis and treatment of type 2 diabetes. Cell 104:517–529

2. Prevention CfDCa (2014) National Diabetes Statistics Report: Estimates of Diabetes and Its Burden in the United States. U. S. Department of Health and Human Services, Atlanta, GA

3. Koenig RJ, Peterson CM, Jones RL, Saudek C, Lehrman M, Cerami A (1976) Correlation of glucose regulation and hemoglobin AIc in diabetes mellitus. N Engl J Med 295:417–420

4. Bennett WL, Maruthur NM, Singh S, Segal JB, Wilson LM, Chatterjee R et al (2011) Comparative effectiveness and safety of medications for type 2 diabetes: an update including new drugs and 2-drug combinations. Ann Intern Med 154:602–613

5. Inzucchi SE, Bergenstal RM, Buse JB, Diamant M, Ferrannini E, Nauck M et al (2012) Management of hyperglycemia in type 2 diabetes: a patient-centered approach: position statement of the American Diabetes Association (ADA) and the European Association for the Study of Diabetes (EASD). Diabetes Care 35:1364–1379

6. Ahmadian M, Suh JM, Hah N, Liddle C, Atkins AR, Downes M et al (2013) PPARgamma signaling and metabolism: the good, the bad and the future. Nat Med 19:557–566

7. Raskin P (2008) Why insulin sensitizers but not secretagogues should be retained when initiating insulin in type 2 diabetes. Diabetes Metab Res Rev 24:3–13

8. Tomkin GH (2014) Treatment of type 2 diabetes, lifestyle, GLP1 agonists and DPP4 inhibitors. World J Diabetes 5:636–650

9. Vivian EM (2014) Sodium-glucose cotransporter 2 (SGLT2) inhibitors: a growing class of antidiabetic agents. Drugs Context 3:212264

10. Puzziferri N, Roshek TB 3rd, Mayo HG, Gallagher R, Belle SH, Livingston EH (2014) Long-term follow-up after bariatric surgery: a systematic review. JAMA 312:934–942

11. Sanghera DK, Blackett PR (2012) Type 2 diabetes genetics: beyond GWAS. J Diabetes Metab 3:6948-6971

12. Yu H, Zheng X, Zhang Z (2013) Mechanism of Roux-en-Y gastric bypass treatment for type 2 diabetes in rats. J Gastrointest Surg 17:1073–1083

13. Baribault H, Majeti JZ, Ge H, Wang J, Xiong Y, Gardner J et al (2014) Advancing therapeutic discovery through phenotypic screening of the extracellular proteome using hydrodynamic intravascular injection. Expert Opin Ther Targets 18:1253–1264

14. Veniant MM, Komorowski R, Chen P, Stanislaus S, Winters K, Hager T et al (2012) Long-acting FGF21 has enhanced efficacy in diet-induced obese mice and in obese rhesus monkeys. Endocrinology 153:4192–4203

15. Foltz IN, Hu S, King C, Wu X, Yang C, Wang W et al (2012) Treating diabetes and obesity with an FGF21-mimetic antibody activating the betaKlotho/FGFR1c receptor complex. Sci Transl Med 4:162ra53

16. Gaich G, Chien JY, Fu H, Glass LC, Deeg MA, Holland WL et al (2013) The effects of LY2405319, an FGF21 analog, in obese human subjects with type 2 diabetes. Cell Metab 18:333–340

17. Wu X, Ge H, Baribault H, Gupte J, Weiszmann J, Lemon B et al (2013) Dual actions of fibroblast growth factor 19 on lipid metabolism. J Lipid Res 54:325–332

18. Wu X, Ge H, Lemon B, Vonderfecht S, Baribault H, Weiszmann J et al (2010) Separating mitogenic and metabolic activities of fibroblast growth factor 19 (FGF19). Proc Natl Acad Sci U S A 107:14158–14163

19. Ridaura VK, Faith JJ, Rey FE, Cheng J, Duncan AE, Kau AL et al (2013) Gut microbiota from twins discordant for obesity modulate metabolism in mice. Science 341:1241214

20. Srinivasan S, Florez JC (2015) Therapeutic challenges in diabetes prevention: we have not found the "Exercise Pill". Clin Pharmacol Ther 98:162–169

21. Townsend KL, Tseng YH (2014) Brown fat fuel utilization and thermogenesis. Trends Endocrinol Metab 25:168–177

22. Halford JC, Harrold JA (2008) Neuropharmacology of human appetite expression. Dev Disabil Res Rev 14:158–164

23. Fulton S (2010) Appetite and reward. Front Neuroendocrinol 31:85–103

24. Yeadon J (2015) Choosing among type II diabetes mouse models. The Jackson Laboratory. https://new.jax.org/news-and-insights/jax-blog/2015/july/choosing-among-type-ii-diabetes-mouse-models#

25. Tam CS, Lecoultre V, Ravussin E (2011) Novel strategy for the use of leptin for obesity therapy. Expert Opin Biol Ther 11:1677–1685

26. Svenson KL, Von Smith R, Magnani PA, Suetin HR, Paigen B, Naggert JK et al (2007) Multiple trait measurements in 43 inbred

mouse strains capture the phenotypic diversity characteristic of human populations. J Appl Physiol (1985) 102:2369–2378

27. Alexander J, Chang GQ, Dourmashkin JT, Leibowitz SF (2006) Distinct phenotypes of obesity-prone AKR/J, DBA2J and C57BL/6J mice compared to control strains. Int J Obes (Lond) 30:50–59

28. Clee SM, Attie AD (2007) The genetic landscape of type 2 diabetes in mice. Endocr Rev 28:48–83

29. Nishikawa S, Yasoshima A, Doi K, Nakayama H, Uetsuka K (2007) Involvement of sex, strain and age factors in high fat diet-induced obesity in C57BL/6J and BALB/cA mice. Exp Anim 56:263–272

30. Luo J, Quan J, Tsai J, Hobensack CK, Sullivan C, Hector R et al (1998) Nongenetic mouse models of non-insulin-dependent diabetes mellitus. Metab Clin Exp 47:663–668

31. Baribault H et al (2014) Advancing therapeutic discovery through phenotypic screening of the extracellular proteome using hydrodynamic intravascular injection. Expert Opin Ther Targets 18(11):1253–1264

32. Kebede M, Alquier T, Latour MG, Semache M, Tremblay C, Poitout V (2008) The fatty acid receptor GPR40 plays a role in insulin secretion in vivo after high-fat feeding. Diabetes 57:2432–2437

33. Buchner DA, Burrage LC, Hill AE, Yazbek SN, O'Brien WE, Croniger CM et al (2008) Resistance to diet-induced obesity in mice with a single substituted chromosome. Physiol Genomics 35:116–122

34. Brommage R (2003) Validation and calibration of DEXA body composition in mice. Am J Physiol Endocrinol Metab 285:E454–E459

35. Gregoire FM, Zhang Q, Smith SJ, Tong C, Ross D, Lopez H et al (2002) Diet-induced obesity and hepatic gene expression alterations in C57BL/6J and ICAM-1-deficient mice. Am J Physiol Endocrinol Metab 282:E703–E713

36. Tinsley FC, Taicher GZ, Heiman ML (2004) Evaluation of a quantitative magnetic resonance method for mouse whole body composition analysis. Obes Res 12:150–160

Chapter 11

Cholesterol Absorption and Metabolism

Philip N. Howles

Abstract

Inhibitors of cholesterol absorption have been sought for decades as a means to treat and prevent cardio-vascular diseases (CVDs) associated with hypercholesterolemia. Ezetimibe is the one clear success story in this regard, and other compounds with similar efficacy continue to be sought. In the last decade, the laboratory mouse, with all its genetic power, has become the premier experimental model for discovering the mechanisms underlying cholesterol absorption and has become a critical tool for preclinical testing of potential pharmaceutical entities. This chapter briefly reviews the history of cholesterol absorption research and the various gene candidates that have come under consideration as drug targets. The most common and versatile method of measuring cholesterol absorption is described in detail along with important considerations when interpreting results, and an alternative method is also presented. In recent years, reverse cholesterol transport (RCT) has become an area of intense new interest for drug discovery since this process is now considered another key to reducing CVD risk. The ultimate measure of RCT is sterol excretion and a detailed description is given for measuring neutral and acidic fecal sterols and interpreting the results.

Key words Cholesterol absorption, Phytosterols, Cholesterol excretion, Reverse cholesterol transport, ACAT, CEL, Inhibitors, Bile acids, Fecal sterols, Dual-label, Obesity, Cardiovascular disease

1 Introduction

Cardiovascular diseases (CVDs) are the leading cause of death in the US and other "Western" societies despite the great success of statin drugs for lowering LDL cholesterol. Thus, finding new targets for pharmaceutical intervention aimed at reducing the risk for CVD has remained a high priority and the focus of decades of research. Since the advent of gene targeting by homologous recombination, a host of mouse models have been generated and widely used for the study of cholesterol and lipoprotein metabolism with the goal of understanding the molecular physiology and etiology of CVD. The result has been identification of numerous proteins, enzymes, and metabolic pathways, other than HMG CoA reductase and the cholesterol synthesis pathway, that are potential drug targets for the treatment and prevention of atherosclerosis and CVD.

Gabriele Proetzel and Michael V. Wiles (eds.), *Mouse Models for Drug Discovery: Methods and Protocols*, Methods in Molecular Biology, vol. 1438, DOI 10.1007/978-1-4939-3661-8_11, © Springer Science+Business Media New York 2016

1.1 Cholesterol Absorption

In this light, cholesterol absorption has received intense focus for several decades. Although the various statins lower LDL by decreasing endogenous cholesterol synthesis, another approach to prevent excess cholesterol accumulation is to reduce absorption of dietary cholesterol. Doing so also prevents reabsorption of biliary cholesterol, which can have a major impact on overall cholesterol metabolism since recirculation of biliary cholesterol represents a large portion of the cholesterol that transits through the intestine. For recent reviews on mechanisms of cholesterol and lipid absorption, *see* Ref. [1–3].

The search for intestinal cholesterol transporters extended for many years, beginning with a debate about whether or not it was even a protein-facilitated process [4, 5]. The pancreatic enzyme carboxyl ester lipase (CEL, also called cholesterol esterase) was believed to be important to this process [6, 7] and several companies devoted considerable resources to the development and testing of compounds to inhibit CEL with mixed results [8–10]. These efforts were abandoned in the mid-1990s, however, after studies with gene-knockout mice demonstrated that the enzyme was important only for absorption of cholesteryl ester [11, 12], which is a minor component of dietary cholesterol and is present at very low levels in bile. Interestingly, CEL is also found in liver where it has been shown to affect HDL metabolism [13]. Thus, it may ultimately play an important role in cholesterol metabolism and may yet prove to be a useful drug target for CVD treatment (Camarota and Howles, unpublished).

Intestinal acyl-CoA:cholesterol acyltransferase (ACAT-2, also present in liver), which esterifies free cholesterol with a palmitic or oleic acid, is another enzyme that was identified early on as a potential target to inhibit cholesterol absorption because most cholesterol in chylomicrons is esterified before being secreted by enterocytes [6, 14]. As for CEL, various inhibitors of this enzyme were also developed and tested with mixed results [10, 15–17]. However, the importance of ACAT-2 was later confirmed by studies of gene-knockout mice, which exhibit markedly reduced cholesterol absorption and atherosclerosis when fed Western diet [18]. Nonetheless, progress in developing effective ACAT inhibitors has been slow, in part because of concerns about the potential for deleterious systemic effects resulting from inhibition of the more widely expressed ACAT-1 [19]. Despite these difficulties as well as mixed results in animal and clinical trials, interest remains for the development of ACAT inhibitors, especially ones that could be restricted to act on ACAT-2 [20–22].

In 1997, Schering-Plough reported dramatic inhibition of cholesterol absorption by a compound, later named ezetimibe, that was developed in the process of modifying known ACAT inhibitors to generate novel compounds with improved pharmacodynamics and pharmacokinetics [23]. Interestingly, while the drug blocked absorption by up to 90%, it was no longer an effective

ACAT inhibitor [23]. Later, this group used database mining and gene-knockout mice to identify that the protein essential for cholesterol absorption is NPC1L1 (Niemann–Pick disease C, type 1-like protein 1), and that this protein is the likely target of ezetimibe [24, 25]. The development of this drug and identification of the putative target gene constituted major breakthroughs in the field, both for research focused on cholesterol and lipoprotein metabolism and for the treatment of hypercholesterolemia and CVD. Ezetimibe (Zetia™) has become widely used clinically, especially for patients whose LDL cholesterol responds poorly to statins or who cannot tolerate the side effects of these drugs [26].

As patent expiration looms for the various statins, major pharmaceutical companies are eagerly seeking new targets and new compounds for lowering plasma cholesterol and reducing CVD risk. Particularly appealing are compounds that would resemble ezetimibe in action and pharmacokinetics. Ezetimibe is glucuronidated by the intestine (and liver), which increases its potency for blocking cholesterol absorption, in part because the glucuronidated form has a high affinity for the intestinal brush border and is poorly reabsorbed after the first pass through the enterohepatic circulation [23]. Thus, compounds with an intestinal site of action that either are not absorbed, or concentrate at the site of action, or have good efficacy at low circulating levels are desired because the risk of off-target effects is reduced (especially if not absorbed).

In this regard, one promising target for drug development is the phospholipase-A2 secreted by pancreatic acinar cells (PLA2) and/or the phospholipase-B made by intestinal epithelial cells. It has been demonstrated that decreasing intestinal PLA2 activity with inhibitors or by gene ablation results in decreased cholesterol absorption [27, 28] and cell culture models have also shown that intact phospholipids inhibit cholesterol absorption apparently by inhibiting micelle formation [29]. Data from knockout mice also suggest that there may be other health benefits of inhibiting PLA2 in the intestine [29]. Thus, this enzyme is an ideal target for decreasing CVD since unabsorbed inhibitors should be efficacious with minimal risk of systemic side effects.

Studies with gene-knockout mice [30] as well as chemical inhibitors [31] have shown that decreasing activity of pancreatic triglyceride lipase (PTL) also decreases cholesterol absorption. Data indicate that this most likely results from incomplete digestion and delayed absorption of triglycerides [30, 31]. Lumenal cholesterol partitions into the oil phase and is carried distally in the small intestine where its absorption is less efficient because of fewer transporter molecules and/or because of reduced bile salt concentration (*see* model in Ref. [1]). Orlistat (tetrahydrolipstatin), currently marketed as alli®, is a potent inhibitor of PTL and is sold as an antiobesity treatment. While decreased cholesterol absorption is a beneficial side effect, targeting lipases is a poor strategy for doing so since steatorrhea can occur, which tends to reduce compliance [32].

After postprandial absorption, lumenal cholesterol is incorporated into chylomicrons (large triglyceride-rich lipoproteins) by enterocytes and secreted into intestinal lymphatics. Secretions into this network collect in the thoracic lymph duct and finally reach the circulation where this duct opens into the left subclavian vein. As chylomicrons are pumped throughout the body, the dietary lipids are rapidly hydrolyzed and taken up by various tissues, as are the remnant particles, with a circulating half-life of only minutes in mice. For this reason, many in vivo investigations of cholesterol absorption and various inhibitor evaluations were performed using rats with cannulated lymph and bile ducts so as to assay intestinal output before it entered the circulation and was metabolized [4, 6, 7, 14, 16, 17, 27]. While extremely powerful, this method has its limitations and questions sometimes arise about how physiological the absorption process is under these conditions. From the perspective of this book, the need for special surgical skills and equipment limits the number of animals per experimental group, so it is not suitable for screening studies. This problem is further magnified because lymph and bile duct cannulation is at least an order of magnitude more difficult in mice (one-tenth the size) than in rats.

These problems were circumvented in most of the studies on knockout and drug-treated mice described earlier, as well as several on rats and other animal models, by using the alternative method(s) for measuring cholesterol absorption described in detail in this chapter. This method was first pioneered and validated by Quintão et al. [33] and remains widely used today because it is simple, accurate, noninvasive, reproducible, inexpensive, and does not require sophisticated equipment or specialized technical or surgical skills.

One of the many difficulties in studying nutrient absorption is that it is sometimes necessary to achieve total recovery of 24 h fecal output. This can be avoided, however, if a nonabsorbable analog of the nutrient in question is available. The method also requires that the marker and nutrient can be obtained with different radiolabels, typically ^{3}H on one and ^{14}C on the other. The marker and nutrient are mixed at a predetermined ratio before being given to the animals. Feces are collected for 24 h and the ratio of excreted marker to excreted nutrient is determined. The percent absorption is calculated from the difference in marker:nutrient ratio between the starting material and the fecal extract. The great advantage of this method is that analysis can be performed on only a portion of the feces, the amount being dictated primarily by the specific activity of the starting material.

While it might seem reasonable to use a generic marker such as polyethylene glycol, which is completely eliminated without absorption (used to verify integrity of the epithelial barrier), it is important for the marker to have physical properties similar to the nutrient in question because of the complexity of the postprandial intestinal milieu—a thick slurry of mixed micelles, oil and water phases, and suspended particles. The marker should partition among the phases

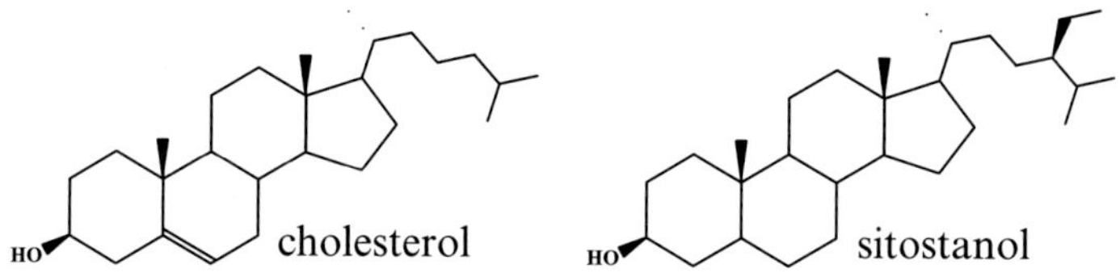

Fig. 1 Structures of cholesterol and a commonly used phytosterol. Adding a branch to the side chain and removing the double bond in the sterol B ring combine to reduce absorption of sitostanol ~20 fold as compared to cholesterol

similarly to the analyte of interest and should have similar intestinal transit times. Thus, sugars must be used to trace sugars, sterols to trace sterols, etc.

For cholesterol absorption, one can take advantage of the fact that plant sterols (phytosterols) are poorly absorbed by mice and humans despite close structural similarity to cholesterol. Beta sitosterol, for example, differs from cholesterol only by the addition of an ethyl group to carbon 24 of the sterol side chain (*see* Fig. 1). This reduces absorption to ~6% [34, 35] as compared to 60–80% for cholesterol. Sitostanol, which has a saturated C5–C6 bond in addition to the ethyl group, is only absorbed at ~3% [34, 35] and is widely used as a nonabsorbed sterol marker (Fig. 1).

1.2 Cholesterol Excretion and Reverse Cholesterol Transport

Recent years have seen increased focus on enhancing reverse cholesterol transport (RCT), as well as on lowering LDL cholesterol levels, as part of strategies for reducing CVD risk. RCT is the HDL-mediated process by which excess cholesterol is transported from peripheral tissues to the liver for disposal as biliary cholesterol and bile salts. In particular, there has been emphasis on finding ways to specifically raise HDL cholesterol based on the premise that doing so facilitates RCT by increasing vehicular capacity for the process. This was one of the goals of the CETP (cholesteryl ester transfer protein) inhibitor Torcetripib developed by Pfizer. While this drug did successfully raise HDL, unacceptable side effects caused it to fail in Phase III clinical trials [36]. Also, related studies yielded equivocal results with respect to whether or not RCT was increased by CETP inhibition [37, 38]. Thus, targets and treatment modalities to increase RCT represent unmet needs of great potential for drug discovery.

Ultimately, what matters is that excess cholesterol be eliminated from the body. Thus, measuring cholesterol excretion has become an important assay for evaluating new potential drug targets and new compounds and a method for doing so is described in detail later. The assay has the advantage of being completely noninvasive to the animals although it is slightly more demanding technically and requires a gas chromatograph. The method is a modification of that described by Post et al. [39] and has been validated and used extensively by us with good success and reproducibility.

In addition, a reasonably physiological method has been recently developed and is now often used to estimate cholesterol flux from macrophage foam cells (as are in atherosclerotic plaque) to plasma HDL, then to liver, bile, and feces [40]. Briefly, the method involves radiolabeling macrophage cells with cholesterol in culture before administering them to mice by intraperitoneal or subcutaneous injection. Movement and excretion of the radiolabel is measured over the course of 3 to 5 days to assay the different steps of RCT. Similarly, the relative contribution of different lipoprotein classes to the RCT process in any given model can be assayed by injecting (i.v.) radiolabeled HDL, LDL, or VLDL and quantitating its appearance in the liver bile and feces over time. Although the experimental details for labeling cells and lipoproteins will not be discussed in this chapter, the final determination of RCT in each of these assays involves measurement of sterol excretion according to the method described here.

2 Materials

2.1 Mice and Diets

C57BL/6 is the most commonly used strain in cardiovascular research so it is the most characterized with respect to cholesterol absorption and excretion. However, mouse strains do differ with respect to cholesterol absorption [41]. Thus, it is very important to have accurate control groups for all absorption experiments. Mice should be matched for age, sex, genetic background, and even diet history (unless that is a parameter being studied). For example, while many investigators will purchase control mice from commercial suppliers or will select them from a "wild-type" colony or collection, the purist will use control mice from the same breeding colony, if not the same cohort, from which the test mice are derived. Inbred mice give the most consistent results and require fewer animals per treatment group (~8) but outbred mice can also be used if group size is increased (at least 12).

Diet effects on cholesterol absorption and excretion measurements are usually not problematic but should be considered. Standard rodent chows from commercial suppliers work well in most instances. However, lot-to-lot variability in some micronutrients, such as phytoestrogens, may complicate interpretation of results from a serial study that spans several months. Semipurified diets avoid this possibility. The caveat with semipurified diets is that most have little or no fiber. Mice eat much less of these diets, have reduced gut motility, and less fecal output.

2.2 Radiolabeled Sterols (See Note 1)

1. ^{3}H-sitostanol or ^{3}H-ß-sitosterol

2. ^{14}C-cholesterol

2.3 Lipids (See Note 2)

1. 10 mg/mL phosphatidylcholine (egg PC) dissolved in ethanol, stored at –20 °C in glass screw-capped vial.

2. 10 mg/mL cholesterol (Sigma-Aldrich) dissolved in ethanol, stored at –20 °C in glass screw-capped vial.

3. Triglyceride: triolein (Sigma-Aldrich) or vegetable oil, such as olive, canola, or safflower oil (grocery brands) stored at –20 °C in glass screw-capped vial to reduce oxidation.

4. Filter sterilized 150 mM sodium taurocholate (Sigma-Aldrich) in water.

2.4 Solvents

Standard solvents, preferably HPLC grade: ethanol, methanol, petroleum ether, chloroform, hexane, diethyl ether.

2.5 Special Supplies

1. Gavage needles: 20 or 22 gauge with 1.25 or 2.25 mm ball, for small or large (>35 g) mice, respectively.

2. Hamilton syringes: 250, 100, and 10 μL with needles.

3. Wire platforms for animal cage bottoms.

4. 35 mL glass centrifuge tubes with screw caps.

5. Ceramic mortar and pestle (2 or more).

2.6 Equipment for Sample Preparation and Analysis

1. Needle manifold for drying samples with a nitrogen stream.

2. Probe sonicator with narrow tip and variable, pulsatile energy delivery.

3. Lyophilizer for drying diet-drug admixtures and feces.

2.7 Optional Equipment for Sample Analysis

1. Silica gel G thin layer chromatography (TLC) plates.

2. TLC tanks.

3. Crystalline iodine (I_2) for staining.

4. Phosphomolybdic acid (10 % in ethanol) solution with an atomizer for staining.

5. Biological oxidizer (e.g., from Harvey instruments).

2.8 Materials Specific to Measurement of Fecal Sterol Mass and RCT

1. Ethanol solutions of 5α-cholestane, lathosterol, coprostanol, desmosterol, sitosterol, stigmasterol, sitostanol.

2. ^{14}C-taurocholate.

3. C18 Bond Elut columns (Varian Inc.).

4. Colorimetric bile acid assay kit (Trinity Biotech PLC).

5. Gas chromatograph (GC) with correct column and configuration to detect neutral sterols.

3 Methods

3.1 Cholesterol Absorption

3.1.1 Preparation of the Sterol Mix

1. Dry the appropriate amounts of each radiolabel (usually supplied in ethanol or toluene) under a stream of N_2 in a glass tube or glass scintillation vial. Typically, the nonabsorbed marker is used at 0.2–0.5 µCi per mouse and the cholesterol at 0.5–1.0 µCi per mouse.

2. Either of two solutions can be used to redissolve the radiolabeled sterols. The first, oil, is simpler to prepare and administer to the mice. However, it is not strictly a physiological representation of dietary cholesterol, which is largely present with phospholipids in cell membranes, and the oil:cholesterol ratio is much greater than would occur in most diets. The second solution, a lipid emulsion, is more tedious to prepare and is less stable but is more physiologically accurate.

   2.1. *Oil*: Dissolve the radiolabeled sterols in a volume (µL) of triglyceride equal to 100–200 times the number of doses being prepared. Because of viscosity and surface tension, the triglyceride is most accurately measured with a syringe or other positive displacement device rather than a pipette or pipettor. After adding the oil to the tube or vial, warm it to 37 °C for 15 min and mix it thoroughly by vortexing or with the syringe. The oil can be pure triolein, or vegetable oil such as olive, canola, or safflower oil.

   2.2. *Emulsion*: For each 1 mL of final suspension, add to the tube or vial from **step 1** previously, 10 µmol of phosphatidylcholine and 5 µmoles of cholesterol from the stock solutions, plus 50 µmol (48 µL) of triolein (or other oil, see earlier) and dry with a N_2 stream at the same time as the radiolabel. The organic solvents dry rapidly leaving the lipids partially dissolved in the oil. Add 1 mL of sterile water or saline to the tube and vortex vigorously to start emulsification. Sonicate for 5–10 min on a 2 s, 50 % duty cycle (1 s on, 1 s off) with the tube suspended in ice for cooling. Adjust burst energy to maximize sonication but *avoid cavitation*. Once prepared, the emulsion should be used within a few hours. Monitor its appearance and resonicate if phase separation becomes evident. Alternatively, the emulsion can be stabilized by adding sodium taurocholate to 10 mM (final concentration). This natural bile salt acts as a detergent to prevent phase separation.

3. Transfer 5 µL of the material prepared in **items 2.1** or **2.2** into a scintillation vial to determine the isotope ratio in the material that will be administered to the animals. For **item 2.2**, it is important to sample when the mixture is thoroughly emulsified.

3.1.2 Animal Preparation and Treatment

1. *Drug Pretreatment*

Either of two approaches can be used for testing drug efficacy. Comparison of a treated group to controls is more common and takes less time, and may be preferred if there are concerns about effects due to prolonged vehicle exposure. The alternative is to measure absorption (or excretion) in the same animals before and after drug treatment. This offers the advantage of requiring fewer animals and is valid as long as drug treatment is not lengthy (2–4 weeks) and the animals are adults (10–12 weeks) at the onset of the experiment.

Before testing for cholesterol absorption or excretion, considerations must be given to how and how long to administer the candidate compound to the mice prior to testing. This decision will be driven largely by the proposed method of action of the drug candidate. If it acts directly on the small intestine or on pancreatic enzymes in the intestinal lumen, acute treatment, 1–3 days may suffice. If the compound acts systemically and a steady-state plasma level must be achieved it may be necessary to pretreat the mice for 1–2 weeks. If the drug is injectable, **steps 1.1** and **1.2** are not relevant. However, oral delivery is preferred for pharmaceuticals (people prefer pills over injections). Also, if the drug is designed to block cholesterol absorption, direct delivery to the site of action, the intestinal lumen and/or the apical surface of enterocytes, is likely to be necessary or at least preferred.

1.1. *Drug delivery by Diet Admix*

In most cases, compounds are supplied as powder or crystal. Therefore, an appropriate amount of diet (chow or semi-purified diet of desired composition) is thoroughly ground (coffee grinder or other kitchen grinder works well). The appropriate amount of compound, ground to fine powder with a smooth-bottom mortar and pestle if at all crystalline, is added and mixed by further grinding. Diet is fed as powder or is repelleted by adding water, compacting with a syringe, and lyophilizing to remove the added water (lyophilize to original weight). The drug:diet ratio is determined by measuring the amount of food eaten per day (2–4 g per day for a 25 g mouse, depending on diet composition) and the desired dose. The latter typically ranges from 1 to 100 mg per kg body weight, but may vary widely. The number of days to feed the compound will depend on its mechanism of action but is typically 3–14 days. Form of the diet, powder or repelleted, should be the same for control mice.

1.2. *Drug delivery by Oral Gavage*

For this method, the compound is dissolved in oil (e.g., olive oil) or water, as appropriate, or can be given as a suspension if care is taken to dispense equal doses. The delivery volume should be 100–200 µL per mouse. This is measured and dispensed using a Hamilton syringe or similar positive displacement device. The concentration of compound should be adjusted so that the entire daily dose is given in a single gavage. Drug should be given to the mice at the same time each day, typically near the end of the dark cycle when their stomachs are relatively empty. The final dose can (or should) be coadministered with the radiolabeled test mixture, depending on mechanism of action.

2. *Acclimation of Mice*

House mice singly in cages with wire platforms and no bedding for 1 or 2 days prior to testing absorption. This allows the animals to acclimate to the stress of the wire platforms (no bedding) as well as to the stress of being singly housed (no nestling with cage mates). The wire platforms greatly reduce caprophagy and simplify fecal collection. Food and water are supplied ad libitum.

3. *Fasting of Mice*

Fast mice during the light cycle on the day of the absorption test. Remove food shortly after the end of the dark cycle. Because this is a short fast (~8 h) during the time when eating is minimal, it does not stress the animals. Fasting is not strictly necessary, but it minimizes stomach content when the test mixture is administered, which aids delivery by gavage.

4. *Treatment of Mice with Radiolabeled Sterol Mix*

The radiolabeled sterol mix is given to the mice 1–3 h before the beginning of the dark cycle. Volume should be 100–200 µL and the dose of isotope as described earlier (Subheading 3.1.1). Sedation is not usually necessary but inexperienced handlers may wish to lightly anesthetize with isoflurane to prevent struggling while inserting the gavage tube. With the head tipped back, the tube should slide down the esophagus and into the stomach with little resistance. It is important to insert the tube into the stomach to prevent aspiration of the lipids. For a 25 g mouse, ~3/4 of the shaft of a 20 gauge, 2 in., gavage needle is inserted in the mouth when the ball is in the stomach.

5. *Feces Collection*

Return the mice to their respective cages and collect feces for 24 h (*see* **Note 3**). Longer collection times have the potential to underestimate absorption since some absorbed cholesterol will end up in HDL and subsequently be taken up by the liver, transported to the bile, and secreted into the intestine. However, semipurified diets are usually very low in fiber, which may reduce intestinal motility. If necessary, sterol transit time is measured in a preliminary experiment where only radiolabeled phytosterol is given and total fecal output is collected on days 1, 2, and 3.

Sterols are quantitatively extracted (*see* Subheading 3.1.3) and the time for 90 % excretion determined.

6. *Tissue Analysis*

   After the 24 h fecal collection, animals are euthanized and any tissues of interest (e.g., for toxicology analysis, such as plasma, liver, kidney, heart, spleen) are harvested. Alternatively, if the mice are precious, they can be used again since the procedure is noninvasive. If the systemic presence of radiolabeled cholesterol does not compromise the second use, little delay is necessary. However, the mice will continue to excrete radiolabeled sterols at detectable levels for 2–3 weeks so bedding must be collected and segregated as contaminated waste.

3.1.3 Extraction of Sterols from Fecal Samples

1. Wire cage bottoms are removed and all fecal output is collected, lyophilized overnight or until dry, and weighed. Normally, 0.8–1.5 g feces can be collected for a 25 g mouse and should be similar for all mice in both control and drug-treated groups. Mice with significantly more or less (e.g., <0.6 or >2.0 g) are noted. If outliers are only in the treated group, consideration is given to side effects or drug toxicity that may affect eating habits or gut motility.

2. Grind the dried material to a coarse powder with mortar and pestle and weigh ~0.5 g from each mouse into conical glass centrifuge tubes for processing. Save the remainder for reanalysis if necessary. Short-term storage (1–2 weeks) of lyophilized material can be at room temperature but long-term storage is best at –20 °C.

3. Add 5–10 mL of water to rehydrate the material. Vortex intermittently and heat at 60 °C for ~30 min to speed the process. If heated, cool tubes to room temperature before proceeding.

4. Add an equal volume (5–10 mL) of hexane or petroleum ether to the tubes and cap tightly. Shake vigorously for 15 s to thoroughly mix the phases. Vortexing will suffice but is less effective.

5. Centrifuge the tubes at $2500 \times g$ for 10 min or until phases are well separated. Recover the organic layer (top) into a scintillation vial or other glass vessel. Take care not to transfer any of the interface or subphase.

6. Repeat the extraction twice and combine organic layers.

7. Transfer 0.5 mL of the combined extract to a fresh scintillation vial, evaporate the solvent, resuspend the residue in 0.25 mL methanol or ethanol, add scintillation cocktail and count using a ^{3}H, ^{14}C dual label program.

8. If cpm values are low (should be several thousand for each isotope), repeat **step 7** using 1 or more mL of the organic fractions. If additional volume is needed, attention should be given to pos-

sible quenching, especially in the ^{3}H channel, due to color in the samples. This is best avoided by including enough isotopes in the gavage dose so small portions ($\leq$0.5 mL) of the extract are sufficient for counting. However, there are three possible solutions:

8.1. If the available scintillation counter does not have an adequate quench correction function, it may be necessary to process a nonradioactive fecal sample as described earlier and add an equivalent amount of the hexane extract to the counting vials for the starting sterol mix (from **item 2.1** or **2.2** earlier) so it is quenched similarly to the fecal samples.

8.2. A second way to resolve quench problems is to separate the sterols from the pigments by TLC. Briefly, samples are concentrated by evaporation and redissolved in a small volume ($\leq$100 μL) of chloroform or hexane and spotted onto a predried, silica gel G, TLC plate along with 5 μg of cholesterol standard. Components are resolved with a mobile phase comprised of petroleum ether, diethyl ether, and acetic acid at a 300:200:1 ratio. Sterols migrate 1/4 to 1/3 of the way to the top of the plate while pigments stay at or near the origin. Spots are visualized with iodine vapor, marked with a pencil, scraped into scintillation vials, and counted directly after soaking overnight in scintillation cocktail.

8.3. A third way to avoid pigment interference is to use a biological oxidizer. This specialized apparatus burns samples (dried onto pieces of filter paper) and collects the ^{3}H and ^{14}C directly into separate scintillation cocktails as water and CO_2, respectively.

3.1.4 Data Analysis

1. Percent absorption is calculated as: $100 \times (1 - (^{14}C/^3H$ feces/ $^{14}C/^3H$ dose)).

2. Cholesterol absorption by control mice is generally reported in the range of 60–80%. Variations within this range depend in part on small differences in method, sample processing, or animal manipulation. If results are notably above or below this range (<40% or >80%) for the control/untreated group, the technique used at each of the above steps should be carefully reviewed to determine where faults might lie. Focus especially on problems with counting efficiency and quenching, as noted in **step 8** of Subheading 3.1.3, and also on sampling of the emulsion in **item 2.2**.

3. Attention should also be given to the standard deviation within groups. A typical value is 10–20% of the group average. Greater variation may indicate inconsistent processing of samples or, for the treated group, variation in drug dosing.

Before accepting the result that absorption is, or is not, different between groups or drug treatments, it may be important to perform one or more control experiments to determine if confounding effects are masking the true result. Three particular areas of concern should be investigated. First, the possibility that the treatment is decreasing activity of the sterol transporter comprised of the two proteins ABCG5 and ABCG8 that are responsible for preventing phytosterol absorption. All sterols are absorbed by enterocytes and can be incorporated into chylomicrons and transported to the circulation. However, the G5/G8 transporter is present on the apical surface of intestinal epithelial cells and pumps plant sterols back into the lumen all along the small intestine so that the net absorption of phytosterols is very low (less than 10 %, as mentioned previously). A simple way to assess this possibility is to calculate the total recovery of radiolabeled phytosterol from 24 and 48 h fecal collections after isotope gavage. This number can be extrapolated from the data generated earlier by correcting for the fraction of extract counted and the fraction of total fecal material extracted. Recovery of the marker should be ~90 % after 24 h and $\geq$95 % by 48 h. The important determination is whether or not there is a difference in recovery between groups or treatments.

If phytosterol recovery differs significantly between groups, an alternative dual label method can be used to measure cholesterol absorption. In this case, a colloidal suspension of ^{3}H-cholesterol (or ^{14}C) is given intravenously (injected into tail, saphenous, or jugular vein) and an oral dose of ^{14}C-cholesterol (or ^{3}H), prepared as described earlier, is given at the same time. Heparinized or citrated blood samples (100 µL) are taken at 48 h and sterols are extracted from whole blood with hexane and counted essentially as described earlier. Absorption is calculated as $100 \times$ (% of oral dose in plasma aliquot/% of injected dose in plasma aliquot). Thus, it is important to accurately measure the exact doses given for each animal. This method is described in detail in the original paper by Zilversmit and Hughes [42]. It has been extensively used and gives results comparable to the method described here. It is technically more demanding, however, because intravenous injections are more challenging in the mouse than in the rat, the species for which the method was originally developed.

A second factor that can give rise to misleading cholesterol absorption data is different rates of biliary cholesterol excretion. This is especially the case if mice are fed a standard chow or a semipurified diet without cholesterol, and/or if the radiolabeled sterols are given in oil rather than as a lipid emulsion. Under these circumstances, the great majority of cholesterol mass in the intestinal lumen is derived from biliary cholesterol. If the latter differs between groups or treatments, the specific activity of the radiolabel (dpm/mg cholesterol) will also differ. Thus, the group having higher biliary cholesterol may appear to have lower absorption because of greater dilution of the radiolabel with unlabeled cholesterol from bile.

If there are reasons to suspect differences in biliary cholesterol secretion, it may be necessary to measure cholesterol concentrations in gall bladder (GB) bile and/or daily neutral sterol excretion. To obtain GB bile, animals are fasted for 10–12 h, typically overnight. Fasting interrupts enterohepatic circulation, resulting in greater volume of bile in the GB. After euthanization, GB is exposed and the bile duct is gripped firmly with surgical forceps (smooth, not serrated) where it enters the GB. The organ is removed by cutting the bile duct and then snipping the ligament and connective tissue that holds it in place. Great care must be taken to not pull to hard on the GB and to not nick it with the scissors or forceps. Surprisingly, the amount of connective tissue varies between strains so that GB are easily harvested from FVB/NJ mice but are more firmly attached in C57BL/6J animals. During and after removal, a firm grip is maintained with the forceps to prevent leakage. The GB is briefly rinsed in a dish of PBS and gently blotted to remove blood and tissue fluids, then suspended over the lip of a small (250 μL) microcentrifuge tube and punctured with a needle to allow the bile to drain into the tube. Store frozen until assay. There is usually enough material to measure lipid composition (bile acids, cholesterol, phospholipids) with standard colorimetric kits (≤ 1 μL needed for each assay). In addition to biliary cholesterol levels, it is important to take note of bile salt concentrations, since these are the detergents which suspend dietary lipids in micelles and deliver them to the intestinal epithelium for absorption by enterocytes. Differences in bile salt concentration alone could lead to differences in cholesterol absorption.

A third consideration is whether or not the treatment has affected fat digestion or absorption. Particular attention should be given to this possibility if bile acid secretion is reduced because micelle capacity will be decreased as mentioned earlier. If undigested or unabsorbed lipids are present in the lumen, sterols will partition into the oil phase, preventing or delaying their absorption [1, 30]. Changes in fat absorption can be measured acutely by giving radiolabeled triolein or tripalmitin (labeled on the fatty acid, not the glycerol) with an oral gavage of olive or other vegetable oil (dosing must be precise). At the same time, mice are also given Poloxamer 407 (1 mg/g body weight; Pluronic F-127, BASF Corp.) by intraperitoneal injection to prevent postprandial hydrolysis and clearance of the absorbed lipid [43]. Plasma levels of radiolabel are monitored hourly for 4–6 h to determine the rate and extent of absorption, and whether or not it is delayed in the treated group. A less invasive and more physiological method for measuring fat absorption is also available which uses sucrose polybehenate (a component of Olestra™)

mixed with the diet as an unabsorbable marker [44]. The fatty acid composition of the diet, including the marker behenic acid, is determined by gas chromatography. After three days of feeding, fatty acid composition of feces is determined and compared to that of the diet using the same equation shown earlier for cholesterol absorption. The power of this method is that analysis requires only a few fecal pellets and timing of their collection is not critical; no isotopes are used so the mice can be used for other studies; data are acquired for individual as well as total fatty acids. The latter point can be important because absorption mechanisms and efficiencies vary according to chain length and saturation of individual fatty acids [45].

3.1.6 Interpretation of Results: Benefit Assessment

While assessing the contributions of alternative scenarios described earlier is important to understanding the mechanism of drug action, assessing the benefit or risk posed by the overall effect is a slightly different process. With respect to the first case described earlier, increased phytosterol absorption due to inhibition of ABCG5/G8 is undesirable since it is likely to cause sitosterolemia—a disease with clinical presentations that includes elevated LDL and accelerated atherosclerosis. If other methods show that cholesterol absorption is decreased independent of the effect on phytosterols, it may be possible to alter drug structure to be more selective.

Effects of drug treatment on bile composition could be beneficial or deleterious. For example, if biliary cholesterol secretion is increased in conjunction with a decrease or no change in cholesterol absorption, the overall effect may be positive. This result would suggest that RCT is increased by the treatment. This would be a very desirable effect as explained earlier. Increased bile salt secretion could also be beneficial for similar reasons. On the other hand, decreased biliary cholesterol or bile salt concentrations are cause for concern since either could be indicative of some degree of hepatotoxicity and/or cholestasis. At the very least, they would indicate decreased RCT and increased CVD risk. As mentioned earlier, if drug treatment decreases cholesterol absorption independently of effects on biliary lipids, structural alterations of the compound may improve specificity and/or decrease its absorption so its effect is limited to the intestine.

While not altogether desirable, it is not necessarily deleterious if drug treatment decreases fat digestion or absorption as well as cholesterol absorption. As described previously, this is the mechanism of action of the antiobesity product alli®. However, if decreased cholesterol absorption is due solely to decreased fat absorption, the efficacy is likely to be poor or highly variable and dependent on diet composition. The primary concern with this outcome is that fat-soluble vitamins will partition into the oil phase and be sequestered, reducing their absorption.

3.2 Quantification of Sterol Excretion and Reverse Cholesterol Transport

3.2.1 Sample Collection and Processing

1. Collect total fecal output for exactly 3 days from mice preacclimated to, and individually housed in, cages with wire platforms as described earlier. Mice should be weighed before and after the collection period.

2. Collect fecal pellets and separate from food particles and other detritus, and lyophilize 12–24 h or until a constant weight is reached. Record exact weight and grind to a fine powder.

3. Measure 0.5 g of powder into tube and add 40 µg 5α-cholestane plus 0.02–0.05 µCi of ^{14}C-taurocholate (see **Note 4**).

4. Add 5 mL methanol and 1 mL 10 N NaOH to each tube and heat at 80–90 °C for 2 h with intermittent vortexing. Except when vortexing, caps should be slightly loose to release pressure. As needed, add methanol to restore original volume.

5. Cool to room temperature, add 10 mL hexane or petroleum ether, tighten caps, and shake vigorously for 15 s. Centrifuge at $1000 \times g$ for 10 min to separate phases and remove the organic layer (top) to a fresh tube or scintillation vial. Repeat the extraction twice and combine organic layers. Save the aqueous layer (bottom) in the original tube for bile acid analysis.

6. Evaporate the organic solvent under a nitrogen stream and redissolve the residue in 1–3 mL hexane. Transfer 100 µL to a GC vial and cap for analysis. The exact volumes for this step have to be determined empirically and are dictated by the sensitivity and injection settings of individual GC instruments.

7. Prepare GC vials to generate standard curves for 5α-cholestane, cholesterol, coprostanol, and lathosterol. Put appropriate amounts of standard from the ethanol stocks into GC vials, dry, and resuspend in 100 µL hexane. Sensitivity varies between instruments but 0.3–10 µg (in half-log steps) is a good initial range. Also prepare vials of the various phytosterols (see Subheading 2.8) to verify peak identification and sufficient separation between peaks to be quantitated (see **Note 5**). Determine area-under-the-curve (AUC) for all peaks of interest.

8. Filter the aqueous layer from **step 5** through Whatman paper and collect the effluent in a glass tube. Wash the original tube and filter 2–3 times with 5 mL methanol, then dry the filtrate with a nitrogen stream.

9. While the material is drying, prewash the C18 BondElut columns with 3 mL methanol followed by 3 mL water.

10. Resuspend the dried filtrates in 6 mL water and load 3 mL onto the columns by gravity flow and wash 3 times with 5 mL water. Centrifuge the columns (with plastic collection tube) for 5 min at $50g$ to remove all the water. Effluents can be discarded.

11. Elute bile acids into a fresh tube with 5 mL methanol. After all 5 mL enters the column, centrifuge as above to maximize sample recovery.

12. Evaporate solvent with nitrogen stream and resuspend in 1 mL methanol. Centrifuge 5 min at $2000g$ to remove any insoluble residue. Count 20–50 μL to determine sample recovery relative to original 0.5 g of fecal powder. Be mindful of potential color quench (see earlier).

13. Measure bile acids with a standard colorimetric kit using 5–20 μL of the solution from 12. Volumes may need to be adjusted due to interference of color or turbidity in the samples or precipitates that may form when reagents and sample are mixed.

3.2.2   Data Analysis

1. The mass of internal standards and analytes of interest in each sample are determined from the appropriate standard curves (mass vs. AUC). The total mass of excreted cholesterol and related sterols is calculated from the 5α-cholestane peak (i.s.) in each sample: sterol mass per 0.5 g feces = sterol mass$_{GC}$ × 40/i.s. mass$_{GC}$. The total excreted per 3 days is then calculated from the total fecal mass collected. Data are finally reported as mass (or moles) excreted per day per gram body weight.

2. The mass of bile acids per 0.5 g feces = mass in 1 mL methanol x dpm ^{14}C-taurocholate added/dpm recovered in 1 mL methanol. As for neutral sterols, the total excreted per 3 days is calculated from the total fecal mass collected and data are reported as mass (or moles) excreted per day per gram body weight.

3.2.3   Interpretation of Sterol Excretion/Reverse Cholesterol Transport Results

Total neutral sterol excretion by mice (male, C57BL/6) fed a chow diet is typically ~80 ± 15 nmol per day per g body weight and bile acid excretion is ~50 ± 10 nmol per day per gram body weight. On semipurified diets, which are low in fiber and reduce gut motility, neutral sterol excretion is reduced as much as 50 % and bile acid excretion can be reduced as much as 5 fold or more (unpublished results). Thus, the diet used in a given study must be chosen carefully. Since human diets have a tendency to be low in fiber, a chow diet may partially mask potential treatment benefits because sterol excretion is already quite high.

GC data should be evaluated carefully. The main sterols of interest are cholesterol and lathosterol, the latter being a late stage intermediate in cholesterol synthesis. Coprostanol is formed by the action of colonic bacteria on cholesterol and may vary considerably between mice. Peaks for coprostanol and other minor animal-derived sterols appear very close to, and may overlap with, that of cholesterol. Lathosterol is usually resolved between cholesterol and the first phytosterol peak. Neutral sterol excretion should be reported as the sum of cholesterol, its precursors, and its derivatives. Evaluation excluding precursors is also appropriate.

If bile acid excretion is different, further experiments should be done to determine if bile acid recovery by the intestine or bile acid pool size is changed. The former can initially be assessed by Western blot detection of the transporter ASBT. Pool size is measured by harvesting liver and intestine from fasted mice and processing them as above starting with 5–10 volumes of methanol + NaOH per gram of tissue. Alternatively, bile acids can be extracted by finely mincing the tissues and soaking them in 5–10 volumes of ethanol at 80 °C for 2–4 h. The resulting solution is then concentrated and assayed as in **steps 8–13** earlier. Radioactive taurocholate should be used as internal standard in either case. A change in bile acid pool size or intestinal transport is interesting mechanistically in that it indicates greater bile acid synthesis. However, the clinically relevant result is change in sterol excretion.

If sterol excretion is increased (neutral or acidic), further investigation is warranted to test for excretion of macrophage cholesterol [40] and prevention or regression of arterial lesions. Particular attention should be given to lathosterol levels in control versus treated groups. If sterol excretion is significantly elevated (>20 %) an even greater difference between groups may be detected for lathosterol, indicating that cholesterol synthesis is elevated in response to treatment. This should be confirmed by direct measurement [46]. If true, the result indicates that the drug is having a physiologically significant impact on cholesterol metabolism that may be beneficial, and that even greater benefit could be achieved by combined therapy with a statin.

4   Conclusion

CVD remains a major cause of morbidity and mortality in the US and elsewhere despite the efficacy of current therapeutic treatments. Thus, new and better targets for pharmaceutical intervention are needed in addition to modification of diet and lifestyle. This chapter has presented methods for analyzing the first and last stages of cholesterol metabolism—absorption and excretion—in part because absorption is the first line of defense against hypercholesterolemia and excretion is the ultimate measure of drug efficacy with respect to RCT. These methods highlight two processes that are receiving greater attention with respect to drug development and target identification. In addition to these assays, lipoprotein and plasma cholesterol dynamics should be routinely evaluated as part of any drug screening process. This includes fractionation of plasma to determine distribution of cholesterol among the different lipoprotein classes; studies to determine rates of turnover of HDL, LDL, and VLDL; and clearance of these particles by liver and other tissues. Key studies would be the use of appropriate animal models to test for effects on plaque deposition or regression. Several widely used methods for such analyses are described throughout the literature.

Genetically engineered and inbred mouse models have been, and remain, central tools in unraveling the complexities of cholesterol and lipoprotein metabolism, and the pathologies associated with abnormalities in these processes, resulting in identification of several realized and potential drug targets. While absorption has been well studied in many mouse models, sterol excretion as a measure of RCT received less attention until recent years. Thus, new drug targets for affecting this process, as well as absorption, may be identified by analyzing available mutants as well as new ones that will be created as these pathways continue to be investigated.

5   Notes

1. The chemical purity of these radiolabeled compounds is very high from most sources. If in question, the purity can be checked and restored by standard TLC and/or HPLC methods. ^{14}C-labeled phytosterols are less available.

2. Lipid stocks in ethanol may precipitate with time at –20 °C and must be completely redissolved at room temperature or 37 °C before use. Triglycerides (triolein, vegetable oils) are also stored at –20 °C to protect them from light and air. If oxidized (color in triolein, "stale" odor in vegetable oils), they should be discarded. All lipids should always be stored in glass, not plastic, if possible. Fresh olive, safflower, and canola oils (grocery) are equally suitable to triolein for the purpose of these methods.

3. Some protocols call for 48 or 72 h feces collection. This is not necessary since ≥ 90 % of the unabsorbed marker sterol is recovered within 24 h by mice eating a chow diet.

4. The cholestane and taurocholate serve as internal standards. They should be dispensed with a Hamilton syringe or other positive displacement device to insure accurate delivery to each sample. Let solvents evaporate before proceeding. The exact amount of taurocholate used should also be dispensed into a scintillation vial to accurately determine the number of dpm added to each sample.

5. There is a lot of plant material in standard rodent chows so peaks from plant sterols will be much larger than those of animal sterols in such samples. Chromatographs from mice fed semipurified diets are simpler and more easily analyzed.

Acknowledgments

This work was supported by NIH grants R01HL078900 and R01DK077170-ARRA.

References

1. Hui DY, Howles PN (2005) Molecular mechanisms of cholesterol absorption and transport in the intestine. Semin Cell Dev Biol 16:183–192

2. Levy E, Spahis S, Sinnett D, Peretti N, Maupas-Schwalm F et al (2007) Intestinal cholesterol transport proteins: an update and beyond. Curr Opin Lipidol 18:310–318

3. Hui DY, Labonté ED, Howles PN (2008) Development and physiological regulation of intestinal lipid absorption III. Intestinal transporters and cholesterol absorption. Am J Physiol Gastrointest Liver Physiol 294:839–843

4. Westergaard H, Dietschy JM (1976) The mechanism whereby bile acid micelles increase the rate of fatty acid and cholesterol uptake into the intestinal mucosal cells. J Clin Invest 58:97–108

5. Thurnhofer H, Hauser H (1990) Uptake of cholesterol by small intestinal brush border membrane is protein-mediated. Biochemistry 29:2142–2148

6. Borja CR, Vahouny GV, Treadwell CR (1964) Role of bile and pancreatic juice in cholesterol absorption and esterification. Am J Physiol 206:223–228

7. Gallo LL, Clark SB, Myers S, Vahouny GV (1984) Cholesterol absorption in rat intestine: role of cholesterol esterase and acyl coenzyme A:cholesterol acyl transferase. J Lipid Res 25:604–612

8. Fernandez E, Borgström B (1989) Effects of tetrahydrolipstatin, a lipase inhibitor, on absorption of fat from the intestine of the rat. Biochim Biophys Acta 1001:249–255

9. McKean ML, Commons TJ, Berens MS, Hsu PL, Ackerman DM et al (1992) Effect of inhibitors of pancreatic cholesterol ester hydrolase (PCEH) on ^{14}C-cholesterol absorption in animal models. FASEB J 6:A1388

10. Krause BR, Sliskovic DR, Anderson M, Homan R (1998) Lipid-lowering effects of WAY-121,898, an inhibitor of pancreatic cholesteryl ester hydrolase. Lipids 33:489–498

11. Howles PN, Carter CP, Hui DY (1996) Dietary free and esterified cholesterol absorption in cholesterol esterase (bile salt-stimulated lipase) gene-targeted mice. J Biol Chem 271:7196–7202

12. Weng W, Li L, van Bennekum AM, Potter SH, Harrison EH et al (1999) Intestinal absorption of dietary cholesteryl ester is decreased but retinyl ester absorption is normal in carboxyl ester lipase knockout mice. Biochemistry 38:4143–4149

13. Camarota LM, Chapman JM, Hui DY, Howles PN (2004) Carboxyl ester lipase cofractionates with scavenger receptor BI in hepatocyte lipid rafts and enhances selective uptake and hydrolysis of choelsteryl esters from HDL$_3$. J Biol Chem 279:27599–27606

14. Chaikoff IL, Bloom B, Siperstein MD, Kiyasu JY, Reinhardt WO et al (1952) C^{14}-cholesterol I: lymphatic transport of absorbed cholesterol-4-C^{14}. J Biol Chem 194:407–412

15. Heider JG, Pickens CE, Kelly LA (1983) Role of acyl CoA:cholesterol acyltransferase in cholesterol absorption and its inhibition by 57–118 in the rabbit. J Lipid Res 24:1127–1134

16. Clark SB, Tercyak AM (1984) Reduced cholesterol transmucosal transport in rats with inhibited mucosal acyl CoA:cholesterol acyltransferase and normal pancreatic function. J Lipid Res 25:148–159

17. Gallo LL, Wadsworth JA, Vahouny GV (1987) Normal cholesterol absorption in rats deficient in intestinal acyl coenzyme A:cholesterol acyltransferase activity. J Lipid Res 28:381–387

18. Buhman KK, Accad M, Novak S, Choi RS, Wong JS et al (2000) Resistance to diet-induced hypercholesterolemia and gallstone formation in ACAT2-deficient mice. Nat Med 6:1341–1347

19. Yagu H, Kitamine T, Osuga J, Tozawa R, Chen Z et al (2000) Absence of ACAT-1 attenuates atherosclerosis but causes dry eye and cutaneous xanthomatosis in mice with congenital hyperlipidemia. J Biol Chem 275:21324–21330

20. Leon C, Hill JS, Wasan KM (2005) Potential role of acyl-coenzyme A:cholesterol transferase (ACAT) inhibitors as hypolipidemic and antiatherosclerosis drugs. Pharm Res 22:1578–1588

21. Rudel LL, Lee RG, Parini P (2005) ACAT2 is a target for treatment of coronary heart disease associated with hyperchoelsterolemia. Arterioscler Thromb Vasc Biol 25:1112–1118

22. Lada AT, Davis M, Kent C, Chapman J, Tomoda H et al (2004) Identification of ACAT1- and ACAT2-specific inhibitors using a novel, cell-based fluorescence assay: individual ACAT uniqueness. J Lipid Res 45:378–386

23. VanHeek M, France CF, Compton DS, McLeon RL, Yumibe NP et al (1997) In vivo metabolism-based discovery of a potent absorption cholesterol inhibitor, SCH58235, in the rat, and rhesus monkey through the identification of the active metabolites of SCH48461. J Pharmacol Exp Therap 283:157–163

24. Altmann SW, Davis HR, Zhu L, Yao X, Hoos LM et al (2004) Niemann-Pick C1 like 1 protein is critical for intestinal cholesterol absorption. Science 303:1201–1204

25. Garcia-Calvo M, Lisnock HG, Bull BE, Hawes DA, Burnett MP et al (2005) The target of ezet-

imibe is Niemann-Pick C1-like 1 (NPC1L1). Proc Natl Acad Sci U S A 102:8132–8137

26. Ziajka PE, Reis M, Kreul S, King H (2004) Initial low-density lipoprotein response to statin therapy predicts subsequent low-density lipoprotein response to the addition of ezetimibe. Am J Cardiol 93:779–780

27. Richmond BL, Boileau AC, Zheng S, Huggins KW, Gramholm NA et al (2001) Compensatory phospholipid digestion is required for cholesterol absorption in pancreatic phospholipase A(2)-deficient mice. Gastroenterology 120: 1193–1202

28. Huggins KW, Boileau AC, Hui DY (2002) Protection against diet-induced obesity and insulin resistance in group 1B PLA$_2$ deficient mice. Am J Physiol Endocrinol Metab 283: E994–E1001

29. Homan R, Hamelehle KL (1998) Phospholipase A2 relieves phosphatidylcholine inhibition of micellar cholesterol absorption and transport by human intestinal cell line Caco-2. J Lipid Res 39:1197–1209

30. Huggins KW, Camarota LM, Howles PN, Hui DY (2003) Pancreatic triglyceride lipase deficiency minimally affects dietary fat absorption but dramatically decreases dietary cholesterol absorption in mice. J Biol Chem 278:42899–42905

31. Mittendorf B, Ostlund RE, Patterson BW, Klein S (2001) Orlistat inhibits dietary cholesterol absorption. Obes Res 9:599–604

32. Drew BS, Dixon AF, Drew JB (2007) Obesity management: update on orlistat. Vasc Health Risk Manag 3:817–821

33. Quintão E, Grundy SM, Ahrens EH (1971) An evaluation of four methods for measuring cholesterol absorption by the intestine in man. J Lipid Res 12:221–232

34. Sanders DJ, Minter HJ, Howes D, Hepburn PA (2000) The safety evaluation of phytosterol esters. Part 6. The comparative absorption and tissue distribution of phytosterols in the rat. Food Chem Toxicol 38:485–491

35. Igel M, Giesa U, Lutjohann D, von Bergmann K (2003) Comparison of the intestinal uptake of cholesterol, plant sterols, and stanols in mice. J Lipid Res 44:533–538

36. Kastelein JJ, van Leuven SI, Burgess L, Evans GW, Kuivenhoven JA, Barter PJ, Revkin JH, Grobbee DE, Riley WA, Shear CL, Duggan WT, Bots ML, RADIANCE 1 investigators

(2007) Effect of torcetrapib on carotid atherosclerosis in familial hypercholesterolemia. N Engl J Med 356:1620–1630

37. Forrester JS, Makkar R, Shah PK (2005) Increasing high-density lipoprotein cholesterol in dyslipidemia by cholesteryl ester transfer protein inhibition. Circulation 111:1847–1854

38. Tchoua U, D'Souza W, Mukhamedova N, Blum D, Niesor E, Mizrahi J, Maugeais C, Sviridov D (2008) The effect of cholesteryl ester transfer protein overexpression and inhibition on reverse cholesterol transport. Cardiovasc Res 77:732–739

39. Post SM, de Crom R, van Haperen R, van Tol A, Princen HM (2003) Increased fecal bile acid excretion in transgenic mice with elevated expression of human phospholipid transfer protein. Arterioscler Thromb Vasc Biol 23:892–897

40. Zhang YZ, Zanotti I, Reilly MP, Glick JM, Rothblat GH et al (2003) Overexpression of apolipoprotein A-I promotes reverse cholesterol transport from macrophages to feces in vivo. Circulation 108:661–663

41. Carter CP, Howles PN, Hui DY (1997) Genetic variation in cholesterol absorption efficiency among inbred strains of mice. J Nutr 127:1344–1348

42. Zilversmit DB, Hughes LB (1974) Validation of a dual-isotope plasma ratio for measurement of cholesterol absorption in rats. J Lipid Res 15:465–473

43. Millar JS, Cromley DA, McCoy MG, Rader DJ, Billheimer JT (2005) Determining hepatic triglyceride production in mice: comparison of poloxamer 407 with Triton WR-1339. J Lipid Res 46:2023–2028

44. Jandacek RJ, Heubi JE, Tso P (2004) A novel, noninvasive method for the measurement of intestinal fat absorption. Gastroenterology 127:139–144

45. LaBonté ED, Camarota LM, Rojas JC, Jandacek RJ, Gilham DE et al (2008) Reduced absorption of saturated fatty acids and resistance to diet-induced obesity and diabetes by ezetimibe-treated and *Npc1l1*$^{-/-}$ mice. Am J Physiol Gastrointest Liver Physiol 295:G776–G783

46. Osono Y, Woollett LA, Herz J, Dietschy JM (1995) Role of the low density lipoprotein receptor in the flux of cholesterol across the tissues of the mouse. J Clin Invest 95: 1124–1132

Chapter 12

Skin Diseases in Laboratory Mice: Approaches to Drug Target Identification and Efficacy Screening

John P. Sundberg, Kathleen A. Silva, Lloyd E. King Jr, and C. Herbert Pratt

Abstract

A large variety of mouse models for human skin, hair, and nail diseases are readily available from investigators and vendors worldwide. Mouse skin is a simple organ to observe lesions and their response to therapy, but identifying and monitoring the progress of treatments of mouse skin diseases can still be challenging. This chapter provides an overview on how to use the laboratory mouse as a preclinical tool to evaluate efficacy of new compounds or test potential new uses for compounds approved for use for treating an unrelated disease. Basic approaches to handling mice, applying compounds, and quantifying effects of the treatment are presented.

Key words Skin, Alopecia areata, Atopic dermatitis, Chronic proliferative dermatitis, Full thickness skin grafts, Hair, Xenograft

1 Introduction

Domestic animals have been incredibly useful for discovering novel approaches to combat major disease problems in humans for centuries. For example, the most notorious infectious disease of all time with a prominent skin lesion was small pox. Descriptions by Barron [1] remind us of the severity of the skin disease, even though the viral infection involved many other organs. Jenner's controversial work, first published in 1798, based on interspecies transmission of cow pox or, more likely, horse pox, from its natural host to human caretakers who subsequently were immune to small pox infection. This observation lead to the development of vaccines (derived from the Latin word *vaccinus*, i.e. relating to, or derived from a cow) [2–4]. Today, we use laboratory mice as a leading biomedical tool, and have amassed a variety of molecular methods to analyze mice with spontaneous or genetically engineered diseases that closely resemble human diseases. These small mammals can effectively be used for preclinical trials for new drugs, identify potential new uses for approved drugs, or using large-scale gene arrays, RNAseq and alike, and other technologies in combination with sophisticated

Gabriele Proetzel and Michael V. Wiles (eds.), *Mouse Models for Drug Discovery: Methods and Protocols*,
Methods in Molecular Biology, vol. 1438, DOI 10.1007/978-1-4939-3661-8_12, © Springer Science+Business Media New York 2016

molecular pathway analysis software to predict the best model for drug testing or drugs to test on a particular mouse model.

The mouse skin represents a somewhat unique organ system to work on. For example with the simple topical application of drugs, effects can easily be seen without special manipulation to rapidly determine if the compound is efficacious. However, with mice they can simply lick the compound off and change the intended treatment from a topical to a systemic effect. As such, education in the use of mouse skin as a model system is of the utmost importance.

There are many reviews on what constitutes good mouse models for human diseases [5,6], as well as models for specific diseases [7–15]. It is beyond the scope of this chapter to discuss the pros and cons of each of these potential models, especially since a recent scan of the public literature and databases revealed hundreds of genetically engineered and spontaneous mouse mutations that have skin disease phenotypes (Mouse Genome Informatics, MGI; http://www.informatics.jax.org/) many of which may serve as models for specific human diseases [15]. Furthermore, we will not describe the usefulness of the mouse to study wound healing, for this the reader is referred to other reviews on the topic [16–18].

In this chapter, we will focus on the use of several specific models illustrating how mouse models for cutaneous drug studies can be effectively utilized to address primary questions and methodologies which will also be applicable to most, if not all mouse skin models. Many so-called single gene-based diseases, including many of the epidermolysis bullosa blistering diseases [19], can be successfully reproduced using mouse models. Yet, even these vary due to background modifier genes [20]. However, similar modifier genes in the mouse genome [20] can be found as modifiers of the homologous human disease (Roopenian, Sproule, and Sundberg, unpublished data), such that while difficult, it is still possible to model even specific subsets of human diseases. Some complex diseases, such as psoriasis, may never be adequately mimicked by mouse models as they lack some of the key anatomic structures or response features found in human patients (Table 1) [13,21]. However, other complex genetic diseases, like alopecia areata, a cell-mediated autoimmune disease that involves the classical lymphocyte co-stimulatory cascade mechanism found in many cell-mediated autoimmune diseases, [22,23] can be modeled in mice and used for screening compounds with a variety of applications [24,25]. Furthermore, many human diseases are actually groups of very similar diseases lumped together based on similar clinical phenotypic features. Now that the molecular bases for many human skin diseases are being unraveled, it is possible to more accurately match them with the specific mouse homolog(s). These mouse models are extremely useful when common molecular targets are the focus of preclinical testing, even where these are not always easy to define.

Table 1
Formulations of commonly used fixative for mouse tissues

Tellyesniczky/Fekete' solution	100 mL 70% EtOH (undenatured, USP) 5 mL Glacial acetic acid 10 mL 37–40% Formalin
Bouin's solution	85 mL, Sat. Ag. Picric Acid 5 mL Glacial acetic acid 10 mL 37–40% Formalin
10% Neutral buffered formalin	100 mL 37–40% Formalin 900 mL dH$_2$O 4 g Sodium phosphate-monobasic 6.5 g Sodium phosphate-dibasic

If the goal is to alter proliferative, scaly skin disease, albeit psoriasis or ichthyosis, panels of mutant mice with this basic phenotype are readily available [26] (https://www.jax.org/jax-mice-and-services) and should be tested and results compared. For many of these mouse models, the specific mutated gene and genetic lesion within that gene is known. Furthermore, many of these models have transcriptome studies completed that can be used to search for specific drug targets that may be dysregulated. If the drug works in one model but not another and the gene is known, this can focus on the best homologous human disease to target [23,27]. If the target is known, for example as with filaggrin (*Flg*) or the closely linked transmembrane protein 79 (*Tmem79*), which are risk factors for atopic dermatitis [28–39], then a mouse with double mutations in *Flg* and *Tmem79*, such as the flaky tail-matted mutant mouse (*Flg*^{ft}, *Tmem79*^{ma}) can be used for testing targeted drugs [40,41].

Gene expression profiling is an approach to both identify dysregulated genes as a potential target for drug treatment and monitor changes in response to drug treatment. If the drug under investigation is known to up- or down-regulate a specific protein, then knowing which mutant mice have down- or up-regulation of transcripts coding for these proteins are logical for initial preclinical screening. Large sets of gene expression profiling data are available through public repositories such as the Gene Expression Omnibus at NIH (http://www.ncbi.nlm.nih.gov/geo/) and the ArrayExpress at EBI (http://www.ebi.ac.uk/microarray-as/ae/). Such databases can be searched for the targeted protein to find candidate mouse models.

References for websites and repositories worldwide for mice models can be found in the *Genetically Engineered Mice Handbook* [42]. Another good source of information on mouse strains is the International Mouse Strain Resource (IMSR) which is available online at MGI (http://www.findmice.org). There are also a number of repository databases that provide summaries on the mouse models and their potential uses. Traditional textbooks continue to serve a useful purpose with detailed images of normal anatomy [43] as well as gross and histologic lesions seen in mutant mice [42,44]. In addition, there are specialized website databases, such as the Mouse Tumor Biology Database (http://tumor.informatics.jax.org/mtbwi/index.do) [45–48], which focuses on cancer models, or more generalized mouse pathology databases such as the European Mouse Pathology Consortium website (http://www.pathbase.net/), whereby one can also access the more specific skin pathology database, see http://skinbase.org [49–51].

With some disease models, even after obtaining them from commercial or collaborative sources it can be difficult to continuously maintain these animals and their desired features, and may be even difficult to generate sufficient sample material for detailed analysis. This can be due to mutant mice not thriving, and/or dying at a young age or developing severe lesions making managing and maintaining a colony difficult. One approach to overcome this problem is to maintain the colony by heterozygous breeding and/or to use full thickness skin grafts, a simple surgical approach, to expand such limited mouse resources [52–54]. Heterozygous mice may be more useful if the mutation causing the problem is semi-dominant. Such mice will develop the phenotype less severely than homozygous mice. Skin grafting can functionally expand from one to many mice. For example, C3H/HeJ mice develop alopecia areata as they age. As with humans, the spontaneous disease will wax and wane [55]. One can perform full thickness skin grafts from one mouse with extensive alopecia areata (alopecia universalis) from one mouse to up to ten young recipients. As C3H/HeJ mice are highly inbred and therefore histocompatible, immunologically intact mice can be used [54,56]. This can also be done for neonatal lethal mutant mice [53] or mice that can be difficult to maintain [52]. Perhaps of greater importance, the most appropriate animal model may not be the mouse but the human. In that case, making mouse xenografts in which human skin [57,58] is grafted onto immunodeficient mice can become the model of choice [59–61]. However, this limits analysis of the immune system if such is a driver of the disease.

In this chapter, we will cover approaches to test compounds on mouse skin and skin disease models in order to identify potential anatomic, metabolic, and genetic changes that correspond to those seen in comparable human diseases. This type of information has proven very useful in identifying and repurposing FDA-approved drugs which are currently available to treat different diseases but with a similar pathogenesis basis. Quantifying changes, as well as selecting the most appropriate changes to measure can be daunting, yet it is a critical aspect of these types of studies. Here, we will show and discuss a variety of approaches and detail how these can all be done.

2   Materials

2.1   Application of Drugs

1. Gavage 20- or 22-gauge feeding tube with 2 mm ball (Instech Laboratories, Inc., Plymouth Meeting, PA).

2. Cherry syrup (can be purchased from local pharmacies anywhere).

3. Sulfatrim (Schein Pharmaceutical, Florham, NJ). The final concentration is ~1 mg/mL (10 mL Sulfatrim [240 mg/ 5 mL]/480 mL water).

4. Elizabethan collars (Harvard Apparatus, Holliston, MA).

5. Hilltop Chambers® (Hill Top Research, Miamiville, OH).

6. 3M™ Coban ™ wrap (3M ™, any supplier of surgical bandaging).

7. Wound clips (9 mm wound clips; Stoelting Co., Wood Dale, IL).

8. Nonstick pads, nonadhering dressing (Johnson & Johnson).

9. ALZET® Osmotic Pumps (Durect Corp., Cupertino, CA).

10. Silastic tubing (Dow-Corning Corp., Midland, MI).

11. Custom-made pelleted implants (Brookwood Pharmaceuticals; http://www.brookwoodpharma.com/drug-loaded-implants.html).

2.2   Quantification of Drug Response

1. Isoflurane.

2. Tribromoethanol.

3. Stick-on type rulers (crime-scene.com/ecpi/references.shtml).

2.2.1   Hair Regrowth/ Repigmentation Evaluation

4. Dermlite (San Juan Capistrano, CA) (http://www.dermlite.com/).

5. TrichoScan Software (Tricholog, Freiburg, Germany) (www.tricholog.de).

2.2.2   Histopathology

1. 10 % neutral buffered formaldehyde, Fekete's acid-alcohol-formalin, Bouin's solution, or other tissue fixative (Table 1).

2. 70 % ethanol for storage of tissues after fixation.

3. Hematoxylin/eosin stain.

4. Toluidine blue stain.

5. Bromodeoxyuridine (5-bromo-2-deoxyuridine, BrdU).

6. Cleaved caspase-3 specific antibody: cleaved caspase-3 (Asp175) antibody (Cell Signaling, cat # 9661, Danvers, MA).

7. Cleaved caspase-9 specific antibody (Novus Biologicals cat# NB100-56118, Littleton, CO).

8. SuperFrost Plus slides (Fisher Scientific, Pittsburgh, PA).

9. MembraneSlides (Zeiss, Oberkochen, Germany).

2.2.3 Scanning Electron Microscopy	1. Fixation buffer: 2.5% glutaraldehyde in 0.1 M cacodylate buffer.
	2. Nylon Mesh (Sefar Filtration, Inc., Depew, NY).
	3. Hitachi S3000N VP Scanning Electron Microscope (Hitachi Science Systems, Japan).
	4. EDAX X-ray microanalysis system (Mahwah, New Jersey).

2.2.4 Transmission Electron Microscopy

1. Extraction buffer: 0.1 mM sodium phosphate buffer (pH 7.9), 2% SDS, 10 mM dithiothreitol.

2. Karnovsky's fixative: 16% paraformaldehyde solution, 50% glutaraldehyde, 0.2 M sodium phosphate buffer.

3. 1% osmium tetroxide.

4. Dehydration solutions: Specimens are transferred through graded solvents (50–60% in distilled water) up to 100% in the solvent. Solvents such as ethanol, methanol, or acetone can be used.

5. Epoxy resin combination: Araldite, Embed 812 (Electron Microscopy Sciences, Hatfield, PA).

6. TEM JEOL JEM-1230 (JEOL).

2.2.5 Laser Capture Microdissection

1. Laser Capture Microdissection system, either Infrared- or Ultraviolet-based systems available; e.g. ArcturusXT LCM (ThermoFisher Scientific, Grand Island, NY), PALM Zeiss UV LCM (Zeiss, Oberkochen, Germany).

2. Fixative: 70% Ethanol or Formalin.

3. Staining Solutions: Cresol Violet.

4. LCM Caps.

2.3 Mice

1. Chronic proliferative dermatitis (*Sharpin^cpdm*/*Sharpin^cpdm*, JAX 007599) mouse.

2. Flaky tail, matted (*Flg^ft*, *Flg^ft*, *Tmem79^ma*/*Tmem79^ma*, JAX 00281) mouse.

3. Immunodeficient mice: e.g. NSG (JAX 00557).

---

3 Methods

3.1 Application of Drugs

There are various ways to administer drugs to rodents in order to test for efficacy and toxicity. The most common approaches are summarized below with references to provide more specific details on how to conduct these types of studies [62].

3.1.1 Oral Routes

Mice can be dosed using a gavage tube or mixing the drug with cherry syrup to allow for measured doses to be delivered to the rodents. Mice should receive sulfamethoxazole in their drinking water for 1 week after surgery, to minimize the risk of infection. Immunodeficient mice receive sulfatrim (final concentration is ~1 mg/mL (10 mL Sulfatrim [240 mg/5 mL]/480 mL water) in their drinking water to minimize complications from *Pneumocystis* spp. infections.

3.1.2 Topical Routes

Although topical applications are the most practical for human patients, it can be problematic as mice depending upon the vehicle, may lick and groom the test compounds off themselves or others in the same box.

Compounds in volatile vehicles (such as acetone) can be applied with a micropipette, allowed to spread over a marked area (site can be tattooed to ensure the compound is repeatedly applied to the same site and spread over the same unit area of skin) and allowed to dry [63]. Small amounts can be applied repeatedly allowing for evaporation between applications to maintain volume in a defined area.

Aqueous or ointment vehicles are more problematic primarily because they cannot be easily contained. We have tried a variety of stick on bubble chambers, compression bandages, and "Elizabethan collars" and rodent jackets for mice that are available from various vendors with variable results. Elizabethan collars, also known as E-collars, are made of plastic, the neck openings are lined with padding, and they close with a Velcro fastener preventing the animal grooming itself (*see* **Note 1**). Hill Top® Chambers are molded plastic chambers that are flexible and conform to the skin. Within the chamber is a pad which holds the test sample. The chambers are secured to the mouse using Coban™ wrap, or other self-adhering bandages, which is secured by 9 mm wound clips. The compounds/drugs are applied with a pipette under the bandages (*see* **Note 2**).

Bandaging the drug application site can be a simple and very effective alternative. A nonstick type pad is placed over the shaved site and a small custom cut vest-shaped elastic bandage is used to hold the pad in place [56]. The compound is applied under the bandage daily and the bandages are changed weekly and untoward effects are noted, e.g. ulcers, swelling, etc. [64].

3.1.3 Injectable Routes

Several routes of administration are available. Frequency will depend on route, volume, and tolerance of the compound, all depending upon Institutional Animal Care and Use Committee (IACUC) approval.

Intravenous Injections

The ventral coccygeal (tail) vein is a common site for intravenous injections. This is located on the ventral side of the mouse tail, running along the midline. The mouse can be placed in a restraint device (plastic box or tube) with a slot at one end the width of the tail. The mouse is allowed to enter the device while the tail is held. Warming the mouse or its tail will help increase the flow of blood.

Small-gauge (30 g) needles are used on syringes through the skin and into the vein. Albino strains are easier to venipuncture than pigmented strains as the red vessel can be seen easily more clearly through the finely haired skin of the tail. Mouse body skin usually lacks interfollicular epidermal pigmentation but this is not the case for the tail skin which also has a very thick epidermis. There are also several alternative but less commonly used sites for intravenous injections including either the femoral or jugular veins, as well as various veins that can be surgically exposed for injection [65,66].

Subcutaneous Injections

The mouse body skin is very loosely attached to the underlying fascia and musculature and can be lifted easily. A small-gauge needle can then be inserted into the tent of skin over the back and neck once it is raised.

Intradermal Injections

It is commonly believed that due to the extremely thin dermis and epidermis of the mouse body skin, injections into the "dermis" are the same as subcutaneous. This can be verified using labeled cells such as those ubiquitously expressing green fluorescent protein (GFP) and tissue sections examined directly by immunofluorescence or indirectly using a GFP-specific antibody and immunohistochemistry [67]. However, it is possible to stretch the dorsal or ventral skin and, using a very small-gauge needle, position it within the layers of the skin instead of through the skin to produce small blisters with the injected material. Recently, dissolving polymer microneedle patches, 25 mm^2 patches housing an array of approximately one hundred 650 µm long needles capable of penetrating 200 µm into the skin, have been developed that would allow for the delivery of compounds to the dermis of mice [68].

Intramuscular Injections

Generally, intramuscular injections are made with small-gauge needles into the epaxial muscles on either side of the lumbar vertebral column or the quadriceps femoris muscle on the ventral side of either rear leg. Due to the small size of mice, very small volumes (e.g. 0.05 mL) should be injected.

Intralesional Injections

Small-gauge needles can be used on syringes to inject drugs into a neoplasm or other raised abnormality affecting a defined area of the skin. The volume used will depend upon the size and number of lesions being treated and what the Institutional Animal Care and Use Committee will allow.

Intraperitoneal Injections

The mouse is manually restrained with the head and body tilted downward. A small-gauge needle is inserted into the caudal left abdominal quarter, thereby avoiding injection into the cecum on the right side.

Retro-Orbital Injections

The mouse is anesthetized and placed in the lateral recumbence position. Gentle downward pressure is applied to skin dorsal and ventral to the eye. A small-gauge needle is placed through the conjunctiva at the medial canthus to the base of the eye and a small volume (≥ 150 µL) is injected.

Unusual Injections

Although rarely used, protocols are available for intracardiac, intrathecal, intrathoracic, and intrarectal injections [65,66].

3.1.4   Osmotic Pumps

ALZET® Osmotic Pumps are miniature, infusion pumps for the continuous dosing of drugs to mice. These minipumps can be surgically implanted intraperitoneally or subcutaneously to transport drugs from 1 day to 6 weeks. Longer durations may be accomplished through serial implantations of pumps. These minipumps provide a convenient and reliable method for controlled agent delivery *in vivo*. The doses are constant and accurate and the variables are minimized thereby producing constant results. By using the minipumps the stress and handling is reduced for the mice and this is especially true where daily doses would be required.

3.1.5   Slow-Release Subcutaneous Implants

Silastic capsules, consisting of 10 mm silastic tubing packed with drug or hormone with sealed ends (insertion of 3 mm glass beads) can be implanted through a small incision over the dorsal thoracic midline into the subcutaneous fat [69,70]. These capsules enable a slow release of encapsulated drug the rate of which is dependent upon formulation. A number of companies will formulate drugs into various pelleted implants and these companies can be found online (e.g. SurModics, http://www.surmodics.com/).

3.2   Quantification of Drug Response

Determination of efficacy of a drug in the mouse models appears to be superficially simple, i.e. the treatment leads to the mouse resolving the abnormal clinical phenotype and does not die due to the treatment. However, to test this unequivocally or to determine a dose–response curve is often, in practice, technically difficult. Although obvious, it is of key importance that the specific goal/s of the study be clearly defined. With skin this can be as simple as hair growth promotion, or even just hair growth on a bald mouse. With wound healing, resolution of scaly skin diseases, etc. needs to be quantified to various degrees. Also it is necessary that one understands the biologic and anatomic differences between mice and humans to appreciate the drug effects (Table 2). This begins with the fact that mice have a prolonged telogen and humans have a prolonged anagen hair cycle, their hair cycles in a wave pattern instead of the human mosaic pattern. Also mouse truncal skin lacks interfollicular pigmentation that is seen in human epidermis. This translates to the fact that mice have pink skin naturally [71], and observed pigment is only seen in actively growing, late anagen stage hair follicles where the hair bulb and shaft are heavily

Table 2
Differences between human and mouse skin

Criteria	Human	Mouse
Hair cycle pattern	Mosaic pattern	Wave pattern
Predominant cycle stage	Anagen	Telogen
Hair follicle size	Large	Small
Hair types: head/trunk	Terminal and vellus	Guard, auchene, awl, and zigzag
Unique hair types	No homologous structure	Vibrissae
Relative hair density	Low	High
Epidermal thickness	High	Low
Rete ridges	Prominent	Nonexistent
Interfollicular epidermal pigmentation	Yes	No
Hair bulb/shaft pigmentation	Yes	Yes

pigmented (Fig. 1). These observations are well-known criteria for monitoring induction and patterning of the hair cycle in pigmented mice. Scoring systems by grey tone intensity are used to monitor hair cycle in the commonly used production strains, C57BL/6J (black) and C3H/HeJ (agouti) for these types of studies [72]. Changing in color from unpigmented (pink) to pigmented (grey to black) indicates hair growth, onset of anagen, and therefore initiation of the hair cycle, which may be the goal of the study.

3.2.1 Hair Regrowth/ Repigmentation Evaluation

While a number of magnification and photography tools are available to dermatologists to visualize and record changes in human skin, these can be more problematic in mice because their relatively small body makes finding flat sites difficult to find, especially when they are very active. To avoid, minimize, or altogether circumvent these issues, a number of approaches can be taken.

Gross Photography

How does one stop a very active mouse from moving? This can be simply done at the end of a study by euthanasia using Institutional Animal Care and Use Committee (IACUC)-approved protocols. Protocols recommended by the American Veterinary Medical Association are available and are regularly reviewed and revised [73] (https://www.avma.org/KB/Policies/Documents/euthanasia.pdf). Currently, the most commonly used and IACUC-approved euthanasia method is carbon dioxide gas asphyxiation. This has the advantage that the mouse dies quickly and a complete necropsy of all tissues can be performed.

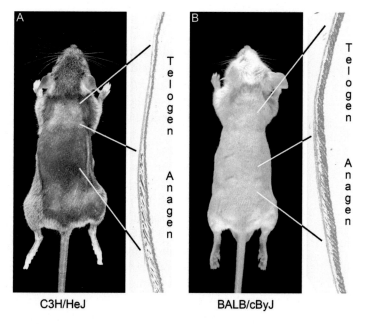

Fig. 1 Hair cycle-related skin pigmentation. Hair cycle stage can be estimated by skin color in pigmented (a, C3H/HeJ) but not lightly pigmented or albino (b, BALB/cByJ) mice. Note the light-colored (in live mice this is *pink* vs. *grey* to *black*) skin correlates histologically with telogen while dark skin is due to hair follicles in late anagen. Unlike humans most of the skin pigmentation in mice is in the hair follicles during the actively growing, anagen stage

Alternative survival methods include anesthesia by inhalation (isoflurane) or injectable (tribromoethanol) anesthetics which can be used to temporarily immobilize the mice for examination and/or photography [74,75]. While isoflurane is commonly used for repeated anesthesias, tribromoethanol is contraindicated [74]. Another approach for partial temporary immobilization of mice includes the use of a 50–50 % O_2/CO_2 mixture in a sealed container (*see* **Note 3**). In an otherwise transparent restraint device mice insert their heads into a black concave area when exposed to bright lights [76]. The side walls of this device are adjustable to keep the mice from moving laterally. The mice are also naturally frightened by bright lights, e.g. flood lamps and will stand still without any anesthesia. Of course this last method will not work in strains that are blind, which is a major issue since many of the commonly used inbred strains carrying mutations causing skin diseases, such as C3H/HeJ and FVB/NJ, also carry the retinal degeneration 1 (e.g. *Pde6b^{rd1}*, phosphodiesterase 6B, cGMP, rod receptor, beta polypeptide) mutation [77]. One surprise with the decision to use C57BL6/N substrains for the international knockout mouse project was that, unlike C57BL/6J mice, these mice carried another single gene mutation (in the *Crb1^{rd8}* gene) that is linked with blindness. Some participants in the project corrected this by genetic engineering [78], emphasizing the importance of knowing the peculiarities of not only the strain but the substrain used. Another approach is to evaluate efficacy on the ventral abdomen of the mice. It is common practice to handle and restrain mice by picking them up by pinching lightly the skin behind their ears and grabbing the tail with the small finger of the same hand. This restrains, immobilizes, and stretches the ventral skin so a photograph can be taken by an assistant.

Many journals prefer to see photographs of live, unrestrained mice and many institutions are implementing policies that discourage the publication of pictures depicting euthanized animals. This can also be accomplished by photographing groups of mice unrestrained in their cages. Individual mice can be placed on the bottom of inverted beakers with a colored background (cloth or art paper).

For whole mouse photographs, a high-quality single lens reflex type digital camera provides ease of use while generating high-quality photographs. A regular 50–55 mm macro lens is adequate for whole body images but a 100 mm macro lens allows closer evaluation of the skin surface. Repeat photographs at regular time intervals can be taken of the same area if the skin is tattooed at the start of the project. A fixed ruler should be placed in the field at the same height as the area of skin being photographed as a fixed internal standard for comparison and morphometric analyses. The animal identification number and date can be written on the ruler, especially if disposable, stick-on type rulers are used. Many modern cameras permit data to be added directly to the images, including date and time (*see* **Note 4**). Several companies sell devices that are designed specifically to photograph human skin in a narrow field with or without magnification. These may have associated software to enable specific types of quantitative analyses to be done in a standardized manner. Dermlite has been used for evaluating melanomas and other skin lesions in humans and has the potential for fine analysis of drug response in a variety of mouse skin disease models [79–81]. This unit attaches to a hand-held camera. Tricholog uses a small cylindrical camera that lays directly on the skin surface but the image is visualized and focused when attached by a cable to a portable computer. The TrichoScan software allows counting and measuring hair shaft size.

3.2.2 Histopathology

Histopathology of mouse skin is preferably done by a board-certified veterinary anatomic pathologist with experience in interpreting mouse skin. There are many examples of mistakes when investigators use "do-it-yourself-pathology" [82]. Especial attention needs to be exercised as misinterpretations of mouse specialized organs that humans do not have occur; for example the preputial and clitoral glands which are modified sebaceous glands in the genital regions, are often misdiagnosed as sebaceous gland tumors [83] or even teratomas [84, 85]. For pathologists experienced with a specific model, normality vs. disease can usually be quickly seen. This is particularly important when strain-specific background diseases confuse the interpretation of results, especially for the skin diseases [86, 87]. These evaluations can be done on routinely fixed and processed hematoxylin and eosin (H&E)-stained paraffin sections (5–6 μm). Quantification or semi-quantitative data on specific structures are often useful to assess drug responses and may need to be developed. Many pathologists use very simple but effective methods based on disease severity. The commonly used adjectives can also have numbers added (normal, 0; mild, 1; moderate, 2; severe, 3; and extreme, 4). Once calibrated or defined, e.g. by a set of images, these can be used effectively when multiple criteria are used to generate a score per criteria as well as a total score. Data are summarized in spreadsheets for all criteria and can be sorted quickly for analysis. This approach can be used to evaluate all organ systems under review [87–90].

Valid quantitative scores can be generated by counting the number of mitotic figures, specific cell types (e.g. mast cells using a toluidine blue stain), per high power field or whatever microscopic field is most appropriate. The area of the field can be obtained easily by measuring the diameter with a micrometer ($A = \pi r^2$). Image analysis programs, such as NIH Image (http://rsb.info.nih.gov/nih-image/), ImageJ (http://rsb.info.nih.gov/ij/), or a calibrated ocular micrometer can be used to rapidly generate quantitative data. It is critical to only choose areas of the slide where the entire lengths of hair follicles are present to serve as an internal standard for orientation. This is extremely important for the epidermis because it is very thin in normal mice. Routine measurements can include the interfollicular epidermal thickness (and each layer, if appropriate), length of the hair follicle, dermal thickness, hypodermal fat layer thickness which normally varies with each hair cycle [26,91,92] (Figs. 2, 3).

Proliferation (mitotic) rates or cell death (apoptosis of keratinocytes in the epidermis) are other useful criteria in looking at compound effects on skin. Both can be evaluated using a simple H&E-stained paraffin section (Fig. 3). Mitotic figures, one criterion for proliferation rate, can be seen in the basal cell layer. We found that hematoxylin alone provides optimal contrast to visualize mitotic figures. In traditional H&E-stained slides of apoptotic keratinocytes are brightly eosinophilic with dark blue to black shrunken nuclei (pycnotic or karyorrhectic nuclei). Proliferation rates can be quantified by injecting mice with 50 μg/g body weight of bromodeoxyuridine (BrdU) intraperitoneally 2 h before necropsy (time interval is critical and can vary based on age and organ under investigation; i.e. normal cell division rates in the tissue under investigation) and then labeling the tissues with an antibody against BrdU by immunohistochemistry or immunofluorescence (Fig. 4). The numbers of positive cells (which are nuclear labeled

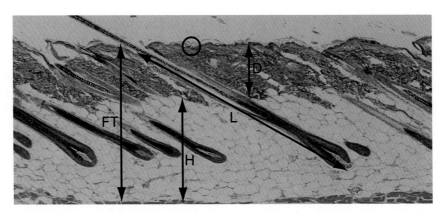

Fig. 2 Quantitative measurements of skin. Routine measurements of skin, in late anagen in this figure, include interfollicular epidermal thickness (*circled area*), hair follicle length (L), dermal thickness (D), hypodermal fat layer thickness (H, normally varies with the hair cycle), and full thickness (FT) from the surface of the stratum corneum to the top of the panniculus carnosus muscle

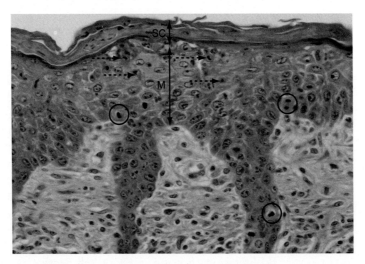

Fig. 3 Epidermal measurements, mitotic figures, and apoptotic keratinocytes in a chronic proliferative dermatitis mutant (*Sharpin^{cpdm}/Sharpin^{cpdm}*) mouse. Routine H&E-stained paraffin histologic sections can be used to determine proliferation rates based on mitotic index (number of mitotic figures, *circled* in the figure, in the stratum basale per 1000 cells) or the presence and numbers of apoptotic epidermal keratinocytes (*dotted arrows*) when present. Epidermal thickness can be measured at high dry magnification (40×) to include the Malphigian, living cell, layer (M), the stratum corneum thickness (SC), or the full thickness of the epidermis (M + SC)

by BrdU incorporation during the S phase of the cell cycle) per millimeter of skin, per 1000 basal cells, or other quantitative criteria can be used [93]. Alternatively, activated caspase-3 or -9 specific antibodies can be used to stain and identify cells undergoing apoptosis [94] (Fig. 5). Lastly, terminal deoxynucleotidyl transferase dUTP nick end labeling (TUNEL) is another approach to evaluate apoptosis, but this method is not specific to apoptosis and can also indicate cells undergoing necrosis.

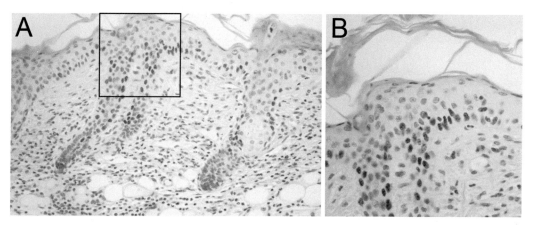

Fig. 4 Bromodeoxyuridine (BRDU) labeling of cells in "S" phase. Routine paraffin sections from mice injected 1 h prior to euthanasia with BRDU have brown nuclei when labeled by immunohistochemistry using diaminobenzidine as a chromogen. These cells were synthesizing DNA at the time of necropsy and therefore incorporated the label. Counting the number of positive nuclei in basal cells per 1000 basal cells yields the proliferation rate. The *boxed area* in (**a**) is enlarged in (**b**) to illustrate the large number of positive (*dark*) nuclei in the skin of an adult chronic proliferative dermatitis mutant mouse (*Sharpin^{cpdm}/Sharpin^{cpdm}*)

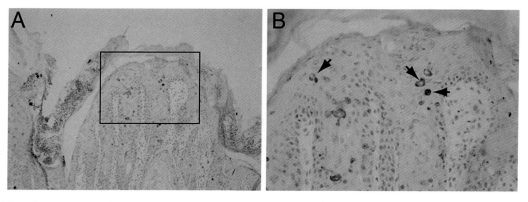

Fig. 5 Determination of apoptosis. Apoptotic keratinocytes can be confirmed using cleaved caspase-3 specific antibodies using immunohistochemistry. The *boxed area* in (**a**) is enlarged in (**b**) to illustrate the large number of positive (*dark*) cells (*arrows*) in the skin of an adult chronic proliferative dermatitis mutant mouse (*Sharpin^{cpdm}/Sharpin^{cpdm}*)

The hair cycle can be roughly graded at the gross level in shaved or alopecic mice by mapping color changes in the skin of pigmented mice from pink (follicles in telogen), to increasing darkening of areas of skin as follicles proceed into the later stages of anagen. The reason for this is described above as illustrated (Figs. 1, 6). At the microscopic level this can be done in two ways. The major hair cycle stages (anagen, catagen, or telogen) can be estimated (as a percentage of the total skin section) to demonstrate major shifts associated with a mutant mouse phenotype or drug treatment [95]. Alternatively, if

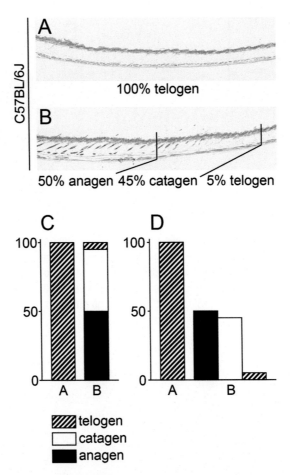

Fig. 6 Estimating changes in hair cycle. A simple means to estimate changes in hair cycle in a semi-quantitative manner is to estimate the percent of hair follicles in anagen, catagen, or telogen in histologic sections, entering the data into an Excel type spreadsheet, and graph results. The section of skin from the mouse in (**a**) is all in telogen. By contrast, the second section (**b**) is approximately 50 % in late anagen, 45 % in catagen, and 5 % in telogen. These data can be graphically presented in various types of bar graphs (**c, d**)

very subtle differences between the test and control groups need to be identified, scoring hairs based on the finely divided grading of all stages and sub-stages of the hair cycle is a complicated but functional approach. This system, developed by the Paus laboratory [96], divides anagen into six stages, catagen into eight stages, and telogen into one stage (Table 3). Although exogen as a stage has been discussed, available defining histologic criteria for the purpose of scoring are very limited [97–100].

All of the types of collected data listed above can be stored in databases for summary and analysis. One such database, the Mouse Disease Information System (MoDIS), is freely available online (https://www.jax.org/research-and-faculty/tools/mouse-disease-information-system). It will automatically spell organ names and diseases processes as word strings are typed which helps

Table 3
Stages of the mouse hair cycle [96, 97]

Anagen	Catagen	Telogen	Exogen
I	I	One stage although more complicated based on hypodermal fat thickness	Not defined anatomically
II	II		
IIIa, b, c	III		
IV	IV		
V	V		
VI	VI		
	VII		
	VIII		

in maintaining nomenclature. This database allows the recording of semi-quantitative histologic scores and can be modified easily for specific tasks. More importantly, when used online and linked to Pathbase (http://www.pathbase.net/), it is possible to confirm the definition of terms used and search for photomicrographs to serve as a virtual second opinion [89,90].

3.2.3 Scanning Electron Microscopy

To evaluate structural integrity of the hair surfaces, scanning electron microscopy (SEM) provides a three-dimensional view of the structure (Fig. 7). The phenotype of many mouse mutations affecting the skin and hair can be due to marked abnormalities of the hair shaft structure [9]. Resolution of these abnormalities can best be seen using SEM. A concurrent qualitative or semi-quantitative follow up can be done, when the SEM is appropriately equipped to perform element analysis on the hair shafts. Sulfur levels are higher in hair shafts and nails than epidermis of the skin due to the presence of the hard (hair and nail) keratins and keratin-associated proteins which are often proteins containing high to ultrahigh sulfur containing amino acids [101]. Hair defects are often due to low sulfur levels, e.g. forms of trichothiodystrophy. Therefore, response to treatment can be validated for hair structural mutants by demonstrating normal hair shafts with normal levels of sulfur.

To perform these assays, it is necessary to collect mouse hair samples. This is easily done from adults where hair follicles are in prolonged telogen and exogen stage. Hair shafts are easily removed without pain to the mouse simply by plucking. Because of the hair cycle stage, it is rare that damage occurs to the shafts no matter how defective they are. However, as human hair is usually in anagen and tightly attached to the scalp, human hairs have to be removed with hemostats to grip the hair which can damage the

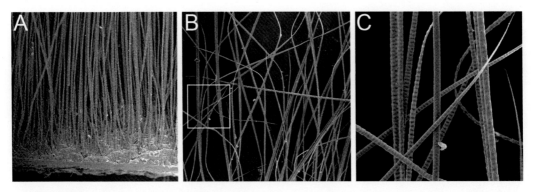

Fig. 7 Scanning electron microscopy reveals details of hair shafts. Normal hairs from an adult C57BL/6J examined as a whole mount (**a**) illustrates density of mouse hairs and the nature of the normal skin surface. Manually plucked hairs illustrate the structural differences between some of the hair shaft types (**b**). Higher magnification of *boxed area* in (**b**) reveals the regular cuticular scale patterns on these hair shafts (**c**). These approaches illustrate details of hair shaft structure and density

hair shafts. As this raises concerns with physicians, an alternative approach utilizes 1 cm^2 biopsies removed from the anatomic site of interest, laid flat on a firm nylon mesh, and fixed routinely in a glutaraldehyde-based fixative. Hair should be removed from the same location on each mouse and the same hair shaft types should be compared at the same location on each hair within a study.

We routinely examine hairs from the dorsal, interscapular region of the trunk. We found sulfur level variations in normal and mutant hairs from three commonly used inbred strains along the length of the hairs and by hair shaft types not significantly detected in normal mice hairs [102].

For SEM, hair shafts are mounted with double-stick tape on aluminum stubs, sputter-coated with a 4 nm layer of gold, and examined at 20 kV at a working distance of approximately 15 mm in our hands on a Hitachi S3000N VP Scanning Electron Microscope [103].

For sulfur content assessment by weight of hair shafts, we suggest using an EDAX X-ray microanalysis system. Samples should be examined for an average of at least 300 live seconds to ensure a comprehensive reading is obtained [101].

3.2.4 Transmission Electron Microscopy

To evaluate the cellular structure within hair shafts, the shafts can be processed for transmission electron microscopy (TEM). Due to hair's hard consistency in plastic blocks, evaluation may be impossible as the hairs are easily lost during sectioning. To avoid this, hair shafts must first be extracted before sectioning:

1. Ten or more hair shafts per mouse are incubated in 2 mL Extraction buffer at room temperature.

2. When the hair became swollen (2–2.5 h), in comparison to a sample incubated in parallel without dithiothreitol, it is immersed in Karnovsky's fixative.

3. Hair is post-fixed with osmium tetroxide overnight at 4 °C.

4. Hair is dehydrated in graded series of ethanol.

5. Hair is embedded in an epoxy resin combination.

6. Blocks are orientated visually to produce longitudinal and cross sections of the hair shafts.

7. 80 nm sections are prepared.

8. Sections are then examined by a high-performance high-contrast TEM (JEOL JEM-1230) [102,104–106].

These TEM studies can be further evaluated by proteomic analysis of hairs to determine specific changes due to various treatments [104–106].

3.2.5 Laser Capture Microdissection

To evaluate single cell or isolated structural effects in tissue at the protein, RNA, and DNA level laser capture microdissection (LCM) can be utilized on tissue sections. LCM allows for fixed, sectioned, and stained samples to be examined under a microscope and have either single cells or small homogenous regions of tissue dissected from the larger sample for analysis [107].

Tissue samples must be fixed though a variety of fixatives with best fit for the following staining protocol and end product, e.g. DNA, RNA, lipid or protein. Commonly 70 % ethanol fixation yields better results than most other fixatives but alternatives such as Tellyesniczky/Fekete (Telly's) Fixative, Formalin, or Bouin's fixative may be used [107–110]. In general tissue samples are sectioned before subsequent staining. Frozen sections can be sectioned using a cryostat, while paraffin-embedded sections may be sectioned on a standard microtome at 5 µm. Sections can be placed on SuperFrost + slides, slides coated with poly-L-lysine, or membrane-coated slides (MembraneSlide).

Tissue sections may then be stained either by standard histologic stains, such as Hematoxylin or Cresyl Violet, or immunohistologically labeled using antibodies specific to a target of interest. Staining of the sections allows easy visualization of the region of interest while dissecting samples [107–110].

The sections may now be placed on the laser capture microdissection platform. There are two dominant platforms on the market that utilize either Infrared (ArcturusXT, Applied Biosystems) or Ultraviolet (PALM MicroBeam, Carl Zeiss Microscopy) lasers to microdissect tissue sections. The infrared platform uses a cap placed over the region of interest. A transfer film is bonded to the top of the cap. After determining the cells and regions to be dissected, the infrared laser interacts with the transfer film to spot-weld the film to the tissue of interest and lifts the dissected material away from the remainder of the slide. The ultraviolet platform also uses a cap, but instead of physically interacting with the tissue the cap is only used as a trap. The ultraviolet laser is used to cut and then float the dissected tissue away from the original section and into the waiting cap.

Lastly, tissue from either platform can be used to extract DNA, RNA, or protein depending on the type of assay needed.

3.3 Alopecia Areata

Alopecia areata is a relatively common autoimmune skin disease that affects humans, mice, rats, horses, dogs, and cattle. There is even a feather form of the disease in chickens [11,25,111–113]. It occurs spontaneously however, mice have a low frequency of this disease. Full-thickness skin grafts from affected C3H/HeJ mice provided a reproducible and predictable model of alopecia areata [54,56,114]. This mouse model has been used to test drugs effective for human alopecia areata applying all methods discussed in this chapter [24].

3.4 Chronic Proliferative Dermatitis

The chronic proliferative dermatitis mutant mouse is one of a number of mouse models proposed for psoriasis. This spontaneous mutant was recently shown to be caused by a mutation in the *Sharpin* gene [115]. This mouse model was used to screen recombinant human cytokines in which recombinant interleukin 12 (IL12) but not interleukin 11 (IL11) effectively corrected the skin disease [27].

4 Notes

1. Elizabethan collars and rodent jackets have not been found to be very effective. In our hands, all mice managed to slip out of them within a 24 h period making these collars and jackets unreliable.

2. Bandages and chambers in our studies were on for 5 days and off for 2 days to allow for any irritation or inflammation of the skin caused by the wound clip to heal.

3. The mice do rapidly recover from these treatments, so the photographer needs to be ready and fast when the mouse is removed from the container.

4. Remember to keep individual identification information (labels with rulers, case number, etc.) off the actual site of interest. This way, the data markers can be cropped out of the images used in the final report.

Acknowledgements

This work was supported by grants from the National Institutes of Health (R01 AR049288, R01 AR056635, R01 CA089713, R01

AR055225, R21 AR063781), Cicatricial Alopecia Research Foundation (CARF), DEBRA International for development of the junctional epidermolysis bullosa mouse model, and The National Alopecia Areata Foundation (NAAF) for support of the alopecia areata mouse model development. The Jackson Laboratory Shared Scientific Services were supported in part by a Basic Cancer Center Core Grant from the National Cancer Institute (CA34196, to The Jackson Laboratory).

References

1. Barron J (1838) Sketch of the history of variola, and variolous inoculation. In: The life of Edward Jenner, with illustrations of his doctrines and selections from his correspondence, vol 1. Henry Colburn, London, pp 217–235

2. Jenner E (1798) An inquiry into the causes and effects of the variole vaccinae, a disease discovered in some of the western counties of England, particularly Gloucestershire and known by the name of the cow-pox. Sampson Low, London

3. Jenner E (1959) An inquiry into the causes and effects of the variola vaccinae, a disease discovered in some of the western countries of England, particularly Gloucestershire, and known by the name of cowpox. In: Camac LNB (ed) Classics of medicine and surgery. Dover, New York, pp 213–240

4. Stedman TL (2005) Stedman's medical dictionary for the health professions and nursing, 5th edn. Lippincott, Williams and Wilkins, London

5. Sundberg JP (1991) Mouse mutations: animal models and biomedical tools. Lab Anim 20:40–49

6. Sundberg JP (1993) Conceptual evaluation of animal models as tools for the study of diseases in other species. Lab Anim 21:48–51

7. Sundberg JP, Beamer WG, Shultz LD, Dunstan RW (1990) Inherited mouse mutations as models of human adnexal, cornification, and papulosquamous dermatoses. J Invest Dermatol 95:62s–63s

8. Sundberg JP, Shultz LD (1991) Inherited mouse mutations: models for the study of alopecia. J Invest Dermatol 96:95S–96S

9. Sundberg JP (1994) Handbook of mouse mutations with skin and hair abnormalities. Animal models and biomedical tools. CRC Press, Inc, Boca Raton, Florida

10. Sundberg JP, Vallee CM, King LE (1994) Alopecia areata in aging C3H/HeJ mice. In: Sundberg JP (ed) Handbook of mouse mutations with skin and hair abnormalities: animal models and biomedical tools. CRC Press, Boca Raton, pp 499–505

11. Sundberg JP, Oliver RF, McElwee KJ, King LE (1995) Alopecia areata in humans and other mammalian species. J Invest Dermatol 104:32s–33s

12. Sundberg JP, HogenEsch H, King LE (1995) Mouse models for scaly skin diseases. In: Maibach HI (ed) Dermatologic research techniques. CRC Press, Inc, Boca Raton, pp 61–89

13. Gudjonsson JE, Johnston A, Dyson M, Valdimarsson H, Elder JT (2007) Mouse models of psoriasis. J Invest Dermatol 127(6): 1292–1308

14. Plikus MV, Sundberg JP, Chuong C-M (2007) Mouse skin ectodermal organs. In: Fox JG, Davisson MT, Quimby FW, Barthold SW, Newcomer CE, Smith AL (eds) The mouse in biomedical research, 2nd edn. Elsevier, Amsterdam, pp 691–730

15. Nakamura M, Schneider MR, Schmidt-Ullrich R, Paus R (2013) Mutant laboratory mice with abnormalities in hair follicle morphogenesis, cycling, and/or structure: an update. J Dermatol Sci 69(1):6–29

16. Pence BD, Woods JA (2014) Exercise, obesity, and cutaneous wound healing: evidence from rodent and human studies. Adv Wound Care (New Rochelle) 3(1):71–79

17. Finnson KW, Arany PR, Philip A (2013) Transforming growth factor beta signaling in cutaneous wound healing: lessons learned from animal studies. Adv Wound Care (New Rochelle) 2(5):225–237

18. Abdullahi A, Amini-Nik S, Jeschke MG (2014) Animal models in burn research. Cell Mol Life Sci 71(17):3241–3255

19. Fine JD, Eady RA, Bauer EA, Bauer JW, Bruckner-Tuderman L, Heagerty H et al (2008) The classification of inherited epidermolysis bullosa (EB): Report of the Third International Consensus Meeting on Diagnosis and Classification of EB. J Am Acad Dermatol 58(6):931–950

20. Sproule TJ, Bubier JA, Grandi FC, Sun VZ, Philip VM, McPhee CG, Adkins EB, Sundberg

JP, Roopenian DC (2014) Molecular identification of collagen 17a1 as a major genetic modifier of laminin mutation induced junctional epidermolysis bullosa in mice. PLoS Genet 10(2):e1004068. doi:10.1371/journal. pgen.1004068

21. Schon MP (2008) Animal models of psoriasis: a critical appraisal. Exp Dermatol 17(8): 703–712

22. Carroll J, McElwee KJ, King LE, Byrne MC, Sundberg JP (2002) Gene array profiling and immunomodulation studies define a cell mediated immune response underlying the pathogenesis of alopecia areata in a mouse model and humans. J Invest Dermatol 119:392–402

23. Xing L, Dai Z, Jabbari A, Cerise JE, Higgins CA, Gong W, Ad J, Harel S, DeStefano GM, Rothman L, Singh P, Petukhova L, Mackay-Wiggan J, Christiano AM, Clynes R (2014) Alopecia areata is driven by cytotoxic T lymphocytes and is reversed by JAK inhibition. Nat Med 20(9):1043–1049

24. Sun J, Silva KA, McElwee KJ, King LE, Sundberg JP (2008) The C3H/HeJ mouse and DEBR rat models for alopecia areata: preclinical drug screening tools. Exp Dermatol 17:793–805

25. King LE, McElwee KJ, Sundberg JP (2008) Alopecia areata. In: Nickoloff BJ, Nestle FO (eds) Dermatologic immunity: current directions in autoimmunity, vol 10. Karger, Basel, pp 280–312

26. Sundberg JP, King LE (2000) Skin and its appendages: normal anatomy and pathology of spontaneous, transgenic and targeted mouse mutations. In: Ward JM, Mahler JF, Maronpot RR, Sundberg JP (eds) Pathology of genetically engineered mice. Iowa State University Press, Ames, pp 181–213

27. HogenEsch H, Torregrossa SE, Boggess D, Sundberg BA, Carroll J, Sundberg JP (2001) Increased expression of type 2 cytokines in chronic proliferative dermatitis (*cpdm*) mutant mice and resolution of inflammation following treatment with IL-12. Eur J Immunol 31:734–742

28. Palmer CNA, Irvine AD, Terron-Kwiatkowski A, Zhao Y, Liao H, Lee SP, Goudie DR et al (2006) Common loss-of-function variants of the epidermal barrier protein filaggrin are a major predisposing factor for atopic dermatitis. Nat Genet 38(4):441–446

29. Palmer CN, Ismail T, Lee SP, Terron-Kwiatkowski A, Zhao Y, Liao H et al (2007) Filaggrin null mutations are associated with increased asthma severity in children and young adults. J Allergy Clin Immunol 120(1):64–68

30. Nomura T, Sandilands A, Akiyama M, Liao H, Evans AT, Sakai K, Ota M, Sugiura H, Yamamoto K, Sato H, Palmer CN, Smith FJ, McLean WH, Shimizu H (2007) Unique mutations in the filaggrin gene in Japanese patients with ichthyosis vulgaris and atopic dermatitis. J Allergy Clin Immunol 119(2):434–440

31. Zhao Y, Terron-Kwiatkowski A, Liao H, Lee SP, Allen MH, Hull PR et al (2007) Filaggrin null alleles are not associated with psoriasis. J Invest Dermatol 127(8):1878–1882

32. Sandilands A, Terron-Kwiatkowski A, Hull PR, O'Regan GM, Clayton TH, Watson RM et al (2007) Comprehensive analysis of the gene encoding filaggrin uncovers prevalent and rare mutations in ichthyosis vulgaris and atopic eczema. Nat Genet 39(5):650–654

33. Sandilands A, Smith FJ, Irvine AD, McLean WH (2007) Filaggrin's fuller figure: a glimpse into the genetic architecture of atopic dermatitis. J Invest Dermatol 127(6):1282–1284

34. Morar N, Cookson WO, Harper JI, Moffatt MF (2007) Filaggrin mutations in children with severe atopic dermatitis. J Invest Dermatol 127(7):1667–1672

35. McGrath JA, Uitto J (2008) The filaggrin story: novel insights into skin-barrier function and disease. Trends Mol Med 14(1):20–27

36. Hamada T, Sandilands A, Fukuda S, Sakaguchi S, Ohyama B, Yasumoto S et al (2008) De novo occurrence of the filaggrin mutation p.R501X with prevalent mutation c.3321delA in a Japanese family with ichthyosis vulgaris complicated by atopic dermatitis. J Invest Dermatol 128(5):1323–1325

37. Betz RC, Pforr J, Flaquer A, Redler S, Hanneken S, Eigelshoven S et al (2007) Loss-of-function mutations in the filaggrin gene and alopecia areata: strong risk factor for a severe course of disease in patients comorbid for atopic disease. J Invest Dermatol 127(11): 2539–2543

38. Saunders SP, Goh CS, Brown SJ, Palmer CN, Porter RM, Cole C et al (2013) Tmem79/Matt is the matted mouse gene and is a predisposing gene for atopic dermatitis in human subjects. J Allergy Clin Immunol 132(5): 1121–1129

39. Sasaki T, Shiohama A, Kubo A, Kawasaki H, Ishida-Yamamoto A, Yamada T et al (2013) A homozygous nonsense mutation in the gene for Tmem79, a component for the lamellar granule secretory system, produces spontaneous eczema in an experimental model of atopic dermatitis. J Allergy Clin Immunol 132(5):1111–1120

40. Presland RB, Boggess D, Lewis SP, Hull C, Fleckman P, Sundberg JP (2000) Loss of

normal profilaggrin and filaggrin in flaky tail (*ft/ft*) mice: an animal model for the filaggrin-deficient skin disease ichthyosis vulgaris. J Invest Dermatol 115:1072–1081

41. Fallon PG, Sasaki T, Sandilands A, Campbell LE, Saunders SP, Mangan NE et al (2009) A homozygous frameshift mutation in the mouse Flg gene facilitates enhanced percutaneous allergen priming. Nat Genet 41(5):602–608

42. Sundberg JP, Ichiki T (2005) Genetically engineered mice handbook. Biomethods for laboratory mice. CRC Press, Boca Raton

43. Treuting P, Dintzis S, Frevert CW, Liggitt D, Montine KS (2012) Comparative anatomy and histology. A mouse and human atlas. Academic Press, Amsterdam

44. Ward JM, Mahler J, Maronpot R, Sundberg JP (2000) Pathology of genetically engineered mice. Iowa State University Press, Ames

45. Krupke D, Begley D, Sundberg J, Bult C, Eppig J (2008) The mouse tumor biology database. Nature Rev Cancer 8(6):459–465

46. Begley DA, Krupke DM, Neuhauser SB, Richardson JE, Schofield PN, Bult CJ, Eppig JT, Sundberg JP (2014) Identifying mouse models for skin cancer using the Mouse Tumor Biology Database. Exp Dermatol 23(10):761–763

47. Begley DA, Krupke DM, Neuhauser SB, Richardson JE, Bult CJ, Eppig JT, Sundberg JP (2012) The Mouse Tumor Biology Database (MTB): a central electronic resource for locating and integrating mouse tumor pathology data. Vet Pathol 49(1):218–223

48. Begley D, Sundberg JP, Krupke DM, Neuhauser SB, Bult CJ, Eppig JT, Morse HC, Ward JM (2015) Finding mouse models of human lymphomas and leukemia's using The Jackson Laboratory Mouse Tumor Biology Database. Exp Mol Pathol 99(3):533–536

49. Schofield PN, Gruenberger M, Sundberg JP (2010) Pathbase and the MPATH ontology: community resources for mouse histopathology. Vet Pathol 47(6):1016–1020

50. Schofield PN, Bard JBL, Booth C, Boniver J, Covelli V, Delvenne P et al (2004) Pathbase: a database of mutant mouse pathology. Nucleic Acids Res 32:D512–D515

51. Schofield PN, Bard JBL, Boniver J, Covelli V, Delvenne P, Ellender M et al (2004) Pathbase: a new reference resource and database for laboratory mouse pathology. Radiat Prot Dosim 112(4):525–528

52. Sundberg JP, Dunstan RW, Roop DR, Beamer WG (1994) Full thickness skin grafts from flaky skin mice to nude mice: mainte-nance of the psoriasiform phenotype. J Invest Dermatol 102:781–788

53. Sundberg JP, Boggess D, Hogan ME, Sundberg BA, Rourk MH, Harris B et al (1997) Harlequin ichthyosis. A juvenile lethal mouse mutation with ichthyosiform dermati-tis. Am J Pathol 151(1):293–310

54. McElwee KJ, Boggess D, King LE, Sundberg JP (1998) Experimental induction of alopecia areata-like hair loss in C3H/HeJ mice using full-thickness skin grafts. J Invest Dermatol 111:797–803

55. Sundberg JP, Cordy WR, King LE (1994) Alopecia areata in aging C3H/HeJ mice. J Invest Dermatol 102:847–856

56. Silva KA, Sundberg JP (2013) Surgical meth-ods for full thickness skin grafts to induce alo-pecia areata in C3H/HeJ mice. Comp Med 63(5):392–397

57. Herbst LH, Sundberg JP, Shultz LD, Gray B, Klein PA (1998) Tumorgenicity of green turtle fibropapilloma-derived fibroblast lines in immu-nodeficient mice. Lab Anim Sci 48:162–167

58. Sundberg JP, King LE, Bascom C (2001) Animal models for male pattern (androgenic) alopecia. Eur J Dermatol 11:321–325

59. Baker BS, Brent L, Valdimarsson H, Powles AV, Al-Imara L, Walker M, Fry L (1992) Is epidermal cell proliferation in psoriatic skin grafts on nude mice driven by T-cell derived cytokines? Brit J Dermatol 126:105–110

60. Nickoloff BJ (1999) Animal models of psoria-sis. Expert Opin Investig Drugs 8(4):393–401

61. Nickoloff BJ (2000) The search for patho-genic T cells and the genetic basis of psoriasis using a severe combined immunodeficient mouse model. Cutis 65(2):110–114

62. Boggess D, Silva KA, Landel C, Mobraaten L, Sundberg JP (2004) Approaches to handling, breeding, strain preservation, genotyping, and drug administration for mouse models of cancer. In: Holland EC (ed) Mouse models of human cancer. John Wiley & Sons, Inc., Hoboken, NJ, pp 3–14

63. Binder RL, Gallagher PM, Johnson GR, Stockman SL, Smith BJ, Sundberg JP, Conti CJ (1997) Evidence that initiated keratino-cytes clonally expand into multiple existing hair follicles during papilloma histogenesis in SENCAR mouse skin. Mol Carcinog 20(1):151–158

64. Freyschmidt-Paul P, JMcElwee K, Happle R, Kissling S, Wenzel E, Sundberg JP et al (2010) Interleukin-10-deficient mice are less susceptible to the induction of alopecia areata. J Invest dermatol 119(4):980–982

65. Hirota J, Shimizu S (2012) Routes of administration. In: Hedrich HJ (ed) The laboratory mouse, 2nd edn. Elsevier, Amsterdam, pp 709–725

66. Fukuta K (2012) Collection of body fluids. In: Hedrich HJ (ed) The laboratory mouse, 2nd edn. Elsevier, Amsterdam, p 727738

67. Liang Y, Silva KA, Kennedy V, Sundberg JP (2011) Comparisons of mouse models for hair follicle reconstitution. Exp Dermatol 20(12):1011–1015

68. Sullivan SP, Koutsonanos DG, DelPilarMartin M, Lee JW, Zarnitsyn V, Choi SO et al (2010) Dissolving polymer microneedle patches for influenza vaccination. Nat Med 16(8):915–920

69. Kumar D, Farooq A, Laumas KR (1981) Fluid-filled silastic capsules: a new approach to a more constant steroidal drug delivery system. Contraception 23(3):261–268

70. Jones DL, Thompson DA, Munger K (1997) Destabilization of the RB tumor suppressor protein and stabilization of p53 contribute to HPV type 16 E7-induced apoptosis. Virology 239(1):97–107

71. Sundberg JP, Silva KA (2012) What color is the skin of a mouse? Vet Pathol 49(1):142–145

72. Uno H (1991) Quantitative models for the study of hair growth in vivo. Ann N Y Acad Sci 642:107–124

73. Leary S, Underwood W, Anthony R, Cartner S, Corey D, Grandin T, Greenacre C, Gwaltney-Brant S, McCrackin MA, Meyer R, Mille D, Sheare J, Yanong R, Golab GC, Patterson-Kane E (2013) AVMA guidelines for the euthanasia of animals: 2013 Edition.1–102

74. Meyer RE, Fish RE (2005) A review of tribromoethanol anesthesia for production of genetically engineered mice and rats. Lab Anim (NY) 34(10):47–52

75. Chu DK, Jordan MC, Kim JK, Couto MA, Roos KP (2006) Comparing isoflurane with tribromoethanol anesthesia for echocardiographic phenotyping of transgenic mice. J Am Assoc Lab Anim Sci 45(4):8–13

76. Binder RL (1996) Nonstressful restraint device for longitudinal evaluation and photography of mouse skin lesions during tumorigenesis studies. Lab Anim Sci 46(3):350–351

77. Smith RS, John SWM, Nashina PM, Sundberg JP (2002) Systematic evaluation of the mouse eye. Anatomy, pathology, and biomethods. Research methods for mutant mice. CRC Press, Boca Raton, FL

78. Low BE, Krebs MP, Joung JK, Tsai SQ, Nishina PM, Wiles MV (2013) Correction of the Crb1rd8 allele and retinal phenotype in C57BL/6N mice via TALEN-mediated homol-ogy-directed repair. Invest Ophthalmol Vis Sci 55(1):387–395

79. Argenziano G, Fabbrocini G, Carli P, DeGiorgi V, Sammarco E, Delfino M (1998) Epiluminescence microscopy for the diagnosis of doubtful melanocytic skin lesions. Comparison of the ABCD rule of dermatoscopy and a new 7-point checklist based on pattern analysis. Arch Dermatol 134(12):1563–1570

80. Argenziano G, Fabbrocini G, Carli P, DeGiorgi V, Delfino M (1999) Clinical and dermatoscopic criteria for the preoperative evaluation of cutaneous melanoma thickness. J Am Acad Dermatol 40(1):61–68

81. Argenziano G, Fabbrocini G, Carli P, DeGiorgi V, Delfino M (1997) Epiluminescence microscopy: criteria of cutaneous melanoma progression. J Am Acad Dermatol 37(1):68–74

82. Ince TA, Ward JM, Valli VE, Sgroi D, Nikitin AY, Loda M et al (2008) Do-it-yourself (DIY) pathology. Nat Biotechnol 26(9):978–979

83. Nakamura Y, Fukami K, Yu H, Takenaka K, Katoka U, Shirakata Y, Nishikawa SI, Hashimoto K et al (2003) Phospholipase C-delta1 is required for skin stem cell lineage commitment. EMBO J 22:2981–2991

84. Fu L, Pelicano H, Liu J, Huang P, Lee CC (2002) The circadian gene Period2 plays an important role in tumor suppression and DNA damage response in vivo. Cell 111:41–50

85. Fu L, Pelicano H, Liu J, Huang P, Lee CC (2002) Erratum. Cell 111:1055

86. Sundberg JP, Taylor DK, Lorch G, Miller J, Silva KA, Sundberg BA et al (2011) Primary follicular dystrophy with scarring dermatitis in C57BL/6 mouse substrains resembles central centrifugal cicatricial alopecia in humans. Vet Pathol 42(2):513–524

87. Sundberg JP, Berndt A, Sundberg BA, Silva KA, Kennedy V, Bronson R et al (2011) The mouse as a model for understanding chronic diseases of aging: the histopathologic basis of aging in inbred mice. Pathobiol Aging Age-related Dis 1(1):7179. doi:7110.3402/pba.v7171i7170.7179

88. Bleich A, Mahler M, Most C, Leiter EH, Lieber-Tenorio E, Elson CO et al (2004) Refined histopathologic scoring systems improves power to detect colitis QTL in mice. Mamm Genome 15:865–871

89. Sundberg JP, Sundberg BA, Schofield PN (2008) Integrating mouse anatomy and pathology ontologies into a diagnostic/phenotyping database: tools for record keeping

and teaching. Mamm Genome 19(6): 413–419

90. Sundberg BA, Schofield PN, Gruenberger M, Sundberg JP (2009) A data capture tool for mouse pathology phenotyping. Vet Pathol 46(6):1230–1240

91. Chase HB, Montagna W, Malone JD (1953) Changes in the skin in relation to the hair growth cycle. Anat Rec 116:75–82

92. Sundberg JP, Peters EM, Paus R (2005) Analysis of hair follicles in mutant laboratory mice. J Investig Dermatol Symp Proc 10(3):264–270

93. Smith RS, Martin G, Boggess D (2000) Kinetics and morphometrics. In: Sundberg JP, Boggess D (eds) Systematic approach to evaluation of mouse mutations. CRC Press, Boca Raton, pp 111–119

94. Potter CS, Wang Z, Silva KA, Kennedy VE, Stearns TM, Burzenski L, Schultz LD et al (2014) Chronic proliferative dermatitis in Sharpin null mice: development of an autoinflammatory disease in the absence of B and T lymphocytes and IL4/IL13 signaling. PLoS One 9(1):e85666

95. Sundberg JP, Rourk M, Boggess D, Hogan ME, Sundberg BA, Bertolino A (1997) Angora mouse mutation: altered hair cycle, follicular dystrophy, phenotypic maintenance of skin grafts, and changes in keratin expression. Vet Pathol 34:171–179

96. Muller-Rover S, Handjiski B, vander Veen C, Eichmuller S, Foitzik K, McKay IA, Stenn KS, Paus R (2001) A comprehensive guide for the accurate classification of murine hair follicles in distinct hair cycle stages. J Invest Dermatol 117:3–15

97. Milner Y, Sudnik J, Filippi M, Kizoulis M, Kashgarian M, Stenn K (2002) Exogen, shedding phase of the hair growth cycle: characterization of a mouse model. J Invest Dermatol 119:639–644

98. Stenn K (2005) Exogen is an active, separately controlled phase of the hair growth cycle. J Am Acad Dermatol 52(2): 374–375

99. Higgins CA, Westgate GE, Jahoda CA (2009) From telogen to exogen: mechanisms underlying formation and subsequent loss of the hair club fiber. J Invest Dermatol 129(9): 2100–2108

100. Tauchi M, Fuchs TA, Kellenberger AJ, Woodward DF, Paus R, Lütjen-Drecoll E (2010) Characterization of an in vivo model for the study of eyelash biology and trichomegaly: mouse eyelash morphology, development, growth cycle, and anagen prolongation

101. Mecklenburg L, Paus R, Halata Z, Bechtold LS, Fleckman P, Sundberg JP (2004) FOXN1 is critical for onychocyte terminal differentiation in nude (*Foxn1nu*) mice. J Invest Dermatol 123:1001–1011

102. Giehl KA, Potter CS, Wu B, Silva KA, Rowe L, Awgulewitsch A, Sundberg JP (2009) Hair interior defect (*hid*) in AKR/J mice maps to mouse Chromosome 1. Clin Exp Dermatol 34:509–517

103. Bechtold LS (2000) Ultrastructural evaluation of mouse mutations. In: Sundberg JP, Boggess D (eds) Systematic characterization of mouse mutations. CRC Press, Boca Raton, pp 121–129

104. Rice RH, Wong VJ, Pinkerton KE, Sundberg JP (1999) Cross-linked features of mouse pelage hair resistant to detergent extraction. Anat Rec 254(2):231–237

105. Rice RH, Rocke DM, Tsai HS, Silva KA, Lee YJ, Sundberg JP (2009) Distinguishing mouse strains by proteomic analysis of pelage hair. J Invest Dermatol 129(9):2120–2125

106. Rice RH, Bradshaw KM, Durbin-Johnson BP, Rocke DM, Eigenheer RA, Phinney BS, Sundberg JP (2012) Differentiating inbred mouse strains from each other and those with single gene mutations using hair proteomics. PLoS One 7(12), e51956

107. Vandewoestyne M, Goossens K, Burvenich C, VanSoom A, Peelman L, Deforce D (2013) Laser capture microdissection: Should an ultraviolet or infrared laser be used? Anal Biochem 439(2):88–98

108. Wang S, Wang L, Gao TZX, Li J, Wu Y, Zhu H (2010) Improvement of tissue preparation for laser capture microdissection: application for cell type-specific miRNA expression profiling in colorectal tumors. BMC Genomics 11:163

109. Cox ML, Schray CL, Luster CN, Stewart ZS, Krytko PJ, Khan KNM, Paulauskis JD, Dunstan RW (2006) Assessment of fixatives, fixation, and tissue processing on morphology and RNA integrity. Exp Mol Pathol 80:183–191

110. Curran S, McKay JA, McLeod HL, Murray GI (2000) Laser capture microscopy. Mol Pathol 53(2):64–68

111. McElwee KJ, Boggess D, Olivry T, Oliver RF, Whiting D, Tobin DJ et al (1998) Comparison of alopecia areata in human and nonhuman mammalian species. Pathobiology 66(2): 90–107

112. McElwee K, Boggess D, Miller J, King L, Sundberg J (1999) Spontaneous alopecia areata-like hair loss in one congenic and seven

by bimatoprost. Br J Dermatol 162(6): 1186–1197

inbred laboratory mouse strains. J Invest Dermatol Symp Proc 4:202–206

113. Sundberg JP, McElwee KJ, Brehm MA, Su L, King LE (2015) Animal models for alopecia areata: what and where? J Invest Dermatol Symp Proc 17:23–26

114. Sundberg JP, McElwee KJ, King LE (2004) Spontaneous and experimental skin-graft-transfer mouse models of alopecia areata. In: Chan LS (ed) Animal models of human inflammatory skin diseases. CRC Press, Boca Raton, pp 429–449

115. Seymour RE, Hasham MG, Cox GA, Shultz LD, Hogenesch H, Roopenian DC, Sundberg JP (2007) Spontaneous mutations in the mouse Sharpin gene result in multiorgan inflammation, immune system dysregulation and dermatitis. Genes Immun 8(5):416–421

Chapter 13

Chronic Myeloid Leukemia (CML) Mouse Model in Translational Research

Cong Peng and Shaoguang Li

Abstract

Chronic myeloid leukemia (CML) is a myeloproliferative disorder characterized by increased proliferation of granulocytic cells without the loss of their capability to differentiate. CML is a clonal disease, originated at the level of Hematopoietic Stem Cells with the Philadelphia chromosome resulting from a reciprocal translocation between the chromosomes 9 and 22t(9;22)-(q34;q11). This translocation produces a fusion gene known as BCR-ABL which acquires uncontrolled tyrosine kinase activity, constantly turning on its downstream signaling molecules/pathways, and promoting proliferation of leukemia cell through anti-apoptosis and acquisition of additional mutations. To evaluate the role of each critical downstream signaling molecule of BCR-ABL and test therapeutic drugs in vivo, it is important to use physiological mouse disease models. Here, we describe a mouse model of CML induced by BCR-ABL retrovirus (MSCV-BCR-ABL-GFP; MIG-BCR-ABL) and how to use this model in translational research.

Moreover, to expand the application of this retrovirus induced CML model in a lot of conditional knockout mouse strain, we modified this vector to a triple gene coexpression vector in which we can co-express BCR-ABL, GFP, and a third gene which will be tested in different systems. To apply this triple gene system in conditional gene knockout strains, we can validate the CML development in the knockout mice and trace the leukemia cell following the GFP marker. In this protocol, we also describe how we utilize this triple gene system to prove the function of Pten as a tumor suppressor in leukemogenesis. Overall, this triple gene system expands our research spectrum in current conditional gene knockout strains and benefits our CML translational research.

Key words Chronic myeloid leukemia (CML), Philadelphia chromosome (Ph+ chromosome), BCR-ABL, Hematopoietic stem cell, Retroviral mouse model, Translational research

1 Introduction

Chronic myeloid leukemia (CML) has been extensively studied and used as a model disease to investigate the molecular basis of leukemia, shedding light on the understanding of other human cancers as well. CML validates the concept that cancer is a genetic disease. CML is derived from the hematopoietic stem cells which harbor the BCR-ABL oncogene and acquire the selective growth advantages over the normal hematopoietic stem cells.

Gabriele Proetzel and Michael V. Wiles (eds.), *Mouse Models for Drug Discovery: Methods and Protocols*,
Methods in Molecular Biology, vol. 1438, DOI 10.1007/978-1-4939-3661-8_13, © Springer Science+Business Media New York 2016

These conditions are drawn based on the transplantation experiments in which peripheral blood (PB) cells, total bone marrow (BM) cells, and primitive cells (CD34+) from the CML patients have been transplanted into irradiated nonobese severe combined immunodeficient (NOD/SCID) mice [1, 2]. Recipient mice transplanted with these PB or BM cells showed engraftment of the human leukemia cells in BM for up to 7 months and those transplanted with CD34+ cells showed a greater engraftment of leukemia cells. Although this xenograft model allows evaluating the capability of transplanted human leukemia cells to initiate and maintain CML disease in recipient mice, the mice with engrafted human leukemia cells did not develop lethal leukemia after 7 months. This calls for further improvement of this xenograft model as it is important to establish a faithful CML mouse model for evaluating promising therapeutic compounds and developing new therapeutic strategies.

During the last 50 years, several milestones in CML research have been reached. In the early 1960s, Peter Nowell and David Hungerford discovered a small abnormal chromosome that was found in almost all human CML samples; later this abnormal chromosome was named as Philadelphia chromosome (Ph+ chromosome) [3]. Ten years later, Janet Rowley proved that this Ph+ chromosome was a product of a reciprocal translocation between chromosome 9 and 22 [4]. In 1984, Groffen et al. found that the t(9;22) translocation resulted in a previously known oncogene on chromosome 9, ABL, becoming fused with a previously unknown gene, BCR, on chromosome 22 [5]. The molecular basis of ph+ chromosome is the formation of the BCR-ABL oncogene. Many signaling molecules have been found to be downstream of BCR-ABL protein (Fig. 1). These downstream molecules include adapter proteins such as GRB2, protein kinase such as MAPK and PI3K, and some new pathways, such as Alox5, Pten, BLK, FOXO, β-Catenin, and Hedgehog, which modulate CML leukemia stem cell and regulate leukemogenesis [6]. BCR-ABL promotes cell proliferation, increases anti-apoptosis activity, and blocks differentiation of specific cell linage through activating key signal molecules/pathways. These molecules serve as promising targets in CML therapies.

Much effort has been made to generate mouse models of Ph+ leukemia. BCR-ABL transgenic models have been made to express BCR-ABL transgene in mice. Different promoters that drive BCR-ABL expression have been tested in these models to express BCR-ABL in different target cells. These promoters include Eμ [7], MPSV-LTR [7], Metallothionein [8], BCR [9], and MSCV-LTR [10]. Although all of these models show the expression of BCR-ABL in mice, there are at least two obvious defects in using these models: (1) not all mice harboring BCR-ABL develop myeloproliferative disorder, with some mice only developing lymphoid leukemia; (2) disease latency is long, restricting the use of these models in developing therapeutic strategies for CML.

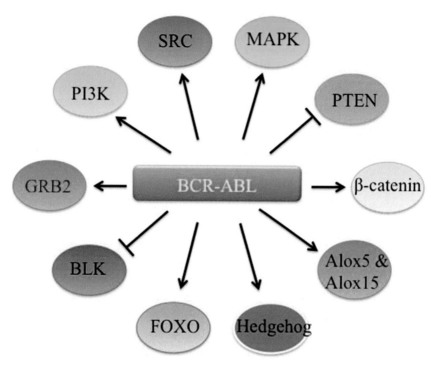

Fig. 1 Major signal pathways activated or inhibited by BCR-ABL. Most pathways listed here have been shown to play a role in functional regulation of LSCs in CML. Note: there are some BCR-ABL-regulated pathways that are not listed in this figure

In contrast, the retroviral transduction/transplantation model is a more faithful model of BCR-ABL induced CML. In 1990, Doley et al. co-cultured mouse bone marrow cells with the retroviral producer cells that produced BCR-ABL expressing retrovirus; they transplanted these infected bone marrow cells into lethally irradiated recipient mice. Three different types of diseases were found in the recipients at up to 5 months post bone marrow transplant. These diseases included CML-like myeloproliferative syndrome, acute lymphoblastic leukemia, and a type of tumor involving macrophages [11]. In the meantime, Kelliher et al. also established a retroviral system in which a JW-RX retrovirus expressing BCR-ABL was used to infect 5-FU pretreated donor mice bone marrow cells. After bone marrow transplantation, more than 90% of recipients developed tumors, with 50% of them developing a myeloproliferative syndrome that shares several features with the chronic phase of chronic myeloid leukemia [12]. Both of these studies proved that BCR-ABL is the primary cause of myeloproliferative syndromes in mice. However, there was more than one type of disease in the recipients, further as not 100% of mice developed CML with similar disease latency; hence it was still difficult to conduct drug testing experiments using these models. To overcome these deficiencies, improvements on the model system have been made including modified construct, transient retroviral packaging system, and changes of virus infection conditions.

At present, we can induce CML in mice with high efficiency, shown by 100% induction of CML in mice [13]. The same CML disease could be induced in most of the inbred mouse strain including C57BL/6, BALB/c, and viable gene knockout mice strains [14]. Because all recipients develop CML with a short latency (about 3 weeks), this provides an excellent model for evaluating therapeutic agents for CML treatment [15]. As CML is derived from the hematopoietic stem cells which harbor BCR-ABL oncogene, CML leukemia stem cells can also be studied in this model [15]. In conclusion, this retroviral model system provides a powerful tool for studying the CML disease mechanism and performing translational research (see **Note 1**).

2   Materials

2.1   Mice

1. C57BL/6J or BALB/cJ mice (The Jackson Laboratory, Bar Harbor, Maine, USA), 4–10 week old. The donor and recipient have to be the same mouse strain.

2. Gamma irradiator: JL Shepherd 2000 Ci Mark I Animal Irradiator.

2.2   Generation of Retroviral Stocks

1. Cell lines: 293T cells (ATCC, Cat#CRL-11268) and NIH3T3 cells (ATCC, Cat#CRL-1658).

2. 293T culture medium: DMEM (Gibco, Bethesda, MD) supplemented with 10% fetal bovine serum (Gibco, Cat# 26140-079), 1% penicillin-streptomycin solution (Gibco, Cat# 15140-122), 1% L-glutamine solution (Gibco, Cat# 25030-081), and 1% MEM non-essential amino acid solution (Sigma, Cat# M7145).

3. NIH3T3 culture medium: DMEM (Gibco, Bethesda, MD) supplemented with 10% new born calf serum (Lonza, Cat# 14-416F), 1% penicillin-streptomycin solution (Gibco, Cat# 15140-122), 1% L-glutamine solution (Gibco, Cat# 25030-081).

4. Calcium chloride ($CaCl_2$) solution: prepare 2 M solution in sterile water then filtered in hood and stored at room temperature.

5. 2× HBS solution: Weigh out 8 g NaCl, 0.37 g KCl, 106.5 mg Na_2HPO_4, 1 g dextrose (Gibco, Cat#15023-021), 5 g HEPES (American Bioanalytical, Cat#AB00892-00500), and add sterile water to 500 mL and adjust pH to 7.05 with 10 N Sodium Hydroxide.

6. Polybrene (Hexadimethrine bromide): (Sigma, Cat: 028K3730) make 8 mg/mL stock solution in H_2O, and keep at −20 °C.

7. 1× Trypsin-EDTA solution (Cellgro, Cat#25-052-Cl).

2.3 Bone Marrow Cell Collection and Medium

1. Micro-dissecting scissor (Roboz, Cat# RS-5925), micro-dissecting forceps (Roboz, Cat# RS-51900).

2. Precision glide needle 27G1/2 (BD, Product Number: 301230) and 10 mL BD Luer-Lok™ tip syringe (BD, Product Number: 309604).

3. Acrodisc Syringe Filters with Super Membrane (Pall, cat#4184).

4. DMEM (Gibco, Bethesda, MD) supplemented with 10 % fetal bovine serum (Gibco, Cat# 26140-079), 1 % penicillin-streptomycin solution (Gibco, Cat# 15140-122), 1 % L-glutamine solution (Gibco, Cat# 25030-081), and 1 % MEM non-essential amino acid solution (Sigma, Cat# M7145).

2.4 Bone Marrow Cell Culture and Transduction Medium

1. Bone marrow cell first time stimulation medium: 77 %(v/v) DMEM, 15 % (v/v) heat inactivated FBS, 5 % (v/v) WEHI-3B conditioned medium, penicillin/streptomycin, 1.0 mg/mL ciprofloxacin, 200 mM L-glutamine, 6 ng/mL recombinant murine IL-3 (Peprotech, Cat#213-13), 10 ng/mL recombinant murine IL-6 (Peprotech, Cat#216-16), and 50–100 ng/mL recombinant murine stem cell factor (SCF; Peprotech, Cat#250-03). The total volume is 10 mL for each sample. For the second round stimulating medium, the volume is 4 mL for each sample.

2. First time transduction medium: 50 % retroviral supernatant, 27 % (v/v) DMEM, 15 % (v/v) heat inactivated FBS, 5 % (v/v) WEHI-3B conditioned medium, penicillin/streptomycin, 1.0 μg/mL ciprofloxacin, 200 mM L-glutamine, 6 ng/mL recombinant murine IL-3 (Peprotech, Cat#213-13), 10 ng/mL recombinant murine IL-6 (Peprotech, Cat#216-16), and 50–100 ng/mL recombinant murine stem cell factor (SCF; Peprotech, Cat#250-03), 1 % (v/v) HEPES and 20 μg/mL polybrene. The total volume is 4 mL for each sample.

3. Second time transduction medium: 2 mL retroviral supernatant, 20 μg/mL polybrene, and 1 % (v/v) HEPES.

4. Red blood cell (RBC) lysis buffer: 150 mM NH_4Cl, 10 mM $KHCO_3$, 0.1 mM EDTA (pH to 7.4).

5. Flow cytometry buffer (FACS buffer): PBS supplied with 1 % BSA

6. 5-fluorouracil (5-Fu, Sigma, Cat#6627) solution: 10 mg/mL in sterile PBS (*see* **Note 6**).

7. 1× HBSS buffer (Fisher, Cat# 21-022-CV).

3 Methods

3.1 Generation of MIG-BCR-ABL (MSCV-BCR-ABL-IRES-GFP) Virus Supernatant

1. Culture 293T cells in 15 cm tissue culture dish (there are about 1×10^8 cells in confluent plates).

2. When the 293T cells reach 90% confluence in the 15 cm dish, remove the medium and wash cells once with 1× PBS. Remove PBS, add 3 mL of Trypsin-EDTA solution, and stop the reaction by adding 20 mL 293T medium. Collect cells carefully in 50 mL centrifuge tube and spin at $514 \times g$, 10 min at room temperature. The 293T cells are passaged to 6 cm dish at 4×10^6 cells/dish at the day before transfection (*see* **Note 3**).

3. Change 4 mL fresh 293T medium to each dish before transfection.

4. In a 15 mL tube, add 10 μg MIG-BCR-ABL plasmid, 5 μg Ecopack plasmid [13], 62 μL 2 M $CaCl_2$, and sterile water to 500 μL total volume. Briefly vortex.

5. Add 500 μL 2× HBS to the tube and mix by vortexing for 10 s (*see* **Note 4**).

6. Gently and quickly drop the DNA/HBS solution onto 293T cells.

7. Rock the dishes forward and backward a few times to achieve even distribution of DNA/$Ca_3(PO_4)_2$ particles.

8. After 24 h, remove the old medium and add 4 mL fresh 293T medium.

9. After 48 h post-transfection of 293T cells, collect the supernatant by 10 mL BD syringe and filter the supernatant through 0.45 μm syringe filter.

10. Aliquot virus supernatant in 4 mL/tube and store at –80 °C (*see* **Note 5**).

3.2 Testing Viral Titer by Flow Cytometry

1. Culture NIH3T3 cells with NIH3T3 cell medium in 10 cm dish.

2. When the NIH3T3 cells reach 90% confluence, remove the medium and wash cells once with 1× PBS. Remove PBS, add 1 mL Trypsin-EDTA solution, and stop the enzymatic reaction by adding 10 mL NIH3T3 medium. Collect cells carefully in 10 mL centrifuge tube and spin at $514 \times g$, 10 min at room temperature. The NIH3T3 cells are passaged to 10 cm dish at 0.6×10^5 cells/dish at the day before infection.

3. At the day of infection, remove the NIH3T3 cell medium and add virus supernatant serially diluted in 293T medium as 1:2, 1:8, and 1:16. Polybrene is added as 80 μg/mL into retroviral supernatant.

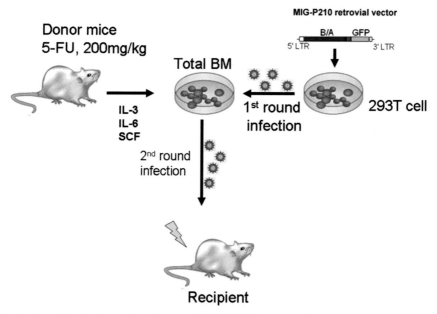

Fig. 2 Retroviral transduction/transplantation model of BCR-ABL induced CML. Donor mice are pretreated with 5-FU, and bone marrow cells are stimulated with cytokines in vitro. After infected twice with MIG-BCR-ABL retrovirus, donor bone marrow cells are transplanted into lethally irradiated recipients for induction of CML

4. After 3 h infection at 37 °C, remove the virus supernatant, and change to 10 mL NIH3T3 medium.

5. After 48 h culture, collect cells into 4 mL FACS buffer. Take 300 μL cells to do the flow cytometry analysis for percentage of GFP-expressing cells. Normally, the good retroviral supernatant means the GFP% can reach to 90–95 % at the 1:2 dilution, 75–85 % at 1:8 dilution, and 60–70 % at 1:16 dilution (*see* **Note 2**).

3.3 Bone Marrow Cells Transduction and Transplantation (See Fig. 2)

3.3.1 Priming of Donor Mice at Day 0

1. Have donor mice ready (e.g., C57BL/6J or BALB/cJ).

2. Freshly suspend the 5-FU powder in PBS.

3. Incubate in 37 °C water bath for 10–30 min, vortex the solution to help to dissolve the powder.

4. Inject 5-FU to donor mice via tail vein (200 mg/kg, 0.6 mL for 30 g mouse).

3.3.2 Bone Marrow Cell Collection of Donor Mice at Day 4

1. Kill primed donor mice with CO_2.

2. Sterilize the skin of the mice with 70 % ethanol.

3. Collect femurs and tibias and place them in cold PBS, clip off the end of the bone, and briefly clean the muscle.

4. Flush out bone marrow cells with 293T medium.

5. Blow the cells with 10 mL pipette up and down to suspend cells. Normally, $2–3 \times 10^7$ total bone marrow cells can be harvested from 10 donors.

6. Spin down cells at $514 \times g$ for 10 min.

7. Resuspend the cell pellet with the first time stimulation medium ($<3 \times 10^7$ cells/dish in 10 mL medium).

8. Incubate cells at 37 °C for 24 h.

3.3.3 First Time Transduction at Day 5

1. Collect the cells and spin cell at $514 \times g$ for 10 min at room temperature.

2. Generate the first time transduction medium containing MIG-BCR-ABL retrovirus: 2 mL MIG-BCR-ABL virus is mixed with 2 mL of the first time transduction medium.

3. Add 4 mL of the transduction medium to suspend the cells and then transfer the cells to a 6-well plate.

4. Spin the cells in the 6-well plate at $1,160 \times g$ at room temperature for 90 min.

5. Incubate the cells at 37 °C, 5 % CO_2 for 3–4 h.

6. Remove the supernatant from each well, and add 4 mL second time stimulation medium, then incubate the cells at 37 °C overnight.

3.3.4 Second Time Transduction at Day 6

1. Remove 2 mL of supernatant from each well carefully, and then add 2 mL of the second time transduction medium.

2. Spin the cells at $1,160 \times g$ for 90 min.

3. Incubate the cells at 37 °C, 5 % CO_2 for 3 h.

4. Collect the cells and spin at $514 \times g$ for 10 min.

5. Suspend the cell with 5 mL of HBSS solution.

3.3.5 Injection of BCR-ABL Transduced Donor Bone Marrow Cells into Lethally Irradiated Recipient Mice at Day 6

1. Recipient mice are treated by two doses of 550-cGy gamma (C57BL/6J) or 450-cGy gamma (BALB/cJ), separated by 3 h.

2. Adjust the cell concentration to 1.25×10^6/mL in HBSS solution then inject 0.5×10^6 cells (0.4 mL) to each mouse via tail vein.

3.4 Monitoring Leukemia Development

After transplantation, recipient mice are evaluated daily for signs of morbidity, weight loss, failure to thrive, and splenomegaly. Depending on the individual animal, hematopoietic tissues and cells are used for several applications, including histopathology, in vitro culture, FACS analysis, secondary transplantation, genomic DNA preparation, protein lysate preparation, or cell lineage analysis.

3.4.1 Monitoring White Blood Cell Count

1. Eye bleed the mice to collect about 100 μL peripheral blood.

2. Add 2 μL peripheral blood to 18 μL RBC buffer, mix and lyse red blood cells on ice for 10 min.

3. Add 20 μL trypan blue, mix well, and drop 15–20 μL cells under the cover slip on a hemocytometer and count the live cells using a microscope. (Dead cells will take up trypan blue, so they are darker than the live cells.)

3.4.2 Preparing Peripheral Blood Cells for FACS

1. Eye bleed the mice to collect about 100 μL peripheral blood into a 0.5 mL Eppendorf tube.

2. Add 1 mL of RBC buffer to the tube, mix well, and lyse the red blood cell on ice for 10 min.

3. Spin cells at 3,300 × g at 4 °C for 5 min.

4. Remove the supernatant and resuspend the cell pellet in 1 mL of PBS to wash the cells.

5. Spin cells at 3,300 × g for 5 min again and remove the supernatant. Resuspend the cell pellet in FACS buffer, adjust cell concentration to 1×10^7 cells/mL.

6. Add PE or APC labeled (or other fluorescent labeled cell surface marker) antibody to 50 μL of cells. Mix well, and stain the cells at 4 °C for 30 min.

7. Add 1 mL of FACS buffer to the cells and wash cells once by spinning at 3,300 × g, 4 °C for 5 min.

8. Resuspend the cell pellet in 300 μL of FACS buffer, and then perform flow cytometry analysis.

3.5 Example of Translation Research Using Ph+ CML Mouse Model

The MIG-BCR-ABL retroviral transduction/transplantation model of CML provides an excellent system not only for identifying critical therapeutic targets downstream of BCR-ABL but also for testing the efficiency of therapeutic agents. Previous studies have shown the effectiveness of imatinib in treating CML in our mouse disease model; i.e., imatinib (Gleevec, Novartis) can significantly prolong the survival of CML mice [14, 15]. Another example of using this mouse model in translational research is to test the heat shock protein 90 (HSP90) as a critical therapeutic target in CML treatment [15]. Below we briefly introduce how to use this model to test the effectiveness of HSP90 inhibitor IPI-504 (retaspimycin hydrochloride; Infinity Pharmaceuticals) in treating CML in mice.

3.5.1 CML Disease Induction in Mice

1. Induce CML as described in Subheadings 3.1–3.3.

3.5.2 Drug Treatment of Mice

1. IPI-504 was dissolved in a solution containing 50 mM citrate, 50 mM ascorbate, 2.44 mM EDTA, pH 3.3.

2. Imatinib was dissolved in water.

3. The drugs were given orally in a volume of less than 0.5 mL by gavage (50 or 100 mg/kg, every other day for IPI-504, and 100 mg/kg, twice a day for imatinib) beginning at 8 days

after bone marrow transplantation, and continuing until the morbidity or death of the leukemic mice.

4. Placebo is a solution containing 50 mM citrate, 50 mM ascorbate, 2.44 mM EDTA, pH 3.3.

3.5.3 Flow Cytometry Analysis

1. Hematopoietic cells were collected from peripheral blood and bone marrow of the CML mice 10 days after bone marrow transplantation.

2. Red blood cells were lysed with NH_4Cl red blood cell lysis buffer (pH 7.4).

3. The cells were washed with PBS, and stained with B220-PE for B cells, Gr1-APC for neutrophils, and Sca1-APC/c-kit-PE for hematopoietic stem cells.

4. After staining, the cells were washed once with PBS and subjected to FACS analysis.

Treatment with HSP90 inhibitor (IPI-504) resulted in the reduced peripheral leukemia cell count, spleen weight, and prolonged survival of mice with induced CML for wild-type or T315I BCR-ABL. Mice with wild-type (WT) or T315I-BCR-ABL transduced bone marrow cells from 5-FU-treated BALB/cJ donor mice were treated with a placebo, the Hsp90 inhibitor IPI-504, or imatinib alone, or the two agents in combination. All placebo-treated mice developed and died of CML within 3 weeks after bone marrow transplant (BMT) (Fig. 3a). As expected, imatinib treatment was effective in treating WT-induced CML but not CML induced by T315I (Fig. 3a). In a dose-dependent manner, treatment with IPI-504 alone significantly prolonged survival of mice with WT CML, but even more markedly prolonged survival of mice with T315I-induced CML (Fig. 3a, $P < 0.001$). Treatment of mice with WT CML with both IPI-504 and imatinib was slightly more effective (but statistically insignificant) than with imatinib alone in prolonging survival of the mice (Fig. 3a), while treatment of mice with BCR-ABL-T315I-induced CML with these two drugs did not further prolong survival of the mice compared with the mice treated with IPI-504 alone (Fig. 3a). Prolonged survival of IPI-504-treated CML mice correlated with decreased peripheral blood BCR-ABL-expressing (GFP-positive) leukemia cells during therapy (Fig. 3b, $P < 0.001$) and less splenomegaly at necropsy (Fig. 3c). As lung hemorrhage caused by infiltration of mature myeloid leukemia cells is a major cause of death of CML mice [13], we further evaluated the therapeutic effect of IPI-504 on CML by examining the severity of lung hemorrhages at day 15 after BMT. Compared with placebo-treated mice, fewer hemorrhages were observed in the lungs of IPI-504-treated mice with BCR-ABLT315I-induced CML

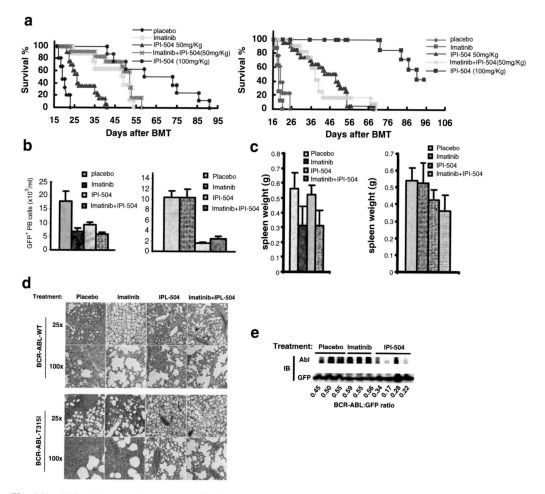

Fig. 3 Hsp90 is a therapeutic target for CML induced by either BCR-ABL-WT or BCR-ABL-T315I. (**a**) Treatment with the Hsp90 inhibitor IPI-504 prolonged survival of CML mice. Mice with BCR-ABL-WT (*left panel*)- or BCR-ABL-T315I (*right panel*)-induced CML were treated with placebo (n = 15 for BCR-ABL-WT; n = 13 for BCR-ABL-T315I), imatinib (100 mg/kg, twice a day by gavage) (n = 8 for both BCR-ABL-WT and -T315I), IPI-504 (50 mg/kg, once every 2 days by gavage) (n = 20 for both BCR-ABL-WT and BCR-ABL-T315I), IPI-504 (100 mg/kg, once every 2 days by gavage) (n = 8 for BCR-ABL-WT; n = 7 for BCR-ABL-T315I), and imatinib + IPI-504 (n = 12 for both BCR-ABL-WT and -T315I), respectively, beginning at day 8 after transplantation. The IPI-504-treated mice with BCR-ABL-T315I-induced CML lived longer than those with BCR-ABL-WT-induced CML (comparing between *left* and *right panels*). (**b**) Flow cytometric evaluation of the leukemic process in IPI-504- or imatinib-treated CML mice. The number of circulating leukemic cells (calculated as percentage of Gr-1+GFP+ cells X white blood cell count) in mice with BCR-ABL-WT (*left panel*)- or BCR-ABL-T315I (*right panel*)-induced CML treated with placebo, imatinib, IPI-504, or the combination of imatinib and IPI-504 was determined on day 14 after transplantation. (**c**) Spleen weights of CML mice treated with placebo, imatinib, IPI-504, and the combination of imatinib and IPI-504. (*Left panel*) BCR-ABL-WT. (*Right panel*) BCR-ABL-T315I. (**d**) Photomicrographs of hematoxylin and eosin-stained lung sections from drug-treated mice at day 14 after transplantation. (**e**) Western blot analysis of spleen cell lysates for degradation of BCR-ABL in IPI-504-treated CML mice. IB indicates immunoblot. Adapted from [16]

(Fig. 3d). Western blot analysis of spleen cell lysates from the treated CML mice showed that IPI-504 reduced the levels of BCR-ABL protein in CML mice (Fig. 3e).

In the BMT model, imatinib prolongs survival of mice with BCR-ABL induced CML, but does not lead to a cure partially due to the inability of imatinib to completely eradicate leukemia stem cells [16]. In CML mice, BCR-ABL-expressing hematopoietic stem cells (HSCs) have been identified as leukemia stem cells (LSCs) [16]. These LSCs are resistant to imatinib treatment and capable of transferring CML to recipient mice. To investigate whether HSP90 is also a major therapeutic target in LSCs, CML mice were treated with IPI-504 (a HSP90 inhibitor) and LSCs monitored by FACS in vivo. Specifically, mice with BCR-ABLT315I-induced CML were treated with a placebo, imatinib, and IPI-504, respectively, for 6 days, and bone marrow cells were analyzed by FACS for LSCs (GFP+Lin-c-Kit+Sca-1+). Imatinib treatment did not lower the percentage and number of LSCs compared to placebo-treated mice, whereas IPI-504 treatment had a dramatic inhibitory effect on LSCs. To determine whether IPI-504 had an effect on normal HSCs in mice, WT mice were treated with IPI-504 or placebo for 2 weeks. Analysis of bone marrow from these mice showed that there was no change in the level of Lin-c-Kit+Sca-1+ cells (Fig. 3b), indicating that IPI-504 treatment did not inhibit survival of normal HSCs.

3.6 A Triple Gene Expression System in CML Mouse Model

To avoid premature death caused by the early developmental defects in conventional gene knockout (KO) mice, conditional KO mice have been generated and used widely in biomedical research. In most conditional KO models, the loxP sites are placed at both ends of the gene of interest to allow the tissue-specific Cre gene to induce the deletion of the gene in a particular tissue by crossing the conditional KO strain with mice expressing the tissue-specific Cre. To utilize these conditional KO strains in CML study, we generated a triple gene expression system that simultaneously expresses BCR-ABL, Cre and GFP genes in the retroviral MSCV vector. Thus, BCR-ABL-expressing wild-type cells or cells lacking the gene of interest can be monitored by detecting GFP expression via FACS. Here is an example for how to use this triple gene expression system in CML mouse model. The tumor suppressor gene phosphatase and tensin homolog (Pten) is inactivated in many types of human cancers, including leukemia stem cells in human AML [17, 18] and CML [19]. In our previous study, we showed that Pten was downregulated by BCR-ABL in LSCs in CML and that Pten deletion causes acceleration of CML development [19]. In addition, overexpression of PTEN delays the development of CML and prolongs survival of leukemia mice. We utilized our triple gene expression system to study the role of Pten in CML leukemogenesis and regulation of LSC function using B6.129S4-$Ptentm1Hwu^{/J}$ (Pten$^{fl/fl}$) conditional strain (JAX 006440).

3.6.1 Construction of MSCV-BCR-ABL-Cre-GFP Triple Gene Coexpression Vector

The original MSCV-IRES-GFP vector was first modified to add new cloning sites for the restriction enzymes MfeI, NotI, and MluI. To do so, the internal ribosomal entry segment (IRES) sequence was first amplified by the murine stem cell virus (MSCV) primer (CGTCTCTCCCCCTTGAACCTCCTCG) and the IRES-MfeI primer (CATGCCATGGCAATTGAGCGGCCGCTT GTGGCCATATTATCATC), which contains the new MfeI, NotI, and the existing NcoI sites. This step allowed us to synthesize a new IRES fragment containing the MfeI, NotI, and NcoI sites.

To replace the original IRES sequence in the original MSCV-IRES-GFP vector with the newly synthesized IRES, this vector was cut with EcoRI and NcoI, and then the new IRES fragment was cloned into the MSCV-IRES-GFP cut with EcoRI and NcoI, forming a new MSCV-IRES-GFP vector that contains two additional sites, MfeI and NotI. To add the MluI site to the new MSCV-IRES-GFP vector, an IRES-GFP fragment was amplified from this vector by the MSCV primer and the GFP-MluI primer (CCATCGATACGCGTAAGCTTGGCTGCAGGTCGA), which contains the existing ClaI and the new MluI sites. The synthesized IRES-GFP fragment was digested with EcoRI and ClaI and then cloned into the new MSCV-IRES-GFP vector between the EcoRI and ClaI sites to generate the final MSCV-IRES-GFP vector. Compared with the original MSCV-IRES-GFP vector, this final MSCV-IRES-GFP vector contains additional sites MfeI, NotI (before the GFP sequence), and MluI (after the GFP sequence). To clone the BCR-ABL cDNA into this final MSCV-IRES-GFP vector, BCR-ABL was cloned into it at the EcoRI site.

To make MSCV-BCR-ABL-Cre-GFP construct, the Cre open reading frame (ORF) was amplified by Cre-MfeI (CGCAATTGATGGTGCCCAAGAAGAAGAGG), and Cre-ClaI (CCATCGATTCAGTCCCCATCCTCGAGCAG) by using pBOB-CAG-iCre-SD (Addgene) as a template. The Cre ORF was cloned into the MSCV-BCR-ABL vector between NotI and MluI sites, and the IRES-GFP fragment was cloned at the MluI site after the Cre ORF (Fig. 4a).

To make the MSCV-BCR-ABL-PTEN-GFP construct, total RNA was isolated from C57BL/6 mice liver tissue to synthesize the Pten cDNA by reverse-transcription polymerase chain reaction (RT-PCR). The Pten cDNA was amplified by PTEN-NotI (5′-AGCGGCCGCATGACAGCCATCATCAAAGAG-3′) and PTEN-MluI (5′-CGACGCGTTCAGACTTTTGTAATTTGTG-3′) primers. The cDNA was sequenced from both ends to confirm the sequence. The Pten cDNA was cloned into the MSCV-BCR-ABL-GFP vector between NotI and MluI sites. The IRES-GFP fragment was amplified by MSCV-MluI (cgacgcgtAATTCC-GCCCTCTCCCTC) and GFP-MluI (ccacgcgtTAAGCTTG-GCTGCAGGTCGA) primers by using MSCV-GFP as a template, and the IRES-GFP fragment was inserted after the Pten sequence at the MluI site (Fig. 4b).

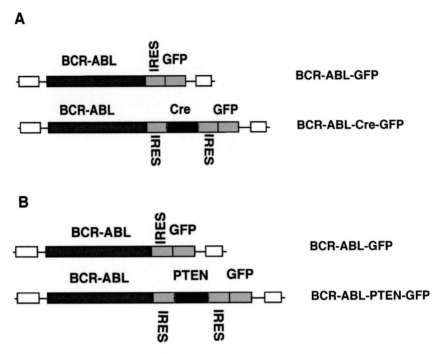

Fig. 4 Schematic structures of triple gene expression and control vectors (**a**) The MSCV vector co-expressing BCR-ABL, Cre, and GFP. (**b**) The MSCV vector co-expressing BCR-ABL, PTEN, and GFP

3.6.2 PTEN Deletion Causes Acceleration of CML Development

To test whether Pten functions as a tumor suppressor in CML development by using Pten conditional knock mice (Pten^fl/fl), we transduced the bone marrow from Pten^fl/fl mice with BCR-ABL-Cre-GFP retrovirus or BCR-ABL-GFP retrovirus as a control. Western blot analysis showed expression of Cre and a significant decrease of the Pten protein level (Fig. 5a), indicating that the Pten gene was deleted from the cells. To test whether deletion of Pten affects CML development, we transduced bone marrow cells from Pten^fl/fl mice with BCR-ABL-Cre-GFP or BCR-ABL-GFP retrovirus, followed by transplantation of the transduced cells into lethal irradiated recipient mice. Mice receiving donor bone marrow cells transduced with BCR-ABL-Cre-GFP developed CML much faster than those receiving bone marrow cells transduced with BCR-ABL-GFP (Fig. 5b; $P < 0.05$). In these CML disease mice, the majority of GFP cells were Gr1^+ leukemia cells. The accelerated death of CML mice in the absence of Pten correlated with a greater percentage of GFP^+Gr1^+ myeloid leukemia cells (Fig. 5c) and a greater number of leukemia cells in peripheral blood of the mice. Accelerated CML development in the absence of Pten also correlated with more severe infiltration of leukemia cells in the lungs (Fig. 5d). These results demonstrated that Pten is a potent tumor suppressor in BCR-ABL-induced CML.

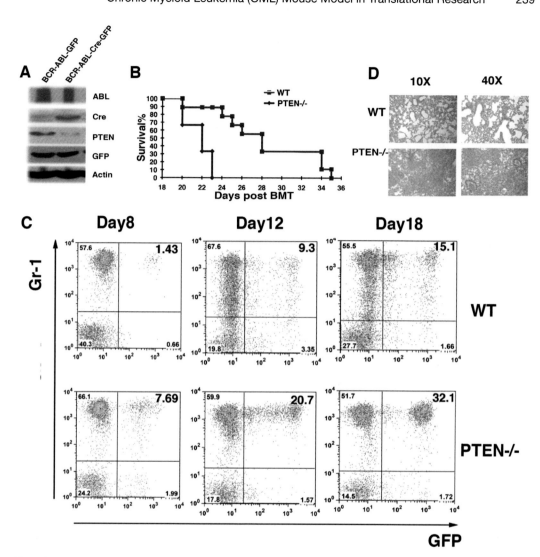

Fig. 5 Pten deletion accelerates CML development. (**a**) BCR-ABL-GFP and BCR-ABL-Cre-GFP retrovirus transduced bone marrow cells from PTENfl/fl mice were cultured under for 1 week. Protein lysates were analyzed by Western blotting with the antibodies indicated. Cre-induced deletion of the Pten gene resulted in the removal of Pten protein. (**b**) Kaplan-Meier-style survival curves for recipients of BCR-ABL-Cre-GFP-transduced bone marrow cells from wild-type (WT; $n = 6$) or PTENfl/fl (PTEN; $n = 9$) mice ($P < 0.001$). (**c**) The percentage of leukemia cells (GFP+Gr1+) in recipients of BCR-ABL-Cre-GFP-transduced bone marrow cells from Ptenfl/fl mice was greater than that in recipients of BCR-ABL-Cre-GFP-transduced bone marrow cells from wild-type mice. (**d**) Photomicrographs of hematoxylin and eosin-stained lung sections from recipients of bone marrow cells from PTEN-deficient CML mice (Pten−/−) showed more severe infiltration of the lungs with myeloid leukemia cells than recipients of bone marrow cells from wild-type mice (WT) at day 14 after BMT

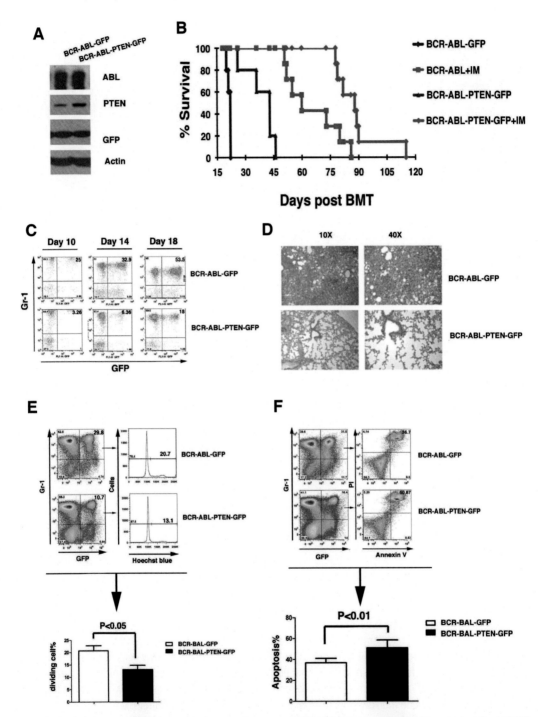

Fig. 6 Overexpression of Pten delays CML development. (**a**) Western blot analysis shows expression of BCR-ABL, PTEN, and GFP from BCR-ABL-PTEN-GFP retrovirus. NIH3T3 cells were transduced with BCR-ABL-GFP or BCR-ABL-PTEN-GFP retrovirus for 3 h. Then, 2 days later, protein lysates were analyzed by Western blotting by the use of the antibodies indicated. (**b**) Overexpression of Pten alone or in combination with imatinib treatment prolongs survival of CML mice. Mice with CML induced with BCR-ABL-GFP (*n*=20) or BCR-ABL-PTEN-GFP (*n*=20)Fig. 6 (continued) were treated with a placebo (*n*=7) or imatinib (*n*=7, 100 mg/kg, twice a day by gavage), beginning at day 8 after transplantation. (**c**) Flow cytometry analysis showed a slower accumulation

3.6.3 PTEN Overexpression Delays CML Development

We also examined whether overexpression of PTEN delays CML development. Western blot analysis showed our BCR-ABL-PTEN-GFP triple-gene retroviral construct allowed overexpression of Pten in cells (Fig. 6a). We next transduced donor bone marrow cells from WT mice with BCR-ABL-PTEN-GFP or BCR-ABL-GFP retrovirus, followed by transplantation of the transduced cells into recipient mice. CML development was significantly slower in mice receiving bone marrow cells transduced with BCR-ABL-PTEN-GFP than in those receiving bone marrow cells transduced with BCR-ABL-GFP (Fig. 6b, $P < 0.001$), indicating that Pten overexpression caused a delay of CML development. The delayed CML development correlated with a less percentage and number of leukemia cells in peripheral blood (Fig. 6c) and infiltration of leukemia cells in the lungs (Fig. 6d). These results further support the role of Pten as a tumor suppressor in CML development. To evaluate whether PTEN overexpression in BCR-ABL-expressing cells synergizes with the therapeutic effect of imatinib on CML, we treated mice receiving bone marrow cells transduced with BCR-ABL-PTEN-GFP or BCR-ABL-GFP retrovirus with imatinib. As expected, imatinib treatment prolonged survival of CML mice receiving bone marrow cells transduced with BCR-ABL-GFP (Fig. 6b; $P < 0.001$). However, imatinib-treated CML mice receiving bone marrow cells transduced with BCR-ABL-PTEN-GFP lived significantly longer than those not treated with imatinib (Fig. 6b; $P < 0.001$). To explain how Pten reduced proliferation of leukemia cells, we performed the DNA content analysis to examine the effect of Pten overexpression on cell cycle progression of these cells. We showed that the percentage of leukemia cells in the S+G2M phase was much lower in leukemia cells with Pten overexpression than in those without Pten overexpression (Fig. 6e; $P < 0.01$), indicating that Pten inhibits the proliferation of leukemia cells by inducing a cell cycle arrest. Furthermore, we examined whether Pten induces apoptosis of leukemia cells by staining the cells with PI and Annexin V. Corresponding to the result in the cell cycle analysis, apoptosis in leukemia cells with Pten overexpression was more severe than in those without Pten overexpression (Fig. 6f; $P < 0.05$).

Fig. 6 (continued) of GFP+Gr1+ leukemia cells in peripheral blood of recipients of BCR-ABL-PTEN-GFP-transduced bone marrow cells than that in recipients of BCR-ABL-GFP-transduced bone marrow cells. (**d**) Photomicrographs of hematoxylin and eosin-stained lung sections from mice with CML induced with BCR-ABL-GFP or BCR-ABL-PTEN-GFP at day 20 after transplantation. (**e**) At day 20 after BMT, peripheral blood cells were stained with Gr1 and Hoechst blue. The S+G2M phase of leukemia cells (GFP+Gr1+) was represented by the percentage of Hoechst blue-positive cells. Mean percentage for each cell population ($n = 3$) was shown. (**f**) At day 20 after BMT, peripheral blood cells were stained with Gr1, Annexin V, and propidium iodide (PI). Apoptotic leukemia cells were represented by the GFP+Gr1+AnnexinV+PI+ population. Mean percentage for each cell population ($n = 3$) was shown

4 Notes

1. In this retroviral transduction/transplantation CML model system, inbred mouse strains mice are highly recommended, as these mice have identical genetic background. This avoids modifier effects which would occur when using mixed, undefined, or outbred mouse strains. This genetic uniformity also assists in the development of consistent disease type with and similar disease latency. Most of human cancers are a type of genetic diseases. Somatic mutations, chromosome deletions, or translocations could cause different type of cancers. Inbred mice with stable genetic backgrounds are the best choice to be used as donors and recipients as secondary genetic events can be avoided.

2. Making high-titer BCR-ABL retrovirus is critical. The 2× HBS is the most important reagent, and its pH needs to be adjusted exactly to 7.05, which can be kept at room temperature up to couple of months.

3. For making retrovirus using 293T cells, we recommend to make sure that the confluence of the cells is about 90%; this degree of cell confluence is critical to making high-titer virus.

4. During mixing the 2× HBS with DNA and $CaCl_2$, gently vortex and drip the solution evenly onto the cultured cells.

5. After collecting the virus supernatant, store them at −80 °C immediately. This will help retain the effectiveness of the virus for up to 1 year.

6. 5-FU has low solubility, so a 37 °C water bath is recommended with gently vortexing to improve the solubility.

References

1. Ren R (2005) Mechanisms of BCR-ABL in the pathogenesis of chronic myelogenous leukaemia. Nat Rev Cancer 5:172–183

2. Wang JC et al (1998) High level engraftment of NOD/SCID mice by primitive normal and leukemic hematopoietic cells from patients with chronic myeloid leukemia in chronic phase. Blood 91:2406–2414

3. Nowell PC, Hungerford DA (1960) Chromosome studies on normal and leukemic human leukocytes. J Natl Cancer Inst 25:85–109

4. Rowley JD (1973) Letter: a new consistent chromosomal abnormality in chronic myelogenous leukaemia identified by quinacrine fluorescence and Giemsa staining. Nature 243:290–293

5. Groffen J et al (1984) Philadelphia chromosomal breakpoints are clustered within a limited region, bcr, on chromosome 22. Cell 36:93–99

6. Chen Y, Peng C, Sullivan C, Li D, Li S (2010) Critical molecular pathways in cancer stem cells of chronic myeloid leukemia. Leukemia 24:1545–1554

7. Hariharan IK et al (1989) A bcr-v-abl oncogene induces lymphomas in transgenic mice. Mol Cell Biol 9:2798–2805

8. Heisterkamp N, Jenster G, Kioussis D, Pattengale PK, Groffen J (1991) Human bcr-abl gene has a lethal effect on embryogenesis. Transgenic Res 1:45–53

9. Castellanos A et al (1997) A BCR-ABL(p190) fusion gene made by homologous recombi-

nation causes B-cell acute lymphoblastic leukemias in chimeric mice with independence of the endogenous bcr product. Blood 90:2168–2174

10. Inokuchi K et al (2003) Myeloproliferative disease in transgenic mice expressing P230 Bcr/Abl: longer disease latency, thrombocytosis, and mild leukocytosis. Blood 102:320–323

11. Daley GQ, Van Etten RA, Baltimore D (1990) Induction of chronic myelogenous leukemia in mice by the P210bcr/abl gene of the Philadelphia chromosome. Science 247:824–830

12. Kelliher MA, McLaughlin J, Witte ON, Rosenberg N (1990) Induction of a chronic myelogenous leukemia-like syndrome in mice with v-abl and BCR/ABL. Proc Natl Acad Sci U S A 87:6649–6653

13. Li S, Ilaria RL Jr, Million RP, Daley GQ, Van Etten RA (1999) The P190, P210, and P230 forms of the BCR/ABL oncogene induce a similar chronic myeloid leukemia-like syndrome in mice but have different lymphoid leukemogenic activity. J Exp Med 189:1399–1412

14. Hu Y et al (2004) Requirement of Src kinases Lyn, Hck and Fgr for BCR-ABL1-induced B-lymphoblastic leukemia but not chronic myeloid leukemia. Nat Genet 36:453–461

15. Peng C et al (2007) Inhibition of heat shock protein 90 prolongs survival of mice with BCR-ABL-T315I-induced leukemia and suppresses leukemic stem cells. Blood 110:678–685

16. Hu Y et al (2006) Targeting multiple kinase pathways in leukemic progenitors and stem cells is essential for improved treatment of Ph+ leukemia in mice. Proc Natl Acad Sci U S A 103:16870–16875

17. Yilmaz OH et al (2006) Pten dependence distinguishes haematopoietic stem cells from leukaemia-initiating cells. Nature 441:475–482

18. Zhang J et al (2006) PTEN maintains haematopoietic stem cells and acts in lineage choice and leukaemia prevention. Nature 441: 518–522

19. Peng C et al (2010) PTEN is a tumor suppressor in CML stem cells and BCR-ABL-induced leukemias in mice. Blood 115:626–635

Chapter 14

Murine Model for Colitis-Associated Cancer of the Colon

Ashley J. Snider, Agnieszka B. Bialkowska, Amr M. Ghaleb, Vincent W. Yang, Lina M. Obeid, and Yusuf A. Hannun

Abstract

Inflammatory bowel disease (IBD), including ulcerative colitis (UC) and Crohn's disease (CD), significantly increases the risk for development of colorectal cancer. Specifically, dysplasia and cancer associated with IBD (colitis-associated cancer or CAC) develop as a result of repeated cycles of injury and healing in the intestinal epithelium. Animal models are utilized to examine the mechanisms of CAC, the role of epithelial and immune cells in this process, as well as the development of novel therapeutic targets. These models typically begin with the administration of a carcinogenic compound, and inflammation is caused by repeated cycles of colitis-inducing agents. This review describes a common CAC model that utilizes the pro-carcinogenic compound azoxymethane (AOM) followed by dextran sulfate sodium (DSS) which induces the inflammatory insult.

Key words Inflammatory bowel disease, Colitis-associated cancer, Azoxymethane, Dextran sulfate sodium, Murine model, AOM/DSS model

1 Introduction

In many human pathologies inflammation and cancer development are known to be intertwined [1, 2]. Ulcerative colitis and Crohn's disease are two types of chronic IBD that increase the risk of development of CAC of the colon [3]. Both disorders are characterized by hyper-activation of the immune response, though they differ in the sites and specifics of the pathological features. The risk of developing CAC depends on multiple factors including the severity and longevity of the underlying inflammation as well as various known and unknown genetic predispositions. Moreover, the exact molecular mechanisms that underlie the transition from inflammation to cancer are not fully understood. The current body of evidence suggests that multiple factors play roles in CAC development: immune response, activation of oncogenes, inhibition of tumor suppressors, modifications to normal microRNA expression patterns, alterations to the epigenetic landscape, as well as commensal

Gabriele Proetzel and Michael V. Wiles (eds.), *Mouse Models for Drug Discovery: Methods and Protocols*,
Methods in Molecular Biology, vol. 1438, DOI 10.1007/978-1-4939-3661-8_14, © Springer Science+Business Media New York 2016

microbiota, and the corresponding inflammatory response of the epithelial cells [4].

Several animal models of CAC have been developed in rodents (mice, rats, and hamsters). The best studied chemically inducible model requires a combination of single injection of a carcinogen [azoxymethane (AOM), a pro-carcinogen that is metabolized to methylazoxymethanol (MAM), the active agent, in the liver] that is followed by an inflammatory insult using dextran sodium sulfate (DSS) [5]. The AOM/DSS model produces a pathology manifested by severe colitis with loss of body weight and bloody diarrhea that is followed by development of multiple colon tumors. The exact location of the tumors along the length of the colon varies depending on the mouse strain and background [4]. The carcinogenic process of this system has a pathological progression from normal intestinal crypts to the formation of foci harboring aberrant crypts with crypt fission and finally to the emergence of microadenomas [4, 6, 7]. These steps recapitulate the sequence of CAC formation in humans from inflammation through dysplasia to carcinoma.

Furthermore, the molecular perturbations governing CAC development and progression in the AOM/DSS model also mimic those observed in human patients. The molecular hallmarks of CAC development following AOM/DSS treatment are described in detail in Table 1. Studies from this animal model have shown increased activity of the Wnt signaling pathway due to mutational activation of β-catenin resulting in its accumulation in the nuclei [5, 7, 8]. The increased Wnt signaling activity along with an enhanced inflammatory immune response (e.g., interleukin 6 (IL-6)) in this model has been shown to result in elevated levels of c-myc, an activator of cell cycle progression and known oncogene [9]. It has been shown that chronic inflammation changes the

Table 1
Effectors of the AOM/DSS animal model

Factor	Effects	Selected reference
Inflammatory cells (macrophages, lymphocytes, plasma cells)	Elevated number	
Nuclear factor κB	Increased activity Increased nuclear localization	[11, 12]
JAK/STAT3 pathway	Increased activity	[14]
Pro-inflammatory cytokines (TNFα and IL-6)	Increased levels	[13, 14]
β-Catenin	Mutations in codons: 32, 33, 34 Increased nuclear localization	[5, 7, 8]

pattern of microRNA expression resulting in the activation of the phosphatidylinositide 3-kinase (PI3K) signaling pathway, aiding in the transition from inflammation to cancer [10]. Many key inflammatory components, nuclear factor of Kappa light polypeptide gene enhancer in B-cells (NF-κB), Janus kinase (JAK), signal transducer and activator transcription 3 (STAT3), pro-inflammatory cytokines (TNF alpha and IL-6), cyclooxygenase 2 (Cox-2; PTGS2), and inducible nitric oxide synthase (iNOS), have been demonstrated to be increased and activated during CAC formation [3, 7, 8, 11–14]. Upon chronic inflammation the immune system displays elevated levels of lymphocyte, plasma cell, and macrophage infiltration into the site of injury, all of which have the potential to modify the inflammatory microenvironment of the intestine and promote tumor establishment [6]. Additionally, recent studies demonstrated that AOM/DSS treatment causes a significant alteration in the diversity of the microbiome, promoting its dysbiosis and nurturing cancer development by promulgating the inflammatory environment [15]. Related to this, it is worth noting that some studies use *ApcMin/+* mice [16].

As with any model, the AOM/DSS-induced CAC model has its limitations. For example, *Kras* or *p53* mutations are typical in human, but have not been detected in this mouse model [17]. In contrast, *Kras* mutations have been observed in rats using the colon cancer model of AOM alone [18, 19]. Another factor is diet that significantly affects the outcome of AOM/DSS-induced CAC. Some of the altered effects may be due to changes in microbiota, an area of rapidly expanding research. The role of microbiota in this process is not well understood; however, alterations in host microbiota may account for inter-institutional variability. Helicobacter is also an additional factor that should be considered, as this can manifest and exacerbate the colitis symptoms of this model.

Rodent models for CAC serve as powerful tools for the investigation into mechanisms by which tumors develop in an inflammatory setting. The samples collected from these models can be analyzed for pathology [20, 21], immune infiltration, and immunohistochemistry for signaling pathways, specific immune cell populations, and genes of interest. Molecular biology techniques can be used to examine gene expression by quantitative RT-PCR and RNA sequencing, as well as by microarray and DNA sequence analysis. Protein profiling can be complemented by western blotting and other advanced proteomic techniques. Chromosomal and microsatellite instability and epigenetic changes can also be examined from tissues collected using this model. AOM/DSS-induced CAC can be utilized in combination with potential novel therapeutic modalities in order to determine and validate novel therapeutic targets.

In summary, the AOM/DSS animal model closely recapitulates histological, pathological, and molecular features of CAC in humans. Thus, it is a suitable model for studying the development, progression, and chemoprevention of this disease.

2 Materials

1. Male or female C57BL/6 mice (C57BL/6J, The Jackson Laboratory or Charles River ex-USA), 8–12 weeks old (*see* **Note 1**). All animal experiments have to be done in accordance with institutional and national guidelines and regulations.

2. Azoxymethane (AOM) (Sigma, cat. #: A5486-100MG).

3. Dextran sulfate sodium (DSS), 36–50 kDa (MP Biomedicals, cat #: 160110 (SKU 0216011050)) (*see* **Note 3**).

4. Acetic acid 100 %.

5. Ethanol 100 %.

6. 10% Buffered formalin.

7. Modified Bouin's fixative (prepared in-house; *see* Subheading 3.2).

8. 1 mL Syringes.

9. 10 mL Syringe.

10. Round-tip gavage needle.

11. 28 × 3/4 Needles.

12. Dissection tools.

13. Dissecting microscope.

14. Wooden toothpicks.

15. Eppendorf tubes.

16. Chamber connected to a gas source of CO_2 or isoflurane.

17. Ruler (in centimeters).

18. Reagent, such as Hemoccult SENSA kit (Beckman Coulter, cat. # 64151), to detect occult blood in stool.

19. Weighing scale sensitive enough to weigh mice (e.g., sensitivity range 0.01–100 g).

20. Clean 150 mm Petri dishes.

21. Phosphate-buffered saline 1× (PBS).

3   Methods

3.1   Preparation of AOM for Injection

Reconstitute AOM to a final concentration of 10 µg/µL by adding 10 mL sterile deionized water (dH$_2$O) to the AOM vial. Vortex until dissolved. Aliquot about 500 µL in Eppendorf tubes and store in -20 °C until use (*see* **Note 2**).

3.2   Preparation of Modified Bouin's Fixative

To prepare 1 L use 1 L measuring cylinder and add 500 mL ethanol 100%, plus 50 mL acetic acid 100%, and then add 450 mL dH$_2$O. Final concentration of ethanol and acetic acid is 50% and 5%, respectively. Stir to mix. Store at room temperature. (This modified fixative contains no picric acid or formaldehyde.)

3.3   Preparation of AOM and Injection in C56BL/6 Mice

Weigh mice, and then per mouse, prepare 10 µg AOM/g mouse in a total volume of 100 µL to be injected (*see* **Notes 2** and **4**). Example: for a 20 g mouse take 20 µL of the 10 µg/µL AOM stock plus 80 µL sterile dH$_2$O, i.e., a total of 100 µL. To compensate for losses due to dead needle volume, per mouse take 60 µL of the AOM stock + 240 µL sterile dH$_2$O = 300 µL, and inject 100 µL i.p. using a 28G needle and 1 mL syringe.

3.4   Preparation and Administration of DSS

1. One week after AOM injection, prepare 2.5% DSS by dissolving 2.5 g DSS per 100 mL dH$_2$O (*see* **Note 3**). For each experiment prepare a total of 300–400 mL of 2.5% DSS to be used per mouse cage. Weigh the mice on the first day of DSS administration, and every day after that until the experiment is terminated (*see* **Note 4**).

2. Leave the mice with 2.5% DSS in drinking water for 5–6 days, followed by 2 weeks of recovery, replacing the DSS water with regular water.

3. After the recovery period start a second cycle of 2.5% DSS for 5 days followed by 2 weeks of recovery, replacing the DSS water with regular water.

4. Euthanize the mice at the end of the last of recovery, 2–6 weeks after the last round of DSS, which varies by background strain and genetic modification (*see* **Note 5**).

3.5   Assessment of Colitis Induction in Treated Mice

1. By days 5–6 of the first DSS treatment, as an indicator of induction of colitis mice will have lost approximately 10% of their starting body weight (*see* **Note 4**).

2. If blood stains are not obvious around the anal region, then collect stool from mice to determine the presence of occult blood by using the Hemoccult kit. Working one mouse at a time, lift the mouse from its tail to elevate its hind legs. Hold the mouse in this position with one hand to induce the mouse to defecate. With the other hand hold a wooden spatula

(provided with the kit) and collect on it any faecal material the mouse produces. Assess the consistency of the stool and determine the stool score (*see* Table 2). Follow instructions from the manufacturer to test for occult blood in stool.

3. By the end of the first recovery period, mice weight should return to approximately the level that it was before DSS treatment.

4. By days 5–6 of the second DSS treatment, mice should have lost about 10–15 % of their starting body weight as an indicator of colitis (*see* **Note 4**).

5. Test for occult blood in stool as described above in **step 2**.

6. By the end of the second recovery period mice weight should return to approximately the level that it was before DSS treatment.

3.6 Assessment of CAC Following AOM/DSS Treatment

1. At the end of the second recovery period (2–6 weeks following the last cycle of DSS), euthanize mice (one mouse at a time) using CO_2 or isoflurane asphyxiation. Put the mouse in the chamber connected to the gas source. Allow the gas to flow gas into the chamber until the mouse is unconscious and all movement has ceased. Take the mouse out and place it on a flat surface with abdominal side down and perform cervical dislocation to ensure and confirm death of the mouse. This is best done by using one hand to hold and place a pair of forceps on the neck of the mouse right behind the skull and then apply slight pressure against the neck. With the other hand hold the tail firmly and pull backwards while keeping pressure on the neck with the forceps. You should hear/feel the cervical spine dislocate.

2. Using dissection forceps and scissors, cut an incision in the skin of the abdominal side. With a pair of forceps, hold the skin at the incision and pull gently away from the abdominal muscle tissue. Use a pair of scissors to cut the abdominal skin and expose the abdominal muscles.

Table 2
Assignment of clinical scores of colitis in C57BL/6 mice [23, 24]

Parameter/score	Weight loss (%)	Stool consistency	Occult/gross bleeding
0	None	Normal	Negative
1	1–5	–	–
2	5–10	Loose stool	Hemoccult positive
3	10–20	–	–
4	>20	Diarrhea	Gross perianal bleeding

3. Hold and pull up the peritoneum with the forceps. Make sure not to be holding and pulling at the intestines. Carefully make an incision in the peritoneal tissue, and continue cutting away to expose the intestines.

4. Identify the colon and trace to the distal end where it joins the rectum/anus. With a pair of scissors, cut as close to the anal opening as you can.

5. With one hand or forceps hold the distal end of the colon, and using the other hand gently unravel the entire length of the colon from any mesenteric connective and/or fat tissue.

6. Identify the cecum (small pouch between the small and large intestine), and cut at where the cecum and the colon join (the proximal end of the colon) to free the colon.

7. Using a ruler, measure to the nearest mm and record the length of the colon.

8. Fill a 10 mL syringe with modified Bouin's fixative and attach a gavage needle to it. Insert the needle about half a centimeter in the anterior opening of the colon.

9. With the fingers of one hand hold the needle inside the colon by applying firm pressure on the colon; with the other hand holding the syringe apply gentle but consistent pressure to flush the contents of the colon using modified Bouin's fixative. This step allows simultaneous cleaning of the colon and immediate fixation. Use a petri dish to collect the flow-through waste. Fixation can be observed by the colon color turning opaque (see **Note 6**).

10. Using scissors, cut open the colon lengthwise. Hold with a pair of forceps and rinse briefly in a petri dish containing PBS.

11. Use the top of a petri dish to place the cleaned and opened colon. Place the colon with the luminal side facing upwards. The luminal side can be easily identified by the variegations/ridges present at the proximal end of the colon. Place the lid with the colon under dissection microscope. Identify any tumor growth along the length of the colon and record the number of tumors (see **Note 7**).

12. Keep the colon flat open and pull it with forceps from its proximal end towards the edge of the petri dish. Keep luminal side facing up and hold the proximal end with the forceps with one hand and with the other hand hold a toothpick. Wrap the edge of the proximal end around the toothpick using the forceps and slightly pinch the wrapped edge against the toothpick to hold it in place. Gently and slowly start rolling the toothpick with your fingers to roll the colon around the toothpick to form a swiss-roll. Once the entire colon length has been rolled up, use a pair of forceps to carefully slide the colon swiss-roll

off the toothpick and into a tissue processing/embedding cassette. Place the cassette in 10 % buffered formalin for processing.

4 Notes

1. While mice from both sexes are susceptible to AOM/DSS-induced CAC, based on author's observations, female mice tend to yield more consistent and reproducible results. An exception to this is for studies involving dietary manipulations, including high-fat diets; these studies typically utilize male mice.

2. Stock solutions of AOM should not be used after repeated freeze-thaw cycles, and should be replaced after 1 year.

3. The source of DSS is very important as DSS from different manufacturers or even different lots can yield significantly different results. It is recommended to utilize the same lot from the same manufacturer for each experiment. DSS should also be prepared fresh for each cycle of administration, as the solution may become turbid after several days of administration.

4. Body weight should be used as an indicator of overall animal health and DSS doses may need to be modified.

5. This chapter describes a model of CAC used primarily in C57BL/6 mice (C57BL/6 mice from any commercial vendor seem to exhibit similar sensitivity). This strain of mice is commonly used for this model as they are susceptible to AOM/DSS-induced CAC [22]. Suzuki et al. evaluated strain differences among murine models of AOM/DSS-induced CAC and determined that Balb/c mice were most susceptible to tumor development. In this study, Balb/c mice demonstrated the highest tumor incidence (100 %) followed by C57BL/6 (80 %), C3H/HeN (29 %), and DBA/2 N mice (20 %) [22]. Of note, this study was conducted using a single injection of AOM (10 mg/kg) as described here; however, mice received 1 % DSS in drinking water for 4 days (only one cycle) and were euthanized at 18 weeks. This study demonstrated that the background strain of mice used significantly affected tumor development. Listed below are some modifications that can be used with this model.

    (a) DSS doses can be increased (or decreased) based on the background strain of mice or genetic modification in mice for mice that are less (or more) susceptible to colitis. Mice that are very susceptible to colitis may need ≤1 % DSS.

    (b) The duration of the experimental model can be manipulated to increase or decrease tumor development. Cycles of

2.5% DSS (or lower) and water may be repeated more than twice to decrease the duration of the model.

(c) Additional carcinogens have also been used to induce the initiation of tumors including 1,2-dimethylhydrazine (DMH; AOM is a metabolite of DMH), and/or methylazoxymethane (MAM). DMH has long been used to study colorectal cancer and adenomas induced by this carcinogen often invade into the submucosa and muscularis, while tumors induced by MAM and AOM do not. When administered with rounds of DSS, DMH yields similar tumor incidence to AOM [23]. MAM is commonly used in combination with 1-hydroxyanthraquinone in rat models of CAC (reviewed in [24]).

6. Make sure not to apply too much pressure during flushing; otherwise the colon might burst open.

7. Under a dissecting microscope, examine the colon for the presence of tumors which should appear as abrupt outgrowth relative to the luminal surface of the colon, and with a relatively denser cellular composition. Size determination is done by measuring the longest and shortest surface diameters of the tumors and calculating the average surface diameter. This step is best done if the dissecting microscope is equipped with an ocular micrometer. Size (average diameter) will vary from ≤ 1 to ≥ 3 mm. The length of the experiment and the particular mouse genotype and background are variables that will determine the tumor number and size.

Acknowledgments

This work was supported by a Veterans Affairs Merit Award (LMO), as well as NIH Grants CA084197 and DK052230 (VWY), and CA172517 and CA097132 (YAH).

References

1. Elinav E et al (2013) Inflammation-induced cancer: crosstalk between tumours, immune cells and microorganisms. Nat Rev Cancer 13(11):759–771

2. Grivennikov SI, Greten FR, Karin M (2010) Immunity, inflammation, and cancer. Cell 140(6):883–899

3. Francescone R, Hou V, Grivennikov SI (2015) Cytokines, IBD, and colitis-associated cancer. Inflamm Bowel Dis 21(2):409–418

4. De Robertis M et al (2011) The AOM/DSS murine model for the study of colon carcinogenesis: from pathways to diagnosis and therapy studies. J Carcinog 10:9

5. Tanaka T (2012) Development of an inflammation-associated colorectal cancer model and its application for research on carcinogenesis and chemoprevention. Int J Inflam 2012:658786

6. Okayasu I et al (1996) Promotion of colorectal neoplasia in experimental murine ulcerative colitis. Gut 39(1):87–92

7. Tanaka T et al (2003) A novel inflammation-related mouse colon carcinogenesis model

induced by azoxymethane and dextran sodium sulfate. Cancer Sci 94(11):965–973

8. Tanaka T (2009) Colorectal carcinogenesis: review of human and experimental animal studies. J Carcinog 8:5

9. Kanneganti M, Mino-Kenudson M, Mizoguchi E (2011) Animal models of colitis-associated carcinogenesis. J Biomed Biotechnol 2011:342637

10. Josse C et al (2014) Identification of a microRNA landscape targeting the PI3K/Akt signaling pathway in inflammation-induced colorectal carcinogenesis. Am J Physiol Gastrointest Liver Physiol 306(3):G229–G243

11. Greten FR et al (2004) IKKbeta links inflammation and tumorigenesis in a mouse model of colitis-associated cancer. Cell 118(3):285–296

12. Paradisi A et al (2009) Netrin-1 up-regulation in inflammatory bowel diseases is required for colorectal cancer progression. Proc Natl Acad Sci U S A 106(40):17146–17151

13. Popivanova BK et al (2008) Blocking TNF-alpha in mice reduces colorectal carcinogenesis associated with chronic colitis. J Clin Invest 118(2):560–570

14. Grivennikov S et al (2009) IL-6 and Stat3 are required for survival of intestinal epithelial cells and development of colitis-associated cancer. Cancer Cell 15(2):103–113

15. Zackular JP et al (2013) The gut microbiome modulates colon tumorigenesis. MBio 4(6): e00692–13

16. Moser AR, Pitot HC, Dove WF (1990) A dominant mutation that predisposes to multiple intestinal neoplasia in the mouse. Science 247(4940):322–324

17. Suzui M et al (1995) No involvement of Ki-ras or p53 gene mutations in colitis-associated rat colon tumors induced by 1-hydroxyanthraquinone and methylazoxymethanol acetate. Mol Carcinog 12(4):193–197

18. Erdman SH et al (1997) Assessment of mutations in Ki-ras and p53 in colon cancers from azoxymethane- and dimethylhydrazine-treated rats. Mol Carcinog 19(2):137–144

19. Takahashi M et al (2000) Altered expression of beta-catenin, inducible nitric oxide synthase and cyclooxygenase-2 in azoxymethane-induced rat colon carcinogenesis. Carcinogenesis 21(7):1319–1327

20. Cooper HS et al (1993) Clinicopathologic study of dextran sulfate sodium experimental murine colitis. Lab Invest 69(2):238–249

21. Kullmann F et al (2001) Clinical and histopathological features of dextran sulfate sodium induced acute and chronic colitis associated with dysplasia in rats. Int J Colorectal Dis 16(4):238–246

22. Suzuki R et al (2006) Strain differences in the susceptibility to azoxymethane and dextran sodium sulfate-induced colon carcinogenesis in mice. Carcinogenesis 27(1):162–169

23. Kohno H et al (2005) Beta-Catenin mutations in a mouse model of inflammation-related colon carcinogenesis induced by 1,2-dimethylhydrazine and dextran sodium sulfate. Cancer Sci 96(2):69–76

24. Tanaka T et al (2000) Colitis-related rat colon carcinogenesis induced by 1-hydroxyanthraquinone and methylazoxymethanol acetate (review). Oncol Rep 7(3):501–508

Chapter 15

Mouse Models for Studying Depression-Like States and Antidepressant Drugs

Carisa L. Bergner, Amanda N. Smolinsky, Peter C. Hart, Brett D. Dufour, Rupert J. Egan, Justin L. LaPorte, and Allan V. Kalueff

Abstract

Depression is a common psychiatric disorder, with diverse symptoms and high comorbidity with other brain dysfunctions. Due to this complexity, little is known about the neural and genetic mechanisms involved in depression pathogenesis. In a large proportion of patients, current antidepressant treatments are often ineffective and/or have undesirable side effects, fueling the search for more effective drugs. Animal models mimicking various symptoms of depression are indispensable in studying the biological mechanisms of this disease. Here, we summarize several popular methods for assessing depression-like symptoms in mice, and their utility in screening antidepressant drugs.

Key words Depression, Animal models, Antidepressant drug screening, Despair, Anhedonia, Chronic stress

1 Introduction

The underlying pathophysiology of depression remains unclear despite the seriousness and prevalence of this disorder [1]. Clinical symptoms of depression manifest at psychological, physiological, and behavioral levels, and include changes in appetite and sleeping patterns, sad or irritable mood, psychomotor agitation, fatigue, anhedonia, poor concentration, feelings of guilt, and recurrent thoughts of suicide or death [2–5]. While the introduction of monoamine-based antidepressants has promoted various neurotransmitter system-based models of depression [1], little is known about their mechanisms of therapeutic action. Additionally, up to 46 % of depressed patients do not fully respond to initial monotherapy antidepressant treatments [6], collectively emphasizing the need for newer and more effective drugs [1, 2, 5, 7].

Animal models are widely used to study the neurobiological mechanisms of depression [8–10]. Ideal animal depression models must be reasonably analogous to the human symptoms, be able to

Gabriele Proetzel and Michael V. Wiles (eds.), *Mouse Models for Drug Discovery: Methods and Protocols*,
Methods in Molecular Biology, vol. 1438, DOI 10.1007/978-1-4939-3661-8_15, © Springer Science+Business Media New York 2016

be monitored objectively, be reversed by the same treatment modalities as humans, and be reproducible between laboratories [3, 5, 11]. Although selected depression symptoms may be irreproducible in animals (e.g., thoughts of suicide), a number of models exhibit considerable construct validity when targeting other clinical endophenotypes of depression [4, 5, 12]. Antidepressant treatment has been shown to affect the behavioral responses in these models (see further), indicating that certain depression paradigms are pharmacologically sensitive, and therefore, can be used in the testing of antidepressant drugs in mice.

A clear distinction must be made between animal models of depression and animal tests (or screens) of antidepressant drugs. Examples of both types of animal paradigms, equally important for further progress in biological psychiatry and drug discovery, are discussed here in detail, based on their common use in behavioral pharmacology research. Finally, automated versions of some of these tests are currently available [13, 14], enabling consistent behavioral measurement, standardization of experimental protocols, and increased throughput and testing.

2   Materials

2.1   Animals

1. Various inbred, selectively bred, and genetically modified (mutant or transgenic [15]) mice (see more details in the chapter 16 on mouse models of anxiety in this volume). We recommend using most of the inbred strains listed in the A-priority list of the Mouse Phenome Project database (www.jax.org/phenome), especially C57BL/6J, 129S1/SvImJ, and BALB/c mice. We also recommend browsing the Mouse Genome Informatics (http://www.informatics.jax.org/) database by the depression phenotype to find appropriate transgenic or mutant strains (e.g., $Disc1^{Rgsc1393}/Disc1^{Rgsc1393}$, Tg(Syn1-ADCY7)11004Btab/0 mice).

2. In general, avoid strains with overt motor or sensory deficits (e.g., vestibular, cardiovascular, and visual) when using tests that may be confounded by these factors (see further). In addition, males and females may exhibit different behavioral reactions to the experimental stimuli [16]. Therefore, the sex is also an important factor to consider when choosing an appropriate animal for experimentation (*see* **Note 1**).

2.2   Housing

1. If mice are obtained from a commercial vendor or another laboratory, allow at least 1 week acclimation from shipping stress. In most cases, a much longer time will be required. Young mice recover more quickly from shipping stress (i.e., 1 week) than adult mice, which may require several weeks to acclimate.

Food and water should be freely available, unless the intake is being controlled for experimental purposes.

2. Utilize plastic, solid-floored cages with sufficient space (e.g., <5 animals per cage). The mouse holding room should be kept at approximately 21 °C, on a 12–12 h light cycle. As mice are nocturnal, the light cycle may be inverted if spontaneous activity measures are needed [17].

3. All experimental procedures (including handling, housing, husbandry, and drug treatment) must be conducted in accordance with national and institutional guidelines for the care and use of laboratory animals.

2.3 Requirements for Experimental Models

1. Sucrose Consumption Test:
   4–10 % Sucrose (*see* **Note 2**).
   Home cage.
   Two drinking bottles, one with pure water and the other with a sucrose solution (add sucrose to the animals' standard drinking water, as a change in water type may dissuade animals from drinking).

2. Forced Swim Test (FST):
   Clean glass cylinder (e.g., height 25 cm, diameter 10–15 cm).
   Water maintained at 23–25 °C.
   Towels to dry animals after swimming.
   Stop watch to calculate the duration of immobility.
   Optional: video-camera for subsequent video-tracking and data analysis (e.g., [13]. For more information on behavioral tracking software, please *see* Table 1). Video tracking software

Table 1
Automated video-tracking system manufacturers

Name	City	Country	Web address
Any-Maze	Wood Dale, IL	USA	www.anymaze.com
Bioseb	Vitrolles	France	www.bioseb.com
CleverSys Inc.	Reston, VA	USA	www.cleversysinc.com
Harvard Apparatus	Holliston, MA	USA	www.harvardapparatus.com
Linton Instrumentation	Diss, Norfolk	England	www.lintoninst.co.uk
Medi Analytika India Pvt. Ltd	Adyar, Chennai	India	www.medianalytika.com
Noldus	Leesburg, VA or Wageningen	USA or Netherlands	www.noldus.com
San Diego Instruments	San Diego, CA	USA	www.sandiegoinstruments.com
TSE Systems	Midland, MI	USA	www.tse-systems.com

Table 2
Selected commercial suppliers of behavioral equipment for depression research

Test apparatus	Manufacturer	Company website
Tail suspension test	Panlab, Barcelona, Spain	www.panlab.com
	Columbus Instruments, Columbus OH, USA	www.colinst.com
	Bioseb, Vitrolles, France	www.bioseb.com
Forced swim test	Panlab, Barcelona, Spain	www.panlab.com
	San Diego Instruments, San Diego CA, USA	www.sandiegoinstruments.com
	Bioseb, Vitrolles, France	www.bioseb.com

requires highly developed algorithmic analysis of input; however, recording typically may be done with a standard video camera. Alternative methods of behavioral tracking include vibration-based (e.g., Bioseb, Vitrolles, France) FST activity monitoring.

3. Tail Suspension Test (TST):

A shelf or tail suspension apparatus to suspend mice. The apparatuses may be wooden or plastic boxes (e.g., $680 \times 365 \times 280$ mm), painted to contrast with mice. The design of the TST apparatus is usually negotiable, given that the animal is securely attached to a solid suspension apparatus, and that this apparatus is at least 35 cm above the nearest surface [18, 19]. Several companies provide behavioral tracking software that is flexible with the variations in experimental design, and would yield reliable data that would translate between designs (Table 1). Additionally, there are also prefabricated apparatuses that can be purchased (*see* Table 2 for details).

Tape measure to determine the height of suspension.

Adhesive tape to secure mice to suspension apparatus (*see* **Notes 13** and **14**).

Optional: automated electromechanical strain gauge device, video tracking system.

In the tail suspension test (TST), mice initially engage in vigorous escape behaviors, but eventually succumb to immobility. Like the FST, longer durations of TST immobility infer a heightened degree of behavioral despair. As such, TST is a commonly used screening method for antidepressant properties of drugs, and is highly sensitive to pharmacological manipulations. Antidepressant drugs generally decrease the duration of TST immobility in mice [14, 20–22].

4. Chronic Mild Stress (CMS):

Supplementary cages for application of stressors.

Various stressors, e.g., soiled rat bedding, confinement tube, or predator sounds (*see* **Note 18**).

3 Methods

3.1 Observations and General Procedures

1. Observers must refrain from making noise or movement, as their presence may alter animal behavior. Assess intra- and inter-rater reliability for consistency. See details in the chapter on animal models of anxiety. Note that strong scents (e.g., perfume) and loud or sudden noises should be avoided in the experimental room.

2. Allow at least 1 h acclimation of mice after their transfer from the animal holding room to the experimental room.

3. After each testing session, clean the equipment (e.g., with a 30% ethanol solution) to eliminate olfactory cues.

3.2 Drug Administration

1. All experimental protocols described here are compatible with testing various antidepressants, administered with a vehicle (e.g., saline). A typical experiment may include one or several drug-treated groups (e.g., several doses or several pretreatment times) compared to a vehicle-treated group of mice. Usually (unless stated otherwise), 10 animals per experimental group will be needed, also providing adequate statistical power (see further). However, if the effects of the drugs are particularly robust, a smaller n (e.g., $n=7-8$) may suffice. For mild effects, a larger number of animals ($n=15-16$) may be required.

2. Common routes of injection include systemic (intraperitoneal (i.p.), intramuscular (i.m.), intravenous (i.v.), peroral (p.o.), subcutaneous (s.c.)) and local (intracerebral (i.c.) or intracerebroventricular (i.c.v) or intranasal (i.n.)). Route of administration, dose, and pretreatment time vary depending on strain sensitivity and the drug being used. Continuous drug infusion (using osmotic pumps, such as Alzet pumps) at a constant rate may be used to improve the availability of the drug, and implantable depots can be used for s.c. drug administration to achieve lasting therapeutic effect.

3.3 Data Analysis

1. Behavioral data may be analyzed with the Mann–Whitney U-test for comparing two groups (parametric Student's t-test may be used only if data are normally distributed), or analysis of variance (ANOVA) for multiple groups, followed by an appropriate post hoc test.

2. Some experiments may require one-way ANOVA with repeated measures, or n-way ANOVA depending upon the number of groups tested (see more details in the chapter on mouse models of anxiety in this volume).

3.4 Sucrose Consumption Test

A core symptom of depression is anhedonia—a decreased interest in pleasurable activities [2]. There are several commonly used tests to assess hedonic deficits in mice. The sucrose consumption test examines anhedonia in a relatively short period of time without the need for expensive equipment or extensive training of the test animals. In this model, a mouse is given free choice between water and a sucrose solution to drink. Usually, healthy mice show a clear preference for the sweetened water, while depressed animals demonstrate markedly less interest. A pure chance would result in animals drinking equally (50 %) from each bottle, and a preference for sucrose of less than 65 % is considered to be an indication of hedonic deficit [23]. Since various antidepressant drugs reverse the anhedonia-like reduction in preference for sucrose, e.g., [24–26], this test is widely used in the screening of antidepressant drugs. As in most experiments, between 8 and 12 mice may be used per group in this test. However, as few as six mice may yield good results, if depression-like phenotypes are robust. C57BL/6J and 129S1/SvImJ mice respond well in this assay (*see* **Note 2**). We recommend finding suitable mouse strains through Mouse Genome Informatics or Mouse Phenome Database based on each laboratory's individual scientific needs.

1. For a set period of time (e.g., 1, 3, or 7 days), allow experimental mice (housed in their standard home cages) access to two freely available water bottles—one containing tap water and the other containing a solution of up to 35 % sucrose. To preclude side preference in drinking, switch the positions of the bottles halfway through the procedure (*see* **Notes 2** and **3**).

2. Measure the volumes of sucrose solution and water consumed. Calculate the preference for the sucrose solution as a percentage of total liquid consumed, and total sucrose intake in mg/g body weight. In addition, commercially available automated lick-counters (lickometers) may be used (e.g., by Lafayette Instrument Co, Lafayette IN, USA or Columbus Instruments, Columbus OH, USA). Assess the number of licks at each bottle for the duration of the test (i.e., 24–72 h) per 100 mg of body weight, and the preference for sucrose as a percentage of total licks [23, 27, 28] (*see* **Notes 4–6**).

3.5 Coat State Assessment

The coat state assessment is a fast and simple qualitative method of assessing mouse depression-like states through observation of the condition of an animal's fur. In rodents, coat state tends to decline with increased depression, similar to depressed patients who frequently exhibit poor hygiene [29–31]. Antidepressants have been shown to improve the coat condition of mice while reducing depression-like symptoms [29–31]. For example, the reduction of corticotropin-releasing factor (CRF) has been associated with improved coat state (and is implicated in depression) [32]. Of importance here, antidepressants (e.g., imipramine) and anxiolytics (e.g., chlordiazepoxide) have been shown to interact with corticotropin-releasing factor [33] (*see* **Note 7**).

1. After removing the animal from the homecage, assess the coat state in each of eight regions: head, neck, forepaws, dorsal coat, ventral coat, hindlegs, tail, and genital region. A coat with a healthy appearance (i.e., unchanged throughout the course of the experiment, normal coat state) should receive a score of 0. Conversely, a coat state that appears damaged or dirty (i.e., noticeably different from a normal coat state) should receive a score of 1. The average of the eight scores for each animal can then be compared among individuals or groups [30, 31]. Due to the subjectivity of this assessment, it is beneficial to have more than one observer score each animal (these results should also be compared for inter-rater reliability). One way to minimize bias is to take the animal in question (i.e., the one with a seemingly dirty coat state) and compare it with another animal with an apparently normal coat state. Another way to observe an abnormal coat is to search for mild to severe piloerection, either generally or on specific body parts [32]. This can either be in addition to, or independent from, a dirty coat appearance. Taken together, these symptoms signify the animal is not grooming normally and has declining hygiene, implicating overt depressive-like symptoms, (*see* **Notes 8** and **9**).

3.6 Forced Swim Test (FST)

1. This test can be performed manually or with automated video/software systems. Manual labor is not as high throughput as automated behavioral tracking software, and the latency of the observer to react can reduce the accuracy or data acquisition. Regardless of manual or automatic observation, the number of mice that can be tested depends on the duration of the test (6 min is enough to obtain reliable data and to determine significance in this test; therefore, 10 mice could be done per hour).

   Although it does not induce experimental depression in mice, the FST is one of the most commonly utilized ethological models of fast high-throughput antidepressant screening. The FST places mice in an inescapable aversive situation and measures their "despair," (learned helplessness) by a measure of increased duration of immobility in the water. Animal FST immobility is markedly reduced by antidepressant drugs. The FST has good predictive validity and is widely used in research investigating acute and chronic effects of antidepressant drugs [20–23, 34] (*see* **Note 10**).

2. Place mice individually into a glass cylinder filled with 10 cm of water for 6 min.

3. As a measure of depression-like behavior, the total duration of immobility and the number of immobility episodes should be recorded. Immobility is defined as the absence of movement,

unless they are necessary for the animal to stay afloat (head above water) (*see* **Notes 11** and **12**).

4. After testing, dry mouse thoroughly with towels and return to their homecages.

3.7 Tail Suspension Test (TST)

1. Mice may be suspended by the tail on the edge of a shelf or in a special apparatus, at least 35 cm above the floor (from the beginning of the tail).

2. The mice should be secured by adhesive tape approximately 1 cm from the tip of the tail for 6 min (*see* **Notes 13** and **14**).

3. Researchers may choose to manually record data through direct observation or automatically collect data using a strain gauge device to detect movements.

4. Mice are considered immobile only when hanging passively and completely motionless (*see* **Note 15**).

3.8 Chronic Mild Stress (CMS)

Chronic mild stress (CMS) presents mice with an unpredictable barrage of stressors to induce (rather than simply measure) a depressed state. CMS reduces sucrose or saccharin intake in mice, a symptom of anhedonia (see above). CMS may also be responsible for decreases in sexual and aggressive behavior, changes in sleeping habits, loss of body weight, pituitary-adrenal hyperactivity, an increased threshold for brain stimulation reward, and an abolishment of place conditioning, making it a valid ethological model of depression. These behavioral deficits can be reduced through chronic treatments of antidepressants, accentuating the pharmacological sensitivity of CMS procedure [12, 35–37] (*see* **Note 16**).

1. Following a random schedule, expose mice to two or more stressors each day for 4–7 weeks (*see* **Note 17**).

2. Typical stressors may include: cage tilting (e.g., 45°), predator sounds, placement in an empty cage, placement in an empty cage with water on the bottom, damp sawdust, inversion of light/dark cycle, lights on during dark cycle, switching cages, food or water deprivation, short-term confinement in a tube, soiled cages with rat odors, and an inescapable footshock [35]. Full experimental design must be published (e.g., degree and duration of cage tilt, dimensions of confinement tube and duration isolated in this tube, quality and duration of predator sounds, size of empty cage, and duration of isolation), so that it can be compared across studies and between laboratories. Time of day during administration of stressors, as well as for assessing depression (sucrose intake, bouts of fighting or aggressive behavior, loss of body weight), must be recorded and standardized when possible. Additionally, duration of exposure to sucrose to measure intake should be provided by each study utilizing this model. The schedule of stressors

should be random; however, they should be recorded and published citing the order in which they occurred to isolate this pattern as potentially manipulating experimental results (*see* **Note 18**).

3. To prevent habituation and enhance the unpredictable nature of the model, stressors should be applied at varying time intervals.

4. Following the period of stress, mice can be tested with behavioral models of depression such as: coat state assessment, sucrose consumption test, FST, or TST (see these protocols above), to determine the effectiveness of the test.

4   Notes

1. Occasionally, mice may have altered cognitive domains that may be easily misinterpreted in models of depression [9]. For example, mice with elevated learning and memory abilities may display active initial locomotion that decreases significantly over time. While this reduction in locomotion may be attributable to heightened learning and habituation, it is often incorrectly assessed as behavioral despair. Likewise, mice with particularly low levels of memory and learning may be misinterpreted as persistently hyperlocomotive. The lack of habituation and decreased sensitivity to repeated stressors may be a result of a reduced learning phenotype, not hyperlocomotion. Similarly, mice displaying hypoactivity and increased sensitivity to repeated stressors (incorrectly categorized as anxious) may be associated with an increased level of depression and enhanced memory. Additionally, sustained hypoactivity coinciding with a decrease in habituation and sensitivity to repeated stressors may not be the result of increased anxiety or decreased despair. Rather, these behaviors may indicate reduced learning and memory, but heightened depression. Overall, cognitive functions may strongly modulate animal performance in ethological models of depression. To diminish the likelihood of incorrect interpretation of behavioral data as depression, it is recommended that mice are carefully tested in memory and learning specific tests [9].

2. The use of a 4–10% sucrose solution will usually generate good results for most mouse strains (e.g., C57BL/6J, 129S1/SvImJ mice). However, some strains (especially mutant or transgenic mice) may have abnormally reduced taste sensitivity, which would make assessment of their hedonic responses in this test difficult. Review Mouse Genome Informatics for mice with abnormal taste sensitivity (e.g., Gnat3tm1Rfm/Gnat3tm1Rfm). Most other mice will respond accurately to

this test. However, always check taste sensitivity prior to performing a sucrose consumption test by using a standard taste sensitivity test (see specific mouse phenotyping literature [8–10] for details). Consider using a different strain if the problem persists. Alternatively, higher concentrations of sucrose (e.g., 20–35 %) may be required.

3. To avoid the confounds of metabolic factors and acute stress, allow food and water *ad libitum* prior to performing the sucrose consumption test. However, some strains may have altered water consumption (e.g., polydypsia), and the sucrose consumption test may not always be suitable for such strains. For example, this test may be unsuitable for diabetic (e.g., $Hk2^{tm1Laak}/Hk2^{+}$) or obese strains with altered water consumption.

4. Although this test can also be performed over a short time period (e.g., 2 h), mice consume so little over such a brief time and errors in measurement can result. Consider lengthening the period of the test to at least 24 h (a 3-day test will be more appropriate in most cases). Note that this protocol must have flexibility and adaptability to be useful as general guidance across laboratories and countries. Rigidly stated specifics can deter novice experimenters from implementing their own research ideas and techniques.

5. Neophobia to the presence of multiple water bottles and to the taste of sucrose may also confound behavioral results in this model. To avoid this problem, acclimate mice by giving them two bottles, each with the sucrose solution, for 72 h before the test, or with one water and one sucrose bottle for 1 h/day for 1 week. Also, consider lengthening the period of the test to at least 24 h. Researchers may choose to utilize video recording to document all water intake, however, endpoint analysis of overall sucrose consumption (as described earlier) is acceptable.

6. Depending on the length of time over which this test is conducted, mice may alternate between active and inactive phases, which demonstrate marked differences in the animals' liquid consumption. When switching the positions of the bottles to avoid side preference, be sure to take shifting activity levels into account so that each bottle is in each position for the same amount of each activity phase.

   Some mouse strains may develop a metabolic syndrome-like phenotype or have pathologically high reward-related phenotype (see Mouse Phenome Project or Mouse Genome Informatix database for these phenotypes of interest: e.g., metabolic syndrome-like phenotypes in $Neil1^{tm1Rsld}/Neil1^{tm1Rsld}$ mice). Thus, their sucrose consumption may be abnormally affected, and alternative methods of depression testing or other mouse strains may be required.

7. Various mouse strains may have different sensitivity in this test. For example, C57BL/6J mice can be somewhat resistant to the deleterious effects of chronic stress on the coat state [38]. Strain differences may result in differing levels of grooming activity. For example, some inbred strains may be inherently poor (e.g., BALB/cJ) or excellent (e.g., A/J) groomers, regardless of stress levels. Some genetically modified mouse strains also display "compulsive" grooming behavior [39] that may mask any alterations in the animal's coat state. Consider a more suitable strain if floor or ceiling effects occur.

8. In socially housed mice, hetero-grooming may confound self-grooming data. Single-housing mice may eliminate this confound, but this practice should be used with caution, as social isolation stress may induce aberrant behavioral effects. Typically use 8–10 subjects per group in order to obtain reliable data. The nature of this specific test is to be used as complementary assessment of depression to supplement other models of depression. This test is relatively simple; however, it may be more useful when performed adjunct to another test, such as the tail suspension test (e.g., in the tail suspension test, before releasing the animal, simply score the coat state in addition to the other endpoints measured in this test). Similarly, coat state can be assessed while animals are still in the home cage, or with little manipulation outside of the home cage.

9. Some mouse strains may display pronounced balding patches due to alopecia [40] or increased auto- or hetero-barbering behavior [41–43], which will make the coat state data less valid. Therefore, this model may not be used in high barbering strains. Likewise, stress *per se* may promote barbering in mice [44], thereby further confounding the coat assessment protocol.

10. Consider strain and individual differences in baseline immobility duration. C57BL/6J, BALB/cJ, and 129/SvEmJ strains have all been shown to provide reliable data, and good sensitivity to pharmacological manipulations in this test [45]. There are a growing number of mouse models with metabolic syndrome-like phenotypes, as well as with altered bone physiology [15]. Mutant strains with calcium or bone deficiency (search Mouse Genome Informatics for specific examples) can potentially confound data in this respect, so this test would not be reliable in an experiment utilizing such mice. Similarly, obese mice may also confound data, as they could either be too buoyant, or simply become exhausted during the test. Consequently, mutant animals with such phenotypes may have affected swimming abilities/buoyancy, and therefore may not be adequately compared in the FST with their wild type littermates.

11. Motor or vestibular deficits may result in poor (abnormal) swimming, including aberrant spinning, turning, and sinking, that may confound FST data. Examine such mice in specific motor or vestibular ability tests. Mice with poor swimming should be excluded from the FST. Also note that some popular inbred mouse strains (e.g., most 129 mouse substrains) are poor swimmers and develop spastic behaviors in FST situations that complicate their swimming.

12. Some mice exhibiting increased levels of FST immobility may be suffering from fatigue rather than depression *per se*. Evaluate the fatigability of animals in separate tests. If mice display high fatigability phenotypes, consider shortening the length of the test.

13. Some mice may fall from the apparatus due to poor fixation by the adhesive tape. Use a cushioned floor for the TST to prevent any damage to the animal and exclude such mice from the experiment.

14. Note that a moderately adhesive tape will be required (preferably, use vinyl tape or medical tape). Since most mice weigh about 25–35 g, duct tape is too strong and would not be required; also, tape of this grade would likely tear hair, skin, and possible part of the tail off the animal.

15. Some strains (e.g., C57BL/6J mice) display specific tail climbing behaviors and may not be an appropriate mouse model for this test [46]. In contrast, BALB/cJ, DBA/2J, and BTBR strains were all shown to be reliable in this test and simultaneously responsive to drug effects (e.g., citalopram) [47]. The growing number of mice with vestibular deficits (e.g., MRL/MpJ, Ce/J, and SJL/J inbred strains; [48]; BDNF knockout mice [49] requires further consideration, since strains with vestibular deficits may show an abnormal "spinning" phenotype in the TST, thereby confounding behavioral data in this model. Consider using other models of depression for testing these mice. Some mutant mice display other specific neurological abnormalities relevant to their TST performance. For example, mutation or deletion of the Pafah1b1 gene of mice on a mixed 129SvEv-NIH Black Swiss background showed a marked increase in "hind leg clutching" behavior [50] whereas hind leg clasping behavior are common in including serotonin transporter knockout mice on 129S1/SvImJ background (own observations). Such phenotypes may result in abnormally high immobility in this test (which can incorrectly be interpreted as low depression). In contrast, spontaneous mild seizures in some mice (see Mouse Genome Informatics database for examples) may lead to reduced TST immobility, again confounding depression-related data.

16. While CMS is a valid model of depression in mice, it is labor intensive, long in duration, and demanding of space. A practical

recommendation for this model is thorough planning of all experiments and consistent completion of the entire CMS battery.

17. Consider strain differences in this paradigm. Some stressors may not affect all strains homogenously, and similarly, some models of depression may not accurately reflect depression in specific strains. For example, C57BL/6J mice are not sensitive to CMS affects on coat state [38].

18. Some suppliers (e.g., Keystone Country Store, www.keystone-countrystore.com) provide electronic devices that emit predator sounds, although rat vocalizations can also be recorded from live specimen. To standardize this, "predator sounds" should be published according to decibel, frequency, pitch, and length of recording. Unfortunately, soiled rat bedding cannot be obtained through a vendor. However, to standardize this stressor it would be possible to: using a metabolic cage, collect and measure the amount of urine and defecation of the rat; and then combine this with fresh rat bedding. A confinement tube is admittedly not descriptive. To isolate the mouse and cause a mild level of anxiety, insert the mouse into a restraint tube. These are typically similar in size and shape, although the exact design and specifications (dimensions) should be published in each manuscript utilizing this stressor. Some manufacturers include ITP (www.intoxproducts.com), AD Research (www.adinstruments.com), and ONARES (onares.com).

Acknowledgements

This work was supported by NARSAD YI Award to AVK, and by Stress Physiology and Research Center (SPaRC) of Georgetown University Medical School. AVK is the President of the International Stress and Behavior Society (ISBS, www.stressandbehavior.com). He is supported by Guangdong Ocean University, St. Petersburg State University (internal grant 1.38.201.2014) and Ural Federal University (Government of Russian Federation Act 211, contract 02-A03.21.0006).

References

1. Wong ML, Licinio J (2004) From monoamines to genomic targets: a paradigm shift for drug discovery in depression. Nat Rev Drug Discov 3:136–151

2. Cryan JF, Holmes A (2005) The ascent of mouse: advances in modelling human depression and anxiety. Nat Rev Drug Discov 4:775–790

3. Cryan JF, Markou A, Lucki I (2002) Assessing antidepressant activity in rodents: recent developments and future needs. Trends Pharmacol Sci 23:238–245

4. Cryan JF, Mombereau C (2004) In search of a depressed mouse: utility of models for studying depression-related behavior in genetically modified mice. Mol Psychiatry 9:326–357

5. Cryan JF, Slattery DA (2007) Animal models of mood disorders: recent developments. Curr Opin Psychiatry 20:1–7

6. Fava M, Davidson KG (1996) Definition and epidemiology of treatment-resistant depression. Psychiatr Clin North Am 19:179–200

7. Malatynska E, Rapp R, Harrawood D, Tunnicliff G (2005) Submissive behavior in mice as a test for antidepressant drug activity. Pharmacol Biochem Behav 82:306–313

8. Kalueff AV, Laporte JL, Murphy DL, Sufka K (2008) Hybridizing behavioral models: a possible solution to some problems in neurophenotyping research? Prog Neuropsychopharmacol Biol Psychiatry 32:1172–1178

9. Kalueff AV, Murphy DL (2007) The Importance of cognitive phenotypes in experimental modeling of animal anxiety and depression. Neural Plast 2007:52087

10. Geyer MA, Markou A (1995) Animal models of psychiatric disorders. In: Kupfer DJ, Bloom F (eds) Psychopharmacology the fourth generation of progress. Raven Press, New York, pp 787–798

11. Frazer A, Morilak DA (2005) What should animal models of depression model? Neurosci Biobehav Rev 29:515–523

12. Willner P (1997) Validity, reliability and utility of the chronic mild stress model of depression: a 10-year review and evaluation. Psychopharmacology (Berl) 134:319–329

13. Crowley JJ, Jones MD, O'Leary OF, Lucki I (2004) Automated tests for measuring the effects of antidepressants in mice. Pharmacol Biochem Behav 78:269–274

14. Juszczak GR, Sliwa AT, Wolak P, Tymosiak-Zielinska A, Lisowski P et al (2006) The usage of video analysis system for detection of immobility in the tail suspension test in mice. Pharmacol Biochem Behav 85:332–338

15. Jackson-Laboratory (2008) Mouse genome informatics. http://www.informatics.jax.org/

16. Palanza P (2001) Animal models of anxiety and depression: how are females different? Neurosci Biobehav Rev 25:219–233

17. Deacon RM (2006) Housing, husbandry and handling of rodents for behavioral experiments. Nat Protoc 1:936–946

18. Steru L, Chermat R, Thierry B, Simon P (1985) The tail suspension test: a new method for screening antidepressants in mice. Psychopharmacology (Berl) 85:367–370

19. Kos T, Legutko B, Danysz W, Samoriski G, Popik P (2006) Enhancement of antidepressant-like effects but not brain-derived neurotrophic factor mRNA expression by the novel N-methyl-D-aspartate receptor antagonist neramexane in mice. J Pharmacol Exp Ther 318:1128–1136.

20. Bai F, Li X, Clay M, Lindstrom T, Skolnick P (2001) Intra- and interstrain differences in models of "behavioral despair". Pharmacol Biochem Behav 70:187–192

21. Bourin M, Chenu F, Ripoll N, David DJ (2005) A proposal of decision tree to screen putative antidepressants using forced swim and tail suspension tests. Behav Brain Res 164:266–269

22. Hunsberger JG, Newton SS, Bennett AH, Duman CH, Russell DS et al (2007) Antidepressant actions of the exercise-regulated gene VGF. Nat Med 13:1476–1482

23. Strekalova T, Spanagel R, Bartsch D, Henn FA, Gass P (2004) Stress-induced anhedonia in mice is associated with deficits in forced swimming and exploration. Neuropsychopharmacology 29:2007–2017

24. Jayatissa MN, Bisgaard CF, West MJ, Wiborg O (2008) The number of granule cells in rat hippocampus is reduced after chronic mild stress and re-established after chronic escitalopram treatment. Neuropharmacology 54:530–541

25. Xu Q, Yi LT, Pan Y, Wang X, Li YC et al (2008) Antidepressant-like effects of the mixture of honokiol and magnolol from the barks of Magnolia officinalis in stressed rodents. Prog Neuropsychopharmacol Biol Psychiatry 32:715–725

26. Zhao Z, Wang W, Guo H, Zhou D (2008) Antidepressant-like effect of liquiritin from Glycyrrhiza uralensis in chronic variable stress induced depression model rats. Behav Brain Res 194:108–113.

27. Perona MT, Waters S, Hall FS, Sora I, Lesch KP et al (2008) Animal models of depression in dopamine, serotonin, and norepinephrine transporter knockout mice: prominent effects of dopamine transporter deletions. Behav Pharmacol 19:566–574

28. Luo DD, An SC, Xhang X (2008) Involvement of hippocampal serotonin and neuropeptide Y in depression induced by chronic unpredicted mild stress. Brain Res Bull 77:8–12

29. Piato AL, Detanico BC, Jesus JF, Lhullier FL, Nunes DS et al (2008) Effects of Marapuama in the chronic mild stress model: further indication of antidepressant properties. J Ethnopharmacol 118:300–304

30. Yalcin I, Aksu F, Belzung C (2005) Effects of desipramine and tramadol in a chronic mild stress model in mice are altered by yohimbine but not by pindolol. Eur J Pharmacol 514:165–174

31. Yalcin I, Aksu F, Bodard S, Chalon S, Belzung C (2007) Antidepressant-like effect of tramadol in the unpredictable chronic mild stress procedure: possible involvement of the noradrenergic system. Behav Pharmacol 18:623–631

32. Ducottet C, Griebel G, Belzung C (2003) Effects of the selective nonpeptide corticotropin-releasing factor receptor 1 antagonist antalarmin in the chronic mild stress model of depression in mice. Prog Neuropsychopharmacol Biol Psychiatry 27:6

33. Zhang L, Barrett JE (1990) Interactions of corticotropin-releasing factor with antidepressant and anxiolytic drugs: behavioral studies with pigeons. Biol Psychiatry 27:953–967

34. Burne TH, Johnston AN, McGrath JJ, Mackay-Sim A (2006) Swimming behaviour and post-swimming activity in vitamin D receptor knockout mice. Brain Res Bull 69:74–78

35. Harkin A, Houlihan DD, Kelly JP (2002) Reduction in preference for saccharin by repeated unpredictable stress in mice and its prevention by imipramine. J Psychopharmacol 16:115–123

36. Willner P, Moreau JL, Nielsen CK, Papp M, Sluzewska A (1996) Decreased hedonic responsiveness following chronic mild stress is not secondary to loss of body weight. Physiol Behav 60:129–134

37. Pothion S, Bizot JC, Trovero F, Belzung C (2004) Strain differences in sucrose preference and in the consequences of unpredictable chronic mild stress. Behav Brain Res 155:135–146

38. Mineur YS, Prasol DJ, Belzung C, Crusio WE (2003) Agonistic behavior and unpredictable chronic mild stress in mice. Behav Genet 33:513–519

39. Greer JM, Capecchi MR (2002) Hoxb8 is required for normal grooming behavior in mice. Neuron 33:23–34

40. Kalueff AV, Keisala T, Minasyan A, Kuuslahti M, Miettinen S et al (2006) Behavioural anomalies in mice evoked by "Tokyo" disruption of the vitamin D receptor gene. Neurosci Res 54:254–260

41. Garner JP, Weisker SM, Dufour B, Mench JA (2004) Barbering (fur and whisker trimming) by laboratory mice as a model of human trichotillomania and obsessive-compulsive spectrum disorders. Comp Med 54:216–224

42. Kalueff AV, Minasyan A, Keisala T, Shah ZH, Tuohimaa P (2006) Hair barbering in mice: implications for neurobehavioural research. Behav Processes 71:8–15

43. Sarna JR, Dyck RH, Whishaw IQ (2000) The Dalila effect: C57BL/6 mice barber whiskers by plucking. Behav Brain Res 108:39–45

44. Garner JP, Dufour B, Gregg LE, Weisker SM, Mench JA (2004) Social and husbandry factors affecting the prevalence and severity of barbering ("whisker-trimming") by laboratory mice. Appl Anim Lab Sci 89:263–282

45. Lucki I, Dalvi A, Mayorga AJ (2001) Sensitivity to the effects of pharmacologically selective antidepressants in different strains of mice. Psychopharmacology (Berl) 155:315–322

46. Mayorga AJ, Lucki I (2001) Limitations on the use of the C57BL/6 mouse in the tail suspension test. Psychopharmacology (Berl) 155:110–112

47. Crowley JJ, Blendy JA, Lucki I (2005) Strain-dependent antidepressant-like effects of citalopram in the mouse tail suspension test. Psychopharmacology (Berl) 183:257–264

48. Jones SM, Jones TA, Johnson KR, Yu H, Erway LC et al (2006) A comparison of vestibular and auditory phenotypes in inbred mouse strains. Brain Res 1091:40–46

49. Rauskolb S (2008) Brain-derived neurotrophic factor: generation and characterization of adult mice lacking BDNF in the adult brain, p 91. University of Basel, Basel, Germany

50. Paylor R, Hirotsune S, Gambello MJ, Yuva-Paylor L, Crawley JN et al (1999) Impaired learning and motor behavior in heterozygous Pafah1b1 (Lis1) mutant mice. Learn Mem 6:521–537

Chapter 16

Experimental Models of Anxiety for Drug Discovery and Brain Research

Peter C. Hart, Carisa L. Bergner, Amanda N. Smolinsky, Brett D. Dufour, Rupert J. Egan, Justin L. LaPorte, and Allan V. Kalueff

Abstract

Animal models have been vital to recent advances in experimental neuroscience, including the modeling of common human brain disorders such as anxiety, depression, and schizophrenia. As mice express robust anxiety-like behaviors when exposed to stressors (e.g., novelty, bright light, or social confrontation), these phenotypes have clear utility in testing the effects of psychotropic drugs. Of specific interest is the extent to which mouse models can be used for the screening of new anxiolytic drugs and verification of their possible applications in humans. To address this problem, the present chapter will review different experimental models of mouse anxiety and discuss their utility for testing anxiolytic and anxiogenic drugs. Detailed protocols will be provided for these paradigms, and possible confounds will be addressed accordingly.

Key words Anxiety, Experimental animal models, Anxiolytic drugs, Anxiogenic drugs, Biological psychiatry, Exploration

1 Introduction

Animal models are widely used for simulating human brain disorders and for providing insight into their neurobiological mechanisms [1–4]. The latter is of great interest in the current neuroscientific community, given the increasing use of laboratory animals for screening various classes of psychotropic drugs [5, 6]. The use of mice has been particularly beneficial, since fine-tuned manipulations of selected genes have led to new animal models relevant to drug discovery [3, 4, 7, 8].

It is important to understand, however, that any animal experiment in the laboratory is an artificial situation, and it may be biologically different from the natural behavior of the animal. Thus, it is crucial to correctly interpret the animal behavior observed in an experiment in order to identify parallels with specific human brain disorders. Although there are many other conceptual and

Gabriele Proetzel and Michael V. Wiles (eds.), *Mouse Models for Drug Discovery: Methods and Protocols*,
Methods in Molecular Biology, vol. 1438, DOI 10.1007/978-1-4939-3661-8_16, © Springer Science+Business Media New York 2016

methodological limitations of working with mice, this species shows much promise for future psychopharmacological research.

In order for animal models to be useful, researchers must follow certain practices and methods which will optimize the translatability of data from animal models to human affective disorders. Here, we will present a broad review of some reliable methods of analyzing mouse anxiety, and their utility for screening for anxiolytic therapeutic agents. All these tests are of a complex nature and we would suggest that the reader explore each system to better understand the variables and sublets of each test. We will also discuss how these protocols can be applied correctly in order to avoid confounding experimental data.

2 Materials

2.1 Animals

1. Various inbred, selectively bred (for specific behavioral/physiological phenotypes), and genetically modified (mutant or transgenic) mice may be used, and some searchable online databases, such as Mouse Phenome Project (www.jax.org/phenome) or Mouse Genome Informatics (www.informatics.jax.org/), may provide appropriate strains for studying mouse anxiety. We recommend using most of the inbred strains listed in the Tier 1 list of the Mouse Phenome Project database, especially C57BL/6J, A/J, and 129S1/SvImJ mice (see http://phenome.jax.org/pub-cgi/phenome/mpdcgi?rtn=docs/pristrains for details) (*see* **Note 1**).

2. Generally, several different models of anxiety that target *different* domains (e.g., locomotion/exploration, risk assessment, defensive responses) are necessary in order to more fully characterize drug effects or a mutant mouse phenotype. The use of a single model, or only models targeting one particular behavioral domain, may not be sufficient.

3. Researchers should also take other factors like age, weight, sex, stage of estrous cycle, diet, and housing situation into account when designing experiments (*see* **Note 2**).

2.2 Housing

1. If mice are obtained from a commercial vendor or another laboratory, allow at least 1 week acclimation from shipping stress. In most cases, a much longer time (e.g., 1 month) may be required for a better acclimation.

2. Housing animals in groups will help avoid social isolation stress/anxiety, but keeping groups small enough (e.g., not more than 5 animals per cage) will be necessary to avoid overcrowding stress. While overcrowding of mice may cause significant levels of stress, single housing is equally detrimental to experimental models. For example, social isolation may lead

to altered neurobiology, increased basal anxiety, reduced exploration, and a profound vulnerability to depression-like behaviors [9, 10].

3. The room in which mice are housed should be kept at approximately 21 °C, on a 12/12-h light cycle. As mice are nocturnal, the light cycle may be inverted if spontaneous activity measures are needed.

4. Food and water should be freely available, unless the intake is being controlled for experimental purposes.

5. Utilize plastic, solid-floored cages with sufficient space for mice to exercise and fully rear up. Note that enrichment items, such as cardboard tunnels, can improve general welfare but may also affect experimental outcomes or increase territorial aggression.

2.3 Drugs

1. All experimental protocols described here are compatible with drug testing. Researchers may choose from various antidepressants, anxiogenics, anxiolytics, or other psychotropic drugs, administered with a vehicle (e.g., saline). A typical experiment may include one or several drug-treated groups (e.g., several doses or several pretreatment times) compared to a vehicle-treated group of mice. Usually (unless stated otherwise), 10 animals per experimental group will be needed, also providing adequate statistical power (see further). However, if the effects of the drugs are particularly robust, a smaller n (e.g., $n = 7–8$) may suffice. For mild effects, a larger number of animals ($n = 15–16$) may be required.

2.4 Observations, Video Recording, and General Procedures

1. Video tracking software:
Ethovision, Noldus, Nijmegen, Netherlands.
Videotrack system, Viewpoint, Lyon, France.
Loco-, Maze- and Top Scan, Clever Sys Inc, Reston, VA, USA.

2. Photobeam-based activity monitoring:
Columbus Instruments, Columbus, OH, USA.
Coulbourn Instruments, Whitehall, PA, USA.

3. Vibration-based activity monitoring:
Laboras/Metris, Hoofddorp, Netherlands.
Bioseb, Vitrolles, France.

2.5 Requirements for Experimental Models

1. Elevated Plus Maze (EPM):
Elevated maze with two open and two closed arms in the shape of a plus, made of steel, fiberboard, or Plexiglas (either transparent or painted matte black), *see* Table 1. Arms are typically 30 cm long and 5 cm wide. The apparatus is usually elevated 40–60 cm on sturdy legs [1, 9, 15].

Table 1
Selected commercial suppliers of behavioral equipment for anxiety research

Test apparatus	Manufacturer	Company website
Elevated plus maze	Panlab, Barcelona, Spain Columbus Instruments, Columbus, OH, USA	www.panlab.com www.colinst.com
Open field	ANY-Maze by Stoelting, Wood Dale, IL, USA Noldus, Wageningen, Netherlands Panlab, Barcelona, Spain San Diego Instruments, San Diego, CA, USA	www.anymaze.com www.noldus.com www.panlab.com www.sandiegoinstruments.com
Startle response	San Diego Instruments, San Diego, CA, USA Columbus Instruments, Columbus, OH, USA	www.sandiegoinstruments.com www.colinst.com
Hole board	Columbus Instruments, Columbus, OH, USA	www.colinst.com
Light-dark box	Panlab, Barcelona, Spain Stoelting, Wood Dale, IL, USA	www.panlab.com www.anymaze.com

2. Open Field.
   Enclosed 50×30 cm wood, plastic, or Plexiglas arena, marked into 10-cm squares (Table 1). Gray or black arenas are typically used. If an arena is not available, a large animal cage marked into squares with indelible ink may be used [10, 16].

3. Marble Burying Test.
   Woodchip bedding (e.g., aspen chips), up to 20 marbles (15 mm in diameter). Animal cages (e.g., large cage 30×20 cm for 20 marbles, smaller cages for 6–8 marbles) [17–20].

4. Defensive Shock-Prod Burying test.
   Familiar test cage or home cage with plentiful bedding and a hole in the wall 2 cm above bedding [6, 21].
   Electrical probe connected to a shock source.
   Ruler for measuring depth to which prod is buried. Optional: Large (e.g., 10 cm) object associated with shock.

5. Grooming Analysis Algorithm.
   Small (e.g., $20 \times 20 \times 30$ cm) transparent observation box.
   Stressors to induce grooming: e.g., novel environment, predator exposure, bright light, or other means of artificially inducing grooming (e.g., water mist).
   Optional: video camera for subsequent frame-by-frame analysis [8].

6. Startle Response.
   Observation box (similar to the open field test).
   Conditioned stimulus: e.g., a light, paired with a footshock, Table 1. Startle stimulus, such as an air puff or loud noise [1, 7, 22].

7. Social Interaction Test.

Low-anxiety version: Test apparatus (similar to the open field test) familiar to the animals, with low illumination.

Mid-low anxiety version: Familiar test apparatus with high illumination.

Mid-high anxiety version: Unfamiliar test apparatus with low illumination.

High-anxiety version: Unfamiliar test apparatus with high illumination [1].

8. Suok Test.

Test apparatus: 2.6-m aluminum tube, 2 cm in diameter (marked into 10-cm segments with indelible ink) with fixed $50 \times 50 \times 1$ cm Plexiglas side walls to prevent escape, elevated 20 cm from a cushioned floor.

Optional (the light-dark version of the test): several 60-W light bulbs suspended 40 cm above one half of the test apparatus [23, 24].

9. Light-Dark Box Test.

Test apparatus: a 2-compartment box, $30 \times 30 \times 30$ cm each; with one black, and one transparent brightly illuminated boxes, separated by a sliding door [5].

10. Stress-Induced Hyperthermia (SIH).

Oiled rectal thermometer with rounded tip, up to 3 mm thick:

Surgilube sterile surgical lubricant by Fougera & Co. (Melville, NY, USA),

K-Y lubricant by Johnson & Johnson (Waltham, MA, USA),

MLT1404 rectal probe for mice by Adinstruments, Inc (Colorado Springs, CO, USA).

Cage or box (as in the open field test) to which mice can be transferred [25].

11. Hole-Board Test.

Test apparatus (similar to the open field test) with hole-board insert (Table 1). The floor has four or more identical holes approximately 3 cm in diameter [26].

12. Rat Exposure Test.

Medium (e.g., $40 \times 30 \times 30$ cm) transparent observation box, with a wire mesh separating the two halves of the box.

Small (e.g., $8 \times 8 \times 12$ cm) black Plexiglas box, serving as the starting placement (home chamber for the mouse).

Transparent Plexiglas tube (e.g., 4.5 cm in diameter, 13 cm in length) connecting the small black box to the medium transparent box [27].

13. Novel Object Test.

Test apparatus similar to the open field test (see above).

Novel objects: e.g., Mega Bloks structures [28].

3 Methods

3.1 Drug Administration

1. Common routes of injection include *systemic* [intraperitoneal (i.p.), intramuscular (i.m.), intravenous (i.v.), per oral (p.o.), subcutaneous (s.c.)] and *local* [intracerebral (i.c.), intranasal (i.n.), or intracerebroventricular (i.c.v)]. Continuous drug infusion (using osmotic pumps, such as Alzet pumps) at a constant rate may be used to improve the availability of the drug, and implantable depots can be used for s.c. drug administration to achieve a lasting therapeutic effect. Though not as commonly used in anxiety research, i.n. drug administration is a rapid, noninvasive method for drug delivery [11]. By this method, drugs can be administered either by pipetting small (6-μl) drops of a drug solution into each nostril once per minute [11] or by placing a 10-μl drop of a drug to be inhaled on the end of the snout [12]. As i.n. administration delivers the drug to the brain directly via the olfactory nerve and olfactory epithelium, this method may be favored for administration of drugs that are rapidly metabolized when given systemically or have difficulty crossing the blood–brain barrier [13]. For example in rats, i.n. dopamine has been shown to decrease grooming activity and increase locomotor activity in the open field at one tenth of the dose needed to observe these effects when the drug is administered i.p. [13]; also see data on its antidepressant effects [14].

2. Route of administration, dose, and pretreatment time vary depending on strain sensitivity and the drug being used.

3.2 Observations, Video Recording, and General Procedures

1. A computer, digital camera mounted above the test apparatus, and video tracking software will aid researchers in the collection of accurate behavioral data.

2. Alternative methods of behavioral tracking include photobeam-based or vibration-based activity monitoring.

3. In addition to automated tracking, an observer with a timer and data sheet to tally behaviors will allow comparison of data if video tracking is unreliable due to poor detectability (from poor angle or bad contrast; e.g., if fur color matches the background).

4. Observers must refrain from making noise or movement, as their presence may alter animal behavior. Assess intra- and inter-rater reliability for consistency.

5. Allow the animals at least 1 h acclimation after their transfer from the animal holding room to the experimental room.

6. Mice should be introduced to the testing environment during their normal waking cycle, to prevent possible confounds. When performing ethological analysis as part of a battery of tests,

consider how effects of these tests (such as habituation) may confound the mouse performance and drug sensitivity.

7. After each testing session, clean the equipment (e.g., with a 30% ethanol solution) to eliminate olfactory cues.

3.3 Data Analysis

1. Behavioral data may be analyzed with the Mann-Whitney U-test for comparing two groups (parametric Student's t-test may be used only if data are normally distributed), or analysis of variance (ANOVA) for multiple groups, followed by an appropriate post hoc test.

2. Some experiments may require one-way ANOVA with repeated measures, while for more complex studies (e.g., those including treatment, genotype, sex, and/or stress) n-way ANOVA may be used.

3. Analysis of statistical power is becoming particularly important in animal research. Effect size (the difference between means of the two groups) in neurobehavioral research can be small (0.2), moderate (0.5), or large (0.8). Statistical power is the probability of finding statistical significance for a true hypothesis, and its common value used in behavioral research is 0.8. Factors that can affect statistical power include the experimental design (independent/dependent groups), 1- or 2-tailed hypotheses, statistical test chosen, effect size, and sample size. In order to determine an effect size, the researcher may rely on effect size reported in similar studies, or can decide on it based on the goals of the study (e.g., use large effect size for robust phenotypes, and small effect size for less profound differences). The researcher can then use statistical power-analyzing software to determine the sample size required for the level of power decided upon. The choice of sample size, based on power calculations, is increasingly important for Institutional research ethics committees, to prove that neither too few nor too many subjects are used in the proposed research.

3.4 Elevated Plus Maze (EPM) Test

Possessing good face-, construct-, and predictive validity, the EPM is a reliable and pharmacologically sensitive paradigm based on the conflict between innate rodent desire to explore and the fear of open, elevated areas [3, 5]. Anxious mice generally have a lower ratio of open arm entries to total arm entries, and display fewer explorative measures such as rears, wall leans, or head dips [3, 5]. Anxiety also increases EPM freezing and stretch-attend postures. After administration of anxiolytic drugs (e.g., diazepam, chlordiazepoxide, ethanol), mice generally display more exploratory behaviors, a greater number of open arm entries, and an increased duration of time spent on open arms [9, 15]. Anxiogenic drugs (e.g., pentylenetetrazole, picrotoxin) produce the opposite behavioral effects in this model.

1. Place rodent on the central platform of the EPM facing either an open arm or a closed arm consistently at the start of each trial (*see* **Notes 1–3**).

2. The open arm, closed arm, and total (open + closed arm) activity can be recorded for 5–10 min using a video tracking system, while the researchers simultaneously document the number of arm entries (all four paws are on the arm) and time spent on each open arm (*see* **Notes 4** and **5**).

3.5 Open Field Test (OFT)

This test is based on the balance between the animal's natural drive to explore novelty and its aversion to open illuminated areas. Measuring exploratory behaviors and generalized motor activity, the open field test is simple and the most frequently used model of mouse anxiety [2]. In general, anxious mice exhibit more freezing, less time spent and a lower percentage of ambulation in the center of the arena (thigmotaxis), and fewer exploratory behaviors (*see* **Note 6**). Anxiolytic drugs generally increase exploration and reduce freezing and thigmotaxis [16].

1. After the apparatus is divided into central and peripheral zones, mice are placed consistently in a corner, or the center of the open field arena, and allowed to explore for 5–10 min.

2. Behavioral measurements can be recorded automatically with appropriate software, and include: time spent in the central area, distance traveled in the center as a ratio of total distance traveled, ambulation duration, time spent immobile (freezing), defecation score, and vertical activity such as rearing and wall leans [16], (*see* **Notes 7** and **8**).

3.6 Marble Burying Test

While not a direct model of anxiety per se, this simple test represents a pharmacologically sensitive method assessing digging activity—a species-typical response to anxiogenic stimuli [17, 19, 30]. Digging behavior is attenuated by low (nonsedative) doses of anxiolytic benzodiazepines and other ligands [18, 20]. Control mice can be expected to bury roughly 75 % of marbles, whereas drug-treated mice show a marked decrease in digging activity [31, 32]. Mouse strains with high basal anxiety levels (e.g., 129S1/SvImJ) should be avoided when testing anxiogenic drugs, to avoid a ceiling effect. Likewise, mice with very low basal anxiety may show poor results when examining the effects of anxiolytic drugs (*see* **Note 9**).

1. Cages should be filled with wood chip bedding approximately 5 cm deep. The bedding must be flattened to create an even surface. Use the same volume (e.g., 300 ml per 26 × 16 cm cage) of bedding in each cage. Although wood chips and shavings are most commonly used, sawdust has also been used with similar effectiveness in this model [33, 34]. There are many suppliers of these types of bedding, including Shavings-Direct

(www.shavings-direct.com), Doctors Foster and Smith (www.drsfostersmith.com), and local pet stores (e.g., Petsmart, www.petsmart.com). The important thing is to be consistent within experiments.

2. Mouse cage dimensions vary, but this test has been used effectively with 26×16 cm (for 6–8 marbles), or 30×20 cm (for 20 marbles) [18]. Marbles should be placed on the surface of the bedding in a regular pattern, roughly 4 cm apart.

3. Place one mouse in each cage. After 30 min, count the number of buried marbles. Any marble covered 2/3 of its depth with bedding is considered "buried" [18]. Alternatively, count fully covered (1/1) and partially covered (2/3) marbles separately, also calculating the sum of the latter two categories (*see* **Notes 9–11**).

4. Use a new clean cage with fresh bedding for each animal. Wash the marbles with ethanol after each test.

3.7 Defensive (Shock-Prod) Burying Test

Similar to the marble burying test, this paradigm is another pharmacologically sensitive method to assess rodent anxiety. Mice usually bury noxious stimuli posing an immediate threat (e.g., electrified shock-prod). The test has pharmacological validity, as benzodiazepines and the serotonergic anxiolytics potently suppress shock-prod burying in a dose-dependent manner, whereas anxiogenic drugs have been proven to increase this behavior [6].

1. In a standard-sized mouse cage (see above) with bedding 5 cm deep, insert a wire-wrapped prod (6–7 cm long) through a hole 2 cm above the bedding surface.

2. After the initial contact with the bare wires and the subsequent shock, record the behavior of the animal for 10–15 min. Behavioral measures of activity may include: prod-directed burying, burying latency, height of pile at prod base, prod contacts (number, duration), prod contact latency, and stretch-attend postures directed at the prod [6] (*see* **Note 12**).

3.8 Grooming Analysis Algorithm

Anxious mice tend to display a disorganized behavioral sequencing of grooming (higher percentage of incorrect transitions, more interrupted bouts) and a longer duration of this behavior. In contrast, anxiolytic benzodiazepines normalize mouse grooming sequencing by significantly reducing interrupted bouts and incorrect transitions [8] (*see* **Notes 13** and **14**).

1. Induce grooming through exposure to novelty or a stressor. Alternatively, mist the animal with water using a spray bottle. Place the animal in a small transparent observation box for 5 or 10 min.

2. If using a video camera, begin recording. With a stopwatch, record cumulative measures of grooming activity, such as:

latency to onset, time spent grooming, and total number of bouts. A new bout takes place after an interruption of greater than 6 s; bouts containing pauses of less than 6 s are deemed "interrupted." Additionally, record the patterning of each bout using the following scale: 0—no grooming, 1—paw licking, 2—nose/face/head wash, 3—body grooming, 4—leg grooming, 5—tail/genitals grooming.

3. There are several types of incorrect transitions, including skipped (e.g., 1–4, 3–5), reversed (e.g., 3–2, 5–3), prematurely terminated (e.g., 3–0, 4–0), and incorrectly initiated (e.g., 0–3, 0–5) transitions. Calculate the percentage of interrupted bouts and the percentage of incorrect transitions; see [8] for details (*see* **Notes 15** and **16**).

3.9 Startle Response

The startle response test pairs a conditioned stimulus (sound, light) with a footshock to induce an anxiogenic "startle" response in mice. While the sensitivity of this test to many drugs is yet to be established, benzodiazepine and serotonergic anxiolytics have been effective in reducing the startle response [1]. Since this model seems to be unaffected by motor phenotypes, activity levels, or neurological deficits, this test (unlike many other anxiety models discussed here) allows researchers to study mouse anxiety without these confounding factors (*see* **Note 17**).

1. In a conditioning trial, a conditioned stimulus (usually a light: e.g., 15 W) is paired with a footshock. The timing of the conditioned stimulus and foots.hock can be controlled by the data acquisition software for consistency [7].

2. In a separate trial 24 h later, the animals are presented with a startle stimulus (e.g., loud noise 70–80 db, or air puff) and their activity is recorded as a baseline. The startle stimulus can be presented in 4 blocks of 5 startles each, with 30–35 s between each startle stimulus [7]. Peak and amplitude of the startle response can be recorded (e.g., using a piezoelectric accelerometer) and digitized [22].

3. 24 h later in testing trials, the conditioned stimulus is displayed immediately prior to the startle stimulus, and the observed response is compared to the baseline startle response. Stimuli should be presented when the animal is quiet and inactive [7] (*see* **Notes 18** and **19**).

3.10 Social Interaction Test

The social interaction test is a useful drug-sensitive approach to assessing anxiety in mice, subjected to the apparatus similar to the OFT. There are four testing conditions which introduce varying levels of stress: (1) familiar test apparatus and low illumination; (2) familiar test apparatus and high illumination; (3) unfamiliar test apparatus and low illumination; and (4) unfamiliar test apparatus and high illumination. The level of anxiety across these conditions ranges from low to high, respectively. Overall, the duration and frequency of social interactions negatively correlate with anxiety. Because this test successfully isolates levels of anxiety (high vs. low) in the subjects, it has been used for pharmacological screening of both anxiolytic and anxiogenic drugs in their effectiveness for increasing or decreasing social interaction, respectively [1].

1. The test environment should be the same in all conditions, except for the test apparatus (familiar or unfamiliar) and the lighting (low or high illumination).

2. Introduce the two mice into the test environment for 5 or 10 min, recording the duration and frequency of all social interactions (e.g., sniffing, following, chasing, touching, and biting). All behavioral endpoints mentioned should be recorded, including the frequency (number of bouts) and duration (cumulative of all bouts per trial). The best way would be to use a video tracking system with either automatic or manual scoring, although it is possible for the observers to record these behaviors manually in real time.

3. After obtaining baseline data for each condition, administer an anxiolytic or anxiogenic drug to the mice. The same test (**step 2**) can be conducted and analyzed relative to baseline data. Alternatively, compare drug-treated with saline-treated groups (only one animal in the interacting pair receives the drug) (*see* **Notes 20–22**).

3.11 Suok Test

The Suok test simultaneously examines anxiety, vestibular, and neuromuscular deficits by combining an unstable rod with novelty. To analyze anxiety, the threats of height, loss of balance, and novelty are presented and animal exploration is recorded. Anxiolytic or anxiogenic drugs will increase or decrease animal exploration, respectively. Risk assessment and vegetative behaviors are generally higher in anxious mice. The model is also sensitive to anxiety-evoked balancing deficits, since administration of anxiogenic drugs increases the number of falls and missteps, while anxiolytics generally improve balancing [23, 24]. A light-dark modification of the test may also be employed, as the illuminated environment will represent an additional stressor. We recommend reviewing the Mouse Phenome and Mouse Genome Informatics Databases for examples of anxious mouse strains, mice displaying low motor or vertical activity, hyperactive mice, and mice with sensory deficits or mouse strains with vestibular difficulties. Researchers should easily find mice that fit their specific experimental needs.

1. Place individual mice in the center of the apparatus facing either end (or, in the light-dark modification, orient the animal facing the dark end) (*see* **Note 23**).

2. Standing approximately 2 m away from the apparatus, record the following behavioral measures (for 5–10 min per animal): horizontal exploration activity (latency to leave central zone, number of segments visited with four paws, distance traveled, number of stops, time spent immobile, average inter-stop distance, number of stops near border separating light-dark areas of the apparatus), vertical exploration (number of vertical rears or wall leans), directed exploration (head dips, side looks), risk assessment behavior (stretch-attend postures), vegetative responses (number of defecation boli and urination spots), and vestibular/motor indices (number and latency of hind-leg slips and falls from rod). If the animal falls, replace it in the same position from where it fell (*see* **Notes 24** and **25**).

3.12 Light-Dark Box Test

This ethological model of anxiety measures the activity and time spent in brightly lit vs. dark compartments of the apparatus, and is based on the animal's innate desire to explore novel areas [5, 16]. Anxious mice exhibit a profound preference for the dark area and display fewer exploratory behaviors (e.g., horizontal activity, vertical rears, or wall leans) in the light. Increased duration of time spent in the light area and more exploratory behaviors can be seen following anxiolytic drug administration.

1. Place one animal into the dark compartment of the box for 5 min for acclimation

2. Lift the shutter to allow the mouse to move freely between the dark and light compartments for 5 min.

3. Measure the latency to initial transition into the light box. Record the duration of time spent in each compartment, the number of transitions between them, and the distance traveled in each box. Additional indices may be vertical rearing, wall leans (in the light compartment), and the number of defecation boli (*see* **Note 26**).

3.13 Stress-Induced Hyperthermia (SIH)

This test relies on the evolutionarily important role of hyperthermia, a rise in body temperature upon encountering stressful stimuli which occurs across many species, including humans. In mouse SIH test, the insertion of a rectal thermometer records a 0.5–1.5 °C increase in body temperature (*see* **Note 27**). SIH is reduced or prevented by different anxiolytic drugs; however, it seems to be unable to detect anxiogenic and antidepressant effects [25].

1. Animals should be put in individual cages the day before testing to avoid effects of acute isolation stress (*see* **Note 28**).

2. Baseline body temperature should be recorded. To test mouse rectal temperature, carefully insert a probe with a rounded tip (up to 3 mm thick) after dipping it in any kind of oil for lubrication. For example, use surgilube sterile surgical lubricant by

Fougera & Co. (Melville, NY, USA), K-Y lubricant by Johnson & Johnson (Waltham, MA, USA), and MLT1404 rectal probe for mice by Adinstruments, Inc (Colorado Springs, CO, USA). The probe should be inserted consistently, approximately 2–2.5 cm for 10 s (*see* **Note 29**).

3. Present the mouse with a stressor, such as a novel cage, and document the change in internal temperature (*see* **Note 30**).

4. After testing, mice may be re-socialized in grouped housing. They may be retested after in 1-week intervals. Typically, 10–15 mice per group are sufficient to observe significant effects [38] (*see* **Note 31**).

5. It is also possible to use an implanted temperature sensor to monitor temperature remotely, and without using this type of invasive measurement. For example, microchip transponders (ELAMS, BioMedic Data Systems, Inc., Seaford, DE, USA) have been shown to reliably monitor temperature without significant difference from rectal temperature measurements [39].

3.14 Hole-Board Test

Conceptually similar to the open field test (OFT), the hole-board test focuses on specific head dipping behaviors. Head dipping, an indication of directed exploration, can be vigorously affected by various drug classes, including anxiolytic and anxiogenic drugs. Due to its short duration and quantifiable behavioral measures, this test is a readily available method for the testing of classic or novel drugs [26]. Place mice individually in hole-board apparatus and record behavior for 5–15 min, documenting traditional exploratory behaviors (as in the OFT, see above) and the number of head dips (*see* **Notes 7, 8**, and **32**).

3.15 Rat Exposure Test

This test utilizes the natural defensive "avoidance" behavioral response of mice to signs of potential danger, such as a natural predator (e.g., rat). Defensive behaviors include stretch-attend posture, stretch approach, freezing, burying, and hiding, and are measured as a function of risk assessment. This test has proven useful to determine strain differences in defensive behaviors and relative levels of anxiety in response to predators (*see* **Note 33**). Additionally, the defensive behaviors measured are sensitive to anxiolytics, making this paradigm useful in pharmacological screening [27].

1. Introduce the mouse into the small black box, which will serve as a "home chamber" (safe environment). The Plexiglas tunnel should allow free movement between the home chamber and the observation box.

2. On the first 3 days of testing, allow the mouse to explore the observation box for 10 min to become familiar with the environment. In these sessions, there should be no rat present.

3. On the fourth day, insert the mouse into the home chamber and the rat into the observation box for 10 min. The rat should be placed in the opposite side of the cage, isolated from the mouse by the wire mesh.

4. In every testing session, record the number of stretch-attend postures, stretch approaches, freezes, and number of times the mouse retreats to the home chamber. Also, measure the amount of time spent in the home chamber and observation box, as well as time in contact with the wire mesh.

3.16 Novel Object Test

This model investigates the approach-avoidance behaviors of mice in response to novel stimuli. Typically, mice tend to explore a novel object longer than a familiar one, and prior exposure to a stimulus increases consecutive approach behavior and decreases avoidance behavior. This robust behavior, as well as the simplicity of this model, makes this test particularly useful for measuring anxiety in a battery of tests [28]. Anxiolytics have been shown to increase exploratory behavior of mice in novel environments [41], suggesting that the use of anxiolytics would similarly increase this behavior with novel objects.

1. On Day 1, introduce the mouse into the test apparatus, allowing it to explore the environment for 30–60 min.

2. On testing day, insert the novel object into the center of the testing apparatus prior to introducing the mouse. Record the frequency and duration of exploratory behavior, such as approaches, sniffing, physical contacts (e.g., touching, licking, biting), wall leans, vertical rears, head dipping, and time spent near the novel object, for 10–30 min. Also record amount of avoidance behavior as time spent in the perimeter.

3. Any video tracking system (*see* Table 3 for details) may be useful for measuring amount of movement and position within the test apparatus. Conversely, the apparatus may be sectioned off, and duration in each section can be recorded, comparing the perimeter sections to the novel object section [28] (*see* **Note 34**).

4   Notes

1. As genetic background greatly influences behavioral phenotypes in mice, comparative studies must take this into account. The use of inbred mice substantially decreases within-subject variation, and also provides valuable insight into the neurobiological mechanisms that modulate specific phenotypes. The most updated detailed nomenclature for mouse strains should be used, and mice should be obtained from certified vendors or other reliable sources to ensure comparability of results

between laboratories. In contrast, outbred mice do not seem to present similar benefits, and therefore, may yield more confounded results.

2. Age, gender, and strain differences affect Elevated Plus Maze (EPM) performance. Young females generally spend less time on the open arms than males, although this varies with the estrous cycle. Pro-estrus rodents spend significantly more time on open arms than di-estrus females (or male mice) [9].

3. All experimental procedures (including handling, housing, husbandry, and drug treatment) must be conducted in accordance with national and institutional guidelines for the care and use of laboratory animals.

4. Since lighting can affect behavior in the maze, make sure it is consistent on all arms. Red light in a darkened room is preferable, as mice cannot see red light. To avoid excessive freezing, testing environment should be kept quiet without disruptions. If the mouse freezes for more than 30% of the total test time, researchers should note of this abnormality, but continue testing. In case of unexpected or loud noises or other disruptions, the data should be discarded from analyses.

5. As some mice may fall off an open arm of the EPM, the data from these animals must be excluded from further analyses.

6. In some cases, reduced anxiety can be mistaken for hyperactivity. A minute-by-minute analysis of exploration and activity may aid in distinguishing these two different domains [29].

7. It is important not to misinterpret reduced locomotion due to high habituation as an anxiogenic response (see Table 2 for details). The mouse learning/memory phenotypes should be assessed in separate tests [29]. If mice have poor habituation, this may result in increased "exploration" that should not be misinterpreted as decreased anxiety [29]. Consider testing mouse cognitive functions in a separate study.

8. Mice with altered motor, vestibular, neurological, memory, or depression domains may need additional screening before use in the OFT. Variable aged ranges may be used, but all mice should be tested at the same age in a comparative experiment.

Table 2
Potential combinations of emotional and cognitive phenotypes [29]

Phenotype	Elevated anxiety	Reduced anxiety
Elevated memory	Increased initial anxiety, increased habituation	Low initial anxiety, increased habituation
Reduced memory	Increased initial anxiety, decreased habituation	Low initial anxiety, decreased habituation

Table 3
Examples of video tracking manufacturers

Name	City	Country	Company website
CleverSys Inc.	Reston, VA	USA	www.cleversysinc.com
Noldus	Leesburg, VA, or Wageningen	USA, or Netherlands	www.noldus.com
San Diego Instruments	San Diego, CA	USA	www.sandiegoinstruments.com
TSE Systems	Midland, MI	USA	www.tse-systems.com
Biobserve GMBH	Bohn	Germany	www.biobserve.com

9. Some strain differences are apparent in digging behaviors. Slow or inactive mouse strains (e.g., 129S1/SvImJ) may be replaced with more active strains (e.g., C57BL/6J) to achieve recordable amounts of burying data. It has been observed that younger mice (2–4 months old) tend to show enhanced digging behaviors as opposed to mice over 1 year old [18].

10. If mice continue to display low burying activity, it may be useful to assess the environment for confounding factors. Unnecessary noise or stress should be eliminated and mice should be undisturbed throughout the experiment. Testing on cage-cleaning days may also cause mice to be less responsive to the new bedding [18].

11. Some strains with low burying/digging activity may require a longer (e.g., 45–60 min) testing time that may help reveal their phenotype.

12. Troubleshooting is the same as in the marble burying test.

13. In choosing a proper mouse strain for testing, it is important to consider possible motor and sensory deficits. For example, C57BL/6J strain is widely used as it has relatively no deficits in these domains and is sensitive to drug and behavioral testing.

14. Abnormally high grooming activity may be due to a strain-specific compulsive-like phenotype (consider using a more appropriate strain), or due to unintended stress in the animal facility (which may be assuaged by improved husbandry or enrichment).

15. High baseline or transfer anxiety may lead to unusually low-grooming activity. This may be alleviated by using smaller observation boxes and dimmer lighting, as well as by improving handling techniques and lengthening acclimation time. Reduced grooming activity may also be due to a strain-specific low-grooming phenotype (e.g., due to abnormal neurological/vestibular/motor phenotype) or overall inactivity of the stain being tested.

16. Detection of different stages of grooming behavior may sometimes be difficult. If using a video camera, replaying in slow motion will make the detection of transitions and interruptions much easier.

17. If the startle stimulus is auditory, some mouse strains may be insensitive to this test because of hearing deficits which can also be age related (e.g., C57BL/6J mice have a progressive hearing with onset after 10 months of age). To rule out this possibility, mice should be tested for hearing problems. If the mouse strain shows abnormally poor hearing, consider using a physical startle stimulus (e.g., air puff or bright light) or a different strain.

18. If the mouse does not show a heightened response to the startle stimulus in the testing trials, it may have cognitive deficits. Memory should be examined in separate, specific tests to ensure accurate data interpretation.

19. C57BL/10J and FVB/NJ strains have high startle responses and 129S1/SvImJ mice have low startle responses, whereas BALB/cJ and C57BL/6J strains have more moderate responses [35]. Some animals may show an abnormally high startle response as a result of brain pathological overexcitation, and this abnormality should be investigated further. Also, consider baseline brain activity as well when administering drugs. For example, due to the floor/ceiling effect, anxiogenic drugs can be tested on mice with a low baseline startle response, whereas anxiolytics would yield clearer results if tested on mice with high baseline responses. Review literature for drug efficacy and concentrations.

20. Some mice display particularly high levels of social interaction, including FVB and C57BL/6J [36]. Certain strains may be more likely to engage in social interaction because of their high sociability phenotype (which may be unrelated to their anxiety or emotionality profile per se). In this case, consider using other strains for this test. Low levels of social interaction may occur with the spontaneous deletion of the Dtnbp1 gene, leading to social withdrawal [37], or in some inbred mouse strains, for example, A/J, BALBcByJ, and BTBR T+ tf/J mice [36].

21. Thus, the use of some strains should be avoided as their autism-like behavior may prevent the relevance of this test as a model of anxiety. In performing the social interaction test for screening anxiolytic and anxiogenic drugs, it is suggested that the same two mice are not re-introduced into the same environment together, as this may eliminate the social novelty of the condition, and will affect their test performance.

22. Mice with abnormally poor or abnormally good cognitive abilities may produce aberrant behavior in this test (e.g., increased or decreased social interaction, respectively). To rule out this

possibility, consider testing mice in some additional memory paradigms. Memory tests, such as the Morris Water Maze and OFT habituation, may be performed to assess cognitive functions in any abnormally behaving mouse.

23. Low motor or vertical activity may be a strain-specific phenotype. Inactive strains will produce less activity overall, and may not be suitable for this model. Likewise, hyperactive strains generally display less non-horizontal exploration and may have difficulties with balance. A narrower apparatus will encourage the animal to show less horizontal activity, enabling it to focus on other behavioral responses. Differences in mouse size should also be addressed. Use animals of similar size, age, and weight to accurately compare between groups.

24. If the mouse displays abnormally high transfer anxiety, gently support it for approximately 5 s to facilitate a solid grip. If the animal continues to display high transfer anxiety, exclude it from the experiment. A dimly lit experimental room may help reduce anxiety.

25. Some strains have difficulties balancing on an aluminum rod, and a more textured surface (e.g., wood) may help stabilize the animal. Increasing the diameter of the rod is another possible solution. If mice continue to struggle with balance or motor abilities, assess motor and vestibular functions separately as these behaviors may be due to a neuromuscular or motor coordination problem unrelated to vestibular deficits or anxiety.

26. Certain strains of mice may be less inclined to explore the test environment, such as mice with anxiety- or depression-like phenotypes (see Mouse Phenome and Mouse Genome Informatics Databases for details). Allow a longer acclimation and/or test time (e.g., 10 or 15 min) to reduce this factor. Some mouse strains (e.g., many albino mice) display visual deficits, and may not be a suitable model for this test. Consider other mouse strains for the light-dark box testing.

27. While most strains respond consistently to this paradigm due to its independence from motor activity, specific mouse strains (e.g., FVB/N) have considerably higher baseline body temperatures, and should be avoided in this model [25]. However, this procedure has been shown to effectively induce hyperthermia to varying degrees in all inbred and outbred strains tested [40], and is also an effective indicator of stress in genetically modified animals [38].

28. In the group-housed mice, the last mice to be tested show an increase in body temperature (compared with the first mice) due to anticipatory anxiety. Therefore, animals should be tested individually, with at least 10 mice in each experimental group [25].

29. Temperature measurements should be performed at the same time due to circadian rhythm. Baseline body temperature is significantly higher during the night. If testing occurs during the dark phase, there may be an interference with the amplitude of hyperthermia when the stressor is presented [25].

30. Mice should be kept undisturbed before the experiment with proper handling and opening of cages to ensure accurate results.

31. The above protocol may be modified for the testing of anxiolytic drugs. Sixty minutes prior to the first temperature measurement, inject the mouse with the desired drug. The first temperature measurement serves as an acute stressor, and is followed after 10 min with a second temperature measurement [38].

32. Certain drugs (e.g., ethanol) are known to be strain-dependent in their effects and may not produce consistent results in the hole-board test. Many commonly used drugs (e.g., fluoxetine) have pronounced dose-dependent effects on head dipping behavior, and therefore, dosing should be carefully considered; see review in [26].

33. Certain mouse strains (e.g., 129S1/SvImJ or BALB/cJ) may not be useful in this test due to their hypoactivity and/or high anxiety phenotypes. If the mouse tested is very inactive and anxious, it may not even leave the home chamber, and this test will not work. In this case, use a milder stressor, such as an anesthetized rat, a toy rat, or rat odor. However, it may also be recommended to use a different mouse strain. Although this test is very useful for comparing defensive behaviors between mouse strains, some strains are not suitable for this test. For example, mice with sensory deficits (e.g., poor vision or olfaction) or with particular cognitive problems (e.g., poor working memory) will not provide reliable data in this paradigm. As mentioned above, it may help to check online mouse databases for selecting an appropriate mouse strain.

34. Similar to some other previously described tests, mouse strains with sensory or cognitive deficits may not provide reliable data in this model. In addition, some mice can exhibit strong neophobia, which would also confound behavioral data. Test mice prior to this experiment to screen for such defects, and consider using alternate strains and/or extending the observation time.

Acknowledgements

This work was supported by the NARSAD YI Award to AVK, and by Stress Physiology and Research Center (SPaRC) of Georgetown University Medical School. AVK is the President of the International Stress and Behavior Society (ISBS, www.stressandbehavior.com).

He is supported by Guangdong Ocean University, St. Petersburg State University (internal grant 1.38.201.2014) and Ural Federal University (Government of Russian Federation Act 211, contract 02-A03.21.0006).

References

1. Warnick JE, Sufka KJ (2008) Animal models of anxiety: examining their validity, utility, and ethical characteristics. In: Kalueff AV, LaPorte JL (eds) Behavioral models in stress research. Nova Biomedical Books, New York, pp 55–71

2. Flint J (2003) Animal models of anxiety and their molecular dissection. Semin Cell Dev Biol 14:37–42

3. Ohl F (2005) Animal models of anxiety. Handb Exp Pharmacol 1:35–69

4. Sousa N, Almeida OF, Wotjak CT (2006) A hitchhiker's guide to behavioral analysis in laboratory rodents. Genes Brain Behav 5(Suppl 2):5–24

5. Borsini F, Podhorna J, Marazziti D (2002) Do animal models of anxiety predict anxiolytic-like effects of antidepressants? Psychopharmacology (Berl) 163:121–141

6. De Boer SF, Koolhaas JM (2003) Defensive burying in rodents: ethology, neurobiology and psychopharmacology. Eur J Pharmacol 463:145–161

7. Falls WA, Carlson S, Turner JG, Willott JF (1997) Fear-potentiated startle in two strains of inbred mice. Behav Neurosci 111:855–861

8. Kalueff AV, Aldridge JW, LaPorte JL, Murphy DL, Tuohimaa P (2007) Analyzing grooming microstructure in neurobehavioral experiments. Nat Protoc 2:2538–2544

9. Walf AA, Frye CA (2007) The use of the elevated plus maze as an assay of anxiety-related behavior in rodents. Nat Protoc 2:322–328

10. Deacon RM (2006) Housing, husbandry and handling of rodents for behavioral experiments. Nat Protoc 1:936–946

11. Martinez JA, Francis G, Liu W, Pradzinsky N, Fine J et al (2008) Intranasal delivery of insulin and a nitric oxide synthase inhibitor in an experimental model of amyotrophic lateral sclerosis. Neuroscience 157(4):908–925

12. Ito N, Nagai T, Oikawa T, Yamada H, Hanawa T (2008) Antidepressant-like effect of l-perillaldehyde in stress-induced depression-like model mice through regulation of the olfactory nervous system. Evid Based Complement Alternat Med 2011:512697

13. De Souza Silva M, Topic B, Huston J, Mattern C (2008) Intranasal dopamine application increases dopaminergic activity in the neostriatum and nucleus accumbens and enhances motor activity in the open field. Synapse 62:176–184

14. Buddenberg TE, Topic B, Mahlberg ED, de Souza Silva MA, Huston JP et al (2008) Behavioral actions of intranasal application of dopamine: effects on forced swimming, elevated plus-maze and open field parameters. Neuropsychobiology 57:70–79

15. Hagenbuch N, Feldon J, Yee BK (2006) Use of the elevated plus-maze test with opaque or transparent walls in the detection of mouse strain differences and the anxiolytic effects of diazepam. Behav Pharmacol 17:31–41

16. Karl T, Duffy L, Herzog H (2008) Behavioural profile of a new mouse model for NPY deficiency. Eur J Neurosci 28:173–180

17. Archer T, Fredriksson A, Lewander T, Soderberg U (1987) Marble burying and spontaneous motor activity in mice: interactions over days and the effect of diazepam. Scand J Psychol 28:242–249

18. Deacon RM (2006) Digging and marble burying in mice: simple methods for in vivo identification of biological impacts. Nat Protoc 1:122–124

19. Nicolas LB, Kolb Y, Prinssen EP (2006) A combined marble burying-locomotor activity test in mice: a practical screening test with sensitivity to different classes of anxiolytics and antidepressants. Eur J Pharmacol 547:106–115

20. Njung'e K, Handley SL (1991) Evaluation of marble-burying behavior as a model of anxiety. Pharmacol Biochem Behav 38:63–67

21. Mikics E, Baranyi J, Haller J (2008) Rats exposed to traumatic stress bury unfamiliar objects—a novel measure of hyper-vigilance in PTSD models? Physiol Behav 94:341–348

22. Halberstadt AL, Geyer MA (2008) Habituation and sensitization of acoustic startle: opposite influences of dopamine D(1) and D(2)-family receptors. Neurobiol Learn Mem 92(2):243–248

23. Kalueff AV, Keisala T, Minasyan A, Kumar SR, LaPorte JL et al (2008) The regular and light-dark Suok tests of anxiety and sensorimotor integration: utility for behavioral

characterization in laboratory rodents. Nat Protoc 3:129–136

24. Kalueff AV, Tuohimaa P (2005) The Suok ("ropewalking") murine test of anxiety. Brain Res Brain Res Protoc 14:87–99

25. Bouwknecht JA, Olivier B, Paylor RE (2007) The stress-induced hyperthermia paradigm as a physiological animal model for anxiety: a review of pharmacological and genetic studies in the mouse. Neurosci Biobehav Rev 31:41–59

26. Kliethermes CL, Crabbe JC (2006) Pharmacological and genetic influences on hole-board behaviors in mice. Pharmacol Biochem Behav 85:57–65

27. Yang M, Augustsson H, Markham CM, Hubbard DT, Webster D et al (2004) The rat exposure test: a model of mouse defensive behaviors. Physiol Behav 81:465–473

28. Powell SB, Geyer MA, Gallagher D, Paulus MP (2004) The balance between approach and avoidance behaviors in a novel object exploration paradigm in mice. Behav Brain Res 152:341–349

29. Kalueff AV, Murphy DL (2007) The importance of cognitive phenotypes in experimental modeling of animal anxiety and depression. Neural Plasticity 2007:52087

30. Broekkamp CL, Rijk HW, Joly-Gelouin D, Lloyd KL (1986) Major tranquillizers can be distinguished from minor tranquillizers on the basis of effects on marble burying and swim-induced grooming in mice. Eur J Pharmacol 126:223–229

31. Bruins Slot LA, Bardin L, Auclair AL, Depoortere R, Newman-Tancredi A (2008) Effects of antipsychotics and reference mono-aminergic ligands on marble burying behavior in mice. Behav Pharmacol 19:145–152

32. Bespalov AY, van Gaalen MM, Sukhotina IA, Wicke K, Mezler M et al (2008) Behavioral characterization of the mGlu group II/III receptor antagonist, LY-341495, in animal models of anxiety and depression. Eur J Pharmacol 592(1-3):96–102

33. Gordon CJ (2004) Effect of cage bedding on temperature regulation and metabolism of group-housed female mice. Comp Med 54:63–68

34. Li X, Morrow D, Witkin JM (2006) Decreases in nestlet shredding of mice by serotonin uptake inhibitors: comparison with marble burying. Life Sci 78:1933–1939

35. Paylor R, Crawley JN (1997) Inbred strain differences in prepulse inhibition of the mouse startle response. Psychopharmacology (Berl) 132:169–180

36. Bolivar VJ, Walters SR, Phoenix JL (2007) Assessing autism-like behavior in mice: variations in social interactions among inbred strains. Behav Brain Res 176:21–26

37. Feng YQ, Zhou ZY, He X, Wang H, Guo XL et al (2008) Dysbindin deficiency in sandy mice causes reduction of snapin and displays behaviors related to schizophrenia. Schizophr Res 106(2-3):218–228

38. Olivier B, Zethof T, Pattij T, van Boogaert M, van Oorschot R et al (2003) Stress-induced hyperthermia and anxiety: pharmacological validation. Eur J Pharmacol 463:117–132

39. Kort WJ, Hekking-Weijma JM, TenKate MT, Sorm V, VanStrik R (1998) A microchip implant system as a method to determine body temperature of terminally ill rats and mice. Laboratory Animals 32(3):260–269

40. Bouwknecht JA, Paylor R (2002) Behavioral and physiological mouse assays for anxiety: a survey in nine mouse strains. Behav Brain Res 136:489–501

41. Klebaur JE, Bardo MT (1999) The effects of anxiolytic drugs on novelty-induced place preference. Behav Brain Res 101:51–57

Chapter 17

Repetitive Behavioral Assessments for Compound Screening in Mouse Models of Autism Spectrum Disorders

Stacey J. Sukoff Rizzo

Abstract

Treatments for repetitive behaviors in Autism Spectrum Disorders (ASD) and other neurodevelopmental disorders remain an unmet medical need. Mouse models are highly valuable tools for investigating the underlying pathophysiology, genetics, and neurocircuitry of ASD, and can also be characterized for ASD-related phenotypes including repetitive behaviors and stereotypies. This chapter describes methods that can be employed for the assessment of repetitive behavior phenotypes and the evaluation of the ability of test compounds to attenuate repetitive behaviors in mouse models of ASD.

Key words Autism, Behavioral testing, Repetitive behaviors, Stereotypy, Repetitive grooming, Repetitive digging, Repetitive jumping

1 Introduction

The DSM-V Diagnostic Criteria for Autism Spectrum Disorders (ASD) includes persistent deficits in social interactions and communication, and the manifestation of restricted, repetitive patterns of behavior including motor stereotypies and restricted, fixated interests or resistance to change in routine [1]. Examples of some repetitive behaviors which are often complex motor stereotypies observed in ASD and other neurodevelopmental disorders such as Tourette, Rett, or Fragile X syndromes include observations of hand flapping or clapping, shoulder shrugging, eye blinking, grimacing, repetitive sniffing, and vocal tics among others (*reviewed in* [2]).

The underlying neurobiology by which repetitive behaviors occur in ASD patients or other disorders remains unknown, although basal ganglia and corticostriatal circuitry have been implicated [3, 4]. Although a number of pharmacological interventions have been employed to treat the repetitive behaviors observed in ASD and other disorders with varied success, there remains no evidence-based drug treatment that can successfully target the spectrum of repetitive behaviors. Hence this remains an unmet medical need [5–7].

Gabriele Proetzel and Michael V. Wiles (eds.), *Mouse Models for Drug Discovery: Methods and Protocols*,
Methods in Molecular Biology, vol. 1438, DOI 10.1007/978-1-4939-3661-8_17, © Springer Science+Business Media New York 2016

Mice exhibit a variety of spontaneous motor stereotypies, with the most commonly reported being excessive self-grooming, repetitive circling, jumping, and digging behaviors [8–10]. Interestingly, a number of these repetitive behaviors have been demonstrated as exacerbated in genetically modified mouse models related to ASD (*reviewed in* [11]). Such models are highly valuable tools for investigating the underlying pathophysiology, genetics, and neurocircuitry of repetitive behaviors and are now also being used to evaluate the potential therapeutic utility of various classes and mechanisms of action of test compounds. Although the observation of repetitive behaviors can easily be quantified and does not necessarily require expensive instrumentation or computer software, to make them ideal parameters for drug screening it is critical that this is approached systematically, utilizing optimized testing environments and proficient personnel trained in recognizing and quantifying repetitive behaviors in mice (*see* **Note 1**). Given the inherent variability of behavior, a rigorous experimental design which is appropriately powered and includes blinding, randomization, counterbalancing, and the inclusion of relevant controls are important considerations to minimize any potential bias in the experiment. All of these parameters combined will lend to the reproducibility of the study [12–15]. Furthermore, an understanding of the test compound's pharmacokinetic and pharmacodynamics properties is needed to aid in the selection of relevant and target specific dose ranges, route of administration, and pretreatment time for testing [16, 17]. Moreover, having knowledge of a test compound's therapeutic window, as defined by the dose range that produces the specific behavioral response (e.g., attenuation of repetitive behavior) relative to the dose range at which adverse or neurotoxic effects occur (e.g., hypoactivity, sedation, ataxia), is crucial to appropriately interpreting the resulting data and its translational utility [17, 18]. For example, a compound which has been described as a potential treatment for repetitive behaviors in ASD due to its ability to attenuate excessive repetitive behavior (e.g., grooming or jumping) in a mouse model should do so in the absence of eliciting any confounding motor effects at that dose. Further, it should be in line with data demonstrating appreciable brain exposure and target engagement activity at the time point that the behavioral data were generated.

2 Materials

2.1 Animals and Housing

1. A number of inbred strains of mice endogenously demonstrate repetitive behaviors and serve as excellent models of repetitive behaviors and for validating assay conditions (*see* **Notes 2** and **3**):

   - BTBR mice (BTBR *T⁺ Itpr3tf*/J; www.jax.org; stock #002882) demonstrate repetitive self-grooming and digging phenotypes [10, 19–22].

- C58 mice (C58/J; www.jax.org, stock # 000669) demonstrate repetitive jumping [10, 19–21].

2. Temperature, humidity, and lighting levels in the housing room and testing rooms should be in accordance with the institution's Animal Care and Use Committee and should be recorded as part of the experimental details as should the details of the housing environment.

2.2 Test Compounds

1. Compounds preferably in powder form and formulated fresh daily, dependent on compound properties to be tested (*see* **Notes 4** and **5**).

2. Vehicle Solvent (*see* **Notes 6** and **7**).

3. Syringes and needles for compound administration.

2.3 Testing Arenas

70 % Ethanol.

Paper towels.

Timer with the audio feature disabled (Fisher Scientific; cat # 14-649-85).

Optional: video camera, mirror placed behind the arena to facilitate viewing of the test animals behavior when it is facing away from the observer.

1. Repetitive Grooming.

A clear, clean, familiar home cage (typical dimensions of $7 \times 11 \times 5$ in or similar home cage) containing no food, water, bedding, or wire cage lid (*see* **Note 8**).

Subsequent clean test cages without food, water, or bedding (one cage per test animal).

A flat aerated lid in place of any wire cage lid to minimize climbing.

Stop watches with the audio feature disabled ([23]; *see* **Note 9**).

2. Repetitive Digging.

A clear, clean home cage without food or water (see above), containing familiar home cage bedding at a defined depth (e.g., 5 cm).

Subsequent clean test cages with clean bedding and without food or water (one cage per test animal).

A flat aerated lid in place of any wire cage lid to minimize climbing.

Stop watches with the audio feature disabled (see above).

3. Repetitive Jumping.

An open field arena with elevated wall heights (e.g., $40 \times 40 \times 40$ cm) (*see* **Note 10**).

Hand tally counter or cell counter (Fisher Scientific; cat# S90189).

3 Methods

3.1 Animal Set Up and Housing

1. Depending on the model and strain background, age and sex of the animals need to be selected accordingly (*see* **Notes 2** and **3**).

2. Cohort sizes need to be adequate, either informed by historical data based on the minimum sample size required to achieve statistical significance or statistically through power calculations (*see* **Note 11**).

3. When ordering from a commercial vendor, it is important to record not only the vendor and the strain and sub-strain information (i.e., C57BL/6J, C57BL/6NJ, C57BL/6NTac) but also the location the mice have been reared, and if possible the specific colony room (*see* **Note 12**).

4. In particular for pharmacology studies, cohorts of behaviorally and drug-naïve mice are recommended for the assessment of a drug effect (*see* **Note 13**).

5. For studies of repetitive behaviors, the inclusion of environmental enrichment in home cages should be minimized given the published data demonstrating that the presence of home cage enrichment can attenuate stereotypies in mice [24, 25]. Other variables such as environmental enrichment and background music in housing rooms need to be recorded and standardized (*see* **Note 14**).

6. Mice obtained from different locations should be acclimated to the test laboratory for about 5 days (*see* **Notes 15** and **16**).

7. A minimum 3-day period between cage change and testing is recommended to avoid the usual increase in novelty induced behaviors typically observed following a cage change, i.e., such behaviors can include grooming and digging (*see* **Note 17**).

8. The light:dark cycle and time of day with respect to lights (on:off) or whether testing was conducted under an inverted light cycle or under red lights should be documented (*see* **Note 18**).

3.2 Testing Environment

1. Low ambient lighting in the testing room (~20–50 lux) is recommended to avoid inhibiting exploratory behavior in the presence of bright light. Lighting levels should be standardized for consistency and should be recorded (*see* **Notes 18** and **19**).

2. Background noise should be maintained at a consistent level and its source and nature recorded. Intermittent disruptions to the testing environment need to be minimized, for example telephones, including cell phones should be silenced, audio features on timers and stopwatches should be disabled (*see* **Notes 9** and **20**).

3. Between test animals, clean the testing apparatus, especially of any urine and fecal boli. The generous use of a sanitizing agent which minimizes odors (e.g., 70 % ethanol) will help eliminate scent cues. Sanitation should be followed by drying the testing apparatus prior to placing the next animal into the apparatus (*see* **Note 21**).

3.3 Acclimation to the Testing Environment

1. Establish room conditions prior to introducing the animals into the testing environment.

2. If an adjacent anteroom is available with identical lighting and background noise conditions to the testing room, this can be used to acclimate and pretreat mice prior to introducing them to the test room for the actual behavioral test (*see* **Note 22**).

3. When selecting mice for the acclimation, weigh and check for overall appearance (i.e., gross health) (*see* **Note 23**).

4. When about to begin leave animals undisturbed for 30–60 min prior to testing in order to acclimate.

3.4 Compound Dosing

1. Compounds for use in testing should be blinded before use (*see* **Note 24**).

2. Prior to dosing, randomize test animals and pre-assign each to a treatment group (*see* **Note 24**).

3. Depending on the compound, the route of administration should be determined, e.g., based on pharmacokinetics. This will be based on time for pretreatment and relative adsorption (*see* **Note 25**).

4. If multiple compound injections are required, e.g., for pharmacodynamic agonist-antagonist experiments, first validate the test with identical multiple injection procedures to understand the impact that multiple injections may have on the behavior test (*see* **Note 26**).

5. Design a spreadsheet before testing starts, which includes the date, test animal ID, the assigned treatment group or dose, body weight, dose time, start and end time of the experiment, time to clean the equipment between animals, and dose time of the next sequential test animal (Fig. 1; *see* **Notes 27** and **28**).

6. Use separate needles and syringes for each test animal. Label and pre-fill syringes before the experiment starts (*see* **Note 28**).

3.5 Repetitive Grooming Assessment

Self-grooming behavior in mice is recognized not only as a primary hygiene function but also as a component of social interaction and has been shown to be sensitive to stressors [26, 27]. Self-grooming is defined as the cleaning, licking, or washing of the limbs, tail, and body surface areas, typically from a head to tail direction, but excludes bouts of scratching. Grooming behavior observations can

Date: _____

Subject Info: (Strain, DOB, arrival date, last cage change date) _____

Testing Environment: Lighting levels (lux), background noise (dB) _____

Time Subjects were habituated to testing room: _____

Test Arena	subject ID	sex	Body Weight	Treatment Group	30 min pretreat time	10 min habituation to test cage	10 min observation time	Cumulative Grooming Time (sec)	Notes/comments
1	1	m		A	9:00	9:30 - 9:40	9:40 - 9:50		
2	2	m		B					
3	3	m		C	9:10	9:40 - 9:50	9:50 - 10:00		
4	4	m		D					
5	5	m		B	9:20	9:50 - 10:00	10:00 - 10:10		
6	6	m		C					
7	7	m		D	9:30	10:00 - 10:10	10:10 - 10:20		
8	8	m		A					
				10 min time to clean testing arenas					
1	9	m		D	10:00	10:30 - 10:40	10:40 - 10:50		
2	10	m		C					
3	11	m		B	10:10	10:40 - 10:50	10:50 - 11:00		
4	12	m		A					
5	13	m		C	10:20	10:50 - 11:00	11:00 - 11:10		
6	14	m		A					
7	15	m		B	10:30	11:00 - 11:10	11:10 - 11:20		
8	16	m		D					
				10 min time to clean testing arenas					
1	17	m		C	11:00	11:30 - 11:40	11:40 - 11:50		
2	18	m		D					
3	19	m		A	11:10	11:40 - 11:50	11:50 - 12:00		
4	20	m		B					
5	21	m		D	11:20	11:50 - 12:00	12:00 - 12:10		
6	22	m		B					
7	23	m		A	11:30	12:00 - 12:10	12:10 - 12:20		
8	24	m		C					
				10 min time to clean testing arenas					
1	25	m		B	12:00	12:30 - 12:40	12:40 - 12:50		
2	26	m		A					
3	27	m		D	12:10	12:40 - 12:50	12:50 - 1:00		
4	28	m		C					
5	29	m		A	12:20	12:50 - 1:00	1:00 - 1:10		
6	30	m		D					
7	31	m		C	12:30	1:00 - 1:10	1:10 - 1:20		
8	32	m		B					

Fig. 1 Example of a spreadsheet for recording of repetitive grooming observations. Note the blinded treatments, randomizing treatment order, counterbalancing treatments across arenas, and precise time points for dosing, habituation, and observations. Additional columns should be added for subsequent genotype (blinded)

be performed manually using a stopwatch with the researcher situated approximately 2–3 feet from the arena, or via closed circuit camera, which can be scored live or recorded and scored at a later time [10, 21]. Highly experienced operators are often capable of observing and scoring two animals simultaneously. However, when multiple measures per test animal are required this is not recommended.

1. Following the designated pretreatment period, place the test animal in the center of the empty observation arena (*see* **Note 29**).

2. Mice are initially allotted an acclimation period to explore the arena, typically 10 min.

3. Evaluate the test animal for cumulative time spent grooming (sec) over the course of 10 min.

    Optional: With a video camera positioned in front of the cage as opposed to overhead and a mirror placed behind the cage, grooming behavior can be recorded. Behavioral tracking software may assist the data analysis (*see* **Note 30**). Video playback allows quantification of additional measures such as total bouts of grooming and latency to groom (sec).

4. Do not return test animals to their home cage until all animals within one home cage have been tested (*see* **Note 31**).

3.6 Repetitive Digging Assessment

Digging behavior is defined as the displacement of bedding, typically using the nose and forepaws [28]. Observations can be performed manually using a stopwatch with the researcher situated approximately 2–3 feet from the arena, or via closed circuit camera and recorded and scored at a later time point as needed [28]. When experienced, operators are capable of observing and scoring two test animals simultaneously for a measurement of cumulative duration spent digging; however additional measurements, e.g., bouts, latency etc., will require video recording.

1. Following the designated pretreatment period, place the test animal in the center of the observation arena (*see* **Note 29**).

2. For cumulative duration of digging (sec) score via stopwatch for 5 min.

    Optional: With a video camera positioned in front of the cage as opposed to overhead and a mirror placed behind the cage, the digging behavior can be recorded. Behavioral tracking software may assist the data analysis (*see* **Note 30**). Video playback allows quantification of additional measures such as total bouts of digging and latency to dig (sec).

3. Do not return test animals to their home cage until all animals within one home cage have been tested (*see* **Note 31**).

3.7 Repetitive Jumping

Jumping behavior is defined as a vertical hindlimb jump where both hindlimbs are displaced off of the arena floor and are typically initiated by the test animal rearing against the arena wall [21]. This can be recorded manually with the researcher situated about 2–3 feet away from the arena using a counter, or via a closed circuit camera, or be recorded and tracking software used to score this at a later time [21]. When experienced, operators are capable of observing and scoring two test animals simultaneously for a measurement of cumulative number of jumps; however additional measurements, e.g., bouts, latency, etc, will require video analysis.

1. Following the designated pretreatment period, place the test animal in the center of the observation arena (*see* **Note 29**).

2. Mice are initially allotted an acclimation period to explore the arena, typically for 10 min.

3. The total number of jumps is scored using a hand counter for a 10 min period.

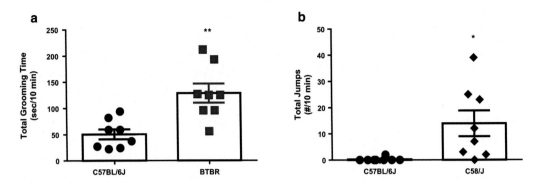

Fig. 2 Representative data for repetitive grooming observations in adult male BTBR mice relative to C57BL/6J (**a**) and repetitive jumping observations in adult male C58/J mice relative to C57BL/6J (**b**). T-test: $*p > 0.05; **p > 0.01$

Optional: A video camera can be used to record the jumps. Standard behavioral tracking software will assist with the data output and analysis (*see* **Note 30**).

4. Do not return test animals to their home cage until all animals within the home cage have been tested (*see* **Note 31**).

3.8 Data Analysis

1. Data should be presented and analyzed as raw values and illustrated in graph form with the distribution of the data points if feasible to assist in interpretation (Fig. 2).

2. Select the appropriate statistical analysis. For example, when comparing two genotypes with a single endpoint use a t-test. Use an analysis of variance (ANOVA) to analyze multiple genotypes or treatment groups (e.g., dose-response curve) for a single behavioral endpoint with an appropriate post hoc test (e.g., Dunnett's post hoc test with vehicle as control).

3. To maintain impartially the researcher/data should only be un-blinded after data have been analyzed.

4. Do not combine several small underpowered experiments, for example by adding additional animals to treatment groups retrospectively if after initial analysis the experiments did not yield statistical significance [12–15, 29, 30]. While pilot studies are helpful to inform dose-response relationships and understand experimental limitations, as well as supportive for power calculations for follow-up experiments, additional appropriately powered experiments may need to be planned and executed to confirm positive or negative outcomes in an independent cohort.

4   Notes

1. A prerequisite for sensitive behavioral tests is a well-trained researcher capable of consistently reproducing published data sets (i.e., positive controls) under blinded conditions.

For example, under optimized environments, compared to C57BL/6J mice, BTBR mice will demonstrate increased levels of grooming behavior, while C58/J mice will demonstrate increased repetitive jumping behaviors [21]. Failure of a researcher to reproduce such positive controls may be due to either the lack of an optimized test environment or an absence of the appropriate proficiency. If this occurs it should caution the investigator on proceeding with phenotyping a novel animal model, or the evaluation of a novel reagent. It should be remembered that the investment in training and running control studies is more than recuperated by achieving reliable and consistent data. Further, it should be stressed that assay and testing conditions need to be optimized and validated before proceeding with testing of novel reagents.

2. Regardless of the mouse model used, it is critical to understand how the genetic background at the selected age, and both sexes, perform in the behavioral task selected. In general, it is helpful to perform pilot experiments in the selected animal model to ensure that the selected assays are appropriate and robust. Further, such pilot studies may inform about the sensitivity for a change of response in the presence of test compound under these test conditions.

3. Include the sex of the animal as an independent variable and whenever possible test both sexes. Irrespective of the behavioral changes associated with estrous, female and male mice have divergent behaviors in many behavioral assays and may also vary in their responses to drug treatment and dosage. However, where both sexes are included within a test, extra caution should be taken to minimize residual female scent in the test equipment which may result in increased sniffing and reduced exploration/performance levels of test males. To help alleviate this it is suggested that provision is made for staggering the testing of females separately from males.

4. Procurement of test compounds in powder form is preferred over pre-packaged solutions as often the vehicle constituents and their precise concentrations are not undisclosed, or may contain solvents that alter behaviors in their own right (*see* **Note 19**). Caution should be taken in preparing compounds with unknown or high instability (e.g., peptides); these may need to be prepared as specified by manufacturer provided that the manufacturer does not recommend a vehicle that alters behavior.

5. Test compound concentrations should be calculated as the activeor parent compound relative to the percentage of the molecule that is a salt (% active moiety), and serial dilutions used for testing. Compounds are often synthesized as a salt to aid in solubility; however it is critical to record how the drug concentration is calculated, as misunderstanding can lead to

reporting dosages that are incorrect as the contribution of a salt varies significantly with the varied formulations (e.g., HCl, HBr). For example, the drug scopolamine (M.W. 303.353) is commercially available in several salt formulations and it is rarely reported whether it is formulated as the freebase or as the salt. As an HCl salt (CAS# 55-16-3), the molecular weight of the molecule is increased to 339.81 of which 89.3 % of the molecule is active compound and the remaining 10.7 % is HCl salt. However, if the hydrobromide trihydrate salt version is used for drug testing (CAS# 6533-68-2), the molecular weight of the molecule is increased to 438.31, of which only 69.2 % of the molecule is active compound, with the remaining 30.8 % being HBr salt. Therefore it is not only critical to report the test compound and source used, but also the manner in which it was calculated, formulated, and which stereoisomer was used. Whenever possible keep a test sample after the compound has been administered for possible subsequent analytics.

6. The choice of solvents for drug formulation and their concentrations should be taken into consideration as the excipient may have its own, possibly concentration-dependent impact on the specific behavior being evaluated [31–34]. We suggest pre-testing any novel vehicle for unexpected behavioral effects prior to testing the drug in the test.

7. To minimize within experiment variability, drug concentrations should be formulated as serial dilutions within the diluent used to formulate the highest concentration as well as used for the vehicle control group. Furthermore, the pH should be titrated to an acceptable range (typically pH 5–8) and made consistent across treatments including the vehicle control. Writhing is a behavior often observed with highly acidic compounds and can confound the behavior being evaluated. Such irritation at the site of injection may well increase grooming and could confound the interpretation of the data. In most cases, test compounds can be titrated to neutral levels with negligible concentrations of base (e.g., NaOH) or acid (e.g., HCl) without impacting solubility.

8. Several vendors offer various home cage sizes for laboratory mice, and their use varies across laboratories. For testing of repetitive and digging behaviors, the mouse's familiar home cage is preferred to minimize the novelty of the cage itself. To ensure that the repetitive behaviors are recordable, validating the assay with animal models with known repetitive phenotypes such as the BTBR and C58/J mice is useful to validate the environment including the home cage used.

9. In our laboratory the phone does not ring, there are restrictions in entering during the dark cycle, and cage changes are done consistently on the same day of the week. All visitors are asked

to silence their cell phones and minimize conversation. Further, the door latches to testing rooms have been adjusted to prevent any latch clicking which could startle the test animals. Wall clocks and timers have been selected to not have any associated audio cues (either beeping or clicking of the secondhand on the clock). When a timer or stopwatch is used for scoring during the test, such need to be silent (i.e., any audio feature disabled) so not startle the test animals (see [23] *for methods for removing the audio feature of commonly used stopwatches*).

10. Selection of equipment and material is important. While acrylic is a fairly inexpensive material that can be readily used to create mazes and arenas, it is highly susceptible to crazing and cracking when exposed to ethanol, an often used sanitizing agent. We recommend polycarbonate as a better alternative to acrylic. To minimize the need to change background color for contrasting against varying coat colors and the additional variability associated with that, an infrared camera paired with an infrared reflecting background systems (e.g., www.Noldus.com; Noldus Information Technology, The Netherlands) can be used which not only eliminates the need to change background but also minimizes glare for camera tracking.

11. Depending on the behavioral test, typical animal sample groups range from 8 to 12 per sex, per treatment group. For chronic studies or sensitive animal models (e.g., models with potential handling induced seizure), additional test animals need to be at hand to supplement for potential attrition.

12. It is important to be aware of behavioral differences across substrains [35] and within a mouse strain, as divergent behaviors may occur as a result of variations in rearing environment. To minimize this variability, when ordering from a commercial vendor, cohorts should be acquired from not only the same vendor but also the same geographical location and optimally from the same colony room, as some vendors have multiple production facilities across the world. For example, background noise level in a husbandry room with ventilated cages may be higher than a husbandry room that uses static caging; such differences may contribute to increased variability in behavioral tests that employ audio stimuli.

13. Habituation to an environment upon re-exposure is a well-reported behavioral phenomenon and should be considered in experimental design, especially when longitudinal studies are being planned. In addition to behavioral habituation, depending on the compound or mechanism of action evaluated, it may also not be known if an experimental compound itself results in tachyphylaxis or tolerance following repeated dosing. We suggest to plan independent cohorts of mice for each specific assay whenever possible, thus avoiding any potential for

drug or behavioral tolerance. If this is not possible, then evaluate multiple test compounds and multiple mechanisms of action with an appreciable washout period between test compounds (e.g., 1 week).

14. Environmental enrichment is a variable and needs to be made consistent across the cohort of all test animals. For example, if laboratory guidelines require environmental enrichment for individually housed mice, but not for group housed mice, then when selecting test animals identical enrichment needs to be provided to all cages housing test animals. Furthermore, exclude mice with wounds or other gross abnormalities from the study to minimize additional uncontrollable factors, such as immune response or veterinary treatments.

15. Provide mice with adequate acclimation time to a new facility so that physiologic changes are normalized prior to the start of an experiment [36]. Interestingly, it has been reported that even brief alterations in the dark cycle by light stimuli can alter behavioral responses in mice, and require several days to re-entrain [37]. For longer shipping times that include crossing time zones, extended acclimation times should be considered. In our laboratory, it is standard procedure to acclimate mice for a minimum of 5 days, even for transfers within the campus when the housing rooms have a different light:dark cycles, e.g., from a 10:14 light:dark cycle to a 12:12 light:dark cycle.

16. Behavioral responses can vary with housing density. Assay validation under different housing densities may be required to understand whether housing density is a contributing variable to the behavior. For example, social behavior may be impacted when mice are housed individually, group housed within mixed genotypes, or housed within same genotype groups [38]. Further, aggressive behaviors in male mice may result in the need to separate cage mates mid-study or exclude test animals due to wounds which would add additional variability to the experiment.

17. Experiments should be planned to avoid being initiated on the same day as a cage change. In some cases, such as social or repetitive grooming behaviors it is recommended that there is a minimum of 3 days between cage change and testing. Importantly, the testing protocol and reported methods should mark the cage change day relative to the test day and this be maintained across studies. While some institutions utilize forceps to transfer mice during cage changes, it is recommended that this is not done for behavioral testing, as this can count as an additional stimulus and would interfere. For example in the case of grooming, tail grooming may be a response to forceps pressure on the tail during transfers.

18. Time of day relative to the light cycle is a critical detail as behavior varies with respect to the time of day; e.g., mice increase activity at times closer to the start of the dark cycle. Although some repetitive behaviors can be more prominent during the animal's active cycle (dark cycle), animal models with true repetitive behavior phenotypes are consistently observed during the light cycle, such that there is not necessarily a requirement to conduct these assessments during the dark phase. If the test day requires an extended window of time to complete all tests, this should be counterbalanced across treatment groups to ensure that matched groups of treatment and controls are tested throughout the various times of day.

19. Commercially available light (lux) meters to measure lighting levels should be purchased. A lux meter should be used to measure lighting in the testing environment and once established should be maintained for each assay as part of its protocol. Lighting terminology such as "dim" or "high" is ambiguous and the lux or lumen level should be reported for both the light levels at the testing equipment where the animal is exposed, as well as the levels for the testing room and any adjacent holding or acclimation areas the mice are maintained in prior to being tested. The type of lighting should also be noted, i.e., LED, fluorescent, and halogen each emit different wavelengths. Lighting systems that emit heat should be avoided both so as to not damage the testing equipment and also to eliminate a heat source for which the mice may gravitate to. Examples of details on lighting level considerations may include reducing light to minimize glare during automated video tracking or to facilitate exploratory behavior under ambient lighting.

20. A decibel sound meter should be purchased and used to measure the background noise of the testing facility. Interference from electrical equipment and other undefined sources which may produce sound in frequency ranges not evident to the human ear but within the ultrasonic range of the mouse should be investigated (a bat detector or other commercially available ultrasonic microphone can be used to identify any issues which could unknowingly interfere with the testing). Background white noise (typically <70 dB) may be used as ancillary background noise to mask out random noise from activity in adjacent areas to the testing room. In this case, commercially available white noise generators can be purchased and should be adjustable appropriately as high levels of background noise may also induce anxiety.

21. Similarly to the testing equipment being cleaned between subsequent test animals, the testing arena also needs to be cleaned prior to placing the first test animal in the arena (with, e.g., 70 % ethanol). This ensures not only that the equipment

is properly cleaned from any previous test, but also makes for the consideration that all animals are treated identically and exposed to identical odors (e.g., 70 % ethanol).

22. Tests that use audio or visual cues are a prime example of systems that should use anteroom holding so as to not inadvertently habituate the test animals to the stimuli prior to measuring. Optimal anterooms allow manipulations such as dosing and are away from the testing area where sensitive observations are being conducted. This will help reduce any disruption by adjacent activities. To minimize visual cues, a curtain can be used to surround the perimeter of the testing arena. This can also help eliminate the visual presence of the experimenter moving around the testing room during the observation period.

23. Body weights should be taken for individual mice prior to dosing, preferably just prior to the start of the experiment; they *should not be* recorded as a mean for the cage. In addition to weighing each mouse, to minimize identification errors, a marker pen (e.g., Sharpie®) can be used to label the tails of the individual animals so as to readily identify which animal with which drug. Alternatively ear notches can be used however, ear notching may be stressful and should be done at least one week prior to testing (ear tags are not in general, recommended). It is common practice in our laboratory to handle each mouse and label its tail prior to the start of the acclimation period for the testing environment. This makes for a consistent practice between tests and researchers. Regardless whether the intention of the experiment is to dose the mice or to conduct phenotyping studies without a compound, this preliminary allows the researcher to evaluate the health and well-being of the test animals prior to dosing and to exclude test animals if necessary, based on predetermined exclusion criteria (e.g., bite wounds, low or excess body weight, excessive barbering).

24. Prior to initiating dosing, randomize test animals and pre-assign each to a treatment group, with attention to counterbalancing treatment groups. Ideally this is done by a researcher familiar with the study, but not responsible for conducting the study. Blinding of drug vials can be facilitated by simple coding (e.g., "A," "B," "C," "D"). Importantly the code should be maintained until the data analysis has been completed. If a second researcher is not available to assist with blinding, then the drug vials can be wrapped (e.g., in aluminum foil), placed in random order, and self-coded by the study manager.

25. The route of administration can impact the outcome of the assays. Some compounds are not readily soluble and can result in a drug suspension precluding certain routes of administration (i.e., intravenous, intracerebroventricular, subcutaneous),

but as a homogenous suspension would be suitable for oral administration, assuming the compound has a good oral bio-availability. Chronic oral dosing has been associated with high attrition rates. Drug-laced water or drug milled into food is in general not precise enough to establish any dose dependency [39]. For pharmacology studies, precise dosing is preferred and based on weight (e.g., 10 mg/kg) versus bolus doses (e.g., 200 μL/mouse). In cases where solubility becomes challenging at the highest concentrations, the concentration can be lowered to achieve the desired dosage; however the dose volume may need to increase, e.g., half the concentration would require twice the dose volume. If this approach is chosen, then all treatment groups need be treated identically, i.e., all dose levels are administered with the same correction for concentration (e.g., by half) and dose volumes including the vehicle control (from 10 mL/kg to 20 mL/kg).

26. Multiple injections either within an experiment or for chronic dosing studies should be administered at independent injection sites and the site of injection should be recorded and consistent across all test animals, e.g., "subcutaneous at nape of neck," "left side intraperitoneal injection," etc.

27. We suggest developing a spreadsheet for each experiment. This will ease the work flow and documentation (including recording mishaps such as bad injection, noise disruption during trial, etc.). Detailed documentation will also help identify possible spurious data during the analysis. Dosing of a subsequent test animal should be avoided while another test animal is being observed so as to avoid changes in behavior of the test animal in response to disruptions induced by movement of cages or vocalizations associated with restraint.

28. It is advisable to pre-plan as much of an experiment as possible to minimize errors during complex testing that requires multi-tasking of the researcher. During the period of time that the test animals are acclimating to the test area, the researcher should prepare test compounds, e.g., pre-fill syringes ahead of dosing. If the compound is a suspension, syringes should not be pre-filled as settling will occur. In these cases, it may be important for the drug to remain on a stir plate to maintain the suspension and syringes should not be filled until immediately prior to dosing. Importantly, it is strongly recommended that each individual test animal is assigned a new syringe and needle for each injection. It is poor practice to preload a single syringe with the intent to inject multiple animals. Not only do needles dull significantly after multiple uses which may result in variable responses of the animal, particularly in its grooming behavior at the injection site, but this practice of "eyeballing"

dosing volume from one test subject to the next negates the precision required for carrying out accurate dosing.

29. Place the test animals consistently in a similar location within the test apparatus (e.g., "in the center," "at the back of the cage"); this should be noted and recorded as part of the standard procedure.

30. While behavioral tracking software is very helpful and helps minimizes additional stimuli due to the researcher being in close proximity to the test animals, it is important to ensure that the behavior being evaluated is what is being captured/ analyzed by the software. For example, recent advances in behavioral tracking technologies have resulted in several vendors offering grooming recognition software. However, in our recent review of a number of these software applications it was determined that the algorithms developed for recognizing grooming behavior by the software may have solely been developed (trained) based on the grooming behavior of a single strain (e.g., C57BL/6J). This training does not necessarily accurately track for the differences in grooming repertoire of other mouse models (e.g., BTBR) and hence cannot be generalized across mouse models to provide an accurate cumulative grooming score. We recommend to first score by hand and compare the data to the automated scoring to ensure the behavior is being captured correctly for each strain or mouse model, and then adjust the tracking software as needed.

31. Have cages ready for returning test animals, either back to a group housed setting at the conclusion of testing for all the cage mates, or alternatively in a separate holding cage until all cage mates have completed testing.

References

1. American Psychiatric Association (2013) Diagnostic and statistical manual of mental disorders, 5th edn. American Psychiatric Association, Washington, DC

2. Goldman S, Wang C, Salgado MW, Greene PE, Kim M, Rapin S (2009) Motor stereotypies in children with autism and other developmental disorders. Dev Med Child Neurol 51(1):30–38

3. Lewis M, Kim S-J (2009) The pathophysiology of restricted repetitive behavior. J Neurodev Disord 1(2):114–132, PMCID: PMC3090677

4. Langen M, Durston S, Kas MJ, van Engeland H, Staal WG (2011) The neurobiology of repetitive behavior: ...and men. Neurosci Biobehav Rev 35(3):356–365. doi:10.1016/j.neubiorev.2010.02.005

5. Bodfish JW (2004) Treating the core features of autism: are we there yet? Ment Retard Dev Disabil Res Rev 10:318–326

6. Boyd BA, McDonough SG, Bodfish JW (2012) Evidence-based behavioral interventions for repetitive behaviors in autism. J Autism Dev Disord 42(6):1236–1248

7. Carrasco M, Vokmar FR, Bloch MH (2012) Pharmacologic treatment of repetitive behaviors in autism spectrum disorders: evidence of publication bias. Pediatrics 129(5):e1301–e1310

8. Hart PC, Bergner CL, Dufor BD, Smolinsky AN, Egan RJ, LaPorte JL, Kalueff AV (2010) Analysis of abnormal repetitive behaviors in experimental animal models. In: Kalueff AV, Warwick JE (eds) Translational neuroscience

and its advancement of animal research ethics. Nova Science, Inc, New York, pp 71–82

9. Lewis MH, Tanimura Y, Lee LW, Bodfish JW (2007) Animal models of restricted repetitive behavior in autism. Behav Brain Res 176: 66–74

10. Silverman JL, Yang M, Lord C, Crawley JN (2010) Behavioural phenotyping assays for mouse models of autism. Nat Rev Neurosci 11(7):490–502

11. Kas MJ, Glennon JC, Buitelaar J, Ey E, Biemans B, Crawley J, Ring RH, Lajonchere C, Esclassan F, Talpos J, Noldus LP, Burbach JP, Steckler T (2014) Assessing behavioural and cognitive domains of autism spectrum disorders in rodents: current status and future perspectives. Psychopharmacology (Berl) 231:1125–1146

12. Kilkenny C, Parsons N, Kadyszewski E, Festing MFW, Cuthill IC, Fry D, Hutton J, Altman DG (2009) Survey of the quality of experimental design, statistical analysis and reporting of research using animals. PLoS One 4(11), e7824. doi:10.1371/journal.pone.0007824

13. Kilkenny C, Browne WJ, Cuthill IC, Emerson M, Altman DG (2010) Improving bioscience research reporting: the ARRIVE guidelines for reporting animal research. PLoS Biol 8, e1000412. doi:10.1371/journal.pbio.1000412

14. Landis SC, Amara SG, Asadullah K, Austin CP, Blumenstein R, Bradley EW, Crystal RG, Darnell RB, Ferrante RJ, Fillit H, Finkelstein R, Fisher M, Gendelman HE, Golub RM, Goudreau JL, Gross RA, Gubitz AK, Hesterlee SE, Howells DW, Huguenard J, Kelner K, Koroshetz W, Krainc D, Lazic SE, Levine MS, Macleod MR, McCall JM, Moxley RT, Narasimhan K, Noble LJ, Perrin S, Porter JD, Steward O, Unger E, Utz U, Silberberg SD (2012) A call for transparent reporting to optimize the predictive value of preclinical research. Nature 490(7419):187–191. doi:10.1038/nature11556

15. Oswald S, Balice-Gordon R (2014) Rigor or mortis: best practices for preclinical research in neuroscience. Neuron 84(3):572–581. doi:10.1016/j.neuron.2014.10.042

16. Rizzo SJ, Edgerton JR, Hughes ZA, Brandon NJ (2013) Future viable models of psychiatry drug discovery in pharma. J Biomol Screen 18(5):509–521. doi:10.1177/1087057113475871

17. Onos KD, Sukoff Rizzo SJ, Howell GR, Sasner M (2015) Toward more predictive genetic mouse models of Alzheimer's disease. Brain Res Bull 122:1-11. doi: 10.1016/j.brainresbull.2015.12.003.

18. Schellinck HM, Cyr DP, Brown RE (2010) How many ways can mouse behavioral experiments go wrong? Confounding variables in mouse models of neurodegenerative diseases and how to control them. In: Jane Brockmann H (ed) Advances in the study of behavior, vol 41. Academic Press, Burlington, pp 255–366

19. Moy SS, Nadler JJ, Young NB, Perez A, Holloway LP, Barbaro RP, Barbaro JR, Wilson LM, Threadgill DW, Lauder JM, Magnuson TR, Crawley JN (2007) Mouse behavioral tasks relevant to autism: phenotypes of 10 inbred strains. Behav Brain Res 176(1):4–20

20. Ryan BC, Young NB, Crawley JN, Bodfish JW, Moy SS (2010) Social deficits, stereotypy and early emergence of repetitive behavior in the C58/J inbred mouse strain. Behav Brain Res 208:178–188

21. Silverman JL, Smith DG, Rizzo SJ, Karras MN, Turner SM et al (2012) Negative allosteric modulation of the mGluR5 receptor reduces repetitive behaviors and rescues social deficits in mouse models of autism. Sci Transl Med 4(131):131ra51

22. Smith JD, Rho JM, Masino SA, Mychasiuk R (2014) Inchworming: a novel motor stereotypy in the BTBR T+ Itpr3tf/J mouse model of autism. J Vis Exp 5(89). doi:10.3791/50791

23. Yang M, Silverman JL, Crawley JN (2011) Automated three-chambered social approach task for mice. Curr Protoc Neurosci. Chapter 8:Unit 8.26. doi: 10.1002/0471142301.ns0826s56

24. Muehlmann AM, Edington G, Mihalik AC, Buchwald Z, Koppuzha D, Korah M, Lewis MH (2012) Further characterization of repetitive behavior in C58 mice: developmental trajectory and effects of environmental enrichment. Behav Brain Res 235(2):143–149. doi:10.1016/j.bbr.2012.07.041

25. Reynolds S, Urruela M, Devine DP (2013) Effects of environmental enrichment on repetitive behaviors in the BTBR T+tf/J mouse model of autism. Autism Res 6(5):337–343. doi:10.1002/aur.1298

26. Spruijt BM, van Hooff JA, Gispen WH (1992) Ethology and neurobiology of grooming behavior. Physiol Rev 72:825–852

27. Kalueff AV, Aldridge JW, LaPorte JL, Murphy DL, Tuohimaa P (2007) Analyzing grooming microstructure in neurobehavioral experiments. Nat Protoc 2(10):2538–2544

28. Deacon RMJ (2006) Digging and marble burying in mice: simple methods for in vivo identification of biological impacts. Nat Protoc 1(1):122–124

29. Steckler T (2015) Preclinical data reproducibility for R&D—the challenge for neuroscience.

Psychopharmacology (Berl) 232(2):317–320. doi:10.1007/s00213-014-3836-3

30. Unger EF (2008) All is not well in the world of translational research. J Am Coll Cardiol 50(8):738–740

31. Castro CA, Hogan JB, Benson KA, Shehata CW, Landauer MR (1995) Behavioral effects of vehicles: DMSO, ethanol, Tween-20, Tween-80, and emulphor-620. Pharmacol Biochem Behav 50(4):521–526

32. Colucci M, Maione F, Bonito MC, Piscopo A, Di Giannuario A, Pieretti S (2008) New insights of dimethyl sulphoxide effects (DMSO) on experimental in vivo models of nociception and inflammation. Pharmacol Res 57(6):419–425. doi:10.1016/j.phrs.2008.04.004

33. Lin HQ, Burden PM, Johnston GAR (1998) Propylene glycol elicits anxiolytic-like responses to the elevated plus-maze in male mice. J Pharm Pharmacol 50(10):1127–1131

34. Rivers-Auty J, Ashton JC (2013) Vehicles for lipophilic drugs: implications for experimental design, neuroprotection, and drug discovery. Curr Neurovasc Res 10(4):356–360

35. Bryant CD, Zhang NN, Sokoloff G, Fanselow MS, Ennes HS, Palmer AA, McRoberts JA (2008) Behavioral differences among C57BL/6 substrains: implications for transgenic and knockout studies. J Neurogenet 22(4):315–331. doi:10.1080/01677060802357388

36. Obernier JA, Baldwin RL (2006) Establishing an appropriate period of acclimatization following transportation of laboratory animals. ILAR J 47(4):364–369. doi:10.1093/ilar.47.4.364

37. Loh DH, Navarro J, Hagopian A, Wang LM, Deboer T, Colwell CS (2010) Rapid changes in the light/dark cycle disrupt memory of conditioned fear in mice. PLoS One. 5(9). pii: e12546. doi: 10.1371/journal.pone.0012546

38. Yang M, Perry K, Weber MD, Katz AM, Crawley JN (2011) Social peers rescue autism-relevant sociability deficits in adolescent mice. Autism Res 4:17–27

39. Gonzales C, Zaleska MM, Riddell DR, Atchison KP, Robshaw A, Zhou H, Sukoff Rizzo SJ (2014) Alternative method of oral administration by peanut butter pellet formulation results in target engagement of BACE1 and attenuation of gavage-induced stress responses in mice. Pharmacol Biochem Behav 126:28–35. doi:10.1016/j.pbb.2014.08.010

Chapter 18

Better Utilization of Mouse Models of Neurodegenerative Diseases in Preclinical Studies: From the Bench to the Clinic

Christopher Janus, Carolina Hernandez, Victoria deLelys, Hanno Roder, and Hans Welzl

Abstract

The major symptom of Alzheimer's disease is dementia progressing with age. Its clinical diagnosis is preceded by a long prodromal period of brain pathology that encompasses both formation of extracellular amyloid and intraneuronal tau deposits in the brain and widespread neuronal death. At present, familial cases of dementia provide the most promising foundation for modeling neurodegenerative tauopathies, a group of heterogeneous disorders characterized by prominent intracellular accumulation of hyperphosphorylated tau protein. In this chapter, we describe major behavioral hallmarks of tauopathies, briefly outline the genetics underlying familial cases, and discuss the arising implications for modeling the disease in transgenic mouse systems. The selection of tests performed to evaluate the phenotype of a model should be guided by the key behavioral hallmarks that characterize human disorder and their homology to mouse cognitive systems. We attempt to provide general guidelines and establish criteria for modeling dementia in a mouse; however, interpretations of obtained results should avoid a reductionist "one gene, one disease" explanation of model characteristics. Rather, the focus should be directed to the question of how the mouse genome can cope with the over-expression of the protein coded by transgene(s). While each model is valuable within its own constraints and the experiments performed are guided by specific hypotheses, we seek to expand upon their methodology by offering guidance spanning from issues of mouse husbandry to choices of behavioral tests and routes of drug administration that might increase the external validity of studies and consequently optimize the translational aspect of preclinical research.

Key words Tauopathy, Alzheimer's disease, Transgenic models, Phenotype, Behavioral tests, Drug delivery, Handling

Abbreviations

Aβ   β-Amyloid peptide
AD   Alzheimer's disease
ApoE   *Apolipoprotein E*
ApoE-ε4   One of the three major alleles of ApoE gene
APP   Amyloid-β precursor protein

Gabriele Proetzel and Michael V. Wiles (eds.), *Mouse Models for Drug Discovery: Methods and Protocols*,
Methods in Molecular Biology, vol. 1438, DOI 10.1007/978-1-4939-3661-8_18, © Springer Science+Business Media New York 2016

A2M   α-2 Macroglobulin
BVAAWF   British Veterinary Association Animal Welfare Foundation
CLU   Clusterin
CNS   Central nervous system
CR   Complement receptor
DLB   Dementia with Lewy body
ES   Embryonic stem
FAD   Familial Alzheimer's disease
FTD   Frontotemporal dementias
FTDP-17 Frontotemporal dementia and parkinsonism linked to chromosome 17
FRAME   Fund for the Replacement of Animals in Medical Experiments
GLM   General linear model
GWAS   Genome-wide association studies
IDE   Insulin-degrading enzyme
LB   Lewy body
LRP   Lipoprotein receptor-related protein
MAPT   Microtubule-associated tau protein
MCI   Mild cognitive impairment
MSA   Multiple system atrophy
MWM   Morris water maze
NFT   Neurofibrillary tangles
PHF   Paired helical filaments
PS1, PS2 Presenilin 1, presenilin 2
RSPCA   Royal Society for the Prevention of Cruelty to Animals
TREM2   *T*riggering *r*eceptor *e*xpressed on *m*yeloid cell *2* protein
UFAW   Universities Federation of Animal Welfare

A man is but what he knoweth,

Sir Francis Bacon [1].

---

1 Introduction

One hundred years ago, Alois Alzheimer wrote a seminal paper describing the behavioral symptoms of his patient, Auguste Deter, who suffered from a mental illness [2] (*see* ref. 3 for English translation of the original paper). He observed that, "*[S]he developed a rapid loss of memory. She was disoriented in her home, [...] She is completely disoriented in time and space. Her memory is seriously impaired. If objects are shown to her, she names them correctly, but almost immediately afterwards she has forgotten everything.*" After her death on the 8th of April 1906, an autopsy revealed dense deposits outside and around nerve cells and twisted strands of fiber inside dead neurons in her brain. Today, these two pathological hallmarks of Alzheimer's disease (AD) are known to be extracellular plaques made largely of β-amyloid peptide (Aβ) and intracellular neurofibrillary tangles composed of hyperphosphorylated microtubule-associated protein tau gene (MAPT) [4, 5].

The current understanding of the disorders is that almost all neurodegenerative disorders present proteinopathies that are associated with the aggregation and accumulation of misfolded proteins [6]. The process, as in the case of AD, may involve intra- and extracellular deposits of MAPT and Aβ or only intra-neuronal aggregates of MAPT [7, 8], which are implicated in the loss of neurons and atrophy of the brain [9–11]. This class of neurodegenerative diseases, which includes Alzheimer's disease among others, certain forms of Parkinson's disease with dementia, and frontotemporal dementia (all together 21 diseases [7]), are characterized neuropathologically by ubiquitous accumulation of abnormal intraneuronal filaments formed by the MAPT and are labeled as neurodegenerative tauopathies [7, 12]. The profound degeneration of neurons in tauopathies leads to a progressive and accelerated decline in mental function. AD is one of the most devastating and so far one of the most intensively studied tauopathies in which a patient's memory and ability to learn are initially compromised and eventually completely destroyed. Although the behavioral phenotypes of tauopathies overlap to a certain extent, they manifest separate clinical entities due to a specific combination of neuropathological changes, variations in the form of abnormally hyperphosphorylated tau, and individual spatiotemporal expression of neurodegeneration in the brain (reviewed in ref. 7).

2   Tauopathies: Hallmarks and Characteristics

2.1   Clinical Diagnosis of Dementia

Tauopathies are characterized by a progressive decline in cognitive abilities and abnormalities in other behavioral systems that cannot be attributed to normal aging [13, 14]. Signs of mild cognitive impairment (MCI) precede overt dementia, as is the case in AD [15]. In practice, individuals will only be classified as cognitively impaired when their ability to perform everyday tasks is compromised to the point that they are no longer able to function at home and in their community [16]. MCI diagnosis does not always correctly predict the development of AD-type dementia. Data derived from neuroimaging, screening for genetic risk factors [17], or detection of increased levels of tau protein in cerebrospinal fluid [18, 19] might provide false indications of the onset of dementia. Even significant hippocampal atrophy is not always a reliable marker since it might occur due to depression, Parkinson's disease (PD), or vascular dementia [17]. Diagnosis of AD is also complicated since amyloid plaques and neurofibrillary tangles (NFTs) comprised of abnormally hyperphosphorylated tau are also present, to some degree, during normal aging. Furthermore, depending on the type of tauopathy, patients may manifest a variety of other cognitive symptoms including delusions, amnesia, executive dysfunction, apathy, agitation, and aggressive behavior [8, 20–22].

2.2 Neuropathology

Postmortem analysis reveals that the main pathological hallmark of tauopathies includes neuronal dysfunction and loss that shows distinct, but also partially overlapping, spatiotemporal distribution of lesions in the brain (reviewed in ref. 8). This neuropathological variability underlines the complex and varying clinical phenotypes observed in tauopathies. In AD, progressive neuronal damage and death appear in brain regions critical for learning and memory (i.e., neocortex, hippocampus, amygdala, anterior thalamus, basal forebrain, and subcortical nuclei including the nucleus basalis of Meynert [23–28]). In AD, the highest atrophy is seen in the entorhinal cortex and hippocampus [29] that positively correlates with the degree of clinically evaluated dementia [30]. The progression of neuropathology is paralleled by a selective decrease in functionality of forebrain and brainstem monoaminergic and cholinergic systems [4, 31–33].

2.3 Intra- and Extracellular Protein Inclusions

As aforementioned, the progressing accumulation of abnormally hyperphosphorylated tau protein in tangles with paired helical filaments (PHF), twisted ribbons, and/or straight filaments [7, 34] is the ubiquitous hallmark of all tauopathies. Excessive intra-neuronal deposition of tau protein is also a key feature of dying neurons during normal aging, albeit at smaller rates [35, 36]. Historically, it has been demonstrated that both tangles and neuritic plaques in the brains of AD patients contain dense accumulation of PHF*s* [37], and in the mid-1980s tau was identified to be the principal component of neurofibrillary lesions in AD [34, 38, 39]. Consequently, similar intracellular tau lesions were also found in other diseases including corticobasal degeneration [40], amyotrophic lateral sclerosis/parkinsonism dementia complex of Guam [41, 42], Down syndrome [43], progressive supranuclear palsy, Pick's disease, argyrophilic degeneration (reviewed in ref. 7), myotonic dystrophy [44], and some forms of frontotemporal dementias (FTD) [45]. Another group of disorders shows additional intracellular inclusions of α-synuclein-containing Lewy bodies (LB) that are comprised of fibrils assembled from the presynaptic α-synuclein protein [46]. These disorders, known collectively as synucleinopathies, include PD, dementia with Lewy body (DLB), rapid-eye-movement sleep disorders, progressive autonomic failure, and multiple system atrophy (MSA). From this, AD is a special form of tauopathy additionally characterized by extracellular deposits of Aβ in the parenchyma and in cerebral blood vessels [47]. Amyloid plaques consist mainly of a 40–42-residue β amyloid peptide ($A\beta_{40}/A\beta_{42}$) cleaved from the amyloid precursor protein (APP), and surrounded by dystrophic nerve cells. The longer $A\beta_{42}$ species is normally present in small, soluble fractions in biological fluids [48] but it is found to be elevated and deposited in cases of familial Alzheimer's disease (FAD) [49, 50]. The increase in levels of both $A\beta_{40}$ and $A\beta_{42}$ correlates with progress in cognitive decline and the increase in $A\beta_{40}$ peptide was detected in 40 % of AD patients before overt amyloid plaques could be detected [51]. Although $A\beta$ pathology in AD coexists with tau pathology, the causal link between the two pathologies has not been yet established and since tau pathology can occur in the absence of $A\beta$, other causes of NFT formation are certain to exist [51].

2.4 Genetics

The first mutation identified as implicated in familial AD was a missense mutation in the gene encoding amyloid-β precursor protein (*APP*) on chromosome 21 [52–54], followed by mutations in presenilin 1 (*PS1*) on chromosome 14 [55, 56] and presenilin 2 (*PS2*) on chromosome 1 as identified in Volga German and Italian families [57, 58]. Studies using assays of $A\beta$ levels in plasma and cultured fibroblasts from patients harboring these mutations revealed a selective enhancement in the production of $A\beta$ which was considered to be a hallmark of the disease [59]. At present, significant evidence supports a hypothesis that elevated production of $A\beta$ plays a pivotal role in the pathogenesis of AD [60–62]. A fourth FAD locus is believed to be located on chromosome 12 with the α-2 macroglobulin (*A2M*) and its receptor, the low-density lipoprotein receptor-related protein (LRP1), being under consideration [63–66]. Despite extensive genetic screens, investigators have identified that mutations in *APP, PS1,* and *PS2* genes remain rare in the disease. In the case of AD, about 33 mutations have been reported for *APP*, 185 for *PS1,* and only 13 for *PS2* [67]. The recent list of all known mutations and nonpathogenic coding variations in the genes related to AD, Parkinson's disease (PD), and frontotemporal dementia (FTD) can be found in the constantly updated Alzheimer Disease and Frontotemporal Dementia Mutation Database (http://www.molgen.vib-ua.be/ADMutations) curated by Dr. Marc Cruts [67].

Epidemiological studies also revealed that 40–50% of all AD patients were carriers of the ε4-allele of the *apolipoprotein E* gene (*ApoE*, located on chromosome 19) in contrast to only 15 % in the normal and healthy population [68]. Since no direct causal links or inheritance patterns are evident, the presence of *ApoE-ε4* identifies an allele-specific risk factor of late-onset AD. The presence of both *ApoE-ε4* alleles leads to an earlier onset of AD in comparison to patients with only one allele or without the alleles [69]. Interestingly, there is evidence suggesting that its effect might vary amongst different races [70]. The presence of *ApoE-ε4* was found to increase the risk of developing AD among white Americans but have no significant effect among African-Americans or Hispanics [70, 71]. The reasons for this differential risk factor remain unknown, but linkage disequilibrium between *ApoE-ε4* and another allele or an unknown environmental or genetic factor present in some races but not in others was suggested as one possibility [70, 72]. Together, these observations indicate that both environmental and genetic factors trigger a chain of events which converge at a final pathogenic pathway and lead to the stereotypic neuropathology [61]. The inheritance of AD follows a dichotomous pattern. Rare mutations in *APP, PS1,* and *PS2* guarantee early-onset (<60 year) FAD which is characterized by Mendelian inheritance and represents about 5% of AD [73]. The *ApoE-ε4* allele represents a risk-modifying factor for late-onset AD. Together, *APP, PS1, PS2,* and *ApoE-ε4* genes account for 30–50% of genetic variance in AD [73, 74]. In addition to the *ApoE* locus, two other loci at *CLU* (also called *APOJ*), encoding clusterin or apoliprotein J, on chromosome 8 and *CR1* encoding the complement component receptor 1 on chromosome 1 were identified as potentially increasing the risk of AD [75, 76].

However, most cases of AD are sporadic (idiopathic) of unknown etiology, which can be strongly influenced by variants in other genes and/or environmental factors. Recent genome-wide association studies (GWAS) identified variants of a gene encoding insulin-degrading enzyme (*IDE*) [77, 78] susceptibility loci (*CD2AP, MS4A6A/ MS4A4E, EPHA1,* and *ABCA7*) implicated in ATP regulation and influx of lipids [79, 80] and a loss-of-function mutation in the *TREM2* gene, which encodes the *t*riggering *r*eceptor *e*xpressed on *m*yeloid cell 2 protein [81, 82] as candidate genes that increase risks of late onset of AD. Although all of these identified susceptibility variants confer only small effects of risk, it is important to search for additional rare variants in other genes that predispose AD. Such findings would considerably broaden our understanding of possible targets in an effort to develop novel strategies for preventing AD.

Although no mutations that are directly associated with tau have been identified in AD, the accumulation of various pathologically hyperphosphorylated forms of the microtubule-associated protein tau in NFTs presents the most ubiquitous pathological hallmark of AD. Many pathological and functional studies point to disturbances involving tau processing as a likely cause, which leads to the malfunction of axonal transport and consequently death [83–86]. Over 30 mutations in *MAPT* were discovered in patients with FTDP-17, an autosomal dominant inherited tauopathy [87–89]. Also, the clinical evidence indicates that excessive intra-neuronal deposition of tau protein is a key feature of dying neurons during normal aging which occurs most often without amyloid deposits [5], suggesting that tau dysfunction may be both necessary and sufficient to cause neuronal death, thus making this gene a particularly interesting candidate for models of neurodegeneration.

2.5   Conclusions

The identification of gene mutations implicated in tauopathies opened a way to model human disorders in a mouse. The translational research paradigm focuses on the production of experimental mouse models that reproduce the essential pathological and phenotypic hallmarks of human tauopathies including the formation of disease-specific intra- and/or extracellular inclusions of protein aggregates, progressing neuronal degeneration, and corresponding compromised behavior. Consequently, an ideal model should have the following characteristics: (1) be specific to a given disorder, intra- and/or extracellular deposition of misfolded proteins, (2) manifest age-progressing dementia or impairment in disease-specific non-mnemonic behaviors, (3) exhibit a disturbance in behavioral systems not directly related to cognitive function but which are observed at specific stages of a disease should be present and stressed in phenotypic characterizations, and (4) exhibit coinciding neuronal loss in disease-specific brain regions and cytoarchitecture.

3 Mouse Models of Neurodegeneration

The first successful mouse model replicating major hallmarks of AD was characterized two decades ago by Games and colleagues [90]. Other models of AD-like amyloidosis followed soon and were proven extremely informative which was chronicled in the number of scholarly reviews [91–102]. Mouse models expressing genes implicated in pure tauopathies were generated around year 2000 (reviewed in refs. 7, 95, 103, 104).

3.1 Criteria of Mouse Models of Tauopathies

At the juncture of a two-decade-long history of using mouse models of neurodegeneration, it seems that a robust and good model should:

1. *Replicate major clinical phenotypes.*

   Since the behavioral decline and region-specific neuronal loss are central to neurodegenerative diseases, the model should recapitulate accurately the main facets of the clinical phenotype.

2. *Manifest key age-progressing phenotypes of human disorder.*

   A credible model should exhibit progressive neuropathology *preceding* cognitive deficits which should be identified in paradigms addressing different memory systems. The extent of progressive behavioral impairment may eventually affect multiple behavioral systems due to significantly increasing brain pathology. Although late severe decline in behavior may raise operational difficulties related to the interpretation of results obtained in cognitive tests, if a disturbance in focal mnemonic domains is confounded by broad impairments in non-mnemonic systems [105], the use of such a model during drug screens may reveal which behavioral disorders could be ameliorated by specific treatment at a given stage of pathology.

3. *Control for the effect of overexpression of mutant pathological proteins.*

Most models involve the expression of pathological mutations and consequently the overexpression of the abnormal, usually human, protein. Therefore, phenotypic changes should be robust and correlated with the presence of familial mutations but should be absent or less overt in the age-matched mice carrying the wild-type (wt, non-mutated) genes expressed at equal (or greater) steady-state levels. These mouse models require minimum variation in their genetic background in order to exhibit subtle changes in a phenotype caused by familial mutation. For that reason, whenever possible, inbred strains having homozygous genomes have been used almost exclusively in biomedical research. Unfortunately, transgenic lines are often created in one strain and later backcrossed to a more suitable for behavioral studies' genetic backgrounds, significantly complicating the control of genetic variability. Wild-derived mouse strains or recombinant strains are usually avoided in transgenic translational research due to unwanted genetic diversity. Similarly, outbred stocks, used predominantly in genetics, toxicology, and pharmacology, are not recommended for transgenic research due to their genetic variability caused by genetic drift, directional selection, and genetic contamination during breeding [106–108].

4. *Robust external validity.*

The independent confirmation and replication of the key facets of the phenotype in independent transgenic lines harboring the same construct should be carried out in independent laboratories [109] including standardization of expertise of technical personnel and differences in handling methods [110].

3.2 Caveats and Pitfalls of Using Mouse Strains

The extent to which the pathology and compromised behavior of a mouse model replicate the phenotype of human disorder often depends on the genetic design of a model and the behavioral systems targeted in tests. The choice of a mouse strain is crucial because many strains routinely used for genetic manipulation are not particularly suitable for behavioral studies due to, but not limited to, deleterious genetic mutations, idiosyncrasies in brain development, or constraints in learning [111, 112]. The genetic background of a mouse strain can also be used as a tool in the analysis of a mutation [113], and for the use of mapping and cloning strategies which allows for the identification of modifier genes existing in different mouse strains [114, 115]. Another noteworthy point is that transgenic mice are usually initially generated on 129 or FVB mouse strain backgrounds since the oocytes of these strains are large. This approach increases the probability of successful injection of the transgene construct. These strains, however, are not particularly suitable for behavioral testing [116] and it is often necessary to transfer the mutation to a more suitable background, usually a C57BL/6 strain, by backcrossing for at least ten generations. While the strategy largely results in the replacement of the donor genetic background with the recipient background, the close region flanking the selected gene remains most likely of donor origin. Thus, the genes in this flanking region are inherited with the transgene, consequently differentiating the transgenic mice from their non-transgenic littermate controls not only by the presence of the transgene in question but also by flanking genes of the original strain. The presence of the latter might conceivably bias the phenotype of a model. The problem of genetic background was partially remedied in the case of knockout models in which genes are targeted directly in C57BL/6-derived embryonic stem (ES) cells (see Knockout Mouse Project (KOMP) http://www.komp.org/ for further information). Many neurodegenerative mouse models, however, are still maintained on mixed and segregated genetic backgrounds; thus even inbred littermates are not genetically identical. Further complication exists in models homozygous for a gene in question. If breeding of homozygous parents is possible, such a breeding system might yield a large number of experimental mice; however, the use of wild-type, non-littermate, controls purchased from commercial suppliers or bred in parallel in-house does not constitute an ideal control in a model. Additionally many strains, including C3H/HeJ, SJL/J, FVB/NJ, MOLF/E, PL/J, SWR/J. BUB/BnJ, CBA/J, or NON/LtJ, are not suitable for behavioral experimentation in tests that rely on visual cues due to the presence of retinal degeneration (rd) (http://eyemutant.jax.org/retinal_degen.html). About 20 % of all inbred mouse strains carry the rd-causing autosomal recessive mutation in *PDE6B* gene [117] that results in the progressing degeneration of rods and cones [118, 119]. Other strains, like A/J, BALB/cByJ, AKR/J, KK/H1J, to mention a few, are albino and could show mild defects in their vision [120, 121]. Other deficits include hearing loss progression with age in A/J, BALB/cByJ, C57BLKS/J, C57L/J, and C57BR/cdJ (but no C57BL/10 J) or a partially developed corpus callosum in 129S1/SvlmJ, BALB/cByJ, or I/LnJ strains (source, http://www.jax.org). In conclusion, taking into account (1) the effect of genetic strain background on behavior, (2) the breeding scheme of a transgenic lines or the generation of multiple transgenic lines, and (3) the presence of retinal degeneration or other possible mutations expressed in a homozygous state, the design of a mouse model has to be carefully planned to avoid potential confounding variables that might affect many behavioral tasks which depend on visual acuity of animals.

Furthermore, the observed pathology in transgenic models may also depend on the choice of promoter used to drive transgene expression. The most common promoters include the *APP* promoter (mouse models of AD-like amyloidosis) [122], the brain-enriched prion protein promoter [123–125], the platelet-derived growth factor b-chain (*PDGFb*) promoter [90] (both PrP and PDGF promoters resulting in a transgene expression also outside of the CNS), and the neuronal specific *Thy-1* promoter [126]. Another problem with transgenic models relates to spontaneous genetic changes that may affect the phenotype of a model. Mice engineered to over-express a transgene can potentially with time change the number of disease-causing transgene copies that lead to possible loss of a phenotype. Without routine checks of transgene expression within and between laboratories, the differences in the transgene copy number can prevent replication of results between laboratories. Awareness of this issue should prompt researchers to periodically check the genetic constitution of their transgenic stocks.

3.3 Modeling Human Dementia

Given the disparities between species, it might not be possible to draw exact and definitive parallels of behavioral profiles obtained in tests between humans and mouse models. In order to make appropriate comparisons, tests of memory in rodent models should target cognitive systems that are clearly identified and conserved across species, including humans, and have a clearly delineated function and a defined neuroanatomy. Assessment of spatial navigation and its dependence on the hippocampus fulfills such assumptions since this memory system is highly conserved anatomically and functionally in mammals [127]. The neuroanatomical structure of the hippocampus and its synaptic plasticity during memory formation [128–135] serves as a well-defined model of memory, which has been extensively studied in rodent species [131, 134–137]. Seminal clinical studies showed, unequivocally, that human subjects with temporal lobe damage demonstrated severe impairment in learning and memory, including the recall of spatial locations and solving spatial maze tasks [138–140], thus confirming the involvement of the hippocampus in spatial memory in humans. Similar findings were reported in AD patients who showed significantly increased atrophy of the hippocampus [141, 142] and impaired performance in spatial navigation tests [143–148].

4   Designing Preclinical Behavioral Studies Using Mouse Models

4.1 Evaluation of Phenotype

The initial evaluation of the phenotype of a mouse model of neurodegeneration should be carried out in a battery of tests characterizing both the overall physical and the motor propensity of mice as well as the characterization of the targeted behavioral system(s) [149].

Since a detailed description of each test is beyond the scope of this chapter, readers should consult specialized textbooks [150, 151], general articles related to behavioral phenotyping [112, 152], articles related to analysis of behavioral strategies [153], memory evaluations [154, 155] in specific tests (e.g., water maze), or methodological descriptions of procedures related to spatial orientation tests [156]. More specialized articles describe experimental approaches which can enhance learning in strains of mice known as poor performers in a specific test [157], or articles comparing the performance of different species (e.g., rats and mice) [158]. The existence of specialized journals, like Nature Protocols (http://www.nature.com/nprot/index.html), that publish detailed experimental protocols in a structured and comprehensive manner, defies any attempt in this chapter to provide a complete guide to the plethora of available behavioral paradigms. These articles not only present the theoretical background underlying each testing paradigm as well as detailed procedural steps and examples of collected data, but also suggest types of equipment, analyses of results, and a list of troubleshooting steps. Instead, we discuss here a few of the most frequently used behavioral tests and provide additional practical information which should serve as a general compendium to help behavioral and molecular researchers successfully launch behavioral screens of available mouse models and facilitate the interpretation of results on the vast background of published literature.

4.2 Evaluation of Mnemonic Function

The prevalence of spatial memory tasks in the characterization of mouse models (Table 1) is justified by high evolutionary conservation of spatial memory across mammalian species. In the case of mouse models of neurodegeneration that over-express genes implicated in human disorders, the main experimental question investigates how a mouse genome copes with the presence of unusually high levels of human protein coded by a transgene. The existence of possible confounding effects due to genetic background of models, modifier genes, compensatory effects, and/or subtle differences in the experimental paradigms, including the strains' response to handling [159], can often yield different or even contradictory results. Therefore, the initial evaluation of the phenotype of transgenic mouse models should also incorporate a characterization of hippocampus-independent memory systems [160–162], evaluation of changes in agitation and aggression levels [163], and assessment of locomotor, exploratory, or stereotypic activity [164]. The results of such studies not only extend our understanding of the effect of the protein(s) coded by transgene(s) on behavior, but also allow us to identify potential confounds in memory tests [165] or recognize the effect of genetic background on the phenotype of a model. Moreover, studying hippocampus-dependent memory in different testing paradigms often provides

Table 1
Evaluation of cognitive phenotypes of mouse models of neurodegeneration

SHIRPA protocol [235]

In most cognitive tests, learning rate and memory strength are inferred from measures of locomotor behavior; therefore any possible effect of a transgene on motor and/or perceptual systems can yield false-positive (impaired learning) results. To this end, a general phenotypic assessment of transgenic mice along with non-transgenic littermates must precede specific cognitive tests in order to eliminate these possible confounds. SHIRPA protocol provides a comprehensive evaluation of mice behavior ranging form the assessment of exploration and activity levels to thermal nociception. This battery of simple tests begins with procedures most sensitive to physical manipulation, like anxiety tests performed in the open-field or elevated plus or zero mazes. Other screens focus on gross phenotyping abnormalities, assessment of sensorimotor deficits (rota-rod), hole-board exploratory activity, and thermal analgesia. Although application of the full battery can be time consuming and requires a well-equipped lab, a subset of simple tests can be carried out and is highly recommended for initial characterization [152].

Morris water maze (MWM) [236, 237]

The MWM test has been the most widely used testing paradigm to study hippocampus-dependent spatial memory in rodent species. Reference memory or place discrimination version of MWM requires mice, trained with repeated trials over several days, to use external visual cues around the testing room to search for the hidden (submerged ~1 cm under water) escape platform in the water maze. Spatial navigation encompasses the development of different search strategies with spatial strategy (reflected by a direct swim to a platform) taking place at the end of this complex learning process [153, 238]. The main dependent variable reflecting learning acquisition is escape latency—the time it takes a mouse to find a platform, or search path, which is less biased by the differences in swim speed. Memory bias is evaluated in trials where a mouse searches a pool where the hidden platform has been removed. Spatial learning is reflected by decreased escape latency or search path, while spatial memory by increased search in areas or quadrants of the pool containing a platform during training. An Annulus Crossing Index (a number of swimming over former platform location adjusted for swims in other three quadrants of a pool [153, 239, 240]) represents an alternative, more stringent measure of memory bias. In cases when more than one probe trial is carried out during training, a mean probe score (the mean percent of time spent in target quadrant during all probes [241]) can be used as a reliable memory evaluation index. Correspondingly, learning impairment is reflected by longer escape latency or search path during training, and memory impairment by displaced or random search, which is reflected by about 25 % of time, spent in each of quadrant of the pool during a probe trial. To address episodic-like memory in mice, a more complex version of the MWM test was developed [242]. In this test, numerous locations of the platform were used and the number of new locations learned during the whole training reflects learning capacity of an individual mouse. In a cued or visible platform version of the MWM test a platform location is marked by a visible cue that mice associate with an escape from water. This version of the test has been often implemented as a control for normal visual acuity, an unimpaired learning of simple association between a proximal cue and an escape platform, or as demonstration of a comparable swim speed between studied genotypes. These control experiments should be used with caution, however, because in contrast to rats, some strains of mice with hippocampal lesions often show also partial impairment in the cue navigation task [243].

OBJECT RECOGNITION (OR) [244, 245]

This test exploits a natural tendency of rodents to explore novel objects and to show an exploratory preference for replaced or displaced objects. The dependence of object recognition memory on the hippocampus is related to the protocol of a test. Short delays between initial exploration phase and a memory test make OR test independent from the hippocampus [162]; however, when longer delays (hours) are implemented, OR memory depends on hippocampus function [246, 247]. Object memory impairment is demonstrated when an animal shows no preference in exploration (close proximity, nose contact) of a new or displaced object.

FEAR CONDITIONING (FC) [248, 249]

The FC paradigm, which is an example of classical Pavlovian associative learning, involves an association of a neutral tone (conditioned stimulus, CS) paired with a brief electric foot shock (unconditional stimulus, US) delivered in a novel context. Mice trained in that manner develop a fear response (conditioned response, CR), expressed as defensive or anti-predatory behavior in a form of freezing (complete cessation of movement) which coincides with autonomic and endocrine response (increased heart beat rate and blood pressure), and sensory alteration (analgesia, potentiated startle). The paradigm may involve two types of conditioning that can be performed simultaneously or independently during a training phase: contextual (CFC), when an animal develops an association between shock and training context (conditioning chamber), and tone fear conditioning when shock (US) is associated with a neutral tone (CS). The tone conditioning is performed either as delay conditioning paradigm when there is a temporal overlap between CS and US (a foot shock is delivered within the last 1–2 s of tone duration), or more demanding trace fear conditioning which requires the association of a CS with an US across an interval of time known as trace interval (a foot shock (US) is delivered after the tone (CS) is turned off). The time between CS and US can vary and an additional temporal processing is required because CS and US are separated; therefore an animal has to retain a trace of CS across this time interval in order to associate it with the US. While delay tone conditioning is hippocampus independent but requires intact amygdala [250, 251], the trace and contextual fear conditioning are sensitive to hippocampal lesions [249, 252]. The sensitivity of the mice to foot shock can be established empirically recording the current thresholds that elicit specific response like flinch and jump [253]. The lowest current eliciting learning (for mice a current of 0.35–0.4 mA is appropriate) should be used. Impairment in FC is evaluated during test phase on the following day after training, and is reflected by reduced freezing time when an animal is placed in familiar chamber (context conditioning) or when the animal is exposed to a conditioned tone in a new environment.

CONDITIONED TASTE AVERSION (CTA) [254–256]

CTA is a special form of classical Pavlovian conditioning, representing an adaptive specialization, which defends an organism against repeated ingestion of toxic foods [254–257]. CTA is well conserved in many different species including humans [257, 258]. When acquiring a CTA response, an animal learns to associate the specific taste of a novel food, usually a saccharine solution (conditioned stimulus, CS) with experimentally induced through i.p. injection of lithium chloride after saccharine intake (unconditional stimulus, US) nausea. Because of one trial pairing between CS and US, a long-lasting avoidance of food with this specific taste develops. The brain areas implicated in the CTA include the agranular insular cortex, the parvicellular thalamic ventral posteromedial nucleus, and the parabrachial nucleus of the pons, which are part of the gustatory pathway [259, 260], and the amygdala [261, 262]. Impairment in CTA is reflected by increased saccharine intake as compared to control mice in choice tests (usually two-bottle test, one containing water, one saccharine).

These testing paradigms represent some of the most commonly used tests employed in behavioral evaluation of mouse models of neurodegenerative disease (reprinted from [95] with permission)

additional information regarding a particular phenotype of a mouse model. For example, when APP Tg2576 mice were tested in T-maze alternation and contextual fear conditioning tasks they showed a significant impairment in T-maze alternation but, surprisingly, the animals were unimpaired in both contextual (hippocampus-dependent) and auditory fear conditioning tests (hippocampus-independent task) [160]. However, these mice showed significantly attenuated contextual discrimination in testing paradigms when the salience of the contextual cues was decreased without changes in tone foreground conditioning. Such detailed validation of the existing mouse models is necessary in order to provide a more powerful experimental framework for behavioral characterization of consequent models used in screens of potential therapeutics. A pragmatic approach would dictate that robust phenotypes obtained in less labile tests (in which data collection is based on motor or strong sensory inputs) would be replicable within tolerable margins, and yield consistent results across different laboratories while more labile phenotypes based on emotional or social behaviors may be strongly affected by differences in laboratory practice [166], especially in animal colonies or laboratories that are less focused on behavioral evaluations of mice.

4.3 Measuring Behavior

We perform experiments in order to test hypotheses that best explain the underlying mechanisms of the observed phenotypes in a model. However, even carefully designed and conceptually sound behavioral experiments might lack sufficient power if sample sizes of animals in experimental groups are too small or the measured behavior in question is too variable. Since the effect of mutation on overall behavior is often not known and difficult to predict a priori, it is advisable to characterize the phenotype of a model in a battery of hypothesis-driven tests along their non-transgenic littermates or if non-transgenic littermates are not available (in the case of breeding homozygous for the transgene mice), wild-type mice (preferably bred alongside a model) that are maintained on the same genetic background. Such initial broader behavioral screen can be helpful in establishing a yardstick for consequent detailed characterizations of specific phenotypes by avoiding floor or ceiling effects in obtained data sets, and might reveal compromise in other behavioral systems, not directly related to modeled behavior. Larger *sample sizes* are desirable in most behavioral studies ($n = 8$–12, or more [167] depending on how robust a focal behavior is); however, care should be executed not to test the mice too long during each day, which may result in their fatigue or the span of tests over different phases of the circadian cycle. Also, one has to be aware that a change in behavior of mice may sometimes be caused not by the experimental treatment but merely by the handling or attention paid to them by the experimenter. The differences in the amount of handling of experimental mice and their familiarity with experimenters are often at the root of discordant results obtained in various laboratories [110]. A common error, which occurs less often in behavioral research but crops up frequently in physiological experiments, refers to treating repeated data points coming from the same subjects as independent measures. This erroneous approach, called *pooling fallacy* [168], leads to inappropriate increases in sample sizes comprised of mixed, dependent, and independent data points, thus violating many assumptions of experimental design and parametric statistical analysis. Problems with independence of data may further arise in less obvious situations, when the obtained measures correlate closely between mice coming from the same litter or between mice housed in the same cage. These *litter* or *cage effects* can be the result of differential maternal care, highly variable housing conditions (mice in cages placed at the bottom shelves of a rack in a densely populated rack room are kept in constant semi-darkness as compared to cages placed on top shelves), and number of mice in a cage; singly housed animals are known to perform worse in cognitive tasks, etc. The above issues often present serious confounding factors, which are difficult to overcome in studies when small numbers of available and oftentimes difficult-to-derive mutant mice are available. Awareness of these potential problems, however, may help during the inspection of raw data and during the exploratory data analyses. For example, if measures generated by mice coming from the same litter or the same cage are on one end of the data distribution, one should consider the replication of the experiment using more careful and balanced assignment of mice to experimental groups.

4.4 Data Analysis

Most experiments, especially those evaluating learning acquisition and memory retrieval, involve repeated training sessions or recall tests. Results are usually presented in blocks of training trials or days. Plotting data in this fashion usually adequately reflects learning processes, but blocking data over too many repeated trials, while reducing variance, may also mask unusual and interesting changes in the patterns of behavior occurring during training (*see* ref. 169). It is advisable, therefore, to inspect the raw data, especially the data generated from first training trials, to check if mutant mice were free from subtle motor or sensory deficits or showed comparable levels of anxiety to that of the control mice. In cases where mutant and control mice show comparable motor and sensory propensity, as well as comparable anxiety to testing situations after being well habituated to experimental handling and testing lab conditions, one can safely assume that the mice of both genotypes should show comparable performance during the first training trial(s) of a mnemonic test. Any cognitive impairment, if not confounded by compromised locomotor or sensory deficiency, should become usually apparent as training progresses but should be absent at the beginning of training. If the severity of cognitive decline impairs the interaction of an animal with the surrounding environment and causes inferior performance by the experimental mice throughout the whole duration of the experiment, it is advisable to evaluate the mice in additional mnemonic tests, and to repeat the tests in cohorts of younger mice to identify the progression of the phenotype. Data obtained in most mnemonic tests are usually analyzed by analysis of variance (ANOVA) with genotype and/or treatment as between-subject factors and training days and/or trials as within-subject factor(s). The data set should meet the criteria of parametric statistics, and in the case of repeated measure or within-subject factor designs an assumption of compound symmetry must be met in order to avoid serious biases in the interpretation of the results. The assumption of compound symmetry refers to a pattern of constant variances on the diagonal and constant covariances off the diagonal in the variance-covariance matrix. In practice, this means that the correlations within the matrix of the repeated factor (days or trials) have to be the same at all distances between measurements. This assumption however is hardly met in the analysis of data obtained in learning experiments; due to improved performance as the mice learn the task over time their performance improves and consequently the variance of scores decreases. A departure from the assumption of compound symmetry is usually evaluated by a slightly more stringent sphericity test (Mauchly sphericity test, SPSS GLM (Statistical Package for Social Sciences, SPSS Inc. Chicago)), and in cases of severe departures degrees of freedom should be adjusted either by Greenhouse-Geisser ε-correction (tends to underestimate, especially when ε is close to 1) or by Huynh-Feldt correction (which tends to overestimate ε) to avoid false-positive results [170].

5   Evaluation of Therapeutics and Behavior of Mouse Models

The successful translation of the results obtained at the preclinical, lab bench, stage to the clinical, bedside, stage will depend not only on the conceptual design of a model and its relevance to human disorder, but also on the choice of behavioral systems in a model. These factors might not necessarily be homologous with human disorder, but might be biologically relevant, robustly expressed, and reliably measurable in a model. Such behavioral models might prove to be sensitive enough to detect not only subtle effects caused by progressing neuropathological events, but also identify changes in the phenotype of a model that may not be related to the effects of treatment. Therefore, in this section we focus on aspects of behavioral experiments that might not be explicitly part of experimental design or even be identified as factors that could significantly change the environment of an experiment and consequently affect the behavior of mice. We discuss only issues that seem to be particularly important for behavioral experiments, including mouse husbandry, conditions of animal facilities, and the effects of pre-experimental handling that is associated with many routes of drug delivery. Emphasis is placed on the awareness of these issues in order to help identify and possibly eliminate potential confounding variables affecting behavior, thus further optimizing the sensitivity of a model. In addition, since the replication of results is at the core of the falsification of hypotheses or theories [171], the replication of the results obtained in preclinical studies across laboratories and in independently derived mouse models of neurodegeneration should be actively pursued in order to obtain the highest experimental validity of a model and translational validity of studied hypothesis.

5.1 Animal Facility and Behavioral Tests

It is not surprising that even the best guide or detailed step-by-step description of procedures and experimental methods pertaining to specific testing paradigms may often yield unpredictable or different results between laboratories despite very careful execution of experiments by researchers experienced in the field [109]. In this section we would like to alert readers to some potential problems and issues related to the quality of husbandry and housing conditions, which may affect the outcome of behavioral experiments and resulting data. The discussion, by no means exhaustive, encompasses problems that are often not formally documented, or issues, which by general consent might be considered trivial or residing outside the Material and Methods sections of most, but mainly specialized, journals.

The characterization of behavioral phenotypes in mice starts with the quality of housing conditions in the animal facility. Animal rooms located in the vicinity of noisy and heavy traffic areas, like cage-washing areas, are less desirable and in some cases may even lower the breeding rate of mice, induce cannibalism, and increase hyper-reactivity and anxiety levels during experimental handling. The adequate housing conditions, which encompass the optimal number of mice in a cage (no more than four mice in a shoebox cage housed over an extended period), and the minimum level of enrichment in the form of pressed cotton nesting material (nestlets), mouse huts or igloos, or pup tents (source http://www.bioserv.com/), can significantly improve breeding success, broaden the behavioral repertoire of mice, and reduce their anxiety in adulthood. Housing mice singly in cages is not recommended due to a heightened rate of developing stereotyping behavior, obesity, and decreased learning ability. When breeding transgenic lines, it is not uncommon that the newly born transgenic pups tend to be smaller than their non-transgenic littermates (for example in the case of APP CRND8 mice, CJ personal observation). Supplementing lactating females and their pups, especially at the age when pups begin to consume solid food (14–15 days), with easily available and more palatable mashed food (e.g., moist powdered mouse Purina chow) facilitates pups' growth and can reduce the weight difference between transgenic and non-transgenic littermates in the CRND8 model (unpublished data). The distance of the housing room to the behavioral testing room(s) is also relevant and the transportation of mice between the two locations might be stressful; therefore longer times of acclimation to testing conditions are recommended. Last but not least, husbandry practices including care and feeding of animals, cleaning of equipment, physical surroundings, and routine checks of the stock health by experienced, well-trained, and well-managed facility staff guarantee good health, growth, reproduction, and survival of mice. Personnel with poor management and/or those who are inexperienced in mouse handling and husbandry methods that are appropriate for behavioral laboratories may adversely affect animal stress level and behavior. Excessive noise produced during cage changing (for example, changing cages under well-ventilated but noisy hoods (required by institutional guidelines) and stacking metal cage lids on the hood's metal surface increase further noise levels, including high levels of ultrasounds which mice are sensitive to) and undetected leaking water bottles or wet cleaning equipment (mops and buckets) left in animal rooms produce difficulty in identifying confounding variables that together might significantly increase stress and anxiety levels of mice, and consequently negatively and variably affect the behavior of mice in tests carried out in well-designed and quiet testing rooms.

As an example, in Fig. 1 we provide the results of the training of two cohorts of mice, maintained on the same C3B6 genetic background (the mice were derived by a number of intercrosses and backcrosses of C3H (C3) and C57BL/6 (B6)) in the spatial reference memory version of a water maze test, in two different animal facilities. Both cohorts were at comparable ages and were trained by the same laboratory assistant, highly experienced with behavioral procedures involving water maze test and mouse husbandry, and certified in laboratory animal medicine. Mice in animal facility A were maintained in a quiet room and highly qualified and well-managed personnel provided husbandry care. The environment in colony B was more stressful and mice were exposed to noisy conditions. The comparison of mice performance between colonies (main between-subject factor) and the analysis of their learning (days as repeated measure or within-subject factor) revealed no significant difference in the average performance between the colonies ($F(1,23) = 1.2$, NS), but also revealed a significant interaction between the colony conditions and learning rate of mice ($F(4,92) = 4.2$, $p < 0.01$, colony × days interaction). Mice in colony A showed a significant and rapid improvement in learning over 5 days in their search for a hidden escape platform

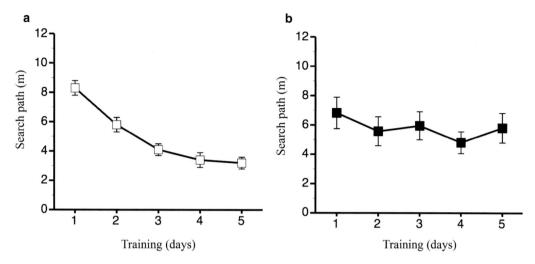

Fig. 1 Learning acquisition in the spatial reference memory version of a water maze test of mice maintained on a mixed C3B6 genetic background in two different animal colonies. (**a**) Mice in colony A were kept in a quiet room with husbandry practices appropriate for behavioral experiments, while (**b**) colony B had increased noise level and suboptimal conditions for behavioral studies. The mice trained in colony A showed a significant improvement in finding a hidden platform location during training (their search path was on average about 5 m shorter at the end of training as compared to at the beginning of training). On the other hand, the mice in colony B showed no improvement (the rate of improvement between days 1 and 5 was about 1 m). See text for further details

($p < 0.001$—simple effect ANOVA with days as a repeated measure, Fig. 1a), while the mice in colony B did not show any signs of improvement over the whole training period (Fig. 1b).

5.2 Routes of Drug Delivery: Does Handling Matter?

Husbandry procedures and the quality and amount of handling of mice change in preclinical studies evaluating therapeutics in mouse models. Depending on the route and frequency of substance delivery, and the total duration of an experiment, mice could be subjected to increased experimental handling often accompanied by restraint and painful or uncomfortable sensations from drug administration [172, 173]. Even routine laboratory husbandry procedures, including handling by lifting an animal, cleaning or moving its cage, or various forms of restraint performed during experiments, result in significant changes in physiologic parameters that correlate with a response to distress (*see* ref. 174 for detailed review of this topic). Although the aversive consequences, including high anxiety and stress responses, caused by routine handling that often involves picking up mice by the tail could be significantly minimized when using open-handed handling (cupped or scoop-up method) or tunnels [175], the administration of substances often involves longer and repeated handling episodes with restraint. Such modifications to experimental designs should be acknowledged, well documented, and recognized as an important part of preclinical studies. Stress associated with repeated restraint or longer immobilization periods employed during drug delivery might exert morphologic changes, biochemical changes, and physiologic changes, and/or disrupt animal homeostasis, including hormonal and immunological responses [176–182], gene expression [182–188], and levels of neurotransmitters or brain activation [189–196]. Prolonged stress may potentially affect pharmacokinetics and/or pharmacology of a drug [178, 197], and such confounding effects might be more potent than the effects of the actual experimental treatment [198]. At the behavioral level, intensive handling (requiring a form of restraint, be it manual in the form of firmly holding the animal or using an apparatus) associated with drug delivery might negatively affect the responses of mice in tests, for example increasing their anxiety levels [199] or increasing stress-induced hypophagia [200], but in the case of strong motivators, for example in Porsolt's swim test, the repeated restraint and drug injection had no measurable effects on behavior [201]. However, recurrent handling, even with restraint, over extended periods of time may also diminish the stress response most likely due to mice becoming habituated to the procedure. In the following section, we provide an example of oral gavage, recognized as one of the most stressful routes of drug administration [172], which is accompanied by frequent handling with restraint, and its potential to actually improve cognitive behavior of mice. The shifts in internal homeostasis due to additional handling related to drug administration might have significant consequences in the field of translational research by generating false-negative results due to the masking of weak effects of a drug, or false-positive results due to synergistic processes between subtle effects of a drug and positive effects of handling.

Substances can be administered to laboratory animals by a wide variety of routes that differ in the amount of associated handling, experimental recommendations and benefits, and potential problems. The administration of substances presents a very broad topic encompassing factors such as absorption, distribution, metabolism, volume, frequency of administration [197, 202], duration of treatment, selection of the vehicle or solvent [172], dosing apparatus, necessary animal restraint, as well as timing in relation to the circadian rhythm [203]. A detailed discussion of these issues goes beyond the scope of this document, and readers are encouraged to consult a number of scholarly articles on the subject [197, 204, 205]. There is a plethora of routes of delivery for any given substance: into the mouth (orally through liquid or solid diet) or directly into the stomach (gastric gavage; per os, PO), into a blood vessel (intravenous, IV), onto, into, under, or across the skin, or into a muscle (epicutaneous (EC), intradermal (ID), subcutaneous (SC), transdermal, and intramuscular (IM), respectively). It may be administered into the peritoneal cavity (intraperitoneal, IP), sprayed into the nose for absorption across the nasal mucous membranes or into the lungs (intranasal), or delivered into the lungs by direct tracheal instillation (intratracheal) or inhalation. More specialized methods encompass applications onto or into the eye (transcorneal or intraocular, respectively), directly into the brain (intracerebral) or the space surrounding the dura mater or the distal spinal cord (epidural and intrathecal, respectively), and directly into the marrow cavity (intraosseous). A comprehensive review of each of these methods, including the use of the technique, summary of the protocol, and potential problems and refinements, is presented in the Report of the BVAAWF/FRAME/RSPCA/UFAW Joint Working Group on Refinement [197], and in an excellent video article discussing the benefits and limitations of each technique, as well as presenting practical guidelines and instructions underlying the protocols of mouse and rat handling and substance administration [206]. The general consensus is that the administration of a substance to an animal should follow the "best practice" principle since preventable mistakes at any stage can cause animal suffering [197]. These guidelines stress reducing or completely avoiding adverse effects, minimizing the number of animals used, and maximizing the quality of results. A strong emphasis is put on the proper animal handing and training of staff who will perform the procedures, in order to maintain high competence and detailed knowledge of what is being done to the animals, ultimately to minimize stress. In some cases the exact cause of stress response might not be easily recognizable such as the type of vehicle and/or the volume used, which can compound stressful restraint during gavage [172]. Alternative methods that do not require handling have been proposed, including a chronically implanted catheter [207] or osmotic minipumps [197].

Enteral (tube) routes, which include oral or nasal gavage, are relatively inexpensive, harmless, and easy to perform, and also mirror the most general way that drugs are administered to patients [202, 208]. These routes are used when systemic exposure is required ensuring good absorption from the gastrointestinal tract. However, this technique causes the most distress to the animal [172–174, 197] and is also technically difficult. In cases when the tube is incorrectly placed, too strong of a force is used, or if the animal moves, the tube may penetrate the trachea, oesophagus, or stomach and lead to serious injury or even the death of an animal. Repeated gavaging may also cause inflammation and ulceration of the oesophagus [197]. Notwithstanding, the technique is frequently used in drug studies, especially when exact doses of a drug have to be administered over long periods of time. Therefore, it is likely that the habituation of mice to repeated restraint and gavage, as well as other delivery routes that require intensive handling, especially in long-lasting experiments with frequent dosing regimens, might reduce the severity of distress or even eliminate it altogether.

5.3 Effects of Long-Term Repeated Handling on Mouse Behavior

It has been well documented that in human neurodegenerative disorders, even in cases with established strong genetic involvement, the onset and severity of neurodegeneration might be significantly influenced by environmental settings [74, 209–211]. Additionally many epidemiological studies confirmed that environmental stimulation might substantially decrease the risk of developing dementia in Alzheimer's disease [212–214]. Similarly, experimental animal studies demonstrated that frequent daily handling could significantly improve hippocampus-dependent learning and memory in rats and mice [215–219], including mice models of neurodegeneration [220]. Therefore, daily handling, which is a necessary component of specific route of a drug administration, might exert a significant effect that can modulate behavior of mouse models of human disorders [221, 222]. Functionally the effect can reflect the influence of increased attention provided by caregivers on a patients' response in clinical studies [223]. It is always informative to elucidate the effect of handling procedures as part of phenotyping characterization of mouse models in order to establish the range of behavioral responses in studies that evaluate therapeutic effect of drugs on mnemonic function in a model. The phenomenon, known in psychology as the Hawthorne effect [224], presents a serious confounding variable in human studies, and pertains to the awareness of research participation and consequent modification of behavior by participants that might significantly contribute to a placebo effect [225].

To evaluate the effect of handling on mnemonic function in a model of neurodegeneration, we performed an experiment that focused on the effect of repeated handling with gavage on spatial reference memory evaluated in the Morris water maze (MWM) test. We used a mouse model of tauopathy, denoted rTg4510 [226], which over-expresses human *P301L* tau mutant gene that is implicated in familial cases of frontotemporal dementia and parkinsonism linked to chromosome 17 (FTDP-17) [88], and displays robust tau brain pathology and neuronal loss in cortical structures within the first 5 months of age [226–228]. The tau pathology at this age is more robust in females but occurs substantially later in males [229]. We focused on spatial reference memory, because it has been well documented that this cognitive system was significantly compromised in this model [226, 229].

In the study, all mice were repeatedly gavaged with water for 3 months, starting at the age of 3 months. At the age of 6 months their spatial learning and memory were evaluated in the MWM test. During that period mice were housed in groups of 3–4 in cages placed in a dedicated room that was accessed by a single experimenter. The mice were gavaged daily in the morning and in the afternoon. During each gavaging session a mouse was removed from its home cage, weighed, and briefly restrained using the dorsum scruff and tail wrap technique. A gavage needle (size $18G \times 2$,

Popper and Sons, Inc.) was gently inserted in the oesophagus and 0.1 mL of water was administered directly to the stomach of the restrained mouse. After removing the needle, the currently handled mouse was returned to its home cage. At the end of the gavaging period, at 6 months of age, all mice underwent the training in the MWM test for 5 days, followed by an evaluation of their spatial memory for the location of the escape platform carried out in a 60-s probe trial with the escape platform absent (*see* ref. 229 and Table 1 for protocol details). All experimental manipulations, handling, gavaging, and the behavioral tests were approved by institutional IACUC and were conducted in accordance with AAALAC guidelines.

Spatial memory for the platform location during a probe trial was evaluated by two measures: the search (% of path) of the target quadrant (a quadrant containing an escape platform during training, TQ), and the Annulus Crossing Index (ACI), a more stringent memory index, which reflects the number of swims over exact platform site location in TQ adjusted for swims over equivalent sites in other quadrants of the pool (see Table 1 for detailed definition of both measures). The obtained results revealed that handled 6-month-old rTg4510 mice showed strong spatial memory, as evaluated by the time spent in the TQ, that was comparable to the memory of the control mice ($F(1,20) = 1.5$, $p = 0.2$, genotype factor). Both memory scores were also significantly higher than chance (25 %) level ($t(9) = 3.6$, $p < 0.01$, and $t(13) = 3.1$, $p < 0.01$ for control and rTg4510 mice, respectively, Fig. 2a). This result was surprising and in stark contrast to our published results that unequivocally demonstrated that briefly acclimated to handling, but otherwise experimentally naïve, 5.5-month-old rTg4510 mice showed significant spatial memory impairment with memory scores no different from the chance level (Fig. 2b, after [229]). However, the analysis of more stringent ACI memory index revealed that gavaged rTg4510 mice showed significantly lower memory scores, which were not significant from a chance level, than control mice ($F(1,20) = 10.6$, $p < 0.01$, genotype effect, Fig. 2c). The ACI scores of the control mice were significantly higher from chance ($t(9) = 3.8$, $p < 0.01$, Fig. 2c). These results were concordant with the ACI memory scores of naïve rTg4510 mice (Fig. 2d, after [229]).

Daily handling with restraint also improved the results of rTg4510 mice's searches for the submerged escape platform during the learning acquisition phase of the test. The overall search path for the platform during MWM training was not significantly different between handled rTg4510 and handled control mice ($F(1,22) = 3.8$, $p = 0.06$). The mice of both genotypes also swam with comparable speeds (data not shown). Interestingly however, the proportion of the search path characterized by thigmotaxic or "wall hugging" swimming of handled mice was greatly reduced

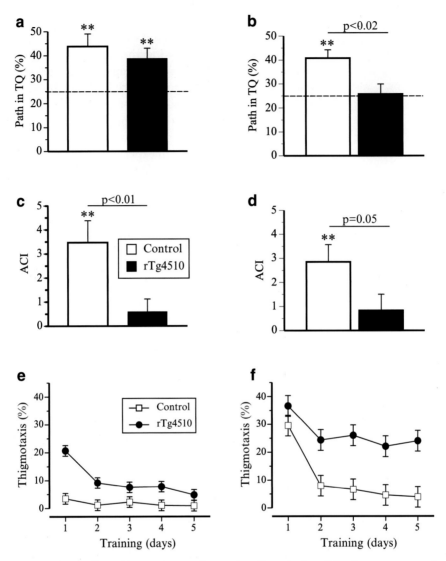

Fig. 2 Repeated daily handling with restraint ameliorates impairment of spatial reference memory in rTg4510 mouse model of tauopathy. Spatial reference memory of mice was evaluated in a probe trial administered at the end of training in the Morris water maze test. Two memory indices were employed: (1) the % of path searching in the target quadrant (TQ) containing an escape platform during training and (2) a more stringent Annulus Crossing Index (ACI), which reflects the number of swims over the platform site in the TQ adjusted for swims over equivalent sites in other quadrants of the pool. (**a**) The spatial memory evaluated by the % of path searching in the TQ was comparable between 6-month-old control and rTg4510 mice that underwent 3-month-long daily handling with restraint (gavage), while (**b**) non-handled 5.5-month-old rTg4510 mice showed significant impairment in their search of the TQ as compared to the controls (modified after [229]). In contrast, the analysis of ACI revealed significantly lower memory indices of rTg4510 mice as compared to the controls after (**c**) 3-month-long intensive handling with gavage, or (**d**) after only brief habituation to experimental handling (modified after [229]). (**e**) Three-month-long daily handling with gavage also significantly affected the behavior of mice during the training phase of the test, reducing thigmotaxic swimming while searching for the platform, in both rTg4510 and control mice. (**f**) In contrast, non-handled mice showed overall higher percentage of thigmotaxic swimming, especially by rTg4510 mice (modified after [229]). See text for details of statistical analyses. $**p < 0.01$ for the comparison of memory scores against chance levels (25 % for %path in TQ and 0 for ACI)

(ranging from ~13% in session 1 to ~3% in session 5, Fig. 2e) with rTg4510 mice showing thigmotaxic behavior at marginally higher proportions than the thigmotaxis of the control mice ($F(1,22) = 4.2$, $p = 0.05$, η^2 (partial eta squared) $= 0.16$). This result contrasted the performance of naïve mice which overall showed higher proportions of thigmotaxic swims during training, ranging on average from ~33% in session 1 to ~16% in session 5 (Fig. 2f). The thigmotaxis of rTg4510 mice was higher than the thigmotaxis of the control mice ($F(1,21) = 43.7$, $p < 0.001$). The high proportion of the thigmotaxic swimming by naïve rTg4510 mice resulted in their significantly longer search paths as compared to their controls ($F(1,21) = 49.0$, $p < 0.001$). The genotype effect accounted for relatively high proportion of variance in thigmotaxic swimming ($\eta^2 = 0.37$).

During thigmotaxic swimming mice follow the wall of the pool as a proximate cue most often during initial stages of the MWM test, and as training progresses they transition to a search encompassing the entire area of the pool. While thigmotaxis should be considered an integral part of the search path, the analysis of this proportion of search path might be informative and should be carried out along the analyses of the whole search path. Longer thigmotaxic swims have often been linked to increased anxiety [230] or to impairments in the choice of optimal search strategies [231]; therefore, the analysis of this behavior might elucidate possible effects of anxiety on the performance of mice in the MWM test. In conclusion, our results indicate that the inclusion of several measures of memory, which differ in the difficulty of task solution and additional measures like thigmotaxis that are related to behavioral plasticity might substantially improve the functional interpretation of results obtained in MWM tests. The most parsimonious explanation of the positive effect of handling on the performance of rTg4510 mice in the MWM test suggests that intensive handling with restraint lowers anxiety levels of mice consequently increased their attention, exploratory motivation, or coping strategies in this testing situation. Mice are generally considered more difficult to motivate to perform than rats to perform in a variety of tasks because of their timidity, but the acclimation to experimental handling often overcomes this problem [232]. Our observations seem to confirm these conclusions, since mice in our experiment became noticeably more docile after being subjected to handling with gavage regimen for the first several days of the experiment.

Since the method of repeated handling with brief restraint during gavage is often a part of routine drug administration in preclinical studies, it needs to be incorporated in experimental designs as a potential confounding factor, which, as our results showed, might solely and significantly ameliorate the behavior of mice modeling human disorder. In such cases the experimental design could include a pre-dosing segment in the overall regimen, where all mice initially receive placebo, and the treatment cohort is only switched to drug treatment when the intended treatment begins. It could also be argued that repeated handling alters the context of an experimental environment that bears some resemblance to well-recognized placebo effects caused by increased attention given to AD patients during drug trials [233].

6   Conclusions

The main goal of the current generation of animal models of human diseases is to better understand their underlying pathology, which should lead to the discovery of new or evaluations of known potential therapeutics. In our review we focused on neurodegenerative diseases that present a complex and age-progressing dementia with rapidly progressing neuronal death. It is probably unrealistic to assume that the full complexity of a human brain disorder could be modeled in a mouse using relatively crude genetic modification. However, as we tried to outline in this review, the interpretation of the results coming from mouse models of neurodegeneration should not follow a "one gene–one disease" paradigm. At present, none of the existing mouse models of tauopathy fully replicate the characteristics of human disorders. Using mouse models to study well-conserved signaling pathways in vivo may be better warranted than attempts to fully replicate the complexity of human dementia. Genetically modified mouse models are an integral part of modern drug discovery, but the interpretation of obtained behavioral results must be carefully administered and carried out within constraints of model and mouse biology. The intensive screens of many compounds would require a systematic, well-controlled, and standardized phenotyping approach. The initial characterization of new models or detailed characterization of specific aspects of existing models should be based on thoughtful experimental design, and encompass not only a larger number of mice in completely randomized experimental designs, but also optimize the conditions of animal facilities and husbandry methods, which ensure maximal expression of the natural behavioral repertoire of mice. Rigor of the experimental design will ensure replication of the results across the labs and between models, adjusting for the differences in genetic backgrounds and transgene(s) expressions. Even seemingly good and robust models of a human disease can yield many false-positive results due to differences in methodology or less rigorously carried out experiments [234]. Our intention is to highlight important aspects of broad experimental design, which are not always identified a priori, but may often generate confounding factors that seriously bias obtained data. We argue that success in the endeavor of modeling human cognitive impairment may often depend on how well we understand the behavior of a mouse. Detailed analysis of the potential and limitations of a model and the interpretation of obtained results within the framework of mouse biology should considerably improve the evaluation of potential therapeutics. By testing specific hypotheses, negative results should be as valuable as positive ones and should be made readily available to the scientific community.

References

1. Dubos R (1968) So human an animal. Charles Scribner's, New York

2. Alzheimer A (1907) Über eine eigenartige Erkankung der Hirnrinde. Allg Z Psychiatrie Psychisch-Gerlichtlich Med 64:146–148

3. Stelzmann RA, Schnitzlein HN, Murtagh FR (1995) An English translation of Alzheimer's 1907 paper, "Uber eine eigenartige Erkankung der Hirnrinde". Clin Anat 8:429–431

4. Braak H, Braak E (1994) Pathology of Alzheimer's disease. In: Calne DB (ed) Neurodegenerative diseases. Saunders, Philadelphia, pp 585–613

5. Braak H, Braak E (1997) Frequency of stages of Alzheimer-related lesions in different age categories. Neurobiol Aging 18:351–357

6. Kosik KS, Shimura H (2005) Phosphorylated tau and the neurodegenerative foldopathies. Biochim Biophys Acta 1739:298–310

7. Lee VM, Goedert M, Trojanowski JQ (2001) Neurodegenerative tauopathies. Annu Rev Neurosci 24:1121–1159

8. Dickson DW (2003) Neurodegeneration: the molecular pathology of dementia and movement disorders. ISN Neuropath Press, Basel

9. Cotman CW, Su JH (1996) Mechanisms of neuronal death in Alzheimer's disease. Brain Pathol 6:493–506

10. Terry RD (2006) Alzheimer's disease and the aging brain. J Geriatr Psychiatry Neurol 19:125–128

11. Davies RR, Hodges JR, Kril JJ, Patterson K, Halliday GM et al (2005) The pathological basis of semantic dementia. Brain 128:1984–1995

12. Iqbal K, Alonso Adel C, Chen S, Chohan MO, El-Akkad E et al (2005) Tau pathology in Alzheimer disease and other tauopathies. Biochim Biophys Acta 1739:198–210

13. Albert MS (1996) Cognitive and neurobiologic markers of early Alzheimer's disease. Proc Natl Acad Sci U S A 93:13547–13551

14. Morgan D (2007) Amyloid, memory and neurogenesis. Exp Neurol 205:330–335

15. Petersen RC, Smith GE, Waring SC, Ivnik RJ, Tangalos EG et al (1999) Mild cognitive impairment: clinical characterization and outcome. Arch Neurol 56:303–308

16. Knopman DS, DeKosky ST, Cummings JL, Chui H, Corey-Bloom J et al (2001) Practice parameter: diagnosis of dementia (an evidence-based review). Report of the Quality Standards Subcommittee of the American Academy of Neurology. Neurology 56:1143–1153

17. Luis CA, Loewenstein DA, Acevedo A, Barker WW, Duara R (2003) Mild cognitive impairment: directions for future research. Neurology 61:438–444

18. Maruyama M, Arai H, Sugita M, Tanji H, Higuchi M et al (2001) Cerebrospinal fluid amyloid beta(1-42) levels in the mild cognitive impairment stage of Alzheimer's disease. Exp Neurol 172:433–436

19. Riemenschneider M, Lautenschlager N, Wagenpfeil S, Diehl J, Drzezga A et al (2002) Cerebrospinal fluid tau and beta-amyloid 42 proteins identify Alzheimer disease in subjects with mild cognitive impairment. Arch Neurol 59:1729–1734

20. Cummings JL (2004) Dementia with lewy bodies: molecular pathogenesis and implications for classification. J Geriatr Psychiatry Neurol 17:112–119

21. Victoroff J, Zarow C, Mack WJ, Hsu E, Chui HC (1996) Physical aggression is associated with preservation of substantia nigra pars compacta in Alzheimer disease. Arch Neurol 53:428–434

22. Pahwa R, Lyons KE (2007) Handbook of Parkinson's disease. Informa Healthcare USA, New York

23. Whitehouse PJ, Price DL, Struble RG, Clark AW, Coyle JT et al (1982) Alzheimer's disease and senile dementia: loss of neurons in the basal forebrain. Science 215:1237–1239

24. Morrison JH, Hof PR (1997) Life and death of neurons in the aging brain. Science 278:412–419

25. Arnold SE, Hyman BT, Flory J, Damasio AR, Van Hoesen GW (1991) The topographical and neuroanatomical distribution of neurofibrillary tangles and neuritic plaques in the cerebral cortex of patients with Alzheimer's disease. Cereb Cortex 1:103–116

26. Hyman BT, Van Hoesen GW, Damasio AR, Barnes CL (1984) Alzheimer's disease: cell-specific pathology isolates the hippocampus formation. Science 225:1168–1170

27. Horn R, Ostertun B, Fric M, Solymosi L, Steudel A et al (1996) Atrophy of hippocampus in patients with Alzheimer's disease and other diseases with memory impairment. Dementia 7:182–186

28. Samuel W, Terry RD, Deteresa R, Butters N, Masliah E (1994) Clinical correlates of cortical and nucleus basalis pathology in Alzheimer dementia. Arch Neurol 51:772–778

29. Karas GB, Burton EJ, Rombouts SA, van Schijndel RA, O'Brien JT et al (2003) A comprehensive study of gray matter loss in patients with Alzheimer's disease using optimized voxel-based morphometry. Neuroimage 18:895–907

30. Jack CR Jr, Shiung MM, Gunter JL, O'Brien PC, Weigand SD et al (2004) Comparison of different MRI brain atrophy rate measures with clinical disease progression in AD. Neurology 62:591–600

31. Jope RS, Song L, Powers RE (1997) Cholinergic activation of phosphoinositide signaling is impaired in Alzheimer's disease brain. Neurobiol Aging 18:111–120

32. Tong XK, Hamel E (1999) Regional cholinergic denervation of cortical microvessels and nitric oxide synthase-containing neurons in Alzheimer's disease. Neuroscience 92:163–175

33. Mattson MP, Pedersen WA (1998) Effects of amyloid precursor protein derivatives and oxidative stress on basal forebrain cholinergic systems in Alzheimer's disease. Int J Dev Neurosci 16:737–753

34. Grundke-Iqbal I, Iqbal K, Quinlan M, Tung YC, Zaidi MS et al (1986) Microtubule-associated protein tau. A component of Alzheimer paired helical filaments. J Biol Chem 261:6084–6089

35. Braak H, Braak E (1997) Diagnostic criteria for neuropathologic assessment of Alzheimer's disease. Neurobiol Aging 18:S85–S88

36. Tomlinson BE, Blessed G, Roth M (1968) Observations on the brains of non-demented old people. J Neurol Sci 7:331–356

37. Kidd M (1964) Alzheimer's disease—an electron microscopical study. Brain 87:307–320

38. Goedert M, Wischik CM, Crowther RA, Walker JE, Klug A (1988) Cloning and sequencing of the cDNA encoding a core protein of the paired helical filament of Alzheimer disease: identification as the microtubule-associated protein tau. Proc Natl Acad Sci U S A 85:4051–4055

39. Kosik KS, Joachim CL, Selkoe DJ (1986) Microtubule-associated protein tau (tau) is a major antigenic component of paired helical filaments in Alzheimer disease. Proc Natl Acad Sci U S A 83:4044–4048

40. Paulus W, Selim M (1990) Corticonigral degeneration with neuronal achromasia and basal neurofibrillary tangles. Acta Neuropathol 81:89–94

41. Elizan TS, Hirano A, Abrams BM, Need RL, Van Nuis C et al (1966) Amyotrophic lateral sclerosis and parkinsonism-dementia complex of Guam. Neurological reevaluation. Arch Neurol 14:356–368

42. Hirano A, Malamud N, Elizan TS, Kurland LT (1966) Amyotrophic lateral sclerosis and Parkinsonism-dementia complex on Guam. Further pathologic studies. Arch Neurol 15:35–51

43. Hof PR, Bouras C, Perl DP, Sparks DL, Mehta N et al (1995) Age-related distribution of neuropathologic changes in the cerebral cortex of patients with Down's syndrome. Quantitative regional analysis and comparison with Alzheimer's disease. Arch Neurol 52:379–391

44. Kiuchi A, Otsuka N, Namba Y, Nakano I, Tomonaga M (1991) Presenile appearance of abundant Alzheimer's neurofibrillary tangles without senile plaques in the brain in myotonic dystrophy. Acta Neuropathol 82:1–5

45. Mott RT, Dickson DW, Trojanowski JQ, Zhukareva V, Lee VM et al (2005) Neuropathologic, biochemical, and molecular characterization of the frontotemporal dementias. J Neuropathol Exp Neurol 64:420–428

46. Goedert M (2001) Alpha-synuclein and neurodegenerative diseases. Nat Rev Neurosci 2:492–501

47. Glenner GG, Wong CW (1984) Alzheimer's disease: initial report of the purification and characterization of a novel cerebrovascular

amyloid protein. Biochem Biophys Res Commun 120:885–890

48. Vigo-Pelfrey C, Lee D, Keim P, Lieberburg I, Schenk DB (1993) Characterization of beta-amyloid peptide from human cerebrospinal fluid. J Neurochem 61:1965–1968

49. Roher A, Wolfe D, Palutke M, KuKuruga D (1986) Purification, ultrastructure, and chemical analysis of Alzheimer disease amyloid plaque core protein. Proc Natl Acad Sci U S A 83:2662–2666

50. Iwatsubo T, Odaka A, Suzuki N, Mizusawa H, Nukina N et al (1994) Visualization of A beta 42(43) and A beta 40 in senile plaques with end-specific A beta monoclonals: evidence that an initially deposited species is A beta 42(43). Neuron 13:45–53

51. Naslund J, Haroutunian V, Mohs R, Davis KL, Davies P et al (2000) Correlation between elevated levels of amyloid beta-peptide in the brain and cognitive decline. JAMA 283:1571–1577

52. Goate A, Chartier-Harlin MC, Mullan M, Brown J, Crawford F et al (1991) Segregation of a missense mutation in the amyloid precursor protein gene with familial Alzheimer's disease. Nature 349:704–706

53. Chartier-Harlin M-C, Crawford F, Houlden H, Warren A, Hughes D et al (1991) Early-onset alzheimer's disease caused by mutations at codon 717 of the ß-amyloid precursor protein gene. Nature 353:844–846

54. Chartier-Harlin MC, Crawford F, Hamandi K, Mullan M, Goate A et al (1991) Screening for the beta-amyloid precursor protein mutation (APP717: Val----Ile) in extended pedigrees with early onset alzheimer's disease. Neurosci Lett 129:134–135

55. Sherrington R, Rogaev EI, Liang Y, Rogaeva EA, Levesque G et al (1995) Cloning of a gene bearing missense mutations in early-onset familial Alzheimer's disease. Nature 375:754–760

56. Campion D, Flaman JM, Brice A, Hannequin D, Dubois B et al (1995) Mutations of the presenilin-1 gene in families with early-onset alzheimer's disease. Human Mol Genet 4:2373–2377

57. Levy-Lahad E, Wasco W, Poorkaj P, Romano DM, Oshima J et al (1995) Candidate gene for the chromosome 1 familial Alzheimer's disease locus. Science 269:973–977

58. Rogaev EI, Sherrington R, Rogaeva EA, Levesque G, Ikeda M et al (1995) Familial Alzheimer's disease in kindreds with missense mutations in a gene on chromosome 1 related to the Alzheimer's disease type 3 gene. Nature 376:775–778

59. Scheuner D, Eckman C, Jensen M, Song X, Citron M et al (1996) Secreted amyloid beta-protein similar to that in the senile plaques of Alzheimer's disease is increased in vivo by the presenilin 1 and 2 and APP mutations linked to familial Alzheimer's disease. Nat Med 2:864–870

60. Golde TE (2003) Alzheimer disease therapy: can the amyloid cascade be halted? J Clin Invest 111:11–18

61. Hardy JA, Higgins GA (1992) Alzheimer's disease: the amyloid cascade hypothesis. Science 256:184–185

62. Selkoe DJ (2002) Deciphering the genesis and fate of amyloid beta-protein yields novel therapies for Alzheimer disease. J Clin Invest 110:1375–1381

63. Blacker D, Wilcox MA, Laird NM, Rodes L, Horvath SM et al (1998) Alpha-2 macroglobulin is genetically associated with Alzheimer disease. Nat Genet 19:357–360

64. Depboylu C, Lohmuller F, Du Y, Riemenschneider M, Kurz A et al (2006) Alpha2-macroglobulin, lipoprotein receptor-related protein and lipoprotein receptor-associated protein and the genetic risk for developing Alzheimer's disease. Neurosci Lett 400:187–190

65. Zappia M, Cittadella R, Manna I, Nicoletti G, Andreoli V et al (2002) Genetic association of alpha2-macroglobulin polymorphisms with AD in southern Italy. Neurology 59:756–758

66. Bertram L, Tanzi RE (2001) Of replications and refutations: the status of Alzheimer's disease genetic research. Curr Neurol Neurosci Rep 1:442–450

67. Cruts M, Theuns J, Van Broeckhoven C (2012) Locus-specific mutation databases for neurodegenerative brain diseases. Hum Mutat 33:1340–1344

68. Strittmatter WJ, Saunders AM, Schmechel D, Pericak-Vance M, Enghild J et al (1993) Apolipoprotein E: high-avidity binding to beta-amyloid and increased frequency of type 4 allele in late-onset familial Alzheimer disease. Proc Natl Acad Sci U S A 90: 1977–1981

69. Corder EH, Saunders AM, Strittmatter WJ, Schmechel DE, Gaskell PC et al (1993) Gene

dose of apolipoprotein E type 4 allele and the risk of Alzheimer's disease in late onset families. Science 261:921–923

70. Evans DA, Bennett DA, Wilson RS, Bienias JL, Morris MC et al (2003) Incidence of Alzheimer disease in a biracial urban community: relation to apolipoprotein E allele status. Arch Neurol 60:185–189

71. Tang MX, Stern Y, Marder K, Bell K, Gurland B et al (1998) The APOE-epsilon4 allele and the risk of Alzheimer disease among African Americans, whites, and Hispanics. JAMA 279:751–755

72. Kim HC, Kim DK, Choi IJ, Kang KH, Yi SD et al (2001) Relation of apolipoprotein E polymorphism to clinically diagnosed Alzheimer's disease in the Korean population. Psychiatry Clin Neurosci 55:115–120

73. Tanzi RE (2012) The genetics of Alzheimer disease. Cold Spring Harb Perspect Med 2

74. Tanzi RE, Bertram L (2001) New frontiers in Alzheimer's disease genetics. Neuron 32:181–184

75. Lambert JC, Heath S, Even G, Campion D, Sleegers K et al (2009) Genome-wide association study identifies variants at CLU and CR1 associated with Alzheimer's disease. Nat Genet 41:1094–1099

76. Harold D, Abraham R, Hollingworth P, Sims R, Gerrish A et al (2009) Genome-wide association study identifies variants at CLU and PICALM associated with Alzheimer's disease. Nat Genet 41:1088–1093

77. Ertekin-Taner N, Graff-Radford N, Younkin LH, Eckman C, Baker M et al (2000) Linkage of plasma Abeta42 to a quantitative locus on chromosome 10 in late-onset alzheimer's disease pedigrees. Science 290:2303–2304

78. Bertram L, Blacker D, Crystal A, Mullin K, Keeney D et al (2000) Candidate genes showing no evidence for association or linkage with Alzheimer's disease using family-based methodologies. Exp Gerontol 35:1353–1361

79. Hollingworth P, Harold D, Sims R, Gerrish A, Lambert JC et al (2011) Common variants at ABCA7, MS4A6A/MS4A4E, EPHA1, CD33 and CD2AP are associated with Alzheimer's disease. Nat Genet 43:429–435

80. Naj AC, Jun G, Beecham GW, Wang LS, Vardarajan BN et al (2011) Common variants at MS4A4/MS4A6E, CD2AP, CD33 and EPHA1 are associated with late-onset alzheimer's disease. Nat Genet 43:436–441

81. Guerreiro R, Wojtas A, Bras J, Carrasquillo M, Rogaeva E et al (2013) TREM2 variants in Alzheimer's disease. N Engl J Med 368:117–127

82. Jonsson T, Stefansson H, Steinberg S, Jonsdottir I, Jonsson PV et al (2013) Variant of TREM2 associated with the risk of Alzheimer's disease. N Engl J Med 368:107–116

83. Mandelkow EM, Mandelkow E (1998) Tau in Alzheimer's disease. Trends Cell Biol 8:425–427

84. Mandelkow EM, Stamer K, Vogel R, Thies E, Mandelkow E (2003) Clogging of axons by tau, inhibition of axonal traffic and starvation of synapses. Neurobiol Aging 24:1079–1085

85. Mattson MP (2004) Pathways towards and away from Alzheimer's disease. Nature 430:631–639

86. Rizzu P, Joosse M, Ravid R, Hoogeveen A, Kamphorst W et al (2000) Mutation-dependent aggregation of tau protein and its selective depletion from the soluble fraction in brain of P301L FTDP-17 patients. Hum Mol Genet 9:3075–3082

87. Clark LN, Poorkaj P, Wszolek Z, Geschwind DH, Nasreddine ZS et al (1998) Pathogenic implications of mutations in the tau gene in pallido-ponto-nigral degeneration and related neurodegenerative disorders linked to chromosome 17. Proc Natl Acad Sci U S A 95:13103–13107

88. Hutton M, Lendon CL, Rizzu P, Baker M, Froelich S et al (1998) Association of missense and 5'-splice-site mutations in tau with the inherited dementia FTDP-17. Nature 393:702–705

89. Spillantini MG, Murrell JR, Goedert M, Farlow MR, Klug A et al (1998) Mutation in the tau gene in familial multiple system tauopathy with presenile dementia. Proc Natl Acad Sci U S A 95:7737–7741

90. Games D, Adams D, Alessandrini R, Barbour R, Berthelette P et al (1995) Alzheimer-type neuropathology in transgenic mice overexpressing V717F beta-amyloid precursor protein. Nature 373:523–527

91. Greenberg BD, Savage MJ, Howland DS, Ali SM, Siedlak SL et al (1996) APP transgenesis: approaches toward the development of animal models for Alzheimer disease neuropathology. Neurobiol Aging 17:153–171

92. Ashe K (2001) Learning and memory in transgenic mice modelling Alzheimer's disease. Learn Mem 8:301–308

93. Ashe KH (2005) Mechanisms of memory loss in Abeta and tau mouse models. Biochem Soc Trans 33:591–594

94. Dodart JC, Mathis C, Bales KR, Paul SM (2002) Does my mouse have Alzheimer's disease? Genes Brain Behav 1:142–155

95. Eriksen JL, Janus CG (2007) Plaques, tangles, and memory loss in mouse models of neurodegeneration. Behav Genet 37:79–100

96. Higgins GA, Jacobsen H (2003) Transgenic mouse models of Alzheimer's disease: phenotype and application. Behav Pharmacol 14:419–438

97. Janus C, Phinney AL, Chishti MA, Westaway D (2001) New developments in animal models of Alzheimer's disease. Curr Neurol Neurosci Rep 1:451–457

98. Price DL, Tanzi RE, Borchelt DR, Sisodia SS (1998) Alzheimer's disease: genetic studies and transgenic models. Annu Rev Genet 32:461–493

99. Seabrook GR, Rosahl TW (1999) Transgenic animals relevant to Alzheimer's disease. Neuropharmacology 38:1–17

100. van Leuven F (2000) Single and multiple transgenic mice as models for Alzheimer's disease. Prog Neurobiol 61:305–312

101. Spires TL, Hyman BT (2005) Transgenic models of Alzheimer's disease: learning from animals. NeuroRx 2:423–437

102. Wong PC, Cai H, Borchelt DR, Price DL (2002) Genetically engineered mouse models of neurodegenerative diseases. Nat Neurosci 5:633–639

103. Hall GF, Yao J (2005) Modeling tauopathy: a range of complementary approaches. Biochim Biophys Acta 1739:224–239

104. Melrose HL, Lincoln SJ, Tyndall GM, Farrer MJ (2006) Parkinson's disease: a rethink of rodent models. Exp Brain Res 173:196–204

105. Le Cudennec C, Faure A, Ly M, Delatour B (2008) One-year longitudinal evaluation of sensorimotor functions in APP751SL transgenic mice. Genes Brain Behav 7(Suppl 1):83–91

106. Cui S, Chesson C, Hope R (1993) Genetic variation within and between strains of outbred Swiss mice. Lab Anim 27:116–123

107. Festing MF (1974) Genetic reliability of commercially-bred laboratory mice. Lab Anim 8:265–270

108. Festing MF (1974) Genetic monitoring of laboratory mouse colonies in the Medical Research Council Accreditation Scheme for the suppliers of laboratory animals. Lab Anim 8:291–299

109. Crabbe JC, Wahlsten D, Dudek BC (1999) Genetics of mouse behavior: interactions with laboratory environment. Science 284:1670–1672

110. Wahlsten D, Metten P, Phillips TJ, Boehm SL 2nd, Burkhart-Kasch S et al (2003) Different data from different labs: lessons from studies of gene-environment interaction. J Neurobiol 54:283–311

111. Banbury Conference on genetic background in mice (1997) Mutant mice and neuroscience: recommendations concerning genetic background. Banbury Conference on genetic background in mice. Neuron 19:755–759

112. Crawley JN, Belknap JK, Collins A, Crabbe JC, Frankel W et al (1997) Behavioral phenotypes of inbred mouse strains: implications and recommendations for molecular studies. Psychopharmacology (Berl) 132:107–124

113. Takahashi JS, Pinto LH, Vitaterna MH (1994) Forward and reverse genetic approaches to behavior in the mouse. Science 264:1724–1733

114. Dietrich WF, Lander ES, Smith JS, Moser AR, Gould KA et al (1993) Genetic identification of Mom-1, a major modifier locus affecting Min-induced intestinal neoplasia in the mouse. Cell 75:631–639

115. Gould KA, Luongo C, Moser AR, McNeley MK, Borenstein N et al (1996) Genetic evaluation of candidate genes for the Mom1 modifier of intestinal neoplasia in mice. Genetics 144:1777–1785

116. Wahlsten D, Cooper SF, Crabbe JC (2005) Different rankings of inbred mouse strains on the Morris maze and a refined 4-arm water escape task. Behav Brain Res 165:36–51

117. Sidman RL, Green MC (1965) Retinal degeneration in the mouse: location of the Rd Locus in Linkage Group Xvii. J Hered 56:23–29

118. Jimenez AJ, Garcia-Fernandez JM, Gonzalez B, Foster RG (1996) The spatio-temporal pattern of photoreceptor degeneration in the aged rd/rd mouse retina. Cell Tissue Res 284:193–202

119. Ogilvie JM, Speck JD (2002) Dopamine has a critical role in photoreceptor degeneration in the rd mouse. Neurobiol Dis 10:33–40

120. Guillery RW (1974) Visual pathways in albinos. Sci Am 230:44–54

121. Rice DS, Williams RW, Goldowitz D (1995) Genetic control of retinal projections in inbred strains of albino mice. J Comp Neurol 354:459–469

122. Lamb BT, Sisodia SS, Lawler AM, Slunt HH, Kitt CA et al (1993) Introduction and expression of the 400 kilobase amyloid precursor protein gene in transgenic mice [corrected]. Nat Genet 5:22–30

123. Chishti MA, Yang DS, Janus C, Phinney AL, Horne P et al (2001) Early-onset amyloid deposition and cognitive deficits in transgenic mice expressing a double mutant form of amyloid precursor protein 695. J Biol Chem 276:21562–21570

124. Hsiao K, Chapman P, Nilsen S, Eckman C, Harigaya Y et al (1996) Correlative memory deficits, Abeta elevation, and amyloid plaques in transgenic mice. Science 274:99–102

125. Hsiao KK, Borchelt DR, Olson K, Johannsdottir R, Kitt C et al (1995) Age-related CNS disorder and early death in transgenic FVB/N mice overexpressing Alzheimer amyloid precursor proteins. Neuron 15:1203–1218

126. Sturchler-Pierrat C, Abramowski D, Duke M, Wiederhold KH, Mistl C et al (1997) Two amyloid precursor protein transgenic mouse models with Alzheimer disease-like pathology. Proc Natl Acad Sci U S A 94:13287–13292

127. Squire LR (1992) Memory and the hippocampus: a synthesis from findings with rats, monkeys, and humans. Psychol Rev 99:195–231

128. Barnes CA, Rao G, McNaughton BL (1996) Functional integrity of NMDA-dependent LTP induction mechanisms across the lifespan of F-344 rats. Learn Mem 3:124–137

129. Bliss TVP, Collingridge GL (1993) A synaptic model of memory: long-term potentiation in the hippocampus. Nature 361:31–39

130. Collingridge GL, Kehl SJ, McLennan H (1983) Excitatory amino acids in synaptic transmission in the Schaffer collateral-commissural pathway of the rat hippocampus. J Physiol (Lond) 334:33–46

131. Eichenbaum H (1996) Learning from LTP: a comment on recent attempts to identify cellular and molecular mechanisms of memory. Learn Mem 3:61–73

132. Fazeli MS, Errington ML, Dolphin AC, Bliss TVP (1988) Long-term potentiation in the dentate gyrus of the anaesthetized rat is accompanied by an increase in protein efflux into push–pull cannula perfusates. Brain Res 473:51–59

133. Malenka RC, Nicoll RA (1993) NMDA-receptor-dependent synaptic plasticity: multiple forms and mechanisms. Trends Neurosci 16:521–527

134. Morris RGM (1989) Synaptic plasticity and learning: Selective impairment of learning in rats and blockade of long-term potentiation in vivo by the N-methyl-d-aspartate receptor antagonist AP5. J Neurosci 9:3040–3057

135. Morris RG, Davis S, Butcher SP (1990) Hippocampal synaptic plasticity and NMDA receptors: a role in information storage? Philos Trans R Soc Lond B Biol Sci 329:187–204

136. O'Keefe J, Nadel L (1978) The hippocampus as a cognitive map. Oxford University Press, Oxford

137. Olton DS, Becker JT, Handelman GE (1979) Hippocampus space and memory. Behav Brain Sci 2:313–365

138. Milner B, Scoville WB (1957) Loss of recent memory after bilateral hippocampal lesions. J Neurol Neurosurg Psychiatry 20:11–21

139. Smith ML, Milner B (1981) The role of the right hippocampus in the recall of spatial location. Neuropsychologia 19:781–793

140. Milner B (1965) Visually-guided maze-learning in man: effects of bilateral hippocampal, bilateral frontal hippocampal lesions. Neuropsychologia 3:317–338

141. Elgh E, Lindqvist Astot A, Fagerlund M, Eriksson S, Olsson T et al (2006) Cognitive dysfunction, hippocampal atrophy and glucocorticoid feedback in Alzheimer's disease. Biol Psychiatry 59:155–161

142. Rodriguez G, Vitali P, Calvini P, Bordoni C, Girtler N et al (2000) Hippocampal perfusion in mild Alzheimer's disease. Psychiatry Res 100:65–74

143. Carlesimo GA, Mauri M, Graceffa AM, Fadda L, Loasses A et al (1998) Memory performances in young, elderly, and very old healthy individuals versus patients with Alzheimer's disease: evidence for discontinuity between normal and pathological aging. J Clin Exp Neuropsychol 20:14–29

144. Ghilardi MF, Alberoni M, Marelli S, Rossi M, Franceschi M et al (1999) Impaired movement control in Alzheimer's disease. Neurosci Lett 260:45–48

145. Kavcic V, Duffy CJ (2003) Attentional dynamics and visual perception: mechanisms

of spatial disorientation in Alzheimer's disease. Brain 126:1173–1181

146. Monacelli AM, Cushman LA, Kavcic V, Duffy CJ (2003) Spatial disorientation in Alzheimer's disease: the remembrance of things passed. Neurology 61:1491–1497

147. Pai MC, Jacobs WJ (2004) Topographical disorientation in community-residing patients with Alzheimer's disease. Int J Geriatr Psychiatry 19:250–255

148. Rizzo M, Anderson SW, Dawson J, Nawrot M (2000) Vision and cognition in Alzheimer's disease. Neuropsychologia 38:1157–1169

149. Janus C, Flores AY, Xu G, Borchelt DR (2015) Behavioral abnormalities in APP/PS1dE9 mouse model of AD-like pathology: comparative analysis across multiple behavioral domains. Neurobiol Aging 36:2519–2532

150. Crawley JN (2007) What's wrong with my mouse?: Behavioural phenotyping of transgenic and knockout mice, 2nd edn. Wiley, New Jersey

151. Whishaw IQ, Kolb B (2005) The behavior of the laboratory rat: a handbook with tests. Oxford University Press, Oxford

152. Crawley JN, Paylor R (1997) A proposed test battery and constellation of specific behavioural paradigms to investigate the behavioural phenotypes of transgenic and knockout mice. Horm Behav 31:197–211

153. Janus C (2004) Search strategies used by APP transgenic mice during navigation in the Morris water maze. Learn Mem 11:337–346

154. Markowska AL, Long JM, Johnson CT, Olton DS (1993) Variable-interval probe test as a tool for repeated measurements of spatial memory in the water maze. Behav Neurosci 107:627–632

155. Spooner RIW, Thomson A, Hall J, Morris RGM, Salter SH (1994) The Atlantis platform: a new design and further developments of Buresova's on-demand platform for the water maze. Learn Mem 1:203–211

156. Dudchenko PA, Goodridge JP, Seiterle DA, Taube JS (1997) Effects of repeated disorientation on the acquisition of spatial tasks in rats: dissociation between the appetetive radial arm maze and aversive water maze. J Exp Psychol 23:194–210

157. Chapillon P, Debouzie A (2000) BALB/c mice are not so bad in the Morris water maze. Behav Brain Res 117:115–118

158. Whishaw IQ, Tomie JA (1996) Of mice and mazes: similarities between mice and rats on dry land but not water mazes. Physiol Behav 60:1191–1197

159. Wahlsten D, Metten P, Crabbe JC (2003) A rating scale for wildness and ease of handling laboratory mice: results for 21 inbred strains tested in two laboratories. Genes Brain Behav 2:71–79

160. Corcoran KA, Lu Y, Turner RS, Maren S (2002) Overexpression of hAPPswe impairs rewarded alternation and contextual fear conditioning in a transgenic mouse model of Alzheimer's disease. Learn Mem 9:243–252

161. Janus C, Welzl H, Hanna A, Lovasic L, Lane N et al (2004) Impaired conditioned taste aversion learning in APP transgenic mice. Neurobiol Aging 25:1213–1219

162. Mumby DG (2001) Perspectives on object-recognition memory following hippocampal damage: lessons from studies in rats. Behav Brain Res 127:159–181

163. Kumar-Singh S, Dewachter I, Moechars D, Lubke U, De Jonghe C et al (2000) Behavioral disturbances without amyloid deposits in mice overexpressing human amyloid precursor protein with Flemish (A692G) or Dutch (E693Q) mutation. Neurobiol Dis 7:9–22

164. Lalonde R, Dumont M, Staufenbiel M, Sturchler-Pierrat C, Strazielle C (2002) Spatial learning, exploration, anxiety, and motor coordination in female APP23 transgenic mice with the Swedish mutation. Brain Res 956:36–44

165. Gerlai R, Fitch T, Bales KR, Gitter BD (2002) Behavioral impairment of APP(V717F) mice in fear conditioning: is it only cognition? Behav Brain Res 136:503–509

166. Wahlsten D, Bachmanov A, Finn DA, Crabbe JC (2006) Stability of inbred mouse strain differences in behavior and brain size between laboratories and across decades. Proc Natl Acad Sci U S A 103:16364–16369

167. Scott S, Kranz JE, Cole J, Lincecum JM, Thompson K et al (2008) Design, power, and interpretation of studies in the standard murine model of ALS. Amyotroph Lateral Scler 9:4–15

168. Machlis L, Dodd FWD, Fentress JC (1985) The pooling fallacy: problems arising when individuals contribute more than one observation to the data set. Zeitschrifte fur Tierpsychologie 68:201–214

169. Billings LM, Oddo S, Green KN, McGaugh JL, LaFerla FM (2005) Intraneuronal Abeta causes the onset of early Alzheimer's disease-related cognitive deficits in transgenic mice. Neuron 45:675–688

170. Stevens J (1990) Intermediate statistics: a modern approach. Lawrence Erlbaum Associates, Hillsdale, New Jersey

171. Popper K (1963) Conjectures and refutations. Routledge and Keagan Paul, London

172. Brown AP, Dinger N, Levine BS (2000) Stress produced by gavage administration in the rat. Contemp Top Lab Anim Sci 39:17–21

173. Walker MK, Boberg JR, Walsh MT, Wolf V, Trujillo A et al (2012) A less stressful alternative to oral gavage for pharmacological and toxicological studies in mice. Toxicol Appl Pharmacol 260:65–69

174. Balcombe JP, Barnard ND, Sandusky C (2004) Laboratory routines cause animal stress. Contemp Top Lab Anim Sci 43:42–51

175. Hurst JL, West RS (2010) Taming anxiety in laboratory mice. Nat Methods 1–2

176. Daftary SS, Panksepp J, Dong Y, Saal DB (2009) Stress-induced, glucocorticoid-dependent strengthening of glutamatergic synaptic transmission in midbrain dopamine neurons. Neurosci Lett 452:273–276

177. Dobrakovova M, Jurcovicova J (1984) Corticosterone and prolactin responses to repeated handling and transfer of male rats. Exp Clin Endocrinol 83:21–27

178. Korte SM (2001) Corticosteroids in relation to fear, anxiety and psychopathology. Neurosci Biobehav Rev 25:117–142

179. Cullinan WE, Ziegler DR, Herman JP (2008) Functional role of local GABAergic influences on the HPA axis. Brain Struct Funct 213:63–72

180. Figueiredo HF, Ulrich-Lai YM, Choi DC, Herman JP (2007) Estrogen potentiates adrenocortical responses to stress in female rats. Am J Physiol Endocrinol Metab 292:E1173–E1182

181. Jankord R, Herman JP (2008) Limbic regulation of hypothalamo-pituitary-adrenocortical function during acute and chronic stress. Ann N Y Acad Sci 1148:64–73

182. Razzoli M, Karsten C, Yoder JM, Bartolomucci A, Engeland WC (2014) Chronic subordination stress phase advances adrenal and anterior pituitary clock gene rhythms. Am J Physiol Regul Integr Comp Physiol 307:R198–R205

183. Abrous DN, Desjardins S, Sorin B, Hancock D, Le Moal M et al (1996) Changes in striatal immediate early gene expression following neonatal dopaminergic lesion and effects of intrastriatal dopaminergic transplants. Neuroscience 73:145–159

184. Choi DC, Nguyen MM, Tamashiro KL, Ma LY, Sakai RR et al (2006) Chronic social stress in the visible burrow system modulates stress-related gene expression in the bed nucleus of the stria terminalis. Physiol Behav 89:301–310

185. Herman JP, Sherman TG (1993) Acute stress upregulates vasopressin gene expression in parvocellular neurons of the hypothalamic paraventricular nucleus. Ann N Y Acad Sci 689:546–549

186. Ostrander MM, Richtand NM, Herman JP (2003) Stress and amphetamine induce Fos expression in medial prefrontal cortex neurons containing glucocorticoid receptors. Brain Res 990:209–214

187. Ostrander MM, Ulrich-Lai YM, Choi DC, Flak JN, Richtand NM et al (2009) Chronic stress produces enduring decreases in novel stress-evoked c-fos mRNA expression in discrete brain regions of the rat. Stress 12:469–477

188. Senba E, Ueyama T (1997) Stress-induced expression of immediate early genes in the brain and peripheral organs of the rat. Neurosci Res 29:183–207

189. Choi DC, Furay AR, Evanson NK, Ostrander MM, Ulrich-Lai YM et al (2007) Bed nucleus of the stria terminalis subregions differentially regulate hypothalamic-pituitary-adrenal axis activity: implications for the integration of limbic inputs. J Neurosci 27:2025–2034

190. Dent G, Choi DC, Herman JP, Levine S (2007) GABAergic circuits and the stress hyporesponsive period in the rat: ontogeny of glutamic acid decarboxylase (GAD) 67 mRNA expression in limbic-hypothalamic stress pathways. Brain Res 1138:1–9

191. Figueiredo HF, Bruestle A, Bodie B, Dolgas CM, Herman JP (2003) The medial prefrontal cortex differentially regulates stress-induced c-fos expression in the forebrain depending on type of stressor. Eur J Neurosci 18:2357–2364

192. Figueiredo HF, Dolgas CM, Herman JP (2002) Stress activation of cortex and hippocampus is modulated by sex and stage of estrus. Endocrinology 143:2534–2540

193. Flak JN, Ostrander MM, Tasker JG, Herman JP (2009) Chronic stress-induced neurotransmitter plasticity in the PVN. J Comp Neurol 517:156–165

194. Herman JP, Flak J, Jankord R (2008) Chronic stress plasticity in the hypothalamic paraventricular nucleus. Prog Brain Res 170:353–364

195. Jankord R, Zhang R, Flak JN, Solomon MB, Albertz J et al (2010) Stress activation of IL-6 neurons in the hypothalamus. Am J Physiol Regul Integr Comp Physiol 299:R343–R351

196. Ziegler DR, Cullinan WE, Herman JP (2005) Organization and regulation of paraventricular nucleus glutamate signaling systems: N-methyl-d-aspartate receptors. J Comp Neurol 484:43–56

197. Morton DB, Jennings M, Buckwell A, Ewbank R, Godfrey C et al (2001) Refining procedures for the administration of substances. Report of the BVAAWF/FRAME/RSPCA/UFAW Joint Working Group on Refinement. British Veterinary Association Animal Welfare Foundation/Fund for the Replacement of Animals in Medical Experiments/Royal Society for the Prevention of Cruelty to Animals/Universities Federation for Animal Welfare. Lab Anim 35:1–41

198. McEwen BS (2000) The neurobiology of stress: from serendipity to clinical relevance. Brain Res 886:172–189

199. Lapin IP (1995) Only controls: effect of handling, sham injection, and intraperitoneal injection of saline on behavior of mice in an elevated plus-maze. J Pharmacol Toxicol Methods 34:73–77

200. de Meijer VE, Le HD, Meisel JA, Puder M (2010) Repetitive orogastric gavage affects the phenotype of diet-induced obese mice. Physiol Behav 100:387–393

201. Hilakivi-Clarke LA (1992) Injection of vehicle is not a stressor in Porsolt's swim test. Pharmacol Biochem Behav 42:193–196

202. Gilbar PJ (1999) A guide to enteral drug administration in palliative care. J Pain Symptom Manage 17:197–207

203. Smolensky MH, Peppas NA (2007) Chronobiology, drug delivery, and chronotherapeutics. Adv Drug Deliv Rev 59:828–851

204. Turner PV, Brabb T, Pekow C, Vasbinder MA (2011) Administration of substances to laboratory animals: routes of administration and factors to consider. J Am Assoc Lab Anim Sci 50:600–613

205. Turner PV, Pekow C, Vasbinder MA, Brabb T (2011) Administration of substances to laboratory animals: equipment considerations, vehicle selection, and solute preparation. J Am Assoc Lab Anim Sci 50:614–627

206. Machholz E, Mulder G, Ruiz C, Corning BF, Pritchett-Corning KR (2012) Manual restraint and common compound administration routes in mice and rats., J Vis Exp

207. Qu WM, Huang ZL, Matsumoto N, Xu XH, Urade Y (2008) Drug delivery through a chronically implanted stomach catheter improves efficiency of evaluating wake-promoting components. J Neurosci Methods 175:58–63

208. Prittie J, Barton L (2004) Route of nutrient delivery. Clin Tech Small Anim Pract 19:6–8

209. Brayne C, Gill C, Huppert FA, Barkley C, Gehlhaar E et al (1998) Vascular risks and incident dementia: results from a cohort study of the very old. Dement Geriatr Cogn Disord 9:175–180

210. Corder EH, Saunders AM, Strittmatter WJ, Schmechel DE, Gaskell PC Jr et al (1995) Apolipoprotein E, survival in Alzheimer's disease patients, and the competing risks of death and Alzheimer's disease. Neurology 45:1323–1328

211. Fratiglioni L, Ahlbom A, Viitanen M, Winblad B (1993) Risk factors for late-onset alzheimer's disease: a population-based, case-control study. Ann Neurol 33:258–266

212. Burns JM, Mayo MS, Anderson HS, Smith HJ, Donnelly JE (2008) Cardiorespiratory fitness in early-stage Alzheimer disease. Alzheimer Dis Assoc Disord 22:39–46

213. Colcombe SJ, Erickson KI, Raz N, Webb AG, Cohen NJ et al (2003) Aerobic fitness reduces brain tissue loss in aging humans. J Gerontol A Biol Sci Med Sci 58:176–180

214. Larson EB (2008) Physical activity for older adults at risk for Alzheimer disease. JAMA 300:1077–1079

215. DeNelsky GY, Denenberg VH (1967) Infantile stimulation and adult exploratory behaviour in the rat: effects of handling upon visual variation-seeking. Anim Behav 15:568–573

216. DeNelsky GY, Denenberg VH (1967) Infantile stimulation and adult exploratory behavior: effects of handling upon tactual variation seeking. J Comp Physiol Psychol 63:309–312

217. Meaney MJ, Aitken DH, van Berkel C, Bhatnagar S, Sapolsky RM (1988) Effect of neonatal handling on age-related impair-

ments associated with the hippocampus. Science 239:766–768

218. Meaney MJ, Aitken DH, Viau V, Sharma S, Sarrieau A (1989) Neonatal handling alters adrenocortical negative feedback sensitivity and hippocampal type II glucocorticoid receptor binding in the rat. Neuroendocrinology 50:597–604

219. Tang AC (2001) Neonatal exposure to novel environment enhances hippocampal-dependent memory function during infancy and adulthood. Learn Mem 8:257–264

220. Tremml P, Lipp HP, Muller U, Ricceri L, Wolfer DP (1998) Neurobehavioral development, adult openfield exploration and swimming navigation learning in mice with a modified beta-amyloid precursor protein gene. Behav Brain Res 95:65–76

221. Adamec RE, Sayin U, Brown A (1991) The effects of corticotrophin releasing factor (CRF) and handling stress on behavior in the elevated plus-maze test of anxiety. J Psychopharmacol 5:175–186

222. Brett RR, Pratt JA (1990) Chronic handling modifies the anxiolytic effect of diazepam in the elevated plus-maze. Eur J Pharmacol 178:135–138

223. Rosenberg MJ (1969) The conditions and consequences of evaluation apprehension. In: Rosnow RL, Rosenthal R (eds) Artifact in behavioral research. Academic, New York

224. McCambridge J, Witton J, Elbourne DR (2014) Systematic review of the Hawthorne effect: new concepts are needed to study research participation effects. J Clin Epidemiol 67:267–277

225. Berthelot JM, Le Goff B, Maugars Y (2011) The Hawthorne effect: stronger than the placebo effect? Joint Bone Spine 78:335–336

226. SantaCruz K, Lewis J, Spires T, Paulson J, Kotilinek L et al (2005) Tau suppression in a neurodegenerative mouse model improves memory function. Science 309:476–481

227. Ramsden M, Kotilinek L, Forster C, Paulson J, McGowan E et al (2005) Age-dependent neurofibrillary tangle formation, neuron loss, and memory impairment in a mouse model of human tauopathy (P301L). J Neurosci 25:10637–10647

228. Spires TL, Orne JD, SantaCruz K, Pitstick R, Carlson GA et al (2006) Region-specific dissociation of neuronal loss and neurofibrillary pathology in a mouse model of tauopathy. Am J Pathol 168:1598–1607

229. Yue M, Hanna A, Wilson J, Roder H, Janus C (2011) Sex difference in pathology and memory decline in rTg4510 mouse model of tauopathy. Neurobiol Aging 32:590–603

230. Treit D, Fundytus M (1988) Thigmotaxis as a test for anxiolytic activity in rats. Pharmacol Biochem Behav 31:959–962

231. Wolfer DP, Stagljar-Bozicevic M, Errington ML, Lipp HP (1998) Spatial memory and learning in transgenic mice: fact or artifact? News Physiol Sci 13:118–123

232. Deacon RM (2006) Housing, husbandry and handling of rodents for behavioral experiments. Nat Protoc 1:936–946

233. Schneider LS, Sano M (2009) Current Alzheimer's disease clinical trials: methods and placebo outcomes. Alzheimers Dement 5:388–397

234. Benatar M (2007) Lost in translation: treatment trials in the SOD1 mouse and in human ALS. Neurobiol Dis 26:1–13

235. Rogers DC, Fisher EM, Brown SD, Peters J, Hunter AJ et al (1997) Behavioral and functional analysis of mouse phenotype: SHIRPA, a proposed protocol for comprehensive phenotype assessment. Mamm Genome 8:711–713

236. Morris R (1984) Developments of a watermaze procedure for studying spatal learning in the rat. J Neurosci Methods 11:47–60

237. Morris RGM (1981) Spatial localization does not require the presence of local cues. Learn Motiv 12:239–260

238. Wolfer DP, Lipp HP (2000) Dissecting the behaviour of transgenic mice: is it the mutation, the genetic background, or the environment? Exp Physiol 85:627–634

239. Gass P, Wolfer DP, Balschun D, Rudolph D, Frey U et al (1998) Deficits in memory tasks of mice with CREB mutations depend on gene dosage. Learn Mem 5:274–288

240. Wehner JM, Sleight S, Upchurch M (1990) Hippocampal protein kinase C activity is reduced in poor spatial learners. Brain Res 523:181–187

241. Westerman MA, Cooper-Blacketer D, Mariash A, Kotilinek L, Kawarabayashi T et al (2002) The relationship between Abeta and memory in the Tg2576 mouse model of Alzheimer's disease. J Neurosci 22:1858–1867

242. Chen G, Chen KS, Knox J, Inglis J, Bernard A et al (2000) A learning deficit related to age

and beta-amyloid plaques in a mouse model of Alzheimer's disease. Nature 408:975–979

243. Logue SF, Paylor R, Wehner JM (1997) Hippocampal lesions cause learning deficits in inbred mice in the Morris water maze and conditioned-fear task. Behav Neurosci 111:104–113

244. Bohut MC, Soffié M, Poucet B (1989) Scopolamine affects the cognitive processes involved in selective object exploration more than locomotor activity. Psychobiology 17:409–417

245. Save E, Poucet B, Foreman N, M-C B (1992) Object exploration and reactions to spatial and nonspatial changes in hooded rats following damage to parietal cortex or hippocampal formation. Behav Neurosci 106:447–456

246. Hammond RS, Tull LE, Stackman RW (2004) On the delay-dependent involvement of the hippocampus in object recognition memory. Neurobiol Learn Mem 82:26–34

247. Vnek N, Rothblat LA (1996) The hippocampus and long-term object memory in the rat. J Neurosci 16:2780–2787

248. LeDoux JE (1993) Emotional memory systems in the brain. Behav Brain Res 58:69–79

249. Phillips RG, LeDoux JE (1992) Differential contribution of amygdala and hippocampus to cued and contextual fear conditioning. Behav Neurosci 106:274–285

250. LeDoux JE (2000) Emotion circuits in the brain. Annu Rev Neurosci 23:155–184

251. Repa JC, Muller J, Apergis J, Desrochers TM, Zhou Y et al (2001) Two different lateral amygdala cell populations contribute to the initiation and storage of memory. Nat Neurosci 4:724–731

252. McEchron MD, Bouwmeester H, Tseng W, Weiss C, Disterhoft JF (1998) Hippocampectomy disrupts auditory trace fear conditioning and contextual fear conditioning in the rat. Hippocampus 8:638–646

253. Bourtchuladze R, Frenguelli B, Blendy J, Cioffi D, Schutz G et al (1994) Deficient long-term memory in mice with a targeted mutation of the cAMP-responsive element-binding protein. Cell 79:59–68

254. Garcia J, Hankins WG, Rusinak KW (1976) Flavor aversion studies. Science 192:265–266

255. Revusky SH, Bedarf EW (1967) Association of illness with prior ingestion of novel foods. Science 155:212–214

256. Rozin P, Kalat JW (1971) Specific hungers and poison avoidance as adaptive specializations of learning. Psychol Rev 78:459–486

257. Garcia J, Kimeldorf DJ, Koeling RA (1955) Conditioned aversion to saccharin resulting from exposure to gamma radiation. Science 122:157–158

258. Bures J, Bermudez-Rattoni F, Yanamoto T (1998) Conditioned taste aversion: memory of a special kind. Oxford University Press, Oxford

259. Rosenblum K, Meiri N, Dudai Y (1993) Taste memory: the role of protein synthesis in gustatory cortex. Behav Neural Biol 59:49–56

260. Kruger L, Mantyh PW (1989) Gustatory and related chemosensory systems. In: Björklund A, Hökfelt T, Swanson LW (eds) Integrated systems of the CNS, Part II. Elsevier Science, Amsterdam, pp 323–411

261. Lamprecht R, Dudai Y (1996) Transient expression of c-Fos in rat amygdala during training is required for encoding conditioned taste aversion memory. Learn Mem 3:31–41

262. Lamprecht R, Hazvi S, Dudai Y (1997) cAMP response element-binding protein in the amygdala is required for long- but not short-term conditioned taste aversion memory. J Neurosci 17:8443–8450

Chapter 19

Neuromuscular Disease Models and Analysis

Robert W. Burgess, Gregory A. Cox, and Kevin L. Seburn

Abstract

Neuromuscular diseases can affect the survival of peripheral neurons, their axons extending to peripheral targets, their synaptic connections onto those targets, or the targets themselves. Examples include motor neuron diseases such as Amyotrophic Lateral Sclerosis, peripheral neuropathies such as Charcot-Marie-Tooth diseases, myasthenias, and muscular dystrophies. Characterizing these phenotypes in mouse models requires an integrated approach, examining both the nerve and muscle histologically, anatomically, and functionally by electrophysiology. Defects observed at these levels can be related back to onset, severity, and progression, as assessed by "Quality of life measures" including tests of gross motor performance such as gait or grip strength. This chapter describes methods for assessing neuromuscular disease models in mice, and how interpretation of these tests can be complicated by the inter-relatedness of the phenotypes.

Key words Motor neuron diseases, Amyotrophic lateral sclerosis (ALS), Spinal muscular atrophy (SMA), Peripheral neuropathies, Charcot-Marie-Tooth diseases, Hereditary motor and/or sensory neuropathies (HSMNs), Congenital myasthenic syndromes, Neuromuscular junction, Muscular dystrophies, Duchenne's disease

1 Introduction

In this chapter, we describe genetic models of neuromuscular diseases in mice and the methods of analysis used to understand the underlying pathophysiology. In discussing these diseases, we are specifically considering those that have their primary pathology either in lower motor neurons (those with cell bodies in the ventral horn of the spinal cord, but including their peripheral axons and synaptic terminals) or in the muscle fibers themselves. Other neurological conditions with exclusively upper motor neuron, basal ganglia, or cerebellar involvement are not discussed. We are also emphasizing diseases that affect the peripheral motor system more than those affecting peripheral sensory systems.

This chapter is reproduced from the previous edition. The methods and approaches described have not changed and therefore the chapter remains relevant; however, additional mouse models are emerging constantly and the lists provided should not be considered comprehensive in 2016

Gabriele Proetzel and Michael V. Wiles (eds.), *Mouse Models for Drug Discovery: Methods and Protocols*, Methods in Molecular Biology, vol. 1438, DOI 10.1007/978-1-4939-3661-8_19, © Springer Science+Business Media New York 2016

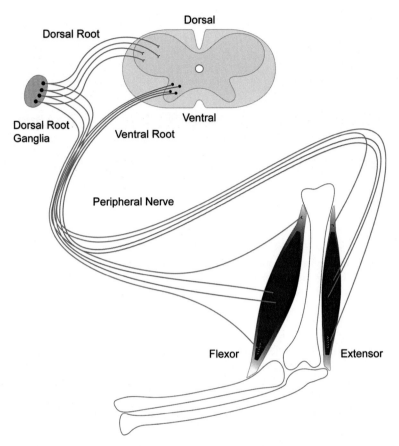

Fig. 1 Anatomy of the neuromuscular system. The neuromuscular system includes the spinal cord, the skeletal muscles, and the motor and sensory nerves that connect them. In cross section, the spinal cord contains motor neuron cell bodies in the ventral horn and sensory neurons in the dorsal horn. Motor axons exit the spinal cord through the ventral roots and join sensory axons in the spinal and peripheral nerves. The cell bodies of the sensory neurons, which are bipolar neurons, are in the dorsal root ganglia and enter the spinal cord through the dorsal roots. In the periphery, motor and sensory fibers are almost always comingled. Major sensory endings relevant to the neuromuscular system include muscle spindles, which sense muscle tension and proprioceptive endings such as Golgi tendon organs. Together with motor neurons, these neurons innervate skeletal muscle, and circuitry within the spinal cord balances force between flexor and extensor muscles; although this is crucial for proper neuromuscular function, it is not discussed in detail here. The diseases discussed in the chapter can generally be grouped into those that affect motor neurons in the spinal cord, those that affect the peripheral axons or synapses (neuromuscular junctions), and those that affect the muscles themselves

The disease models can be grouped into four primary categories (Fig. 1). (1) Motor neuron diseases, in which the death of motor neuron somata in the spinal cord results in denervation of the muscles, progressive flaccid paralysis, and usually premature death. In humans, examples of such diseases would include amyotrophic lateral sclerosis (ALS) and spinal muscular atrophy (SMA). (2) Peripheral neuropathies, in which axonal integrity or conduction are not maintained, resulting in axon degeneration and impaired connectivity of the nervous system and the musculature. These diseases are often less severe than the motor neuron diseases, and although potentially debilitating, they do not necessarily result in shortened lifespan. In humans, these diseases include the Charcot-Marie-Tooth diseases, or hereditary motor and/or sensory neuropathies (HSMNs). These conditions may be the result of demyelination or defects intrinsic to the motor axon itself. (3) Myasthenias, which are diseases of the neuromuscular junction (NMJ). These diseases impair synaptic transmission and can be caused by presynaptic or postsynaptic defects, or by defects in the extracellular matrix of the synaptic cleft, particularly by affecting the localization or activity or acetylcholine esterase (AChE) at the NMJ. In humans, these diseases are the congenital myasthenic syndromes (CMS), which are rare but typically very severe. Adult onset myasthenias, such as myasthenia gravis, tend to be autoimmune and are not considered in this chapter. (4) Muscular dystrophies, which are diseases that affect the muscle fibers themselves, often leading to a loss of sarcolemmal integrity and degeneration of the muscle fibers, although some diseases may progress through other mechanisms. In humans, these diseases include congenital muscular dystrophies, such as Duchenne's, and also limb-girdle muscular dystrophies.

In reality, the diseases in humans and the phenotypes in mice cannot be so cleanly grouped. For instance, ALS results in lower motor neuron degeneration, but also has upper motor neuron involvement, or congenital muscular dystrophy type 1A, which has profound muscular dystrophy, but also severe hypomyelination in the central and peripheral nervous system. In some cases, mouse models fully recapitulate the human disease pathologies, and in other cases there are differences, which are both caveats for disease-oriented research and interesting opportunities for examining disease mechanisms through comparative physiology. However, recognizing the full phenotypic spectrum of these diseases is important for designing effective therapies.

Overtly, the phenotypes of mice with any of these conditions may be quite similar. They are generally smaller, shaky either from weakness or myelination defects, have difficulty fully supporting their weight, and atrophy or wasting, most often in the hindlimbs. The methods we will outline below could be applied towards an existing model, for instance to determine whether a drug treatment or genetic manipulation changes the course of the disease, or to a novel model to determine the basis of an overt neuromuscular phenotype. It should be noted that the overt phenotype is a very sensitive predictor of neuromuscular dysfunction, although it may be subtle, such as in the *mdx* mouse, a model of Duchenne's muscular dystrophy resulting from mutation in the dystrophin gene.

We will first mention useful mouse models in each of the disease categories listed above, and then describe methods for assessing each pathology, as well as general methods, including basic neuromuscular physiology and behavioral tests of gross motor performance.

2   Materials

2.1   Mouse Strains Used in Neuromuscular Disease Research

The tables below list mouse models used in neuromuscular disease research. Strains were included based on their popularity, but also on their uniqueness in cases where a mutation illustrates an apparently distinct disease mechanism. In many cases, the targeted mutations (usually nulls) result in non-neurological phenotypes, in some cases embryonic lethality, and in others, no phenotype. Neuropathy phenotypes in such models are often generated by transgenic expression of human disease alleles, either in wild type or in knockout backgrounds. General caveats in using any of these strains include variation in phenotype in different genetic backgrounds, increased phenotypic variability in hybrid genetic backgrounds, and the existence of multiple alleles for some genes, which may have markedly different characteristics.

1. Motor neuron disease models:
   Table 1: Motor neuron disease models.

2. Peripheral neuropathy models:
   Table 2: Peripheral neuropathy models.

3. Myasthenia models:
   Table 3: Myasthenia models.

4. Muscular dystrophy models:
   Table 4: Muscular dystrophy models.

5. Research tools:
   Table 5: Research tools.

2.2   Histology

1. Hematoxylin and Eosin stain (H&E).
2. Cresyl violet/luxol fast blue stain (CV/LFB).
3. Bouin's fixative.
4. Decalcifying solution (Fisher Cal-Ex).
5. Paraffin.
6. Xylene.
7. Ethanol series (70% to 95% to 100%).

Table 1
Motor neuron disease models

Gene	Cellular function	Tg/KO/knockin/spontaneous	Human disease	Mouse phenotype	Alleles	References
Ighmbp2	RNA/DNA helicase	Sp	DSMA1	Motor neuron degeneration, cardiomyopathy ~8–10 weeks lifespan	*nmd1, nmd2J*	[2, 39, 45, 46]
Smn1	Assembly of spliceosomal snRNP complexes	Tg + KO	SMA	KO is embryonic lethal, addition of Tg-SMN2, SMN2^{Δ7} and/or SMN1^{A2G} produces lifespans ranging from 2 to 227 days	Tg-SMN2, Tg-SMN2^{Δ7}, Tg-SMN1^{A2G}; *Smn1tm1Msd*	[47–50]
Tbce	Tubulin-specific chaperone	Sp	Kenny-Caffey syndrome/HRD	Caudocranial degeneration of motor axons, ~7 weeks lifespan	*pmn*	[51, 52]
Dctn1	p150(glued), component of dynein/dynactin complex	KO/KI, Tg	dSBMA	Tg and KI hets, late-onset (>10 months), slowly progressive weakness and altered NMJ integrity	*Dctn1tm1Cai*; Tg-DCTN1(G59S)	[53–55]
Dctn2	P50(dynamitin) component of dynein/dynactin complex	Tg	Unknown	Late-onset (>10 months) slowly progressive motor neuron degeneration and denervation of muscle	Tg-Dctn2	[56]
SOD1	Dominant gain of toxic function unrelated to dismutase activity	Tg	ALS	Motor neuron degeneration, distal axonopathy, and NMJ denervation	mSod1-G86R; hSOD1-G93A, G85R, G37R, H46R/H48Q	[57–63]

(continued)

Table 1
(continued)

Gene	Cellular function	Tg/KO/knockin/ spontaneous	Human disease	Mouse phenotype	Alleles	References
Cln8	ER localized, lipid synthesis, transport, or sensing	Sp	CLN8; EPMR—epilepsy, progressive with mental retardation	Neuronal ceroid lipofuscinoses and retinal photoreceptor degeneration	*mnd*	[64–66]
Dync1h1	Axonal transport	Sp	Unknown	Hets: sensory neuropathy with muscle spindle deficiency. Hom: Neonatal lethality	*Loa, Cra1, Swl*	[67–69]
Fig4	Phosphoinositide phosphatase	Sp	CMT4J, ALS and PLS	Pale coat color, axonal degeneration-motor and sensory, limited segmental demyelination	*plt*	[70–72]
Vps54	Golgi-associated retrograde protein (GARP) complex of vesicle sorting proteins	Sp, KO	Unknown	Vacuolization and degeneration of brainstem and ventral horn neurons; KO embryonic lethal	*wr, Gr(RRl497)Byg*	[73]

Examples of mouse mutations resulting in the loss of ventral horn motor neurons are given. In many instances, these models are created by the transgenic overexpression of human disease genes, such as *SOD1* alleles that cause ALS. In such cases, the loss-of-function mouse mutation frequently does not cause the same motor neuron disease. The mouse models may also vary in phenotype, not only because of the different alleles introduced, but also because of transgene expression levels (copy number) and integration sites. In other cases, the human and mouse phenotypes simply differ, such as *Tbce*, where human mutations cause congenital hypoparathyroidism, mental retardation, facial dysmorphism, and extreme growth failure [40], whereas the mouse mutations cause motor axon degeneration. The genetics of *Smn1* in the mouse is also complicated. Mice have only one *Smn* gene, while humans have two. To replicate the human SMA condition, mouse models frequently involve the transgenic expression of human *SMN2* and/or variant forms of human SMN1 in an *Smn1* knockout background. Also noteworthy is the wide array of cellular functions these genes normally serve. Some have a logical association with motor neurons, such as those involved in axonal transport, but others are ubiquitous "house keeping" genes that selectively affect motor neurons when mutated

Table 2
Peripheral Neuropathy disease models

Gene	Cellular function	Tg/KO/ spontaneous	Human disease	Mouse phenotype	Alleles	References
Demyelinating, Type 1 CMTs						
Pmp22, Peripheral myelin protein 22 kDa	Peripheral myelin packing	Sp, KO, ENU	CMT1A	Demyelinating neuropathy	Trembler (3)	[74]
Mpz, Myelin Protein Zero	Peripheral myelin component, transmembrane cell adhesion	KO, Tg	CMT1B	Demyelinating neuropathy, hypomyelination	Null, Tg-WT, S63C, S63del	[75–77]
Egr2, early growth response2	Zinc finger transcription factor, immediate early gene	KO, KI	Congenital hypomyelinating neuropathy, CMT1D	Embryonic lethal (null), lack of CNS, PNS myelination	Several, including null and Egr2^{lo}	[78, 79]
Gjb1, Gap junction protein beta1	Connexin32, gap junction component	KO, Tg	X-linked CMT	Late-onset demyelinating neuropathy	Null, Tg-WT and R142W	[80–82]
Axonal, Type 2 CMTs						
Mfn2, Mitofusin2	Mitochondrial fusion/transport	Tg	CMT2A	In Tg, Motor neuropathy, hindlimb weakness	Null, Tg-T105M	[83]
Gars, Glycyl tRNA synthetase	Glycine to tRNA^{gly} aminoacylation	Sp, KO	CMT2D SMA-DV	Loss of large diameter sensory and motor axons in periphery	Nmf249 XM256	[3]

(continued)

Table 2
(continued)

Gene	Cellular function	Tg/KO/ spontaneous	Human disease	Mouse phenotype	Alleles	References
Nefl Neurofilament light chain	KO, Tg	CMT2E	Impaired regeneration (null), motor neuron loss (Tg)	Null, Tg.Nefl[pro]	[84, 85]
Hspb1 Heat shock protein1, 27 kDa	KO	CMT2F	No phenotype (expression pattern?)	Null	[86]
Recessive CMTs						
Prx, periaxin	PDZ-scaffolding protein	KO	CMT4F	Demyelination, neuropathic pain, shortened internodal distance	Null	[4, 87]
Fig4		Sp	CMT4J		Pale tremor	[71]

Examples of demyelinating (type 1), X-linked, axonal (type 2), and recessive (type 4) Charcot-Marie-Tooth diseases are given. The demyelinating and recessive models fairly accurately reproduce the human disease phenotypes caused by mutations in their orthologous genes. For genes such as *Pmp22* and *Egr2*, there are many alleles in mice, with some phenotypic differences. This is particularly the case for *Egr2*, which is involved in numerous developmental pathways, and a demyelinating neuropathy is present only in a partial-loss-of-function allele (*Egr2lo*); null mice have hypomyelination in the CNS and PNS, but do not survive long enough to study. Interestingly, the peripheral neuropathy phenotype is more difficult to reproduce in the axonal models. Motor neuron phenotypes are observed in transgenic mice overexpressing mutant forms of *Mfn2* or *Nefl*, but these phenotypes are not evident in straight loss-of-function alleles. The *Hspb1* mutant mice do not have a peripheral neuropathy, which may again indicate a requirement for expression of a mutant form of the protein, or it may reflect differences in expression pattern and function between mice and humans. The *GarsNmf249* mouse provides a phenotypically accurate axonal CMT model. The allele was identified in a phenotype-driven screen for neuromuscular disease mutations, and no dominant phenotype is observed in mice lacking *Gars* expression at the RNA level. This suggests the mutant form of the protein is required to cause the phenotype and is an argument in favor of phenotype-based approaches to generating accurate disease models. Note that some mutations, such as the aggregate-forming *Neflpro* transgenics and other alleles of *FIG4* in humans, can cause motor neuron loss and not just axonopathy, possibly indicating that some axonal neuropathies may in fact be part of a spectrum of motor neuron disease manifestations. It should also be noted that there are several identified CMT genes in humans that do not yet have mouse models, such as *RAB7*, *GDAP1*, *HSPB8*, and *YARS*

Table 3

Myasthenic syndrome disease models

Gene	Cellular function	Tg/KO/ spontaneous	Human disease	Mouse phenotype	Alleles	References
Presynaptic						
Chat, Choline acetyl-transferase	ACh synthesis	KO	CMS with episodic apnea	No synaptic transmission, neonatal lethal	Conditional null	[88–90]
Vacht, Vesicular acetylcholine transporter	ACh vesicular transport	KO		Myasthenia, behavioral changes	Partial loss of function	[91]
Synaptic						
Colq, Collagenic tail of AChE	Collagen like tail, AChE anchoring	KO	Endplate AChE deficiency	Severe myasthenia, weakness, reduced viability	Null	[92]
Ache, Acetylcholine esterase	ACh degradation	KO		Severe myasthenia, weakness, reduced viability	Null	[93, 94]
Synaptic differentiation/structure, trans-synaptic signaling						
Agrn, Agrin	Nerve-derived differentiation factor	KO		Neonatal lethal, no postsynaptic differentiation	Deletion, isoform specific, conditional	[95–99]
Musk, Muscle-specific kinase	Agrin receptor (with Lrp4), tyrosine kinase receptor	KO	CMS with AChR deficiency	Neonatal lethal, no postsynaptic differentiation	Null (and autoimmune)	[100]
Lrp4, LDL-receptor related protein 4	Agrin receptor	ENU		Neonatal lethal, no postsynaptic differentiation	Null	[101]

(continued)

Table 3
(continued)

Gene	Cellular function	Tg/KO/ spontaneous	Human disease	Mouse phenotype	Alleles	References
Rapsn, Rapsyn	Intracellular scaffolding of AChRs	KO	CMS with AChR deficiency	Neonatal lethal, no AChR clustering	Null	[102]
Dok7, Downstream of Kinase-7	Intracellular adaptor protein	KO	CMS1B, limb girdle	Neonatal lethal, no postsynaptic differentiation	Null	[103]
Lamb2, Laminin-beta2	Extracellular matrix protein	KO	Pierson syndrome (kidney)	Failure of NMJ maturation, death at 4 weeks	Null	[104]
Acetylcholine receptor (AChR) defects (postsynaptic)						
Chrnb1, Acetylcholine receptor beta1	AChR subunit	KI, phosphorylation deficient	CMS with AChR deficiency and slow channel	NMJ morphology changes	3 intracellular tyrosines	[105]
Chrne, Acetylcholine receptor epsilon subunit	AChR subunit (adult)	KO	CMS with AChR deficiency, fast and slow channel	Death at 2–3 months with progressive weakness	Null	[106, 107]
Chrng, Acetylcholine receptor Gamma subunit	AChR subunit (embryonic)	KO, KI	Escobar syndrome	Neonatal lethal (null). Broadened endplate band (KI)	Null, epsilon knockin	[108–110]

Examples of mutations causing myasthenia phenotypes are given, including those with presynaptic, synaptic, trans-synaptic signaling, or postsynaptic functions. In several cases, mouse mutations with relevant phenotypes, or in relevant pathways, are listed in the absence of an established human disease caused by mutations in the orthologous gene (*Vacht, Ache, Agrn, Lrp4*). In addition, the mouse mutations are frequently a complete loss-of-function (knockout), whereas the human mutations are a partial-loss-of-function (*Rapsn, Musk, Dok7*). Such models are useful for understanding gene function, but accurate models of the associated human disease will require more specific alterations, such as generation of knockin alleles recreating the human genetic changes. In some cases, loxP-conditional alleles offer compromise models, where the timing and extent of the loss-of-function can be controlled, even though the alleles created are complete loss-of-function following cre-mediated excision (*Chat, Agrn*). Elimination of *Lamb2* has both pre- and postsynaptic effects in mice, but human mutations are characterized by kidney dysfunction and not myasthenia (Pierson syndrome). Kidney defects are also observed in the *Lamb2* mutant mice [41]. In humans, mutations in nicotinic acetylcholine receptor subunits (*CHRNA, CHRNB1, CHRND, CHRNG, CHRNE*) cause a variety of myasthenic syndromes. The mouse models of these diseases are incomplete largely because the function of these genes is well understood (encoding ligand-gated ion channels), and the biophysical properties of variants can be effectively studied using in vitro and heterologous systems

Table 4

Muscular dystrophy disease models

Gene	Cellular function	Tg/KO/KI/ENU/spontaneous	Human disease	Mouse phenotype	Alleles	References
Sarcolemmal/ ECM						
Dmd	DGC associated, sarcolemmal integrity	Sp, ENU, KO	Duchenne and Becker muscular dystrophy	Dystrophic pathology by 2 weeks, particularly progressive in diaphragm	*mdx, -2Cv, -3Cv, -4Cv, -5Cv; tm1Mok, tm1Khan*	[111–117]
Large	Glycosylation of α-dystroglycan	Sp	MDC1D	Progressive dystrophy and ocular abnormalities	*myd, vls*	[118, 119]
Lama2	ECM	Sp, ENU, KO	MDC1A	Variable severity among alleles, progressive dystrophy and hypomyelination	*dy, -2J, -3J, -6J, -7J; tm1Eeng, tm1Stk*	[120–124]
Dysf	Membrane repair	Sp, KO	LGMD2B, Miyoshi myopathy	Dystrophic pathology by 3 months, progressive necrosis and fatty infiltration	*im* (SJL/J), *prmd* (A/J); *tm1Kcam, tm1Mcho*	[125–127]
Dtna	DGC associated	KO	Left ventricular noncompaction (LVNC)	Dystrophic pathology by 1 month and progressing to 6 months	*tm1Jrs*	[128]
Col6a1	Type VI collagen	KO	Bethlem myopathy and Ullrich congenital muscular dystrophy	Dystrophic pathology as early as 3 days; mitochondrial dysfunction	*tm1Gmb*	[129, 130]
Col15a1	Type XV collagen	KO	Unknown	Dystrophic pathology after 3 months; microvessel damage	*tm1Pih*	[131]
Cav3	Structural protein of caveolae	KO	LGMD1C	Mild myopathic changes; dilated and longitudinally oriented T-tubules	*tm1Mls*	[132]

(continued)

Table 4
(continued)

Gene	Cellular function	Tg/KO/KI/ENU/spontaneous	Human disease	Mouse phenotype	Alleles	References
Cytoskeletal/sarcomeric						
Acta1	Skeletal actin	KO	Nemaline myopathy - 3	Neonatal lethal ~10 days	*tm1Jll*	[133, 134]
Ttn	Sarcomere assembly and passive tension	Sp, ENU, KO	LGMD2J, TMD	Progressive dystrophy, onset by 2 weeks, 8–10 weeks lifespan. KO embryonic lethal	*mdm; sbrn; tm1Her*	[135–137]
Cryab	Alpha-B crystallin	KO	Myofibrillar myopathy	Muscle wasting and dystrophic pathology after 40 weeks	*tm1Wawr*	[138]
Bag3	Regulate Hsp70 family molecular chaperones	KO	Myofibrillar myopathy	Myofibrillar degeneration ~4 weeks lifespan	*Gt(OST16086)Lex*	[139, 140]
Des	Intermediate filament; Z-discs	KO	Desmin-related myopathy	Cardiac, skeletal, and smooth muscle myofibrillar defects	*tm1Cap, tm1Cba, tm1Ltho*	[141–143]
Myot	Z-disc protein	Tg, KO	LGMD1A	Myofibrillar myopathy, KO no pathology	*Tg-Myot(T57I); tm1Moza*	[144, 145]
Other						
Chkb	Phosphatidyl choline synthesis	Sp	Unknown	Rostrocaudal progressive dystrophy with fatty infiltration	*rmd*	[146]
Mtm1	PI(3)P - phosphatase	KO	X-linked myotubular myopathy	Progressive myopathy by 4 weeks with a 6–14 weeks lifespan	*tm1Jman*	[147]
Lmna	Component of nuclear lamina	KI, KO	EDMD2 and EDMD3	Knockin mutations develop later dystrophy, 9 months lifespan; KO alleles lifespan ~8 weeks	*LmnaH222P/H222, tm1Stw*	[148, 149]

Gene	Function	Type	Disease	Phenotype	Allele(s)	Ref.
Cox10	Mitochondrial electron transport	Conditional KO	COX deficiency	Slowly progressive myopathy after 3 months	*tm1Ctm*	[150]
Dmpk	CUG repeat expansion in 3′ UTR	Tg, KO	Myotonic dystrophy 1 (DM1)	Myotonia; myonuclear RNA foci	*Tg-HSALR*, *Tg-DM300*, *Tg-Dm960*, *tm1Rdd*	[151–154]
Mbnl1	CUG RNA binding protein	KO	Myotonic dystrophy	Myotonia; myonuclear RNA foci	*MbnlΔ3/Δ3* (*tm1Sws*)	[155]
Gne	Sialic acid biosynthesis	KO/Tg	Distal myopathy with rimmed vacuoles (DMRV)	Decreased motor performance >30 weeks, inclusion bodies	*tm1Sngi*, Tg-GNE(D176V)	[156]

Examples of mouse mutations exhibiting a muscular dystrophy phenotype are presented. A prevalent mechanism for these mutations is a failure in linking the extracellular matrix to the intracellular cytoskeleton via components of the dystroglycan glycoprotein complex (DGC), resulting in compromised sarcolemmal integrity. Similar dystrophies are caused by defects in membrane repair, which is critical for maintaining mechanically contracting muscle fibers. A second class of genes causing muscular dystrophy includes those associated with the structure of the sarcomere and the cytoskeleton. Finally, a variety of other examples are given that arise through more disparate or poorly understood mechanisms ranging from phospholipid biosynthesis to RNA binding. In many cases, multiple alleles exist for any given gene, often with varying phenotypes that correlate with the severity of the loss-of-function. For example, *Lama2* was first identified as dystrophia muscularis (*dy*) in 1955, an allele that is a severe hypomorph. Additional alleles including milder, partial-loss-of-function (*dy2J*, *dy7*) and a complete null (*tm1Stk*, also called *dy3K*) have been identified or engineered since. The appropriate allele for use in research should be chosen based on the experiment; for example, the severity of a hypomorphic mutation could be either enhanced or suppressed, and in contrast, a null allele has no retained activity so any suppression has to come through other pathways or activation of downstream factors

Table 5
Strains useful for neuromuscular research

Strain name	Jax #	Phenotype/utility	References
Cre transgenic strains			
Nestin-cre, B6.Cg-Tg(Nes-cre)1Kln/J	3771	Pan-neuronal expression of cre from embryonic ages	[157]
HB9-cre, B6.129S1-Mnx1^tm4(cre)Tmj/J	6600	Motor neuron-specific expression of cre from embryonic ages	[158, 159]
HSA-cre	NA	Muscle-specific expression of cre from myoblast stage (including satellite cells in adult)	[160]
Mck-cre	NA	Muscle-specific expression of cre from early myotube stage	[161]
Myf5-cre B6;129S4-Myf5tm3(cre)Sor/J	7845	Knockin allele, skeletal muscle and dermis expression of cre	[162]
P0-cre	NA	Schwann cell specific expression of cre	[163]
SLICK mice, B6.Cg-Tg(Thy1-cre/ESR1,-EYFP)Gfng/J	7606, 7610	Inducible cre (ESR-cre) expression coupled to YFP expression, driven by Thy1 promoter	[164]
Fluorescent protein strains			
Thy1-YFP-16 B6.Cg-Tg(Thy1-YFP)16Jrs/J	3709	YFP expression in all motor neurons embryonic ages	[165]
Thy1-YFP-H, B6.Cg-Tg(Thy1-YFPH)2Jrs/J	3782	YFP expression in sparse subset of motor neurons beginning at 3–4 weeks of age (useful for tracing motor units)	[165]
Thy1-CFP-23 B6.Cg-Tg(Thy1-CFP)23Jrs/J	3710	CFP expression in all motor neurons	[165]
Thy1-Mito-CFP B6;CB-Tg(Thy1-CFP/COX8A)Lich/J	7940, 6614, 6617	Mitochondrially localized CFP in all motor neurons, useful for transport/trafficking studies	[166]

S100-XFP, B6;D2-Tg(S100B-EYFP)1Wjt/J	5620, 5621	YFP of GFP expression in Schwann cells	[167]
HB9-GFP, B6.Cg-Tg(Hlxb9-GFP)1Tmj/J	5029	Motor neuron-specific expression of GFP (also useful differentiation/cell type marker in vitro)	[168]
Brainbow mice B6.Cg-Tg(Thy1-Brainbow1.0)Lich/J B6.Cg-Tg(Thy1-Brainbow2.1)Lich/J	7901, 7910, 7911, 7921	Multicolored neuronal expression	[169]
Other			
Wlds Wallerian degeneration slow	NA	Slowed distal axonal degeneration in response to a variety of insults	[170]

A variety of strains that are useful in neuromuscular disease research are listed. *Cre* transgenic strains mediate the deletion of sequences flanked by loxP sites ("floxed" sequences), the most common construct used in condition mutations in mice. Tissue specificity is achieved by selectively driving cre expression, and both neuron-specific and muscle-specific strains are listed. In addition, temporal control can be achieved by using inducible cre systems such as the SLICK mice. A second class of transgenic mice that greatly simplify analysis is those expressing markers such as fluorescent proteins (largely GFP derivatives) in selected cell types. Many of these strains incorporate the same DNA construct and variable expression occurs as a result of genomic insertion site of the transgene (position effect variegation), for example, many constructs use the Thy1 promoter to drive expression in the nervous system, but the precise expression pattern of each of these strains is distinct. This also highlights a caveat for anyone considering transgenic studies; the expression level and pattern of each transgene must be confirmed before results can be reliably interpreted. Finally, the Wlds allele is an interesting spontaneous genomic rearrangement that creates a fusion of the *Ube4e* and *Nmnat* genes. This mutation suppresses distal axonal degeneration in a number of acute (injury) and genetic models. It appears that the *Nmnat* portion of the fusion may be primarily responsible for conferring this activity, and it is proposed to function as a chaperone and to prevent mitochondrial dysfunction [42–44]. This strain can therefore be useful for exploring pathogenic mechanisms in different neurodegenerative conditions

2.3 Assessing Motor Neuron Loss	1. Strong fixative: 2% paraformaldehyde, 2% glutaraldehyde in 0.1 M cacodylate buffer.
	2. No. 2 forceps.
	3. Toluidine Blue.
2.4 Internodal Distance Assessment	1. No. 5 forceps.
	2. 30-gauge needle.
2.5 Light Microscopy of the NMJ	1. 2% paraformaldehyde (Electron Microscopy grade).
	2. Anti-neurofilament marker, e.g., the monoclonal antibody 2H3 and anti-SV2 (Developmental Studies Hybridoma Bank, University of Iowa, *see* http://dshb.biology.uiowa.edu).
	3. Postsynaptic receptor marker, e.g., fluorescent conjugates of alpha-bungarotoxin (Molecular Probes, Invitrogen).
	4. Schwann cell marker: anti-S100 antibody.
2.6 Electron Microscopy of the NMJ	1. Paraformaldehyde (Electron Microscopy grade).
	2. Fixative: 2% paraformaldehyde, 2% glutaraldehyde in 0.1 Mm cacodylate buffer.
2.7 Evan's Blue Staining	1. Sterile saline solution.
	2. Evan's Blue (Sigma), dissolved in sterile saline at 10 mg/ml.

3   Methods

The protocols below provide an intermediate level of detail. Specifics are provided when they are critical to the success of the experiment, but otherwise should be sought in the references. Our goal is to provide a list of experimental approaches for assessing neuromuscular function, and to illustrate caveats and possible interpretations so that appropriate controls can be used and accurate conclusions can be drawn.

3.1   Motor Neuron Diseases

3.1.1   Histological Analysis of the Spinal Cord

1. Dissect the vertebral column. Note: the cell bodies of lumbar motor neurons are more rostral than their exit points, i.e., L4 cells bodies are in upper lumbar or even lower thoracic vertebrae, even though the spinal nerve exits at the fourth lumbar vertebrae. The more caudal vertebral column contains only the cauda equina, the dorsal and ventral roots of lower lumbar and sacral motor neurons (Fig. 2).

2. Fix the spinal cord in the intact vertebral column using Bouin's fixative, either by immersion after dissection or by transcardial perfusion prior to dissection. The bone of the vertebrae will eventually decalcify in Bouins (requiring 2–4 weeks), or after

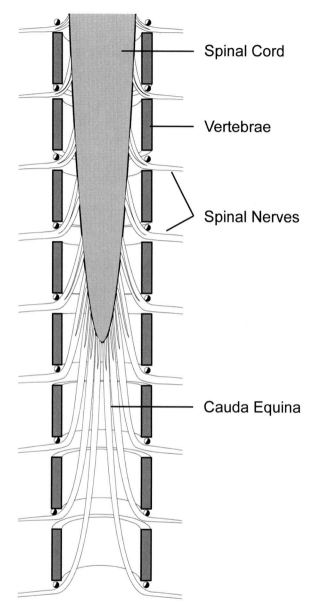

Fig. 2 The spinal cord and cauda equina. The caudal spinal cord does not completely fill the caudal vertebral column. As a result, the dorsal and ventral roots of lumbar and sacral neurons are very long, spanning from the more rostral cell bodies to the more caudal exit points. Therefore the dissected spinal cord resembles a horse's tail. The practical effect of this anatomy is that it becomes very hard to reliably determine the level of the spinal cord that is being studied in any given cross section

24 h, the sample can be transferred to decalcifying solution overnight.

3. Rinse the sample thoroughly (>12 h) in water and process for paraffin embedding (dehydrate through an ethanol series and then to xylene, do not allow nervous system samples to sit longer then necessary in ethanol or white matter tracts will look like Swiss cheese).

4. Embed the samples for cross section (horizontal section).

5. Stain 4–5 μm sections using standard hematoxylin and eosin protocols (H&E) or with cresyl violet/luxol fast blue (CV/LFB). H&E stains protein rich cells (eosin) and counterstains nuclei (hematoxylin). CV/LFB also stains protein, but also stains myelin. If available, an automated tissue processor is recommended for consistency of staining.

6. Interpretation: Dying cells in the ventral horn of the spinal cord may be distinguished by loss of Nissl substance in the cell bodies, by pyknotic nuclei, and by pink cytoplasm in cells that are early in the degeneration process (see **Note 1**).

3.1.2 Assessing Motor Neuron Number (Loss)

In the spinal cord, motor neuron number can be assessed in cross sections as described above. However, there are two primary challenges. First, motor neuron cell bodies need to be accurately identified in the ventral horn. Markers such as HB9-GFP, Thy1.2-YFP, or ChAT can be useful to positively identify motor neurons, but these are often incompatible with strong fixatives, paraffin embedding, or histological counterstains. A second challenge is being sure that comparisons are made at exactly the same level of the spinal cord, which is difficult considering the cord does not fill the full rostral/caudal extent of the vertebral column and the number of motor neurons per cross section varies greatly with level. An easier approach that we prefer is to count axons in the ventral roots as they exit the spinal cord (Fig. 3). The ventral roots can be dissected, embedded, cross-sectioned, and myelinated axons can be counted after staining. A decrease in axon number should reflect a loss in motor neurons in the ventral horn.

1. Perfuse an animal transcardially with strong fixative (2 % paraformaldehyde, 2 % glutaraldehyde in 0.1 M cacodylate buffer, standard electron microscopy fixative).

2. Expose the sciatic nerve dorsally adjacent to the femur and follow it proximally to the sciatic notch where it disappears under the pelvis.

3. Using a pair of fine blunt-point scissors carefully split pelvis in the direction followed by the nerve and gently separate the split bone. Continue to follow the nerve until branches into L6, the L5 and L4 spinal nerves are evident (N.B. L6 is most distal and smallest).

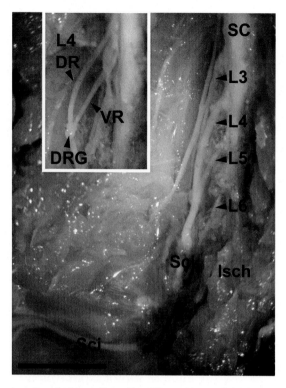

Fig. 3 Dorsal and ventral root dissection. The anatomy of the ventral and dorsal roots during dissection is shown. Mouse is prone (rostral is up) and the left sciatic nerve (Sci) has been exposed along the femur and to its origin at the lumbar spinal nerves. The spinal nerves are labeled (Lumbar 3 through 6). The ischium (Isch) has been removed to follow the sciatic nerve. The inset shows the L4 root separated into its ventral and dorsal roots, note the DRG associated with the dorsal root. The scale bar is 6 mm

4. Perform a hemilaminectomy (removing the dorsal bone of the vertebrae) in the lumbar region using a pair of No. 2 forceps to remove small pieces of bone until spinal cord is exposed from the dorsal midline laterally to the entry point of the spinal nerves.

5. Grasping each spinal nerve, follow it to its bifurcation into the ventral (motor) and dorsal (sensory) roots (*see* **Note 2**).

6. Cut the ventral root free at the bifurcation point and as close to the spinal cord as possible (NB ventral roots enter cord proximal to spinal nerve entry point).

7. Process for plastic embedding (as for electron microscopy).

8. Cut 500 nm cross sections near the middle of the sample to avoid dissection damage and stain with Toluidine Blue.

The most straightforward interpretation is that decreased axon number will reflect death of motor neurons in the spinal cord. *See* Fig. 4 for an example of this analysis. This interpretation is strengthened

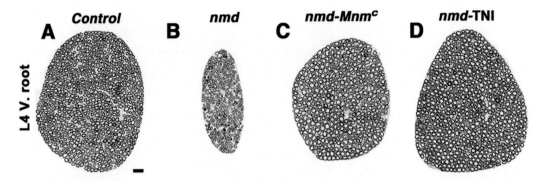

Fig. 4 An example of motor neuron loss reflected in ventral root axon number. The L4 ventral root stained for myelin is shown in cross section in (**a**) a control mouse (**b**) an nmd mouse (Ighmbp2 mutant), (**c**) an nmd mouse carrying a modifier locus from CAST, and (**d**) an nmd mouse with transgenic rescue of the Ighmbp2 gene driven in the nervous system. For details, *see* Maddatu et al. [39]

if combined with spinal cord histology described above. Degenerating axon profiles may also be seen, which may indicate an axonopathy in the absence of evidence for dying cells in the ventral horn. An often-invoked complication to this analysis is that motor axons may branch and this would mask the loss in axon number. Motor axons typically branch only after they have entered the muscle. While a pathological state may cause them to branch at the level of the ventral root, we have not encountered any examples of this, although it is a formal possibility and a caveat to this analysis. Furthermore, if axon loss is seen, the worst-case scenario is that the number is an underestimate due to branching.

Degenerating axons in the peripheral nerves can also be examined as described below, and electrophysiological estimates of motor unit number could be informative. Motor units are defined as a single motor neuron (axon) and the muscle fibers its terminals innervate. In motor neuron diseases, the number of motor units is anticipated to decrease as motor neurons die, but the size of motor units may increase with compensatory sprouting and reinnervation (e.g., [1]). Interpretation can be further confounded by factors such as a change in muscle fiber number or innervation of muscle fibers by multiple motor axons. Therefore, the best interpretation results from corroborating evidence from the spinal cord, nerve, synapse, and the muscle.

3.2 Peripheral Neuropathies

The femoral nerve provides an excellent system for evaluating peripheral neuropathy, provided there is hindlimb involvement. In combination with counts of ventral roots, femoral axon counts can be used to distinguish peripheral neuropathy from motor neuron death. The nerve has a primarily motor branch that innervates the quadriceps, and a primarily sensory branch that becomes the saphenous nerve more distally (Fig. 5).

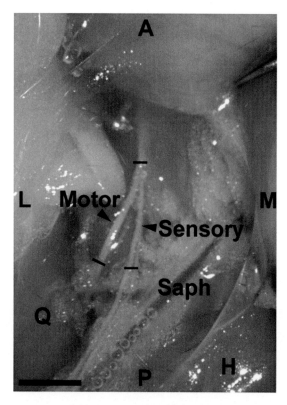

Fig. 5 Femoral nerve dissection. The motor and sensory branches of the femoral nerve are exposed. The mouse is supine and the right hip is shown (forceps are retracting the abdominal wall, A, P, M, L are anterior, posterior, medial, and lateral respectively, H is hamstring muscles). Some adipose tissue has been removed for clarity. The motor branch of the femoral nerve innervates the quadriceps (Q). The sensory branch becomes the saphenous nerve, which runs adjacent to the saphenous vein (Saph) on the medial side of the thigh. Dissecting the nerve where the tick marks provides a reasonable length of nerve to work with. Note the sensory branch sometimes runs as two fascicles and both should be taken to get reproducible counts. The scale bar is 2 mm

Each branch can be easily dissected free (the animals can be transcardially perfused before, or the nerves can be fixed by immersion after dissection if care is taken to be sure they are extended at full length when immersed).

1. Nerves should be plastic embedded and cross-sectioned as above for the ventral roots.

2. Axons can be counted from Toluidine Blue stained sections.

3. The distribution of axon diameters, myelin thickness, and G-ratios (inner/outer diameters, the inner being the axoplasm, the outer including the myelin) can also be determined. This may be done most accurately by low magnification (4000–6000×) transmission electron microscopy.

4. The assessment of axon diameters may reveal general axonal atrophy, or a missing class, such as large diameter, fast motor axons.

5. The assessment of myelin layering and myelin thickness may reveal a demyelinating or hypomyelinating neuropathy.

6. The G-ratio may indicate abnormal reciprocal signaling between the axon and the myelinating Schwann cell, or may highlight thin myelin or conversely, thin axons, since there is normally a rough correlation between axon diameter and myelin thickness.

7. Other peripheral nerve pathologies such as onion bulbs (indicating rounds of demyelination/remyelination), Schmidt-Lanterman Incisures (indicating abnormal myelin packing), myelinating Schwann cells wrapping multiple axons, and bundles of regenerating axons can also be seen in these sections. For examples of these pathologies, *see* Fig. 6 and references such as [2, 3].

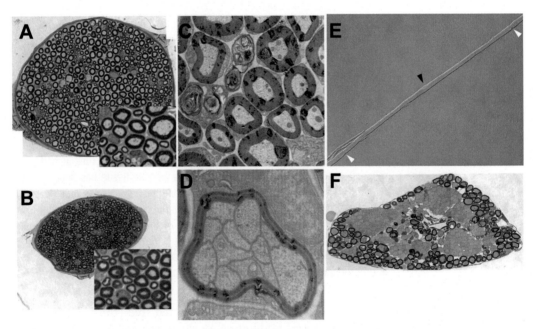

Fig. 6 Peripheral neuropathy phenotypes. A cross section of the motor branch of the femoral nerve in a control mouse (**a**) and a *GarsNmf249/+* model of Charcot-Marie-Tooth 2D (**b**) are shown. The *insets* highlight the varied axon diameters in a control nerve and the almost complete absence of large diameter axons in the mutant nerve. (**c**) The same mutation examined by transmission electron microscopy demonstrates degenerating axon profiles. Note, the irregularities in the myelin are fixation artifact and not pathology, highlighting the need to always process control samples in parallel. (**d**) A myelinating Schwann cell that has ensheathed multiple axons, probably representing a failure in radial sorting during early postnatal development. Note, this is different from a Remak bundle of small sensory axons, in which a non-myelinating Schwann cell enwraps a number of axons in a single basal lamina (*see* the *bottom right corner* of **c**). (**e**) Nodes of Ranvier can be examined by light microscopy in teased nerve preparations. The nucleus of the Schwann cell (*black arrow*) is typically midway between the nodes (*white arrows*). (**f**) An example of hypomyelination in the ventral root of a *Lama2dy/dy* mouse. Normally, 100 % of the ventral root axons are myelinated. In this mutation, bundles of large but unmyelinated axons are evident

| 3.2.1 *Internodal Distance*
Assessment | In addition to axon loss/atrophy and defects in myelination, the intermodal distance can also affect nerve conduction velocities. |

1. To determine intermodal distance, dissect a 1–2 cm segment of peripheral nerve such as the femoral or sciatic, and fix as above (*see* Subheading 3.1.2).

2. Tease the nerve longitudinally to individual fibers using No. 5 forceps or a 30-gauge needle. Keeping the nerve immersed in a drop of PBS, tease the nerve directly on a microscope slide.

3. Coverslip the teased nerves and view using Nomarski-DIC optics.

4. Measure intermodal distances and correlate them with axon diameters.

5. Again, there should be a rough correlation, with larger axons having longer intermodal distances. This analysis requires software calibrated for digital image analysis to determine the distances (Fig. 6).

| 3.2.2 *Nerve Conduction*
Velocities | Nerve conduction velocities are used diagnostically to distinguish type 1 (demyelinating) and type 2 (axonal) neuropathies in humans. Type 1 neuropathies typically have pathologically reduced NCVs (below 30 m/s compared to normal values near 50 m/s in humans). Nerve conduction velocities can also be measured in the mouse as described below. However, significant decreases in axon diameter or internodal distance can also contribute to decreased NCV [3, 4]. Furthermore, NCVs record the fastest (largest) axons present and may therefore miss axonal pathologies that do not cause a marked decrease in these neurons. Therefore, this functional measure should again be combined with an examination of axon diameters, myelination, and intermodal distance to determine the underlying mechanism. |

| **3.3 Myasthenias** | Myasthenias are diseases of the NMJ. This synapse is highly accessible, highly stereotyped in its morphology, and easily visualized by light or electron microscopy. Evaluation of the morphology of the junctions is generally very well correlated with function, although electrophysiology may be required to fully assess synaptic transmission and to determine if defects are pre- or postsynaptic. |

| 3.3.1 *Staining NMJs*
for Light Microscopy | Neuromuscular junctions can be visualized by light microscopy following labeling of the presynaptic nerve terminal and the postsynaptic acetylcholine receptors. For best results, muscles should be prepared for longitudinal sections, and NMJs in an en face orientation can be imaged. In almost all muscles, the endplate band is near the middle of the muscle and this represents the region of interest. |

1. Muscles should be lightly fixed in buffered 2 % paraformaldehyde (2–4 h on ice) (*see* **Note 3**).

2. The muscles can be prepared for staining in a number of ways, but should be oriented for longitudinal sections.

3. Cutting thick, frozen sections using a cryostat (20–40 μm thick sections), or using a vibratome to cut 50 μm sections of unfrozen tissue (remove the tendons to facilitate sectioning) gives good results. Muscle fibers can also be teased directly on slides to obtain individual fibers.

4. Samples are then stained using standard immunocytochemistry procedures.

5. Presynaptic antigens that work well include a cocktail of anti-neurofilament (for instance the monoclonal antibody 2H3) and anti-SV2 to fully label both the axon and the nerve terminal.

6. The postsynaptic receptors are brightly and specifically labeled using fluorescent conjugates of alpha-bungarotoxin.

7. Standard techniques involving application of the cocktail of primary antibodies followed by washes, then application of a cocktail of the secondary antibody and α-bungarotoxin followed by washes, work well provided patience is used (for instance, primary antibodies should be applied overnight followed by several hours of washing the next day).

8. Standard antibody dilution buffers such as PBS with BSA or normal goat serum as blocking agents can be used, provided generous detergent (0.5–1 % triton X-100) is also included.

9. Samples can be viewed on a standard fluorescence scope, but given the large size and 3-dimensional nature of the samples, a confocal Z-series usually gives better results.

10. A number of defects can be readily observed, including partially innervated or completely denervated postsynaptic receptor sites, fragmented or shrunken postsynaptic receptors, atrophied axons or terminals, and swollen or dystrophic axons or terminals. More subtle defects include sprouting nerve terminals and multiple innervation or synaptic sites (single innervation is normally present after 2 weeks of age). Normal NMJs have a contiguous pretzel-like morphology, the AChR staining has a zebra stripe appearance (the junctional folds), and the nerve terminals completely overlap the receptors. Examples are shown in Fig. 7.

Caveats include minor differences in size and shape from muscle to muscle, and fixation artifacts that can eliminate staining, especially presynaptically. In general, defects in the presynaptic terminal, such as partial retraction, are reflected less-precise definition in the postsynaptic receptors.

Additional analyses that can be informative include staining with anti-S100 to visualize Schwann cells (the terminal Schwann cells play an important role in guiding terminal sprouting and reinnervation [5, 6]), and histochemical stains to visualize acetylcholine esterase [7].

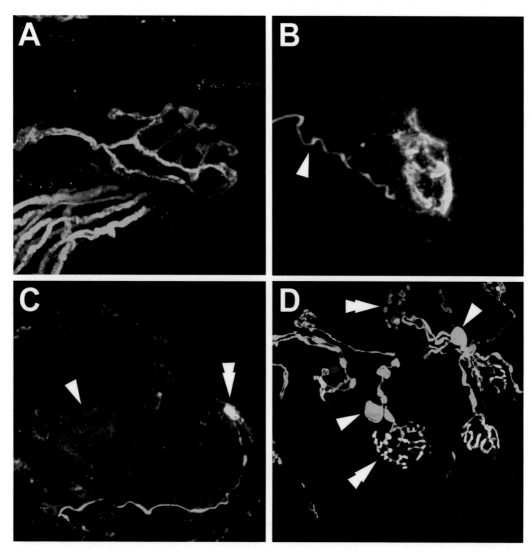

Fig. 7 Neuromuscular junctions, light microscopy. In a control neuromuscular junction (**a**), the nerve (stained green using anti-neurofilament plus anti-SV2, or using a Thy1-YFP transgenic strain) completely overlaps the postsynaptic AChRs (stained red using fluorescent α-bungarotoxin conjugates). (**b**) An example of partial innervation, where an atrophied motor axon and terminal fails to completely cover the AChRs on the muscle. (**c**) An example of frank denervation, where a site of postsynaptic AChRs with no associated nerve (*arrowhead*) is observed near a site of partial innervation (double arrowhead). (**b**) and (**c**) are examples from the *Gars^{Nmf249/+}* mouse model of Charcot-Marie-Tooth 2D [3]. Other NMJ pathologies are evident in an example from an unpublished spontaneous mutation. The axons and terminals have irregular diameters and varicosities (*arrowheads*) and postsynaptic sites are fragments (*double arrowheads*). Such changes are predictive of eventual denervation. These pathologies are also observed in very old mice (greater than 15 months), but are evident in this mutant by 3 months of age

3.3.2 Electron Microscopy of the NMJ

NMJs can also be visualized by transmission electron microscopy for a more detailed look at pre- and postsynaptic anatomies.

1. Muscles should be prepared for electron microscopy using standard techniques, including rapid fixation (preferably by perfusion) with glutaraldehyde-based fixatives.

2. The muscle should be dissected free and trimmed for cross sections at the point where the nerve enters the muscle. The endplate band is only a narrow region near the middle of the muscle (*see* **Note 4**).

3. Samples should be postfixed, osmicated, and embedded in plastic, cross-sectioned, and mounted on EM grids using standard procedures (e.g., [8]).

The challenge to viewing sections is finding NMJs. They can be spotted by scanning the grids at 10–12 K magnification and concentrating on areas where axons, fat, or blood vessels are also present. NMJs are rarely found in areas where the muscle fibers tightly stacked. Detailed images can be obtained at 30–60 K magnification.

Normal NMJs have a generally polarized nerve terminal with accumulations of 40–50 nm small clear vesicles near the presynaptic membrane and mitochondria located farther away (Fig. 8). In mice, the terminal Schwann cell capping the nerve terminal

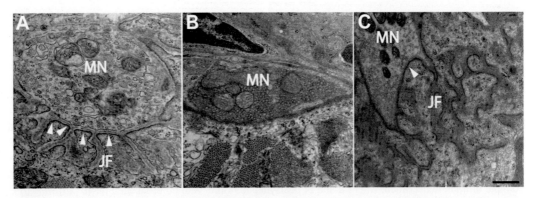

Fig. 8 Neuromuscular junctions analyzed by transmission electron microscopy. (**a**) In wild-type mice, the motor nerve terminal (MN) is depressed into the muscle fiber surface. The terminal is polarized, with small clear vesicles near the presynaptic membrane and mitochondria in the more proximal portion of the terminal. The postsynaptic membrane has deep convolutions (junctional folds, JF) and the membrane near the tops of these folds is very electron dense because of the high density of acetylcholine receptors (*arrowheads*). (**b**) In some myasthenias where the nerve sprouts but remains in contact with the muscle, terminals with mitochondria and vesicles are observed in the absence of any postsynaptic specialization. Presumably these are sprouting terminals that have not established a functional connection. (**c**) Partial innervation of postsynaptic sites is evident as elaborate junctional folds in the muscle membrane with no overlying nerve terminal. In these examples, the interpretations were aided by light microscopy examination of other samples as described in Fig. 8 in parallel with electron microscopy. The mutation shown in (**b**), (**c**) is an unpublished ENU-induced allele of agrin

can be difficult to resolve. The postsynaptic membrane has a series of junctional folds invaginating into the muscle fiber. At the mouth (crest) of each fold, the membrane appears electron dense because of the accumulation of AChRs. The synaptic cleft is pronounced and contains a visible basal lamina.

Pathological deviations include an absence of junctional folds, partial innervation (folds without an overlying nerve terminal), and vacuolated mitochondria. Assessing more subtle defects, such as changes in vesicle number, requires a statistical analysis on many junctions.

3.4 Muscular Dystrophies

3.4.1 Histological Analysis of Muscular Dystrophies

Muscle pathology can be accurately assessed by histology, focusing on similar hallmarks to those used in the diagnosis of human muscular dystrophies (Fig. 9). Muscle weight to body weight ratios can be used to indicate pathology in the muscle that is not simply proportional to decreased body size. Mice with muscular dystrophy or atrophy will typically have lower body weights, but an even greater reduction in muscle weight:body weight.

1. Muscles can be fixed by perfusion or immersion in Bouin's fixative, dehydrated, and embedded in paraffin (record body weights, and muscle weights following dissection, prior to processing).

2. Cross sections should be cut using a microtome and stained using H&E. All of these techniques are standard histological protocols. Care should be taken to collect sections from standardized regions of the muscle (such as the belly) to avoid differences in fiber number, fiber size, or composition that can vary along the longitudinal axis in limb muscles.

3. Samples should be evaluated for muscle fiber diameters (and the uniformity of fiber sizes), centrally located myonuclei (a sign of a regenerated muscle fiber), fibrosis, fatty infiltration, and atrophied muscle fibers.

4. Interpretation: Muscular dystrophies can be distinguished from other neuromuscular conditions based on the loss of muscle fibers, fibrosis, and signs of degeneration/regeneration. Atrophy resulting from denervation may have a similar appearance, but the affected fibers are typically scattered throughout the muscle (reflecting the anatomy of motor units and the pattern of innervation by motor neurons) while dystrophies tend to affect most of the fibers in a particular region of the muscle.

3.4.2 Muscle Fiber Integrity, Evan's Blue Staining

Many muscular dystrophies result in the loss of sarcolemmal integrity. This is true of dystrophies affecting the dystrophin/glycoprotein complex (DGC), including defects in dystrophin, the sarcoglycans, and dystroglycan, or in dysferlin, which is involved in

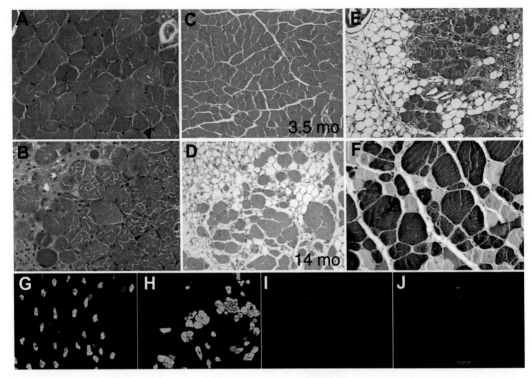

Fig. 9 Dystrophy phenotypes. (**a**) In histological staining of wild-type muscle (H&E), the muscle fibers are closely packed, regular in size, and the nuclei are near the cortex of the muscle fiber. (**b**) In a dystrophic muscle (*Lama2^{dy/dy}* is shown), the fibers are variable in diameter, some are atrophied, fibrotic cells are replacing muscle fibers, and nuclei in the center of fibers indicate regenerated fibers. (**c, d**) Mice lacking dysferlin, which is involved in membrane repair, have a progressive dystrophy. Muscle is histologically normal (**c**) until approximately 8 months of age, by 14 months (**d**), the muscle is severely dystrophic, with a great deal of fatty infiltration. (**e**). The *rmd* mouse mutation also has a severe dystrophy with fatty infiltration into the muscle, but the phenotype is much more severe in the hindlimbs (shown) than the forelimbs. (**f**) Antibody staining for myosin isoforms (fast myosin is shown) can determine if certain fiber types are selectively sensitive. The *myd* mouse is shown, in which both fast and slow fibers show central nuclei and signs of dystrophy, indicated both fiber types are affected. (**g, h**) Fiber type staining can also identify grouping, as seen with slow myosin stains of control (**g**) and *Gars^{Nmf249/+}* muscle (**h**). Such grouping is indicative of denervation and reinnervation and not a dystrophy intrinsic to the muscle. (**i, j**) Muscle fiber integrity can be assessed by Evan's Blue dye infiltration. In control muscle (**i**), Evan's Blue administered intraperitoneally is excluded from the muscle fibers but stains the membranes and connective tissue. In dystrophic muscles in which the sarcolemmal membrane integrity is compromised (*mdx* is shown), the dye stains the entire fiber (**j**)

membrane repair. Other mutations, such as those in titin, do not result in a loss of muscle fiber integrity. Therefore, particularly for the mechanistic evaluation of new models, it is important to determine if membrane integrity is compromised. This is easily done using Evan's Blue, an Azo dye that binds albumin and is normally excluded from healthy cells, but infiltrates into the cytoplasm of compromised cells (Fig. 9).

1. Evan's Blue dye should be dissolved in sterile saline at 10 mg/ml.

2. This solution can then be injected IP using 0.1 ml per 10 g of body weight (100 mg/kg).

3. Within a few hours of injection, exposed skin, such as the feet, tail, and ears, should have a pronounced blue tinge.

4. After 12–18 h (overnight), dissect muscles of interest and view.

5. All tissue will have a bluish cast, but dystrophic muscle will have marked streaks of blue due to the compromised fibers.

6. This can be most readily assessed by embedding the tissue and cutting cryostat cross sections.

7. Under fluorescence (rhodamine filters), the Evan's blue fluoresces red. Healthy muscle will have positively labeled membranes, as well as blood vessels, but muscles with compromised fiber integrity will contain fibers in which the entire cross-sectioned cytoplasm is strongly fluorescent.

This technique is simple and very sensitive. Intraperitoneal injection is straightforward and as effective as intravenous injection. Positive muscle fibers can also be the result of necrosis or injury in the muscle, but this usually results in a few, widely scattered fibers instead of a large number of fibers grouped in a specific region of the muscle.

3.4.3 Caveats

As with all neuromuscular diseases, there can be tremendous variation muscle-to-muscle, and even within a muscle (this is true for neuropathies as well, see for example [9]). For instance, in some dystrophy models the diaphragm is severely affected while in others, it is spared. Therefore, it is very important to evaluate the same muscle in each mouse, and to evaluate more than one muscle. A picture of the anatomy of the lower hindlimb of the mouse is shown in Fig. 10. The underlying basis of these differences is not clear and does not seem to be as obvious as differences in fiber-type composition, activity levels, or force generated. Within a model, the pattern of pathology is usually fairly reproducible, but there can again be animal-to-animal variability that complicates quantification of results.

Other techniques such as serum levels of muscle creatine kinase can also be used to assess dystrophy, but this requires great care in mouse handling. Creatine kinase is released from damaged muscle fibers. If the mouse struggles significantly while the sample is collected (more than a few seconds), values will be artificially high and variable. In addition, hemolysis in the sample will also reduce accuracy. Therefore, while this measurement is outwardly straightforward and valuable, obtaining consistent results can be challenging.

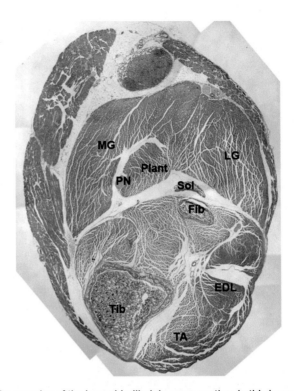

Fig. 10 The muscles of the lower hindlimb in cross section. In this image, anterior is down and medial is left. Abbreviations are as follows: *MG* medial gastrocnemius, *LG* lateral gastrocnemius, *Plant* plantaris, *PN* plantar nerve, *Sol* soleus, *Fib* fibula, *EDL* extensor digitorum longii, *TA* tibialis anterior, *Tib* tibia. The mouse muscles are predominantly fast muscle fibers, but the soleus is valuable for its high percentage of slow fibers. Note, the darker mass on the posterior portion of the leg is a lymph node that provides a convenient landmark when sectioning to establish that reproducible sections are examined in the proximal/distal axis. Also, the peripheral muscles in the section are the hamstrings, which insert along the tibia in the lower leg in the mouse

3.5 Techniques Generally Applicable to Neuromuscular Disease Models

3.5.1 Gross Motor Performance

The disease models described above, motor neuron diseases, neuropathies, myasthenias, and muscular dystrophies, can all benefit from an analysis of gross motor performance. The rotarod is often used to assess neuromuscular function and other movement disorders and has been successfully used to describe defects in many mouse models (e.g., [10–12]). The widespread use of the device is due in part to the apparent simplicity of the testing procedures. However, recent work emphasizes that, as with any behavioral test, there are a number of potential confounding factors [13, 14]. In our experience data collection is more efficient using gait analysis (treadmill). In addition raw video data is often useful on its own and the extensive and varied measurements that are possible offer significant flexibility. Grip strength measures also provide a more direct assessment of muscle force generation that can be combined with electrophysiological techniques described below, and are only minimally confounded by other parameters affecting motor performance such as coordination or balance. Therefore, our preferred methods are grip strength and gait analysis.

3.5.2   *Grip Strength*

Measure of grip strength gives a good indicator of muscle strength and is analogous to grip strength measures in humans, e.g., [15], which reveal weakness as a major presenting symptom of neuromuscular dysfunction.

To measure grip strength, the mouse is prompted to grab a bar connected to a force transducer with either its hind or fore paws (or less often all paws using a grid). Once the mouse achieves a grip the tester typically pulls the mouse horizontally away from the bar until the animal is no longer able to maintain its grip. The peak force registered by the transducer is recorded. This approach, recommended by equipment suppliers, requires careful attention to the rate and direction of force applied by the tester to "break" the animals grip. If this approach is used, reliability should be confirmed for each individual performing the test and between testers if multiple individuals are collecting data.

However, to improve reproducibility for forelimb grip strength we recommend orienting the force transducer vertically (Columbus Instruments, Columbus OH USA) and modifying the test procedure as follows. A weight (100 g) is attached to the base of the mouse's tail using a small plastic clip. The mouse is held by the scruff of the neck with the tail and weight in the tester's other palm. The mouse is advanced toward the bar until it instinctively grasps the bar with both paws. The mouse is slowly lowered to a vertical position and both animal and weight are released. The transducer records the peak force generated by the animal just prior to grip loss. This method insures consistent force and acceleration away from the grip bar, and is not any more traumatic for the mouse. Typically measures from three repeat trials are averaged to represent the value for each animal. This approach can be used to track changes in longitudinal studies and has been used effectively for most standard inbred strains (http://phenome.jax.org).

3.5.3   *Gait Analysis*

Analysis of gait has a long history, and because it is also routinely used in humans, establishing clinical relevance for murine disease models is facilitated. The simplest method for quantification of gait in mice is to paint the feet of the mice and to motivate them to walk across a piece of paper. Manual measurements of footprints can then be made to derive stride length and stance width. This method is particularly useful for quantification in mice with overt, observable movement defects, but may also be sufficient for more subtle phenotypes (e.g., [16]). Practically, the greatest difficulty with this method is in motivating the mice to walk, and investigators have addressed this with a variety of strategies. For example, placing the mice in a lighted "corridor" facing toward a darkened space will both restrict wandering and improve motivation, as mice will seek to "escape" to the dark enclosure. Two additional considerations for this method are that only a limited number of parameters can be measured, and that the speed of locomotion cannot be controlled. Thus, common time domain parameters that divide a stride (step) into component phases of stance and swing are not possible and measurements that are derived need to consider locomotory speed in the interpretation.

We expanded the utility of the basic footprint analysis by employing video recordings of mice walking on a treadmill (Fig. 11). The mice are placed in an enclosure on a treadmill with a clear plastic tread (Columbus Instruments, Columbus, OH). The ventral surface of the mouse is reflected in a mirror placed at 45° under the tread and is recorded by an adjacent digital video camera (Basler, Inc). Typically a video clip of ~10 s provides sufficient data for valid measurement. Interactive analysis software (Clever Sys Inc, Reston, VA) is used to track body position and paw placement of each paw during locomotion at a fixed speed. The software derives the standard time domain gait parameters for each paw (e.g., stride and stance, swing phase times) as well as a variety of additional measures (body angle, foot placement angles, inter-limb phase ratios, etc.).

The variety of measures that can be derived with such a system allow detailed description of gait in many different models and offer the significant advantage of comparing animals at the same speed of locomotion and may be more sensitive to some subtle neuromuscular changes prior to overt movement deficits [17]. However, the treadmill represents a novel context for the mice, and we have noted changes in treadmill gait with repeated trials and age both of which can complicate characterization of progressive changes. Also, different mouse strains vary in their willingness to walk consistently on the treadmill, and compliance may be further reduced with repeated exposures (for details *see* ref. 18). These factors need to be considered carefully for each model in order to optimize experimental design.

3.5.4 *Electrophysiology*   We will describe basic electrophysiological analysis of neuromuscular function, similar to evaluation using electromyography by neurologists. More specialized techniques such as two electrode voltage clamp for monitoring synaptic transmission may also be useful, but are beyond the basic nature of this chapter. We describe briefly techniques for assessment of muscle contractile function including electromyography and nerve conduction velocity. These techniques provide measures of muscle force output and fatigability and insight into contractile dynamics related to energy supply,

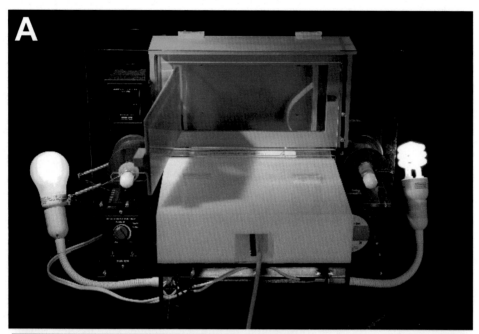

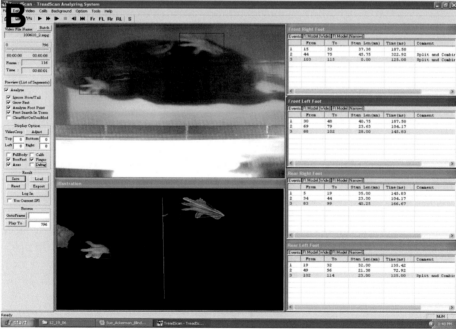

Fig. 11 Gait analysis device. (**a**) The gait analysis device is shown. The mouse is placed on the treadmill in the *green box*. The mouse would be facing *left* and the loop at *right* in the chamber bumps the mouse's tail if they lag on the treadmill. The *white box* in front contains the digital camera that videos the mouse using an angled mirror. (**b**) The automated data analysis is trained to recognize the mouse's paws in the video and gait parameters are calculated, as well as body axis, toe spread, and other data

calcium handling, and excitation-contraction coupling. In addition, motor unit number can be estimated and significant deficits in synaptic function may be detectable.

3.5.5 Muscle Contractile Function

Contractile properties in rodents can be measured either in vitro in a dissected muscle, or in vivo in an intact preparation with an anesthetized animal (e.g., [19–21]). Measurements made under isometric conditions are perhaps most common and use the most straightforward setup. The addition of servomotors for dynamic control of muscle length allows simulation of dynamic conditions (eccentric, isotonic etc.) that may be modified by disease or other processes [22, 23]. For in vitro studies, the muscle is anchored by ligating the tendon (origin) to a support; for in vivo studies, the bone (femur) is clamped to prevent movement. The other tendon (insertion) is then coupled to a force transducer. In both cases, a recording electrode is also placed in contact with the muscle to record the compound action potential, and a stimulating electrode is used to stimulate the nerve or the muscle, as described below.

It should be noted that both in vitro and in vivo preparations present considerable technical challenges and are best developed/established in consultation with an experienced laboratory. In general, an in vitro preparation is somewhat less demanding because surgical preparation is less demanding and extended life support/maintenance of the animal during experimentation is not required.

Muscle force can be elicited by *nerve stimulation* to test both muscle function and the integrity of the nerve-muscle connection or by *direct muscle stimulation* to evaluate only muscle contractile independent of the synapse. The latter measure reflects the total force that the muscle is able to generate, and if it is significantly greater than that obtained by nerve stimulation, it implies denervation, or a failure in conduction or synaptic transmission in the nerves. Two general types of stimuli are typically used, a single brief (100–200 μs) stimulus of intensity sufficient to activate all functional connections, referred to as maximal twitch force (Pt), and trains of stimuli delivered at a frequency (100–200 pps) which induces tetanic fusion of individual contractions to produce maximum tetanic force (P0).

3.5.6 Contractile Measures

Measurement of single twitches provides insight into contraction/relaxation dynamics of the muscle. For example, reduced AChE or altered Ca^{2+} handling in the muscle causes a slower relaxation phase whereas shifts toward "faster" ATPase isoforms will reduce time to reach peak twitch force (Fig. 12).

Stimulus trains of varying lengths (0.5–1.2 s) and frequencies of (10–200 pps) can provide additional information. As stimulus frequency is increased, there is greater tetanic fusion and higher peak forces are produced. A force-frequency curve (F-F) can also be generated to evaluate factors that affect muscle contractile speed (time-to-peak, half-relaxation). If, for example, disease processes

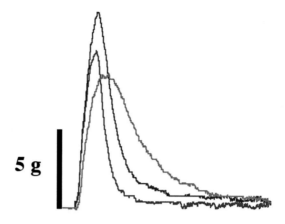

5 g

Fig. 12 Muscle twitch force. The force generated by a control muscle in response to a single 200 μs stimulus (*black trace*) compared to two mutants showing exaggerated force (*red*) or slower relaxation time (*orange*) is shown. This is an unpublished mutation that we interpret as having decreased AChE at the synapse, possible as a secondary compensation for impaired presynaptic neurotransmitter release

lead to slower contraction/relaxation time, more tetanization occurs at lower frequencies and the shape of the F-F curve will be shifted compared to normal muscle.

A variety of protocols exist for measurement of muscle fatigue in mice. In general a series of stimulus trains are repeated at a set frequency for several minutes. The protocol selected depends on the muscle being tested and can vary depending on the model and the experimental objective. Different laboratories have established unique protocols that can be implemented [24–28]. The basis for the difference in protocols is not always evident; different stimulus paradigms may have been selected on a theoretical basis or determined empirically for the purposes of a given laboratory.

Long stimulus trains will also induce "tetanic fade," observed as a decrease in force observed *during* a single contraction. Fade is measured as the ratio of final force to initial peak force. Diseased muscle/nerve may show a lower ratio. Accompanying measurement of EMG can reveal if fade is due to failure of the muscle or the nerve. In purely muscle defects (muscle dystrophy) force will drop without any change in EMG whereas reduced force accompanied by reduced EMG suggests a defect in transmission (NMJ) (e.g., [3]) or excitation-contraction coupling.

3.5.7   Motor Unit Number Estimates (MUNE)

Delivering brief nerve stimuli with gradually increasing amplitudes will evoke an incremental increase in Pt as additional motor units are recruited. If stepwise increases in twitch force or electromyogram amplitudes are recorded, the distinct increments can be counted to provide an estimate of motor unit number or the number of functional motor axons innervating a given muscle (Fig. 13).

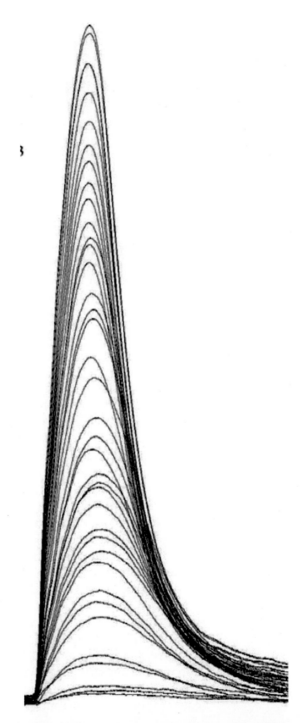

Fig. 13 Motor unit estimation. Stimuli of gradually increasing intensity are applied to the nerve and the force generated by the muscle is recorded. The number of steps of force produced approximately reflects the number of motor units in the nerve

There are a variety of MUNE techniques [29] but no universally accepted standard has emerged. Nonetheless, the method, originally used in humans [30], has been used successfully in other species including mice (e.g., [1]) and recently developed methods can facilitate implementation [31].

3.5.8 Nerve Conduction Velocities (NCV)

In contractile experiments the time from the stimulation of the nerve to the CMAP recorded in muscle provides an estimate of NCV. The length of the nerve from the stimulating electrode to the muscle can simply be measured and divided by the time. However, the time recorded in this way includes the delay for synaptic transmission, which may be increased in models with synaptic defects. If this is a concern or if the only parameter desired is NCV, then the measurement can be obtained noninvasively with a relatively simple setup (e.g., [3]). Using the sciatic nerve NCV can be calculated by measuring the latency of compound motor action potentials recorded in the muscle of a rear paw. Action potentials are produced by subcutaneous stimulation at two separate sites: proximal stimulation at the sciatic notch and distally at the ankle. NCV is then calculated by using the two latencies and conduction distance. Decreases in nerve conduction velocity most often reflect defects in myelination, but may also be the result of changes in intermodal distance, decreased axon diameters, or altered excitability.

3.6 Conclusion

The pathophysiology of neuromuscular diseases is highly integrated, reflecting the extensive reciprocal signaling and functional interdependency of the peripheral nervous system and the skeletal muscles. As a result, changes in the nerves result in changes in the muscles and vice versa, making it a challenge to distinguish primary and secondary effects. This is particularly true when new models are being examined and it is unknown where the causative gene normally functions. As a result, the functions of genes and the mutant phenotypes must be examined comprehensively, and the results must be taken as a whole, in order to understand how the final phenotype arises. In this chapter, we have given examples of mouse models used in neuromuscular disease research, and provided a battery of analyses that can be applied to these models to understand the underlying disease mechanisms. While many of these diseases are now well understood, many remain equally mysterious, and continued efforts to understand both the genetics and physiology are needed to eventually have a complete understanding of the neuromuscular system.

4 Notes

1. Purple condensed cells in H&E can be "Dark Neuron Artifact" caused by deforming the tissue before it was well fixed [32–34].

2. The dorsal root has an enlargement, the dorsal root ganglion containing the sensory cell bodies near the point of bifurcation.

3. The paraformaldehyde should be of high quality (such as that supplied by Electron Microscopy Services), and should be prepared fresh each day. Overfixation or low-grade fixatives will dramatically reduce or eliminate the antigenicity of presynaptic proteins.

4. It is better to lose some NMJs than to waste a lot of effort sectioning regions of the muscle where there are no synapses.

5. There are a number of other useful online resources and reference texts that further explain the genetics, physiology, and clinical features of neuromuscular diseases in both humans and mice. The database "Online Mendelian Inheritance in Man (OMIM)" is particularly useful as a source of genetic information related to all forms of heritable human diseases (http://www.ncbi.nlm.nih.gov/sites/entrez?db=OMIM&itool=toolbar). For a resource of mouse genetics, including mapping data, alleles, molecular characterization and expression information, the Mouse Genome Informatics website provides a comprehensive and current database (http://www.informatics.jax.org/). Other online resources related to specific human diseases also exist, such as the Inherited Peripheral Neuropathies website, providing genetic and molecular information pertaining to inherited diseases of the peripheral nervous system (http://www.molgen.ua.ac.be/CMTMutations/).   Useful reference texts include *Myology* (A. G. Engel, C Franzini-Armstrong, Eds), *Peripheral Neuropathy* (P. J. Dyck and P. K. Thomas, Eds.), *Pathology of Peripheral Nerves* (J. M. Schroder, Ed.), and *Neuromuscular Disorders: Clinical and Molecular Genetics* (A. E. H. Emery, Ed.,). *See* references [35–38].

References

1. Hegedus J, Putman CT, Gordon T (2007) Time course of preferential motor unit loss in the SOD1 G93A mouse model of amyotrophic lateral sclerosis. Neurobiol Dis 28:154–164

2. Grohmann K, Schuelke M, Diers A, Hoffmann K, Lucke B, Adams C et al (2001) Mutations in the gene encoding immunoglobulin mu-binding protein 2 cause spinal muscular atrophy with respiratory distress type 1. Nat Genet 29:75–77

3. Seburn KL, Nangle LA, Cox GA, Schimmel P, Burgess RW (2006) An active dominant mutation of glycyl-tRNA synthetase causes neuropathy in a Charcot-Marie-Tooth 2D mouse model. Neuron 51:715–726

4. Court FA, Sherman DL, Pratt T, Garry EM, Ribchester RR, Cottrell DF et al (2004) Restricted growth of Schwann cells lacking Cajal bands slows conduction in myelinated nerves. Nature 431:191–195

5. Son YJ, Thompson WJ (1995) Nerve sprouting in muscle is induced and guided by processes extended by Schwann cells. Neuron 14:133–141

6. Son YJ, Thompson WJ (1995) Schwann cell processes guide regeneration of peripheral axons. Neuron 14:125–132

7. Enomoto H, Araki T, Jackman A, Heuckeroth RO, Snider WD, Johnson EM Jr et al (1998) GFR alpha1-deficient mice have deficits in the enteric nervous system and kidneys. Neuron 21:317–324

8. Patton BL, Cunningham JM, Thyboll J, Kortesmaa J, Westerblad H, Edstrom L et al (2001) Properly formed but improperly localized synaptic specializations in the absence of laminin alpha4. Nat Neurosci 4:597–604

9. Pun S, Santos AF, Saxena S, Xu L, Caroni P (2006) Selective vulnerability and pruning of phasic motoneuron axons in motoneuron disease alleviated by CNTF. Nat Neurosci 9:408–419

10. Morgan D, Munireddy S, Alamed J, Deleon J, Diamond DM, Bickford P et al (2008) Apparent behavioral benefits of Tau overexpression in P301L Tau transgenic mice. J Alzheimers Dis 15:605–614

11. Stam NC, Nithianantharajah J, Howard ML, Atkin JD, Cheema SS, Hannan AJ (2008) Sex-specific behavioural effects of environmental enrichment in a transgenic mouse model of amyotrophic lateral sclerosis. Eur J Neurosci 28:717–723

12. Turgeman T, Hagai Y, Huebner K, Jassal DS, Anderson JE, Genin O et al (2008) Prevention of muscle fibrosis and improvement in muscle performance in the mdx mouse by halofuginone. Neuromuscul Disord 18:857–868

13. Bohlen M, Cameron A, Metten P, Crabbe JC, Wahlsten D (2009) Calibration of rotational acceleration for the rotarod test of rodent motor coordination. J Neurosci Methods 178(1):10–14

14. Rustay NR, Wahlsten D, Crabbe JC (2003) Influence of task parameters on rotarod performance and sensitivity to ethanol in mice. Behav Brain Res 141:237–249

15. Bautmans I, Mets T (2005) A fatigue resistance test for elderly persons based on grip strength: reliability and comparison with healthy young subjects. Aging Clin Exp Res 17:217–222

16. Fernagut PO, Diguet E, Labattu B, Tison F (2002) A simple method to measure stride length as an index of nigrostriatal dysfunction in mice. J Neurosci Methods 113:123–130

17. Wooley CM, Sher RB, Kale A, Frankel WN, Cox GA, Seburn KL (2005) Gait analysis detects early changes in transgenic SOD1(G93A) mice. Muscle Nerve 32:43–50

18. Wooley CM, Xing S, Burgess RW, Cox GA, Seburn KL (2009) Age, experience and genetic background influence treadmill walking in mice. Physiol Behav 96(2):350–361

19. Brooks SV, Faulkner JA, McCubbrey DA (1990) Power outputs of slow and fast skeletal muscles of mice. J Appl Physiol 68:1282–1285

20. Seburn KL, Gardiner P (1995) Adaptations of rat lateral gastrocnemius motor units in response to voluntary running. J Appl Physiol 78:1673–1678

21. Lynch GS, Hinkle RT, Chamberlain JS, Brooks SV, Faulkner JA (2001) Force and power output of fast and slow skeletal muscles from mdx mice 6-28 months old. J Physiol 535:591–600

22. Brooks SV, Faulkner JA (1988) Contractile properties of skeletal muscles from young, adult and aged mice. J Physiol 404:71–82

23. Grange RW, Gainer TG, Marschner KM, Talmadge RJ, Stull JT (2002) Fast-twitch skeletal muscles of dystrophic mouse pups are resistant to injury from acute mechanical stress. Am J Physiol Cell Physiol 283:C1090–C1101

24. Belluardo N, Westerblad H, Mudo G, Casabona A, Bruton J, Caniglia G et al (2001) Neuromuscular junction disassembly and muscle fatigue in mice lacking neurotrophin-4. Mol Cell Neurosci 18:56–67

25. Danieli-Betto D, Esposito A, Germinario E, Sandona D, Martinello T, Jakubiec-Puka A et al (2005) Deficiency of alpha-sarcoglycan differently affects fast- and slow-twitch skeletal muscles. Am J Physiol Regul Integr Comp Physiol 289:R1328–R1337

26. Grange RW, Meeson A, Chin E, Lau KS, Stull JT, Shelton JM et al (2001) Functional and molecular adaptations in skeletal muscle of myoglobin-mutant mice. Am J Physiol Cell Physiol 281:C1487–C1494

27. Parry DJ, Desypris G (1985) Fatigability and oxidative capacity of forelimb and hind limb

muscles of dystrophic mice. Exp Neurol 87:358–368

28. Swallow JG, Garland T Jr, Carter PA, Zhan WZ, Sieck GC (1998) Effects of voluntary activity and genetic selection on aerobic capacity in house mice (Mus domesticus). J Appl Physiol 84:69–76

29. Shefner JM (2001) Motor unit number estimation in human neurological diseases and animal models. Clin Neurophysiol 112:955–964

30. McComas AJ, Fawcett PR, Campbell MJ, Sica RE (1971) Electrophysiological estimation of the number of motor units within a human muscle. J Neurol Neurosurg Psychiatry 34:121–131

31. Major LA, Hegedus J, Weber DJ, Gordon T, Jones KE (2007) Method for counting motor units in mice and validation using a mathematical model. J Neurophysiol 97:1846–1856

32. Ebels EJ (1975) Dark neurons: a significant artifact: the influence of the maturational state of neurons on the occurrence of the phenomenon. Acta Neuropathol 33:271–273

33. Garman RH (1990) Artifacts in routinely immersion fixed nervous tissue. Toxicol Pathol 18:149–153

34. Jortner BS (2006) The return of the dark neuron. A histological artifact complicating contemporary neurotoxicologic evaluation. Neurotoxicology 27:628–634

35. Schrèoder JM (2001) Pathology of peripheral nerves: an atlas of structural and molecular pathological changes. Springer, Berlin/New York

36. Emery AEH (1998) Neuromuscular disorders: clinical and molecular genetics. Wiley, Chichester; New York

37. Dyck PJ, Thomas PK (2005) Peripheral neuropathy, 4th edn. Elsevier Saunders, Philadelphia, PA

38. Engel A, Franzini-Armstrong C (2004) Myology: basic and clinical, 3rd edn. McGraw-Hill, Medical Pub. Division, New York

39. Maddatu TP, Garvey SM, Schroeder DG, Hampton TG, Cox GA (2004) Transgenic rescue of neurogenic atrophy in the nmd mouse reveals a role for Ighmbp2 in dilated cardiomyopathy. Hum Mol Genet 13:1105–1115

40. Hershkovitz E, Rozin I, Limony Y, Golan H, Hadad N, Gorodischer R et al (2007) Hypoparathyroidism, retardation, and dysmorphism syndrome: impaired early growth and increased susceptibility to severe infections due to hyposplenism and impaired polymorphonuclear cell functions. Pediatr Res 62:505–509

41. Noakes PG, Miner JH, Gautam M, Cunningham JM, Sanes JR, Merlie JP (1995) The renal glomerulus of mice lacking s-laminin/laminin β2: nephrosis despite molecular compensation by laminin β1. Nat Genet 10:400–406

42. Zhai RG, Cao Y, Hiesinger PR, Zhou Y, Mehta SQ, Schulze KL et al (2006) Drosophila NMNAT maintains neural integrity independent of its NAD synthesis activity. PLoS Biol 4, e416

43. Zhai RG, Zhang F, Hiesinger PR, Cao Y, Haueter CM, Bellen HJ (2008) NAD synthase NMNAT acts as a chaperone to protect against neurodegeneration. Nature 452:887–891

44. Press C, Milbrandt J (2008) Nmnat delays axonal degeneration caused by mitochondrial and oxidative stress. J Neurosci 28:4861–4871

45. Maddatu TP, Garvey SM, Schroeder DG, Zhang W, Kim S-Y, Nicholson AI et al (2005) Dilated cardiomyopathy (DCM) in the nmd mouse: transgenic rescue and QTLs that improve cardiac function and survival. Hum Mol Genet 14(21):3179–3189

46. Cox GA, Mahaffey CL, Frankel WN (1998) Identification of the mouse neuromuscular degeneration gene and mapping of a second site suppressor allele. Neuron 21:1327–1337

47. Monani UR, Pastore MT, Gavrilina TO, Jablonka S, Le TT, Andreassi C et al (2003) A transgene carrying an A2G missense mutation in the SMN gene modulates phenotypic severity in mice with severe (type I) spinal muscular atrophy. J Cell Biol 160:41–52

48. Le TT, Pham LT, Butchbach ME, Zhang HL, Monani UR, Coovert DD et al (2005) SMNDelta7, the major product of the centromeric survival motor neuron (SMN2) gene, extends survival in mice with spinal muscular atrophy and associates with full-length SMN. Hum Mol Genet 14:845–857

49. Monani UR, Sendtner M, Coovert DD, Parsons DW, Andreassi C, Le TT et al (2000) The human centromeric survival motor neuron gene (SMN2) rescues embryonic lethality in Smn(-/-) mice and results in a mouse with spinal muscular atrophy. Hum Mol Genet 9:333–339

50. Schrank B, Gotz R, Gunnersen JM, Ure JM, Toyka KV, Smith AG et al (1997) Inactivation

of the survival motor neuron gene, a candidate gene for human spinal muscular atrophy, leads to massive cell death in early mouse embryos. Proc Natl Acad Sci U S A 94:9920–9925

51. Martin N, Jaubert J, Gounon P, Salido E, Haase G, Szatanik M et al (2002) A missense mutation in Tbce causes progressive motor neuronopathy in mice. Nat Genet 32:443–447

52. Bommel H, Xie G, Rossoll W, Wiese S, Jablonka S, Boehm T et al (2002) Missense mutation in the tubulin-specific chaperone E (Tbce) gene in the mouse mutant progressive motor neuronopathy, a model of human motoneuron disease. J Cell Biol 159: 563–569

53. Puls I, Jonnakuty C, LaMonte BH, Holzbaur EL, Tokito M, Mann E et al (2003) Mutant dynactin in motor neuron disease. Nat Genet 33:455–456

54. Chevalier-Larsen ES, Wallace KE, Pennise CR, Holzbaur EL (2008) Lysosomal proliferation and distal degeneration in motor neurons expressing the G59S mutation in the p150Glued subunit of dynactin. Hum Mol Genet 17:1946–1955

55. Lai C, Lin X, Chandran J, Shim H, Yang WJ, Cai H (2007) The G59S mutation in p150(glued) causes dysfunction of dynactin in mice. J Neurosci 27:13982–13990

56. LaMonte BH, Wallace KE, Holloway BA, Shelly SS, Ascano J, Tokito M et al (2002) Disruption of dynein/dynactin inhibits axonal transport in motor neurons causing late-onset progressive degeneration. Neuron 34:715–727

57. Ripps ME, Huntley GW, Hof PR, Morrison JH, Gordon JW (1995) Transgenic mice expressing an altered murine superoxide dismutase gene provide an animal model of amyotrophic lateral sclerosis. Proc Natl Acad Sci U S A 92:689–693

58. Kunst CB, Messer L, Gordon J, Haines J, Patterson D (2000) Genetic mapping of a mouse modifier gene that can prevent ALS onset. Genomics 70:181–189

59. Bruijn LI, Becher MW, Lee MK, Anderson KL, Jenkins NA, Copeland NG et al (1997) ALS-linked SOD1 mutant G85R mediates damage to astrocytes and promotes rapidly progressive disease with SOD1-containing inclusions. Neuron 18:327–338

60. Wang J, Xu G, Gonzales V, Coonfield M, Fromholt D, Copeland NG et al (2002) Fibrillar inclusions and motor neuron degeneration in transgenic mice expressing superoxide dismutase 1 with a disrupted copper-binding site. Neurobiol Dis 10: 128–138

61. Wong PC, Pardo CA, Borchelt DR, Lee MK, Copeland NG, Jenkins NA et al (1995) An adverse property of a familial ALS-linked SOD1 mutation causes motor neuron disease characterized by vacuolar degeneration of mitochondria. Neuron 14:1105–1116

62. Gurney ME, Pu H, Chiu AY, Dal Canto MC, Polchow CY, Alexander DD et al (1994) Motor neuron degeneration in mice that express a human Cu, Zn superoxide dismutase mutation. Science 264:1772–1775

63. Heiman-Patterson TD, Deitch JS, Blankenhorn EP, Erwin KL, Perreault MJ, Alexander BK et al (2005) Background and gender effects on survival in the TgN(SOD1-G93A)1Gur mouse model of ALS. J Neurol Sci 236:1–7

64. Messer A, Flaherty L (1986) Autosomal dominance in a late-onset motor neuron disease in the mouse. J Neurogenet 3:345–355

65. Bronson RT, Lake BD, Cook S, Taylor S, Davisson MT (1993) Motor neuron degeneration of mice is a model of neuronal ceroid lipofuscinosis (Batten's disease). Ann Neurol 33:381–385

66. Ranta S, Zhang Y, Ross B, Lonka L, Takkunen E, Messer A et al (1999) The neuronal ceroid lipofuscinoses in human EPMR and mnd mutant mice are associated with mutations in CLN8. Nat Genet 23:233–236

67. Hafezparast M, Klocke R, Ruhrberg C, Marquardt A, Ahmad-Annuar A, Bowen S et al (2003) Mutations in dynein link motor neuron degeneration to defects in retrograde transport. Science 300:808–812

68. Chen XJ, Levedakou EN, Millen KJ, Wollmann RL, Soliven B, Popko B (2007) Proprioceptive sensory neuropathy in mice with a mutation in the cytoplasmic Dynein heavy chain 1 gene. J Neurosci 27:14515–14524

69. Ilieva HS, Yamanaka K, Malkmus S, Kakinohana O, Yaksh T, Marsala M et al (2008) Mutant dynein (Loa) triggers proprioceptive axon loss that extends survival only in the SOD1 ALS model with highest motor neuron death. Proc Natl Acad Sci U S A 105:12599–12604

70. Chow CY, Landers JE, Bergren SK, Sapp PC, Grant AE, Jones JM et al (2009) Deleterious

variants of FIG4, a phosphoinositide phosphatase, in patients with ALS. Am J Hum Genet 84:85–88

71. Chow CY, Zhang Y, Dowling JJ, Jin N, Adamska M, Shiga K et al (2007) Mutation of FIG4 causes neurodegeneration in the pale tremor mouse and patients with CMT4J. Nature 448:68–72

72. Zhang X, Chow CY, Sahenk Z, Shy ME, Meisler MH, Li J (2008) Mutation of FIG4 causes a rapidly progressive, asymmetric neuronal degeneration. Brain 131:1990–2001

73. Schmitt-John T, Drepper C, Mussmann A, Hahn P, Kuhlmann M, Thiel C et al (2005) Mutation of Vps54 causes motor neuron disease and defective spermiogenesis in the wobbler mouse. Nat Genet 37:1213–1215

74. Suter U, Welcher AA, Ozcelik T, Snipes GJ, Kosaras B, Francke U et al (1992) Trembler mouse carries a point mutation in a myelin gene. Nature 356:241–244

75. Giese KP, Martini R, Lemke G, Soriano P, Schachner M (1992) Mouse P0 gene disruption leads to hypomyelination, abnormal expression of recognition molecules, and degeneration of myelin and axons. Cell 71:565–576

76. Wrabetz L, Feltri ML, Quattrini A, Imperiale D, Previtali S, D'Antonio M et al (2000) P(0) glycoprotein overexpression causes congenital hypomyelination of peripheral nerves. J Cell Biol 148:1021–1034

77. Wrabetz L, D'Antonio M, Pennuto M, Dati G, Tinelli E, Fratta P et al (2006) Different intracellular pathomechanisms produce diverse Myelin Protein Zero neuropathies in transgenic mice. J Neurosci 26:2358–2368

78. Le N, Nagarajan R, Wang JY, Araki T, Schmidt RE, Milbrandt J (2005) Analysis of congenital hypomyelinating Egr2Lo/Lo nerves identifies Sox2 as an inhibitor of Schwann cell differentiation and myelination. Proc Natl Acad Sci U S A 102:2596–2601

79. Topilko P, Schneider-Maunoury S, Levi G, Baron-Van Evercooren A, Chennoufi AB, Seitanidou T et al (1994) Krox-20 controls myelination in the peripheral nervous system. Nature 371:796–799

80. Nelles E, Butzler C, Jung D, Temme A, Gabriel HD, Dahl U et al (1996) Defective propagation of signals generated by sympathetic nerve stimulation in the liver of connexin32-deficient mice. Proc Natl Acad Sci U S A 93:9565–9570

81. Anzini P, Neuberg DH, Schachner M, Nelles E, Willecke K, Zielasek J et al (1997) Structural abnormalities and deficient maintenance of peripheral nerve myelin in mice lacking the gap junction protein connexin 32. J Neurosci 17:4545–4551

82. Jeng LJ, Balice-Gordon RJ, Messing A, Fischbeck KH, Scherer SS (2006) The effects of a dominant connexin32 mutant in myelinating Schwann cells. Mol Cell Neurosci 32:283–298

83. Detmer SA, Vande Velde C, Cleveland DW, Chan DC (2008) Hindlimb gait defects due to motor axon loss and reduced distal muscles in a transgenic mouse model of Charcot-Marie-Tooth type 2A. Hum Mol Genet 17:367–375

84. Lee MK, Marszalek JR, Cleveland DW (1994) A mutant neurofilament subunit causes massive, selective motor neuron death: implications for the pathogenesis of human motor neuron disease. Neuron 13:975–988

85. Zhu Q, Couillard-Despres S, Julien JP (1997) Delayed maturation of regenerating myelinated axons in mice lacking neurofilaments. Exp Neurol 148:299–316

86. Huang L, Min JN, Masters S, Mivechi NF, Moskophidis D (2007) Insights into function and regulation of small heat shock protein 25 (HSPB1) in a mouse model with targeted gene disruption. Genesis 45:487–501

87. Gillespie CS, Sherman DL, Fleetwood-Walker SM, Cottrell DF, Tait S, Garry EM et al (2000) Peripheral demyelination and neuropathic pain behavior in periaxin-deficient mice. Neuron 26:523–531

88. Brandon EP, Lin W, D'Amour KA, Pizzo DP, Dominguez B, Sugiura Y et al (2003) Aberrant patterning of neuromuscular synapses in choline acetyltransferase-deficient mice. J Neurosci 23:539–549

89. Misgeld T, Burgess RW, Lewis RM, Cunningham JM, Lichtman JW, Sanes JR (2002) Roles of neurotransmitter in synapse formation. Development of neuromuscular junctions lacking choline acetyltransferase. Neuron 36:635–648

90. Buffelli M, Burgess RW, Feng G, Lobe CG, Lichtman JW, Sanes JR (2003) Genetic evidence that relative synaptic efficacy biases the outcome of synaptic competition. Nature 424:430–434

91. Prado VF, Martins-Silva C, de Castro BM, Lima RF, Barros DM, Amaral E et al (2006)

Mice deficient for the vesicular acetylcholine transporter are myasthenic and have deficits in object and social recognition. Neuron 51:601–612

92. Feng G, Krejci E, Molgo J, Cunningham JM, Massoulie J, Sanes JR (1999) Genetic analysis of collagen Q: roles in acetylcholinesterase and butyrylcholinesterase assembly and in synaptic structure and function. J Cell Biol 144:1349–1360

93. Adler M, Manley HA, Purcell AL, Deshpande SS, Hamilton TA, Kan RK et al (2004) Reduced acetylcholine receptor density, morphological remodeling, and butyrylcholinesterase activity can sustain muscle function in acetylcholinesterase knockout mice. Muscle Nerve 30:317–327

94. Mesulam MM, Guillozet A, Shaw P, Levey A, Duysen EG, Lockridge O (2002) Acetylcholinesterase knockouts establish central cholinergic pathways and can use butyrylcholinesterase to hydrolyze acetylcholine. Neuroscience 110:627–639

95. Burgess RW, Nguyen QT, Son YJ, Lichtman JW, Sanes JR (1999) Alternatively spliced isoforms of nerve- and muscle-derived agrin: their roles at the neuromuscular junction. Neuron 23:33–44

96. Burgess RW, Skarnes WC, Sanes JR (2000) Agrin isoforms with distinct amino termini: differential expression, localization, and function. J Cell Biol 151:41–52

97. Lin W, Burgess RW, Dominguez B, Pfaff SL, Sanes JR, Lee KF (2001) Distinct roles of nerve and muscle in postsynaptic differentiation of the neuromuscular synapse. Nature 410:1057–1064

98. Harvey SJ, Jarad G, Cunningham J, Rops AL, van der Vlag J, Berden JH et al (2007) Disruption of glomerular basement membrane charge through podocyte-specific mutation of agrin does not alter glomerular permselectivity. Am J Pathol 171(1):139–152

99. Gautam M, Noakes PG, Moscoso L, Rupp F, Scheller RH, Merlie JP et al (1996) Defective neuromuscular synaptogenesis in agrin-deficient mutant mice. Cell 85:525–535

100. DeChiara TM, Bowen DC, Valenzuela DM, Simmons MV, Poueymirou WT, Thomas S et al (1996) The receptor tyrosine kinase MuSK is required for neuromuscular junction formation in vivo. Cell 85:501–512

101. Weatherbee SD, Anderson KV, Niswander LA (2006) LDL-receptor-related protein 4 is crucial for formation of the neuromuscular junction. Development 133:4993–5000

102. Gautam M, Noakes PG, Mudd J, Nichol M, Chu GC, Sanes JR et al (1995) Failure of postsynaptic specialization to develop at neuromuscular junctions of rapsyn-deficient mice. Nature 377:232–236

103. Okada K, Inoue A, Okada M, Murata Y, Kakuta S, Jigami T et al (2006) The muscle protein Dok-7 is essential for neuromuscular synaptogenesis. Science 312:1802–1805

104. Noakes PG, Gautam M, Mudd J, Sanes JR, Merlie JP (1995) Aberrant differentiation of neuromuscular junctions in mice lacking s-laminin/laminin beta 2. Nature 374:258–262

105. Friese MB, Blagden CS, Burden SJ (2007) Synaptic differentiation is defective in mice lacking acetylcholine receptor beta-subunit tyrosine phosphorylation. Development 134:4167–4176

106. Missias AC, Mudd J, Cunningham JM, Steinbach JH, Merlie JP, Sanes JR (1997) Deficient development and maintenance of postsynaptic specializations in mutant mice lacking an 'adult' acetylcholine receptor subunit. Development 124:5075–5086

107. Witzemann V, Schwarz H, Koenen M, Berberich C, Villarroel A, Wernig A et al (1996) Acetylcholine receptor epsilon-subunit deletion causes muscle weakness and atrophy in juvenile and adult mice. Proc Natl Acad Sci U S A 93:13286–13291

108. Takahashi M, Kubo T, Mizoguchi A, Carlson CG, Endo K, Ohnishi K (2002) Spontaneous muscle action potentials fail to develop without fetal-type acetylcholine receptors. EMBO Rep 3:674–681

109. Koenen M, Peter C, Villarroel A, Witzemann V, Sakmann B (2005) Acetylcholine receptor channel subtype directs the innervation pattern of skeletal muscle. EMBO Rep 6:570–576

110. Liu Y, Padgett D, Takahashi M, Li H, Sayeed A, Teichert RW et al (2008) Essential roles of the acetylcholine receptor gamma-subunit in neuromuscular synaptic patterning. Development 135:1957–1967

111. Sicinski P, Geng Y, Ryder Cook AS, Barnard EA, Darlison MG, Barnard PJ (1989) The molecular basis of muscular dystrophy in the mdx mouse: a point mutation. Science 244:1578–1580

112. Chapman VM, Miller DR, Armstrong D, Caskey CT (1989) Recovery of induced

mutations for X chromosome-linked muscular dystrophy in mice. Proc Natl Acad Sci U S A 86:1292–1296

113. Danko I, Chapman V, Wolff JA (1992) The frequency of revertants in mdx mouse genetic models for Duchenne muscular dystrophy. Pediatr Res 32:128–131

114. Cox GA, Phelps SF, Chapman VM, Chamberlain JS (1993) New mdx mutation disrupts expression of muscle and nonmuscle isoforms of dystrophin. Nat Genet 4:87–93

115. Im WB, Phelps SF, Copen EH, Adams EG, Slightom JL, Chamberlain JS (1996) Differential expression of dystrophin isoforms in strains of mdx mice with different mutations. Hum Mol Genet 5:1149–1153

116. Araki E, Nakamura K, Nakao K, Kameya S, Kobayashi O, Nonaka I et al (1997) Targeted disruption of exon 52 in the mouse dystrophin gene induced muscle degeneration similar to that observed in Duchenne muscular dystrophy. Biochem Biophys Res Commun 238:492–497

117. Kudoh H, Ikeda H, Kakitani M, Ueda A, Hayasaka M, Tomizuka K et al (2005) A new model mouse for Duchenne muscular dystrophy produced by 2.4 Mb deletion of dystrophin gene using Cre-loxP recombination system. Biochem Biophys Res Commun 328:507–516

118. Grewal PK, Holzfeind PJ, Bittner RE, Hewitt JE (2001) Mutant glycosyltransferase and altered glycosylation of alpha-dystroglycan in the myodystrophy mouse. Nat Genet 28:151–154

119. Lee Y, Kameya S, Cox GA, Hsu J, Hicks W, Maddatu TP et al (2005) Ocular abnormalities in Large(myd) and Large(vls) mice, spontaneous models for muscle, eye, and brain diseases. Mol Cell Neurosci 30:160–172

120. Kuang W, Xu H, Vachon PH, Liu L, Loechel F, Wewer UM et al (1998) Merosin-deficient congenital muscular dystrophy. Partial genetic correction in two mouse models. J Clin Invest 102:844–852

121. Michelson AM, Russell ES, Harman PJ (1955) Dystrophia muscularis: a hereditary primary myopathy in the house mouse. Proc Natl Acad Sci U S A 12:1079–1084

122. Miyagoe Y, Hanaoka K, Nonaka I, Hayasaka M, Nabeshima Y, Arahata K et al (1997) Laminin alpha2 chain-null mutant mice by targeted disruption of the Lama2 gene: a new model of merosin (laminin 2)-deficient congenital muscular dystrophy. FEBS Lett 415:33–39

123. Patton BL, Wang B, Tarumi YS, Seburn KL, Burgess RW (2008) A single point mutation in the LN domain of LAMA2 causes muscular dystrophy and peripheral amyelination. J Cell Sci 121:1593–1604

124. Sunada Y, Bernier SM, Utani A, Yamada Y, Campbell KP (1995) Identification of a novel mutant transcript of laminin α2 chain gene responsible for muscular dystrophy and dysmyelination in dy2J mice. Hum Mol Genet 4:1055–1061

125. Bittner RE, Anderson LV, Burkhardt E, Bashir R, Vafiadaki E, Ivanova S et al (1999) Dysferlin deletion in SJL mice (SJL-Dysf) defines a natural model for limb girdle muscular dystrophy 2B. Nat Genet 23:141–142

126. Bansal D, Miyake K, Vogel SS, Groh S, Chen CC, Williamson R et al (2003) Defective membrane repair in dysferlin-deficient muscular dystrophy. Nature 423:168–172

127. Ho M, Post CM, Donahue LR, Lidov HG, Bronson RT, Goolsby H et al (2004) Disruption of muscle membrane and phenotype divergence in two novel mouse models of dysferlin deficiency. Hum Mol Genet 13:1999–2010

128. Grady RM, Grange RW, Lau KS, Maimone MM, Nichol MC, Stull JT et al (1999) Role for alpha-dystrobrevin in the pathogenesis of dystrophin-dependent muscular dystrophies. Nat Cell Biol 1:215–220

129. Bonaldo P, Braghetta P, Zanetti M, Piccolo S, Volpin D, Bressan GM (1998) Collagen VI deficiency induces early onset myopathy in the mouse: an animal model for Bethlem myopathy. Hum Mol Genet 7:2135–2140

130. Irwin WA, Bergamin N, Sabatelli P, Reggiani C, Megighian A, Merlini L et al (2003) Mitochondrial dysfunction and apoptosis in myopathic mice with collagen VI deficiency. Nat Genet 35:367–371

131. Eklund L, Piuhola J, Komulainen J, Sormunen R, Ongvarrasopone C, Fassler R et al (2001) Lack of type XV collagen causes a skeletal myopathy and cardiovascular defects in mice. Proc Natl Acad Sci U S A 98:1194–1199

132. Galbiati F, Engelman JA, Volonte D, Zhang XL, Minetti C, Li M et al (2001) Caveolin-3 null mice show a loss of caveolae, changes in the microdomain distribution of the dystrophin-glycoprotein complex, and t-tubule abnormalities. J Biol Chem 276:21425–21433

133. Crawford K, Flick R, Close L, Shelly D, Paul R, Bove K et al (2002) Mice lacking skeletal muscle actin show reduced muscle strength and growth deficits and die during the neonatal period. Mol Cell Biol 22:5887–5896

134. Ilkovski B, Cooper ST, Nowak K, Ryan MM, Yang N, Schnell C et al (2001) Nemaline myopathy caused by mutations in the muscle alpha-skeletal-actin gene. Am J Hum Genet 68:1333–1343

135. Garvey SM, Rajan C, Lerner AP, Frankel WN, Cox GA (2002) The muscular dystrophy with myositis (mdm) mouse mutation disrupts a skeletal muscle-specific domain of titin. Genomics 79:146–149

136. Gotthardt M, Hammer RE, Hubner N, Monti J, Witt CC, McNabb M et al (2003) Conditional expression of mutant M-line titins results in cardiomyopathy with altered sarcomere structure. J Biol Chem 278:6059–6065

137. May SR, Stewart NJ, Chang W, Peterson AS (2004) A Titin mutation defines roles for circulation in endothelial morphogenesis. Dev Biol 270:31–46

138. Brady JP, Garland DL, Green DE, Tamm ER, Giblin FJ, Wawrousek EF (2001) AlphaB-crystallin in lens development and muscle integrity: a gene knockout approach. Invest Ophthalmol Vis Sci 42:2924–2934

139. Homma S, Iwasaki M, Shelton GD, Engvall E, Reed JC, Takayama S (2006) BAG3 deficiency results in fulminant myopathy and early lethality. Am J Pathol 169:761–773

140. Selcen D, Muntoni F, Burton BK, Pegoraro E, Sewry C, Bite AV et al (2009) Mutation in BAG3 causes severe dominant childhood muscular dystrophy. Ann Neurol 65(1):83–89

141. Li Z, Colucci-Guyon E, Pincon-Raymond M, Mericskay M, Pournin S, Paulin D et al (1996) Cardiovascular lesions and skeletal myopathy in mice lacking desmin. Dev Biol 175:362–366

142. Milner DJ, Weitzer G, Tran D, Bradley A, Capetanaki Y (1996) Disruption of muscle architecture and myocardial degeneration in mice lacking desmin. J Cell Biol 134:1255–1270

143. Thornell L, Carlsson L, Li Z, Mericskay M, Paulin D (1997) Null mutation in the desmin gene gives rise to a cardiomyopathy. J Mol Cell Cardiol 29:2107–2124

144. Moza M, Mologni L, Trokovic R, Faulkner G, Partanen J, Carpen O (2007) Targeted deletion of the muscular dystrophy gene myotilin does not perturb muscle structure or function in mice. Mol Cell Biol 27:244–252

145. Garvey SM, Miller SE, Claflin DR, Faulkner JA, Hauser MA (2006) Transgenic mice expressing the myotilin T57I mutation unite the pathology associated with LGMD1A and MFM. Hum Mol Genet 15:2348–2362

146. Sher RB, Aoyama C, Huebsch KA, Ji S, Kerner J, Yang Y et al (2006) A rostrocaudal muscular dystrophy caused by a defect in choline kinase beta, the first enzyme in phosphatidylcholine biosynthesis. J Biol Chem 281:4938–4948

147. Buj-Bello A, Laugel V, Messaddeq N, Zahreddine H, Laporte J, Pellissier JF et al (2002) The lipid phosphatase myotubularin is essential for skeletal muscle maintenance but not for myogenesis in mice. Proc Natl Acad Sci U S A 99:15060–15065

148. Sullivan T, Escalante-Alcalde D, Bhatt H, Anver M, Bhat N, Nagashima K et al (1999) Loss of A-type lamin expression compromises nuclear envelope integrity leading to muscular dystrophy. J Cell Biol 147:913–920

149. Arimura T, Helbling-Leclerc A, Massart C, Varnous S, Niel F, Lacene E et al (2005) Mouse model carrying H222P-Lmna mutation develops muscular dystrophy and dilated cardiomyopathy similar to human striated muscle laminopathies. Hum Mol Genet 14:155–169

150. Diaz F, Thomas CK, Garcia S, Hernandez D, Moraes CT (2005) Mice lacking COX10 in skeletal muscle recapitulate the phenotype of progressive mitochondrial myopathies associated with cytochrome c oxidase deficiency. Hum Mol Genet 14:2737–2748

151. Mankodi A, Logigian E, Callahan L, McClain C, White R, Henderson D et al (2000) Myotonic dystrophy in transgenic mice expressing an expanded CUG repeat. Science 289:1769–1773

152. Seznec H, Agbulut O, Sergeant N, Savouret C, Ghestem A, Tabti N et al (2001) Mice transgenic for the human myotonic dystrophy region with expanded CTG repeats display muscular and brain abnormalities. Hum Mol Genet 10:2717–2726

153. Orengo JP, Chambon P, Metzger D, Mosier DR, Snipes GJ, Cooper TA (2008) Expanded CTG repeats within the DMPK 3′ UTR causes severe skeletal muscle wasting in an inducible mouse model for myotonic dystrophy. Proc Natl Acad Sci U S A 105:2646–2651

154. Reddy S, Smith DB, Rich MM, Leferovich JM, Reilly P, Davis BM et al (1996) Mice lacking the myotonic dystrophy protein kinase develop a late onset progressive myopathy. Nat Genet 13:325–335

155. Kanadia RN, Johnstone KA, Mankodi A, Lungu C, Thornton CA, Esson D et al (2003) A muscleblind knockout model for myotonic dystrophy. Science 302:1978–1980

156. Malicdan MC, Noguchi S, Nonaka I, Hayashi YK, Nishino I (2007) A Gne knockout mouse expressing human GNE D176V mutation develops features similar to distal myopathy with rimmed vacuoles or hereditary inclusion body myopathy. Hum Mol Genet 16:2669–2682

157. Tronche F, Kellendonk C, Kretz O, Gass P, Anlag K, Orban PC et al (1999) Disruption of the glucocorticoid receptor gene in the nervous system results in reduced anxiety. Nat Genet 23:99–103

158. Yang X, Arber S, William C, Li L, Tanabe Y, Jessell TM et al (2001) Patterning of muscle acetylcholine receptor gene expression in the absence of motor innervation. Neuron 30:399–410

159. Arber S, Han B, Mendelsohn M, Smith M, Jessell TM, Sockanathan S (1999) Requirement for the homeobox gene Hb9 in the consolidation of motor neuron identity. Neuron 23:659–674

160. Schwander M, Leu M, Stumm M, Dorchies OM, Ruegg UT, Schittny J et al (2003) Beta1 integrins regulate myoblast fusion and sarcomere assembly. Dev Cell 4:673–685

161. Bruning JC, Michael MD, Winnay JN, Hayashi T, Horsch D, Accili D et al (1998) A muscle-specific insulin receptor knockout exhibits features of the metabolic syndrome of NIDDM without altering glucose tolerance. Mol Cell 2:559–569

162. Tallquist MD, Weismann KE, Hellstrom M, Soriano P (2000) Early myotome specification regulates PDGFA expression and axial skeleton development. Development 127: 5059–5070

163. Feltri ML, D'Antonio M, Previtali S, Fasolini M, Messing A, Wrabetz L (1999) P0-Cre transgenic mice for inactivation of adhesion molecules in Schwann cells. Ann N Y Acad Sci 883:116–123

164. Young P, Qiu L, Wang D, Zhao S, Gross J, Feng G (2008) Single-neuron labeling with inducible Cre-mediated knockout in transgenic mice. Nat Neurosci 11:721–728

165. Feng G, Mellor RH, Bernstein M, Keller-Peck C, Nguyen QT, Wallace M et al (2000) Imaging neuronal subsets in transgenic mice expressing multiple spectral variants of GFP. Neuron 28:41–51

166. Misgeld T, Kerschensteiner M, Bareyre FM, Burgess RW, Lichtman JW (2007) Imaging axonal transport of mitochondria in vivo. Nat Methods 4(7):559–561

167. Zuo Y, Lubischer JL, Kang H, Tian L, Mikesh M, Marks A et al (2004) Fluorescent proteins expressed in mouse transgenic lines mark subsets of glia, neurons, macrophages, and dendritic cells for vital examination. J Neurosci 24:10999–11009

168. Wichterle H, Lieberam I, Porter JA, Jessell TM (2002) Directed differentiation of embryonic stem cells into motor neurons. Cell 110:385–397

169. Livet J, Weissman TA, Kang H, Draft RW, Lu J, Bennis RA et al (2007) Transgenic strategies for combinatorial expression of fluorescent proteins in the nervous system. Nature 450:56–62

170. Coleman MP, Conforti L, Buckmaster EA, Tarlton A, Ewing RM, Brown MC et al (1998) An 85-kb tandem triplication in the slow Wallerian degeneration (Wlds) mouse. Proc Natl Acad Sci U S A 95:9985–9990

Chapter 20

Bright-Field Imaging and Optical Coherence Tomography of the Mouse Posterior Eye

Mark P. Krebs, Mei Xiao, Keith Sheppard, Wanda Hicks, and Patsy M. Nishina

Abstract

Noninvasive live imaging has been used extensively for ocular phenotyping in mouse vision research. Bright-field imaging and optical coherence tomography (OCT) are two methods that are particularly useful for assessing the posterior mouse eye (fundus), including the retina, retinal pigment epithelium, and choroid, and are widely applied due to the commercial availability of sophisticated instruments and software. Here, we provide a guide to using these approaches with an emphasis on post-acquisition image processing using Fiji, a bundled version of the Java-based public domain software ImageJ. A bright-field fundus imaging protocol is described for acquisition of multi-frame videos, followed by image registration to reduce motion artifacts, averaging to reduce noise, shading correction to compensate for uneven illumination, filtering to improve image detail, and rotation to adjust orientation. An OCT imaging protocol is described for acquiring replicate volume scans, with subsequent registration and averaging to yield three-dimensional datasets that show reduced motion artifacts and enhanced detail. The Fiji algorithms used in these protocols are designed for batch processing and are freely available. The image acquisition and processing approaches described here may facilitate quantitative phenotyping of the mouse eye in drug discovery, mutagenesis screening, and the functional cataloging of mouse genes by individual laboratories and large-scale projects, such as the Knockout Mouse Phenotyping Project and International Mouse Phenotyping Consortium.

Key words Bright-field fundus imaging, Optical coherence tomography, Noninvasive ocular imaging, Mouse eye phenotyping, Mouse vision research

1 Introduction

Since it was first described in the 1990s [1–3], noninvasive imaging of the posterior mouse eye (fundus) has been a mainstay of genetic discovery and hypothesis testing in mouse vision research. This region of the eye includes the retina, retinal pigment epithelium (RPE), and choroid, in which pathological changes occur in many blinding diseases. In contrast to indirect ophthalmoscopy [1, 4], which can be used by a highly skilled observer to identify phenotypic differences rapidly, noninvasive imaging techniques

Gabriele Proetzel and Michael V. Wiles (eds.), *Mouse Models for Drug Discovery: Methods and Protocols*,
Methods in Molecular Biology, vol. 1438, DOI 10.1007/978-1-4939-3661-8_20, © Springer Science+Business Media New York 2016

provide an objective permanent record that may be scrutinized more thoroughly and is amenable to quantitative analysis. Noninvasive imaging techniques also facilitate repeated examination at different times, allowing short-term assessment of ocular properties, such as blood flow in the retinal vasculature, and long-term studies, for example longitudinal analysis of ocular development or disease progression in individual mice as they age. The most widely used methods for imaging the posterior eye in live mice parallel those of the human ophthalmology clinic: bright-field or fluorescence fundus imaging, also known as fundus photography [1, 4]; scanning laser ophthalmoscopy [5, 6]; and optical coherence tomography (OCT) [7, 8].

In this chapter, we present our approach to two of these methods, bright-field fundus imaging and OCT. Other recent articles provide alternative protocols for these and related approaches [4, 9]. The protocols, which are presented in detail below, can be summarized briefly as follows: the eyes of mice are dilated and the animals are held (bright-field fundus imaging) or anesthetized and restrained (OCT) on a support stand that can be translated or rotated in up to three dimensions (Fig. 1). The positions of the mouse and/or instrument light path are then adjusted so that the ocular region of interest is optimally oriented, illuminated, and focused. Multi-frame videos (bright-field fundus imaging) or replicate volume scans (OCT) can then be acquired. Finally, the scans are saved in or converted to *.tif format, registered, averaged, and processed digitally to enhance image detail and minimize artifacts.

In bright-field fundus imaging, the pupil is dilated and a specialized microscope is aimed into the mouse eye to provide illumination and imaging capability. In earlier work, researchers used a clinical fundus camera adapted for the mouse eye with an intervening adaptor lens [1–4], and results obtained by this approach continue to be published. However, over the past decade, new ocular imaging systems have become available that are designed for use with a contact liquid or gel between the microscope objective and the mouse corneal surface. These microscopes employ a carefully developed optical system that shines light into the eye and collects the reflected image on a color sensor. The digital signal is acquired in video mode and can be further processed by the manufacturer's software or exported for analysis by other programs that may offer additional functionalities.

The fundamental principle underlying OCT is that light passes at differing rates through the translucent compartments and tissue layers of the eye, which differ in refractive index. In spectral domain OCT, the technology available in most commercial OCT instruments, a near-infrared narrow bandwidth beam is aimed into the dilated eye (typically without an intervening contact liquid) and simultaneously split to a reference mirror. The beam reflected from

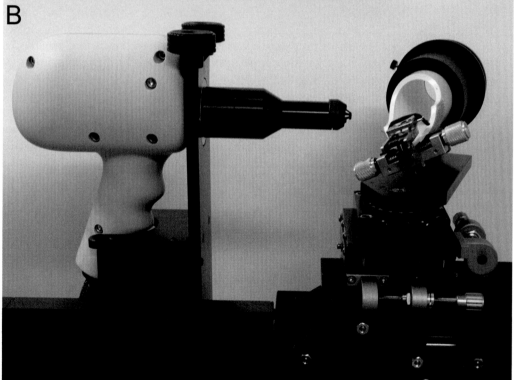

Fig. 1 Noninvasive imaging mouse platforms. (**a**) Camera and stand of a Phoenix Research Laboratories instrument. The Micron III system is shown; the camera and stand of the Micron IV system are very similar. (**b**) Probe and stand of a Bioptigen ultrahigh-resolution (UHR) Envisu R2210 spectral domain OCT (SDOCT) imaging system

the eye interferes with the reference beam, and the resulting spectral domain interference pattern is transformed mathematically into a plot of tissue reflectivity by axial distance (A-scan). The beam is swept in a radial plane to produce the equivalent of a histological cross section (B-scan), and in some instruments can also be advanced in raster fashion to create a dataset (rectangular volume) that captures the three-dimensional structure of ocular tissue. Commercial instruments that offer rapid acquisition rates over large capture areas are among the most widely used by mouse vision researchers. These systems provide software for image processing and analysis, but also allow export of the data in a format compatible for analysis with other programs.

A key aspect of our approach is the use of post-acquisition processing to improve image quality. Ideally, images would be obtained with high signal-to-noise in a single exposure that is short enough to freeze motion artifacts, which arise from eye movement, heartbeat, respiration, and mouse movement. However, as currently available commercial instruments yield relatively noisy data in a single short exposure, image quality can be improved by acquiring multiple exposures, followed by registration (image alignment) and averaging. Digital processing is used further to improve image appearance and ensure uniformity for downstream applications by adjusting orientation, compensating for uneven illumination, sharpening, and optimizing brightness and contrast. For this purpose, we have developed a plug-in and macro that make use of the public domain Java-based software ImageJ [10] (currently released as ImageJ2 [11]), which is bundled with many useful plug-ins in a distributed version of ImageJ named Fiji [11, 12]. Processed images are suitable for downstream analysis, including introduction into automated pipelines for digital survey and statistical image comparison. Our processing tools are available through links provided in this chapter.

The bright-field and OCT imaging methods described here have been used in experiments documenting new mouse lines, including a C57BL/6N-derived strain bearing a targeted correction of the $Crb1^{rd8}$ mutation [13] and several new models of human disease [14–17] identified through the Translational Vision Research Models (TVRM) program [18, 19] at The Jackson Laboratory (JAX). In addition, the bright-field fundus imaging approach has been adopted by the Knockout Mouse Phenotyping Project (KOMP²) pipeline at JAX. The methods aim toward the high-quality end of the quality-quantity continuum, which contrasts the need for data with high information content against the need for examining large numbers of subjects. These considerations are particularly important for large-scale projects that require high throughput, including mutagenesis screens, drug discovery programs, big data analysis, and gene cataloging initiatives, such as

KOMP² and the International Mouse Phenotyping Consortium (IMPC; http://www.mousephenotype.org). While the methods presented here may have suboptimal throughput for certain large-scale projects, they may be easily accelerated if lower image quality is acceptable.

2   Materials

Mice: Desired mouse strain, pigmented or albino, bred and maintained with institutional approval (for example, the ARVO Statement for the Use of Animals in Ophthalmic and Vision Research).

Anesthetic cocktail: 1.6 mL Ketamine (100 mg/mL), 1.6 mL xylazine (20 mg/mL), and 6.8 mL sodium chloride (0.9 % w/v).

Dilating agent: 1 % Atropine.

Contact gel: GenTeal Severe Dry Eye Relief, Lubricant Eye Gel, or Goniovisc Eye Lubricant.

Corneal hydration agent: Systane Ultra Lubricant Eye Drops.

Ocular imaging system: Micron III or IV (Phoenix Research laboratories), Envisu R2210 (Bioptigen), or equivalent imaging systems.

Surgical spears: Sugi, Kettenbach, USA.

Computer: Personal computer (Mac OS X or Windows based) with good graphics capability, 64 bit.

Software: Fiji can be downloaded and installed from http://fiji.sc/Fiji. Once installed, we suggest accepting all updates available except for developer tools, which are not needed for the protocols described here. Plug-ins are necessary and need to be installed to provide for registration of color images [20]. This is done by dragging the Image_Stabilizer.class and Image_Stabilizer_Log_Applier.class files onto the Fiji menu bar.

3   Methods

3.1   Bright-Field Fundus Image Acquisition and Processing

These instructions are based on the use of a Micron IV imaging system (Phoenix Research Laboratories, Pleasanton, CA), using Streampix (NorPix, Montreal, Quebec, Canada) as the video acquisition interface. The instructions also apply to the Micron III imaging system and to the use of Discover (Phoenix Research Laboratories) as the video acquisition software in newer installations, with exceptions noted.

1. Set Streampix so that video files are stored automatically as a *.tiff image stack immediately following acquisition (*see* **Note 1**). Under *Streampix Settings> Recording> When a recording ends> Auto export sequence to*, select TIFF image files. Under *…> General*, select *Save images as TIFF image files*. Under *…> Limits*, select *Limit sequence on disk to X frames* and set frames to the desired number, typically 100. Also set *Loop recording when limit is reached* to *No*. Finally, under *More…> Images*, set *TIFF format* to *Multi-TIFF* (*see* **Note 2**).

2. Adjust the Streampix software so that each video file automatically receives a timestamp and datestamp: under *Streampix Settings> Auto Naming> Auto-naming for EXPORTING to image files*, check the box labeled *Scheme* and complete the default entry in the adjacent box so that it reads (*working-folder)\(customtoken*) (*date*)_(*time*). This ensures that every image video will have a unique name, and that the date and time of each acquisition file will be associated with the file, independently of when it is moved or copied (*see* **Note 3**).

3. Set up the acquisition session by creating a new folder and browsing to select it under *Streampix Settings> Workspace(s)> Default Working Folder*. As a simple organizational aid, include the acquisition date (in *yymmdd* format) at the beginning of the folder name.

4. Create a unique sample identifier for your first sample by entering it under *Autonaming> Default value of the (customtoken)*; this will be updated manually with each additional sample, or automatically if you elect to increment the *customtoken* automatically according to *Edit (increment) token settings*. For recordkeeping, use a unique sample identifier that includes the strain, mouse number, and eye designation, such as *JR0664 1528 OS*.

5. Turn on the lamp. Wave in front of the lens to ensure that all parts of the system are functioning properly; an image of your moving hand should appear on the live display.

6. Set the color balance of the imaging sensor at first use and every few months thereafter. Hold a white card in front of the lens, move the card so that each color channel falls within the 0–255 intensity range on the imaging histogram display, and click on the automatic white balance (AWB) button in the imaging dialog box (*see* **Note 4**). The correction will be completed within a few seconds.

7. Dilate the pupils of mice to be examined. Restrain the mouse firmly in one hand and squeeze a drop of dilating agent directly onto the surface of both eyes (or one, if only one will be examined). Allow 5–10 min for dilation to occur; pupils will remain dilated for several hours when atropine is used. Whiskers

(vibrissae) may be trimmed to prevent them from interfering with imaging, but with practice they may be displaced by the camera nosepiece, so trimming becomes unnecessary (*see* **Note 5**).

8. Scruff the mouse (Fig. 2a) and place a generous drop of contact gel on the eye to be imaged (*see* **Note 6**).

9. Place the mouse on the imaging stand (Fig. 2b). To allow the camera nosepiece to approach without hitting the stand, the eye should be approximately centered on the groove perpendicular to the long axis of the stand (Fig. 2b).

10. Image the head and/or eye position to permit downstream adjustment of ocular rotation. Adjust the camera gain to a setting of 12 and turn the illumination control to its maximum value. To image the head, move the camera forward with the lateral positioner until the head including the eye and nose fill the image frame. Adjust the focus and the illumination intensity to minimize overexposed regions and acquire a snapshot. To image the external eye, move the camera forward with the midline axial knob until the eye and tear duct fill the image frame (Fig. 2c). Make sure that the pupil and the ventral edge of the nictitating membrane (third eyelid) are clearly defined and free of reflections, as these features will be used to measure ocular rotation. Adjust focus and illumination and acquire a snapshot (*see* **Note 7**).

11. Set the exposure time to a minimum. An upgrade to the Streampix software available through Phoenix Research Laboratories allows the user to adjust the exposure time in the *JAI GigE Control* panel, which shows camera settings. Typically, a value of 20 ms is used (*see* **Note 8**).

12. Reset the gain to 0 to ensure the best signal-to-noise ratio.

13. Adjust the mouse position so that the eye is centered on the pupil. Carefully move the camera forward until the retina is visible. Initially adjust the focus on the superficial vasculature (large blood vessels on the retinal surface), which are typically the easiest landmarks to detect.

14. Carefully adjust the mouse position so that the optic nerve head (ONH) is centered on the displayed image and aligned with the optical axis of the camera. The superficial vessels radiating from the ONH resemble spokes on a bicycle wheel with the ONH being the axle. Use the vertical position knob on the mouse stand to move the displayed image of the ONH up or down, coordinately adjusting the tilt knob to keep the ONH axle free of tilt. If needed, move or rotate the base of the stand to center the ONH laterally. The image center can be seen as a dim spot on the displayed image (*see* **Note 9**).

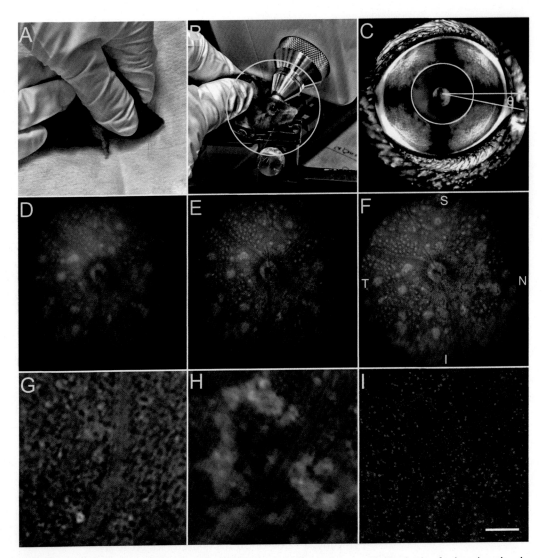

Fig. 2 Bright-field fundus imaging. (**a**) Manual restraint (scruffing) of unanesthetized mice for imaging showing position of right thumb between the mouse ear and front leg. (**b**) Positioning of mouse on the stand for imaging. In this case, the left eye is being imaged and the top of the stand has been swiveled accordingly. Contrast within the area inside the *circle* has been enhanced in Fiji to visualize the mouse. (**c**) External view of a dilated right mouse eye obtained with the Micron IV camera. The superimposed diagram shows an ellipse outlining the pupil (*circle*) and the angle (θ) between horizontal and a line drawn from the center of the ellipse to the ventral edge of the upper nictitating membrane. This angle is applied to bright-field fundus images of the same eye to achieve a uniform orientation. (**d**) Single frame from an unprocessed video acquisition of 100 frames. The right eye of a mouse identified in a chemical mutagenesis screen as part of the TVRM program was imaged. (**e**) Same image as in (**d**) after registration and averaging with *ImageStabilizeMicronStack*. (**f**) Same image as in (**d**) after adjusting orientation using *Image>Transform>Rotate* with an angle of −11.0° determined as in **step 26**, and with *Bilinear* interpolation, and by correcting for nonuniform illumination using *Polynomial Shading Corrector* with *Degree x* and *Degree y* set to 10 and *Regularization percent of peak* set to 60. The image was also adjusted with *Image>Adjust>Brightness/Contrast*. Retinal quadrants are indicated as *S* superior, *N* nasal, *I* inferior, and *T* temporal. (**g**) Detection of the RPE polygonal lattice in processed images (Micron III) of a mutant strain that exhibits retinal degeneration in the form of photoreceptor loss. (**h**) Apparent loss of pigment from RPE cells in processed images (Micron III) of the eye of a mutant strain (*tvrm267*) that shows RPE defects [15]. (**i**) Fluorescence microscopy of an RPE flat mount stained to reveal filamentous actin (rhodamine phalloidin, *red*) and nuclei (DAPI, *blue*) showing close correspondence of RPE lesions with those observed in processed bright-field fundus images. Scale bar, 100 μm

15. Readjust the focus so that the features of interest are clearly defined. For example, lesions in the RPE are typically in a different focal plane than the superficial retinal vessels.

16. Adjust the illumination so that no channel exceeds a value of ~200 on the histogram of pixel intensity. On the Micron IV, the red channel is usually the most intense.

17. Once the ONH is centered and free of tilt, features of interest are focused, and illumination is optimized, initiate a video acquisition by clicking on *Record* with the computer mouse or stepping on the footswitch, if available. Monitor the display to ensure that no abrupt movements, such as blinks, occur during the recording (if observed, simply click *Stop* to abort the recording). If the data acquisition has been pre-set properly, conversion of the video to TIFF (*.tif on Mac platforms) format should complete within a few seconds immediately after the acquisition finishes.

18. Repeat the process several times until you are satisfied that a high-quality video has been acquired.

19. Repeat with **steps 8–18** for the next eye or mouse to be imaged after updating the sample information as described in **step 4**.

20. Once all data have been collected, prepare the files for processing with *ImageStabilizeMicronStack*. Although the plug-in is designed to identify and process only *.tif stacks containing more than one slice, it may be convenient to prepare the output from the recording session. Move all video files in TIFF format to a single folder, named *Raw*. Remove *.seq files, which can only be read by Streampix, as well as any files that are not of full length, such as recordings that were stopped prematurely. Move snapshots in TIFF format to a folder named *Headshots*; these can be identified and sorted by their relatively small size (~2 MB). Create a third folder named *Processed* to hold the processed data.

21. Launch Fiji and drag the *ImageStabilizeMicronStack* macro onto the menu bar to compile it.

22. Use the *ImageStabilizeMicronStack* macro to register, average, and sharpen all of the videos collected during the session. The macro will prompt the user for input and folders (*Raw* and *Processed*, respectively) (*see* **Note 10**).

23. The *ImageStabilizeMicronStack* macro outputs images that are sharpened by applying two rounds of *Process> Filters> Unsharp Mask...* (compare Fig. 2e, d). The rationale for using a registered and averaged stack as input is that sharpening introduces noise if a single unaveraged frame is used. The justification for two rounds of *Unsharp Mask...* is that this approach appears to

reveal true ocular features, such as the RPE cells shown in Fig. 2g, and yields image details that correspond well with microscopic views of the same tissue (compare Fig. 2h, i). However, if desired, the degree of unsharp masking can be reduced simply by decreasing the *Mask Weight* parameter of the *Unsharp Mask...* command within the *ImageStabilizeMicronStack* macro. A helpful analysis of applying unsharp mask to scientific image data has been presented [21].

24. Curate the processed images to identify and retain one with the highest quality. From the *Processed* folder, select all images with the same *customname* (ignoring the time and date stamps) and drag them onto the Fiji menu bar. Use *Window > Tile* to distribute the images over the screen. Close images that show evidence of poor focus, motional blurring, or bright artifacts associated with eyelashes or bubbles. Identify the highest quality image among those that remain.

25. Delete all images and raw datasets except for those that correspond to the high-quality images selected. This approach assures that a minimal number of raw and processed images are archived (that is, reduces storage space).

26. If uniform orientations are desired, use the external eye snapshots to approximate the orientation angle. Draw a line from the center of the pupil to the ventral edge of the upper nictitating membrane as shown in Fig. 2c. The angle of displacement (θ) of this line from horizontal, which is chosen arbitrarily as an orientation reference, appears in the Fiji menu bar. Rotate the fundus image with *Image > Transform > Rotate* by $\theta°$ (right eye) or $\theta + 180°$ (left eye).

27. Image stacks are a convenient form to store, process, review, and analyze large numbers of bright-field images. Fiji may be used to create image stacks simply by dragging a folder of images onto the menu bar and selecting "create image stack" to the dialog box. Individual images within the stack can be reviewed by moving the scroll bar at the base of the image or by pressing the right or left arrow keys. Operations such as contrast adjustment or masking can be performed on the entire stack. The *Image > Stacks > Make Montage...* operation is particularly useful for creating arrays of images for review or publication (Fig. 3).

3.2 OCT Image Acquisition and Processing

These instructions are based on the use of an ultrahigh resolution (UHR) Envisu R2210 spectral domain OCT (SDOCT) imaging system (Bioptigen, Durham, NC), with InVivoVue 1.4.0.4260 software installed on a 32-bit Windows computer as supplied by the manufacturer. The instructions may be adapted to acquire and process OCT data from other instruments as long as data are

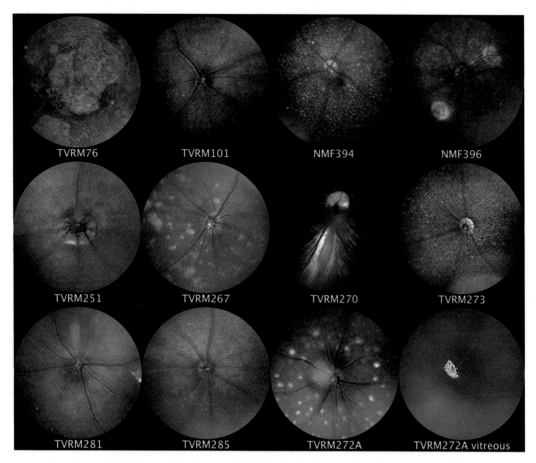

Fig. 3 Montage display of bright-field fundus images obtained with the Micron III imaging system and processed with *ImageStabilizeMicronStack* as described in this protocol, except that no correction was made for uneven image illumination or orientation. Using Fiji operations, processed images were subsequently combined in a stack, adjusted for brightness and contrast, cropped with a circular mask, and displayed in an array. The figure shows the varied phenotypes of mutant and putative mutant mice identified by indirect ophthalmoscopy screens of chemically mutagenized mice for ocular disease phenotypes as part of the TVRM program at JAX [18, 19]

available in *.tif format (required for processing). This capability has been demonstrated with data from an extra-high-resolution (XHR) Envisu R4110 SDOCT with InVivoVue 2.2.20 software on a 64-bit platform (Bioptigen), and the Micron IV imaging system fitted with an OCT accessory (Phoenix Research Laboratories).

1. Turn on the instrument and its light source, and launch the InVivoVue software to prepare for image acquisition. On first use, set up a template to record ten replicate scans for each eye in rectangular volume mode, typically with scans of 1000 A-scans per B-scan × 100 B-scans per rectangular volume × 1 frame (no averaging) over an area of 1.4–1.6 mm. Check the box to designate OD or OS; also uncheck the box labeled

"Save OCU files" to reduce the size of saved data by two-thirds. Replicate scans must be created as a "protocol." Once this protocol has been saved, it is recalled when an exam is initiated (*see* **Note 11**).

2. Enter mouse-specific information for all of the mice to be examined in a single session. The software interface is designed for medical use and therefore retains clinical terminology (patient, doctor, exam) and data-handling strategies (first name, last name, and patient ID number are required entries; patients cannot be deleted once they are associated with data). These parameters must be adapted for mouse studies. Critically, the *New Patient* dialog box requires completing three fields (first name, last name, and ID number), and the ID number is limited by the software to ten characters. For simplicity, copy a unique sample identifier that includes the strain and mouse number, such as *JR0664_528*, into each field. Designation of left or right eye is added at each step of the acquisition (*see* **Note 12**).

3. Dilate the pupils of mice to be examined. Restrain and hold the mouse firmly in one hand, and squeeze a drop of dilating agent directly onto the surface of both eyes (or one eye, if only one will be examined). Allow 5–10 min for dilation to occur; the pupils will remain dilated for at least an hour. Whiskers (vibrissae) may be trimmed to prevent them from interfering with imaging, but with practice they may be displaced by the camera nosepiece, so this is unnecessary (*see* **Note 13**).

4. In the meantime, align the nodal point of the instrument with the metal phantom provided by the manufacturer.

5. Weigh mice that have been treated with dilating agent and record their weight. Anesthetize one mouse with the ketamine/xylazine cocktail, 0.10 mL/20 g body weight. The animal will become sedated with several minutes of injection. Assess anesthetization by the absence of movement when a hind paw is gently pinched.

6. Once the mouse appears sedated, apply a drop of GenTeal Severe gel to both eyes. This will prevent the corneal surface from drying prior to image acquisition.

7. Decide which eye will be imaged and orient the mouse stand (AIM-RAS in Bioptigen nomenclature) so that this eye is closest to the probe.

8. Apply Systane Ultra drops two to three times to wash away the GenTeal Severe gel from the eye being imaged, carefully removing excess liquid by touching a tissue to the corner of the eye while avoiding the corneal surface. Leave a small amount of liquid on the ocular surface to keep it moist during alignment of the eye with the optical path.

9. Place the mouse on the stand using the bite bar and dental cotton rolls or tissue on either side of the body to secure the mouse snugly within the stand. For small mice (10 g or less), additional padding may be required under the lower jaw to position the head and eye properly. Place the Velcro strap provided by the manufacturer over the back of the mouse to hold the mouse and padding in position.

10. Coarsely align the eye with the probe optical path. Place the red aiming tip on the end of the probe; if desired, it may be kept on continuously, although it may sometimes interfere with positioning. By turning the crank on the sample arm, advance the probe head to within several mm of the corneal surface. To move the mouse, twist the white cylindrical tube of the mouse stand and rotate the stand around the vertical axis until the eye is close to the center of the red aiming tip.

11. Initiate a live scan and make fine adjustments of the retinal image. Use the bite bar adjustment knobs and reference and sample arm positions until the optic nerve head is centered and the retinal layers and RPE are relatively free of curvature, horizontal in the left-hand live display, and vertical in the right. The choroid should be closest to the top of the left-hand live display. If necessary, the horizontal and vertical translation knobs on the mouse stand may be used for adjustment, but if they are used, the nodal point will be displaced, and may require recentering before the imaging the next mouse (*see* **Note 14**).

12. Once the eye is properly aligned, use a surgical spear to wipe away remaining eye drop liquid from the ocular surface to ensure a high-quality image (a slight realignment may be required if the eye is moved during this procedure) (*see* **Note 15**).

13. Acquire ten replicate scans in rapid succession. At a setting of $1000 \times 100 \times 1$, each scan takes ~3 s to acquire, and ~17 s to save (if only *.oct data are saved). If each scan is saved immediately following acquisition, a series of ten scans will require ~3 min (*see* **Note 16**).

14. If desired, rotate the mouse in the stand, align the remaining eye with the optical path, and image as described in **steps 10–13**.

15. Remove the mouse from the stand and repeat **steps 5–14** with the next mouse.

16. At the end of the session, transfer the OCT rectangular volume scans to a single folder. The software automatically saves several file types for each recording, but only *.oct files are needed for subsequent processing. Files are saved by default in a folder named *image* with the date as the folder name (*yyyy.*

mm.dd); the *.oct files can be sorted by type and copied to another folder or moved to a server.

17. Use the OCT Volume Averager plug-in in Fiji to register and average all scans from each session. The plug-in will automatically convert files from *.oct format to *.tif format, and perform registration and averaging functions on the *.tif formatted data. The output includes an image stack similar to the original volume and an orthogonal (*en face*) image stack that has been rescaled as needed to compensate for the unequal dimensions of nonisotropic scans. The performance of OCT Volume Averager is illustrated in Fig. 4. A B-scan from a single $1000 \times 100 \times 1 \times 1.4$ mm volume dataset is shown in Fig. 4a, and rescaled *en face* slices at different depths of the image stack are shown in Fig. 4c–f. The corresponding B-scan and *en face* views generated by OCT Volume Averager are shown in Fig. 4b and g–j, respectively. The use of the plug-in improves the signal-to-noise ratio, enhances the detection of detail in both B-scan and *en face* images, and reduces the horizontal banding that arises from eye movement due to the systemic pulse (*see* **Note 17**).

18. Image stacks generated by OCT Volume Averager may be manipulated in Fiji to review lesions and reveal morphological features of the posterior eye. One useful feature is the orthogonal displays generated by applying *Image > Stacks > Orthogonal Views* to *en face* datasets, which allow the user to assess where lesions or morphological features lie with respect to retinal layers, as shown Fig. 5a, b. A second useful feature is the *Image > Stacks > Tools > Grouped Z Project…*, which can project averaged subsets from the full *en face* volume based on user input, as shown in images of a wild-type C57BL/6J mouse shown in Fig. 5c. This approach allows assessment of tissue abnormalities at a glance. For example, dysplastic lesions in the inferior retina are readily identified in a homozygous *Crb1*^{rd8} mutant (compare Fig. 5d with c).

4 Notes

1. The Discover software allows files to be directly written in TIFF format, so additional time is not needed to convert to this format, as is the case in the Streampix software.

2. Fewer than 100 frames can be acquired to decrease raw data file size, but output should be assessed to ensure that image quality is acceptable. Acquisitions of >100 frames may be useful for additional downstream analysis.

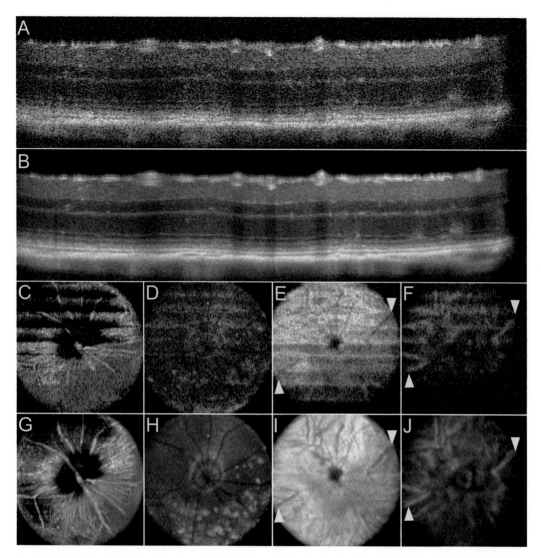

Fig. 4 OCT imaging. (**a**) B-scan from a single-volume dataset obtained from the right eye of a homozygous STOCK *Crb1^rd8* mouse (propagated and rederived from JAX Stock No. 003392). (**b**) B-scan at same position as in (**a**) obtained by registering and averaging 10 volumes with the OCT Volume Averager plug-in for Fiji. The processed image shows improved contrast and detection of detail. (**c–f**) Single *en face* images from a volume dataset at (**c**) the vitreoretinal surface, (**d**) the outer nuclear layer, (**e**) the choriocapillaris just beneath the RPE, and (**f**) the deeper choroid. Horizontal bands result from retinal movement during acquisition due to heartbeat and respiration. (**g–j**) Single *en face* images from the corresponding processed dataset. Details are enhanced and horizontal banding is substantially reduced. The large choroidal vessels near the optic nerve head in panels (**f**) and (**j**) (*arrowheads*), which are also evident as hyporeflective tracts in panels (**e**) and (**i**) (*arrowheads*), may correspond to the long posterior ciliary arteries and may be used to standardize ocular orientation

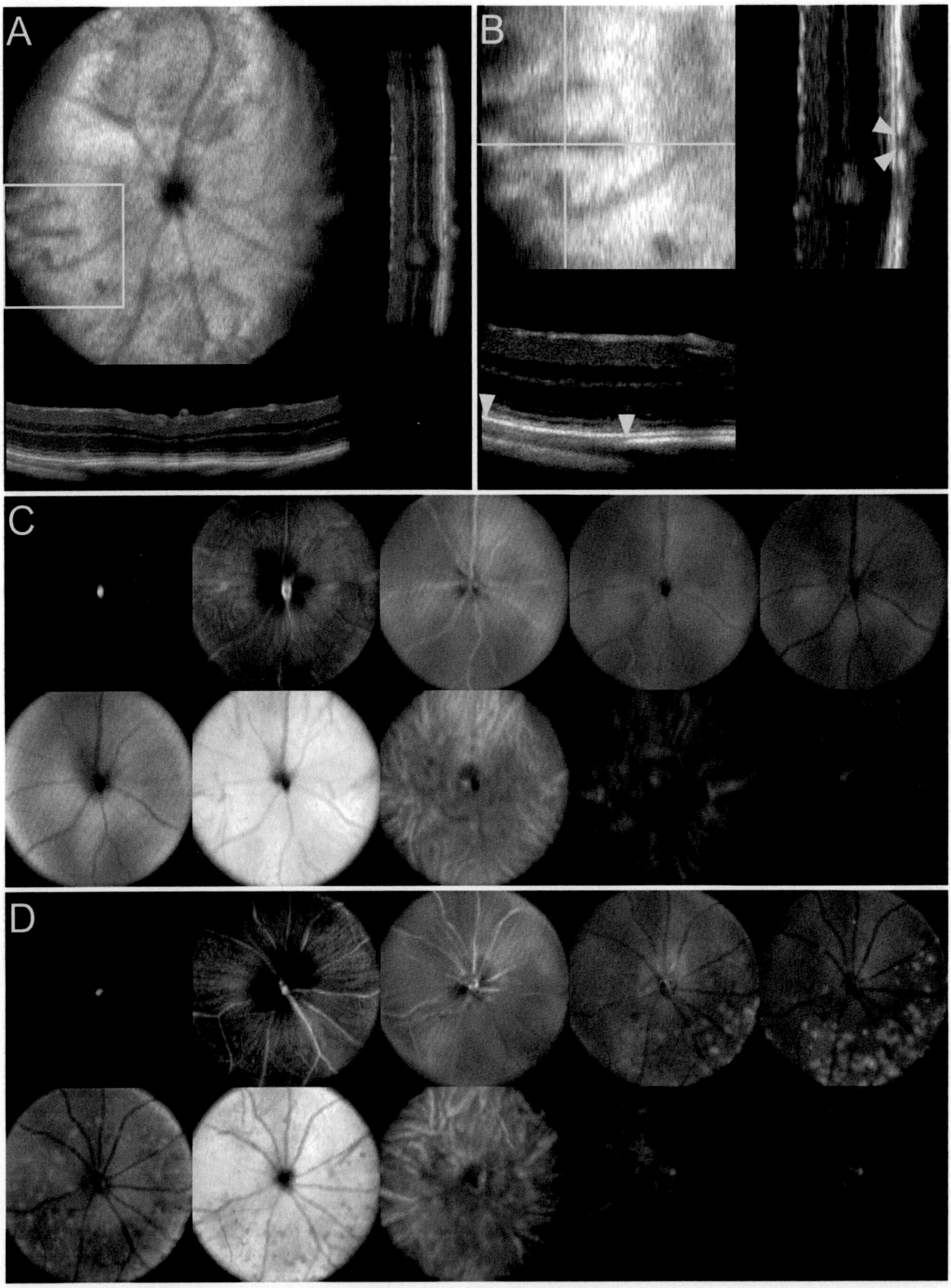

Fig. 5 Presentation of OCT imaging data in Fiji. (**a**) *En face* view of an OCT volume dataset of a mouse eye with orthogonal projections, useful for identifying the localization of lesions and vascular features with respect to retinal layers. In the image shown, the hyporeflective tracts observed at the level of the choroid in the *en face* view correspond to large-diameter tubelike structures in cross section, providing further evidence that these correspond to the long posterior ciliary arteries. (**b**) Detail of (**a**) showing the structure of the temporal long posterior ciliary artery near the optic nerve head. Note the gaps in the choroidal hyperreflective band between

3. The settings in **steps 1–4** will be retained when the program is launched subsequently, so they may be skipped after the program is first used.

4. To reduce the red appearance of the posterior eye in albino mice, the manufacturer recommends performing AWB with a pink card. This approach may also be used to manipulate the color balance of images acquired with the Micron IV camera, which has a sensor that yields images with a greater red contribution than those obtained with the Micron III instrument. A software feature that allowed users to adjust the gain of each channel independently might be the most desirable approach for manipulating color, to ensure the reproducibility of image parameters for downstream analysis. However, as of this writing, this approach is unavailable on the Micron systems.

5. Atropine may be toxic to the mouse if a large amount is accidentally ingested, so be sure to prevent drops from entering the mouth.

6. The mouse may be scruffed and mounted on the stand prior to applying GenTeal Severe gel. It is also possible to anesthetize the mouse prior to imaging, as described in **step 5** of the OCT acquisition and processing protocol. However, image contrast may be reduced in anesthetized animals due to modest opacification of ocular tissues.

7. This step is optional and should only be used if the rotational orientation of the eye is critical for downstream image analysis. Many mouse models of ocular disease, such as $Crb1^{rd8}$ mice [22], show regional bias in the appearance or distribution of lesions in the retina and/or retinal pigment epithelium (RPE). Thus, methods to ensure a uniform ocular orientation may be essential for understanding pathogenic mechanisms.

8. In the Streampix software as originally provided by Phoenix Research Laboratories for a Micron IV installation at JAX,

---

Fig. 5 (continued) the *arrows*, which presumably correspond to areas where the artery displaces pigmented choroidal cells. (**c**) Grouped Z-projection showing features of the right eye of a C57BL/6J mouse (JAX Stock No. 000664) at different levels throughout the image. To generate a stack encompassing the tissue that could be divided into groups of manageable size (30 slices), the original *en face* image stack of 392 slices was cropped to 300 slices by *Image > Duplicate* with *Duplicate Stack* checked and a *Range* of 77–376. The *Projection Method* was then set to *Average Intensity* and *Group Size* was set to 30 for the *Grouped Z Project…* operation. The resulting stack was displayed with *Image > Stacks > Make Montage…* and *Image > Adjust > Brightness/Contrast… Auto* was used to optimize image appearance. This approach can be used to rapidly identify gross abnormalities in the posterior eye. (**d**) Grouped Z-projection of the homozygous $Crb1^{rd8}$ eye shown in Fig. 4 with the same parameters as in (**c**). Note the abundant dysplastic lesions in the lower portion of the image, which corresponds to the inferior retina

exposure time was set to a default value of ~50 ms, the maximum allowed at a video acquisition rate of 20 frames per second (fps). Phoenix Research Laboratories provided greater flexibility in setting exposure times by upgrading their implementation of the Streampix software. Exposure time may be important for the highest quality images. In an unanesthetized mouse, the position of the retina may cycle 5–15 times per second due to the systemic pulse of ~300–900 beats/min, and the eye may undergo additional motion due to respiration at a rate of ~80–230 breaths/min [23]. Even under anesthesia, the heart rate is 300–450 beats/min and breathing is ~55–65 breaths/min [24]. Motional blurring and change in focus due to these physiological processes can be reduced by adjusting the exposure time to the minimum that allows sufficient image brightness. Exposure times of <20 ms are possible, but may result in dim images even if the source illumination is adjusted to its maximum. Gain may be increased to brighten the image, but this introduces noise that is undesirable for downstream processing. The Discover software does not currently allow adjustment of exposure time.

9. The Discover software includes an option to mark the image center onscreen.

10. The macro requires ~1–1.5 min to process a 100-slice TIFF file. The macro is available through the Fiji update site (http://fiji.sc/List_of_update_sites).

11. Bioptigen instruments on 32-bit Windows platforms can allocate sufficient memory to record at maximum a nonisotropic volume of 1000 A-scans per B-scan × 100 B-scans by 1 frame (abbreviated as $1000 \times 100 \times 1$; frame refers to the number of repeat B-scans at the same raster position), or an isotropic volume of around $330 \times 330 \times 1$. The nonisotropic approach allows high resolution in at least one dimension, and is therefore preferred for surveying fundus features, as in the protocol presented here. Newer Bioptigen 64-bit installations can acquire scans of $1000 \times 1000 \times 1$; however, the resulting datasets appear to be too large for the capabilities of the Fiji OCT Volume Averager plug-in. Efforts are in progress to overcome this limitation. In the meantime, it is still possible to work with $1000 \times 100 \times 1$ scans generated on these platforms, which can be exported in *.tif format by InVivoVue 2.2.20.

12. In more recent versions of the Bioptigen software, such as InVivoVue 2.2.20, more than ten characters may be used in the subject name. Nevertheless, the use of ten-character names should be retained, as the OCT Volume Averager plug-in is designed to recognize output from InVivoVue 1.4.0.4260.

13. Atropine may be toxic to the mouse if a large amount is accidentally ingested, so be sure to prevent drops from entering the mouth. If a mouse eye was treated with dilating agent earlier on the same day prior to OCT, an additional treatment may be unnecessary.

14. The procedure describes acquiring OCT data in enhanced depth imaging (EDI) mode, which emphasizes detail in the RPE and underlying choroid.

15. This is perhaps THE most important step to ensure high-quality images. Residual GenTeal Severe gel or drops can blur the OCT image considerably, so efficient removal of the liquid from the corneal surface is essential. The small amount of liquid that remains after the use of surgical spears appears to be sufficient to hydrate the eye during ten repeat rectangular volume scans.

16. Bioptigen InVivoVue 2.2.20 software on a 64-bit platform can acquire and save to disk in *.oct format the same total data volume in ~1.5 min (using a $1000 \times 100 \times 10$ scan), about half the time required by repeat scan protocol described here. Fewer keystrokes are required in the Bioptigen approach, as only one scan is initiated. Although both methods aim to produce an averaged rectangular volume with improved signal-to-noise, they differ in the acquisition schedule. In the approach presented here, the full image frame is acquired in one repeat. In the Bioptigen approach, ten repeats (frames) are obtained for each B-scan before moving to the next. Possible differences in the registered volume generated by these approaches have not yet been examined.

17. The OCT Volume Averager plug-in is available at the Fiji update site (http://fiji.sc/List_of_update_sites), and the corresponding source code at GitHub (github.com). The processing time for ten replicate rectangular volume scans ($1000 \times 100 \times 1$) using the OCT Volume Averager plug-in is ~20 min to create a *.tif image stack. By contrast, Bioptigen InVivoVue software 2.2.20 on a 64-bit platform can register and average a $1000 \times 100 \times 10$ scan and export the registered data in *.tif format in ~4.5 min. However, unlike OCT Volume Averager, the Bioptigen registration feature does not align adjacent slices of the averaged rectangular volume. In addition, the Bioptigen software cannot currently run in batch. Thus, additional hands-on time is required to load files from memory prior to registration and averaging, or to allow processing during the recording session.

References

1. Hawes NL, Smith RS, Chang B, Davisson M, Heckenlively JR, John SW (1999) Mouse fundus photography and angiography: a catalogue of normal and mutant phenotypes. Mol Vis 5:22

2. DiLoreto D, Grover DA, del Cerro C, del Cerro M (1994) A new procedure for fundus photography and fluorescein angiography in small laboratory animal eyes. Curr Eye Res 13:157–161

3. Nakamura A, Yokoyama T, Kodera S, Zhang D, Hirose S et al (1998) Ocular fundus lesions in systemic lupus erythematosus model mice. Jpn J Ophthalmol 42:345–351

4. Chang B (2013) Mouse models for studies of retinal degeneration and diseases. Methods Mol Biol 935:27–39. doi:10.1007/978-1-62703-080-9_2

5. Seeliger MW, Beck SC, Pereyra-Munoz N, Dangel S, Tsai JY, Luhmann UF, van de Pavert SA, Wijnholds J, Samardzija M, Wenzel A, Zrenner E, Narfstrom K, Fahl E, Tanimoto N, Acar N, Tonagel F (2005) In vivo confocal imaging of the retina in animal models using scanning laser ophthalmoscopy. Vision Res 45(28):3512–3519. doi:10.1016/j.visres.2005.08.014

6. Paques M, Simonutti M, Roux MJ, Picaud S, Levavasseur E, Bellman C, Sahel JA (2006) High resolution fundus imaging by confocal scanning laser ophthalmoscopy in the mouse. Vision Res 46(8–9):1336–1345. doi:10.1016/j.visres.2005.09.037

7. Srinivasan VJ, Ko TH, Wojtkowski M, Carvalho M, Clermont A, Bursell SE, Song QH, Lem J, Duker JS, Schuman JS, Fujimoto JG (2006) Noninvasive volumetric imaging and morphometry of the rodent retina with high-speed, ultrahigh-resolution optical coherence tomography. Invest Ophthalmol Vis Sci 47(12):5522–5528. doi:10.1167/iovs.06-0195

8. Fischer MD, Huber G, Beck SC, Tanimoto N, Muehlfriedel R, Fahl E, Grimm C, Wenzel A, Reme CE, van de Pavert SA, Wijnholds J, Pacal M, Bremner R, Seeliger MW (2009) Noninvasive, in vivo assessment of mouse retinal structure using optical coherence tomography. PLoS One 4(10):e7507. doi:10.1371/journal.pone.0007507

9. Alex AF, Heiduschka P, Eter N (2013) Retinal fundus imaging in mouse models of retinal diseases. Methods Mol Biol 935:41–67. doi:10.1007/978-1-62703-080-9_3

10. Schneider CA, Rasband WS, Eliceiri KW (2012) NIH Image to ImageJ: 25 years of image analysis. Nat Methods 9(7):671–675

11. Schindelin J, Rueden CT, Hiner MC, Eliceiri KW (2015) The ImageJ ecosystem: an open platform for biomedical image analysis. Mol Reprod Dev 82(7–8):518–529. doi:10.1002/mrd.22489

12. Schindelin J, Arganda-Carreras I, Frise E, Kaynig V, Longair M, Pietzsch T, Preibisch S, Rueden C, Saalfeld S, Schmid B, Tinevez JY, White DJ, Hartenstein V, Eliceiri K, Tomancak P, Cardona A (2012) Fiji: an open-source platform for biological-image analysis. Nat Methods 9(7):676–682. doi:10.1038/nmeth.2019

13. Low BE, Krebs MP, Joung JK, Tsai SQ, Nishina PM, Wiles MV (2014) Correction of the Crb1^{rd8} allele and retinal phenotype in C57BL/6N mice via TALEN-mediated homology-directed repair. Invest Ophthalmol Vis Sci 55(1):387–395. doi:10.1167/iovs.13-13278

14. Charette JR, Samuels IS, Yu M, Stone L, Hicks W, Shi LY, Krebs MP, Naggert JK, Nishina PM, Peachey NS (2016) A chemical mutagenesis screen identifies mouse models with ERG defects. Adv Exp Med Biol 854:177–183

15. Collin GB, Hubmacher D, Charette JR, Hicks WL, Stone L, Yu MZ, Naggert JK, Krebs MP, Peachey NS, Apte SS, Nishina PM (2015) Disruption of murine Adamtsl4 results in lens zonular fiber detachment and retinal pigment epithelium dedifferentiation. Hum Mol Genet 24(24):6958–6974

16. Zhao LH, Spassieva S, Gable K, Gupta SD, Shi LY, Wang JP, Bielawski J, Hicks WL, Krebs MP, Naggert J, Hannun YA, Dunn TM, Nishina PM (2015) Elevation of 20-carbon long chain bases due to a mutation in serine palmitoyltransferase small subunit b results in neurodegeneration. Proc Natl Acad Sci U S A 112(42):12962–12967. doi:10.1073/pnas.1516733112

17. Saksens NT, Krebs MP, Schoenmaker-Koller FE, Hicks W, Yu M, Shi L, Rowe L, Collin GB, Charette JR, Letteboer SJ, Neveling K, van Moorsel TW, Abu-Ltaif S, De Baere E, Walraedt S, Banfi S, Simonelli F, Cremers FP, Boon CJ, Roepman R, Leroy BP, Peachey NS, Hoyng CB, Nishina PM, den Hollander AI (2016) Mutations in CTNNA1 cause butterfly-shaped pigment dystrophy and perturbed retinal pigment epithelium integrity. Nat Genet 48(2):144–151. doi:10.1038/ng.3474

18. Won J, Shi LY, Hicks W, Wang J, Hurd R, Naggert JK, Chang B, Nishina PM (2011) Mouse model resources for vision research. J Ophthalmol 2011:391384. doi:10.1155/2011/391384

19. Won J, Shi LY, Hicks W, Wang J, Naggert JK, Nishina PM (2012) Translational vision research models program. Adv Exp Med Biol 723:391–397. doi:10.1007/978-1-4614-0631-0_50

20. Li K (2008) The image stabilizer plugin for ImageJ. http://www.cs.cmu.edu/~kangli/code/Image_Stabilizer.html

21. Sedgewick J (2012) Scientific imaging: to sharpen or obscure? http://www.americanlaboratory.com/913-Technical-Articles/121695-Scientific-Imaging-To-Sharpen-or-Obscure/

22. Mehalow AK, Kameya S, Smith RS, Hawes NL, Denegre JM, Young JA, Bechtold L, Haider NB, Tepass U, Heckenlively JR, Chang B, Naggert JK, Nishina PM (2003) *CRB1* is essential for external limiting membrane integrity and photoreceptor morphogenesis in the mammalian retina. Hum Mol Genet 12(17):2179–2189. doi:10.1093/hmg/ddg232

23. Mouse Facts. http://www.informatics.jax.org/mgihome/other/mouse_facts1.shtml

24. Ewald AJ, Werb Z, Egeblad M (2011) Monitoring of vital signs for long-term survival of mice under anesthesia. Cold Spring Harb Protoc 2011(2):pdb prot5563. doi:10.1101/pdb.prot5563

Chapter 21

Mouse Models as Tools to Identify Genetic Pathways for Retinal Degeneration, as Exemplified by Leber's Congenital Amaurosis

Bo Chang

Abstract

Leber's congenital amaurosis (LCA) is an inherited retinal degenerative disease characterized by severe loss of vision in the first year of life. In addition to early vision loss, a variety of other eye-related abnormalities including roving eye movements, deep-set eyes, and sensitivity to bright light also occur with this disease. Many animal models of LCA are available and the study them has led to a better understanding of the pathology of the disease, and has led to the development of therapeutic strategies aimed at curing or slowing down LCA. Mouse models, with their well-developed genetics and similarity to human physiology and anatomy, serve as powerful tools with which to investigate the etiology of human LCA. Such mice provide reproducible, experimental systems for elucidating pathways of normal development, function, designing strategies and testing compounds for translational research and gene-based therapies aimed at delaying the diseases progression. In this chapter, I describe tools used in the discovery and evaluation of mouse models of LCA including a Phoenix Image-Guided Optical Coherence Tomography (OCT) and a Diagnosys Espion Visual Electrophysiology System. Three mouse models are described, the *rd3* mouse model for LCA12 and LCA1, the *rd12* mouse model for LCA2, and the *rd16* mouse model for LCA10.

Key words Leber's congenital amaurosis (LCA), Mouse models, Retinal degeneration, Fundus, Electroretinography (ERG), Transgenic

1 Introduction

1.1 Leber's Congenital Amaurosis in Human

Leber's congenital amaurosis (LCA; MIM 204000), a severe form of inherited retinal dystrophy, affects 1 in approximately 30,000–80,000 people in the general population, accounting for more than 5 % of all retinopathies and 20 % of blindness in school-age children [1, 2]. It is an early-onset retinal degenerative disease with clinical features usually appearing within the first year of life and is one of the most common causes of blindness in children. The disease is characterized by nystagmus, fundus changes, severe congenital vision loss, and minimal or no detectable electroretinogram (ERG) response [3]. Leber's congenital amaurosis is genetically

Gabriele Proetzel and Michael V. Wiles (eds.), *Mouse Models for Drug Discovery: Methods and Protocols*,
Methods in Molecular Biology, vol. 1438, DOI 10.1007/978-1-4939-3661-8_21, © Springer Science+Business Media New York 2016

heterogeneous and mutations in 17 different genes have been described (Table 1), with the subtypes being distinguished by their genetic cause, patterns of vision loss, and related eye abnormalities. The LCA causative genes encode proteins with a wide variety of retinal functions in diverse cellular pathways crucial for photoreceptor cell structure, function and survival, including phototransduction, vitamin A metabolism, vesicle trafficking, protein assembly, ciliary structure and transport, photoreceptor development and morphogenesis, guanine nucleotide synthesis and outer segment (OS) phagocytosis [4, 5].

Table 1
Leber's congenital amaurosis (LCA) in human and mouse

| Type | OMIM# | Gene | Locus | | Mouse mutant | Mouse model |
			Human	Mouse		
LCA1	204000	GUCY2D	17p13.1	Chr. 7	Gucy2dtm1Sdm	
LCA2	204100	RPE65	1p31	Chr. 3	Rpe65rd12	B6(A)-Rpe65^{rd12}/J
LCA3	604232	SPATA7	14q31.3	Chr. 3	Spata7tm1(KOMP)Vlcg	
LCA4	604393	ALPL1	17p13.1	Chr. 11	Aipl1tm1Mad	
LCA5	604537	LCA5	6q11-16	Chr. 9	Lca5tm1a(EUCOMM)Wtsi	
LCA6	605446	RPGRIP1	14q11	Chr. 14	Rpgrip1nmf247	C57BL/6J-Rpgrip1^{nmf247}/J
LCA7	602225	CRX	19q13.3	Chr. 7	CrxRip, Crxtm1Clc	B6.129-Crx^{tm1Clc}/J
LCA8	604210	CRB1	1q31-32	Chr. 1	Crb1rd8	STOCK Crb1^{rd8}/J
LCA9	608553	NMNAT1	1p36	Chr. 4	Nmnat1tm1.1Lcon	
LCA10	610142	CEP290	12q21.33	Chr. 10	Cep290rd16	B6.Cg-Cep290^{rd16}/Boc
LCA11	146690	IMPDH1	7q31.3-32	Chr. 6	Impdh1tm1Bmi	
LCA12	610612	RD3	1q32.3	Chr. 1	Rd3rd3	B6.Cg-Rd3^{rd3}/Boc
LCA13	612712	RDH12	14q31.3	Chr. 12	Rdh12tm1Kpal	
LCA14	613341	LRAT	4q32.1	Chr. 3	Lrattm1Bok	
LCA15	613843	Tulp1	6p21.31	Chr. 17	Tulp1tm1Pjn	B6.129X1-Tulp1^{tm1Pjn}/Pjn
LCA16	614186	KCNJ13	2q37.1	Chr. 1	Kcnj13tm1(KOMP)Vlcg	
LCA17	615360	GDF6	8q22.1	Chr. 4	Gdf6tm1Lex	

1.2   Mouse Models of LCA

Mouse models have been discovered or generated for all of the 17 human LCA causing genes (Table 1). Mouse models offer the advantages of low cost, disease progression on a relatively rapid time scale, with the ability to perform precision genetic manipulation. The ability to target and alter a specific gene(s) is an important and necessary tool leading to mouse models with mutations in the genes of choice. Creating these model models with mutations in genes (known as knockout or transgenic) is termed "reverse genetics," in contrast "forward genetics" approaches initiate as spontaneous/induced mutations that are discovered as a result of the overt phenotypes and the underlying mutation is subsequently identified. Studies of mouse models of LCA are important for the understanding of the pathophysiology, as well as the etiology, of these diseases. Using these mouse models much progress has been made in elucidating gene defects underlying retinal disease, understanding disease mechanisms, and providing tools for translational research, and developing gene-based therapies to interfere with the progression of disease [5–9]. Three examples are the *rd3* mouse model for LCA12 and LCA1, the *rd12* mouse model for LCA2, and *rd16* mouse model for LCA10.

1.2.1   The Retinal Degeneration 3, rd3, Mouse Model Used in Studies for LCA12 and LCA1

The *rd3* is a spontaneous autosomal recessive mutation that arose in the RBF/DnJ strain (Stock No: 000726) at The Jackson Laboratory and it causes an early onset retinal degeneration. Genetic mapping placed the *rd3* mutation on mouse Chromosome 1 and its human homolog on Chromosome 1q32 [10, 11]. The *Rd3rd3* allele was backcrossed onto C57BL/6J (Stock No: 000664) from STOCK In(5)30Rk/J (Stock No: 000852), by five cycles of backcross-intercross breeding to reach an incipient congenic strain B6.Cg-*Rd3rd3*/Boc (Stock No: 008627). Using a combination of recombinational mapping and positional candidate gene approach, a C→T substitution in a novel gene, *Rd3* was identified that encodes an evolutionarily conserved protein of 195 amino acids. The substitution mutation in the *Rd3* gene is predicted to cause a stop codon after residue 106. To explore the possibility that mutations within the homolog of the *Rd3* gene caused similar retinal diseases in humans, DNA from a sister and brother from a consanguineous Indian family diagnosed with Leber's congenital amaurosis (LCA12), previously mapped to Chromosome 1q32 was sequenced, and a homozygous alteration in the invariant G nucleotide of the *RD3* exon 2 donor splice site was identified. Both affected siblings had poor vision since birth. Nystagmus and atrophic lesions in the macular area with pigment migration were found upon examination [12].

The 23 kDa RD3 protein, of unknown function, encodes by a gene associated with photoreceptor degeneration in humans with Leber's congenital amaurosis type 12 (LCA12), the rd3 mouse, and rcd2 collie, it also colocalizes and interacts with guanylate cyclases, GC1 and GC2, in rod and cone photoreceptor cells of normal mice.

In *rd3* mouse deficient in RD3, GC1 and GC2 are undetectable in photoreceptors by immunofluorescence microscopy. Therefore, RD3 appears to be important for the stable expression of guanylate cyclase in photoreceptor cells. Cell expression studies show that RD3 mediates the export of GC1 from the endoplasmic reticulum to endosomal vesicles, and that the C terminus of GC1 is required for RD3 binding. Perhaps not surprisingly, mutations leading to C terminal truncation of RD3 leads to *LCA12*, *rd3*, and *rcd2*. Further, mutations in GC1 are known to cause Leber's congenital amaurosis type 1 (LCA1) [13]. It is also possible that impaired association of retinal degeneration-3 (RD3) with guanylate cyclase-1 (GC1) and guanylate cyclase-activating protein-1 (GCAP1) can lead to Leber's congenital amaurosis type 1 [14].

To explore gene therapy as a potential treatment for *LCA12*, the adeno-associated viral vector (AAV8) with an Y733F capsid mutation, containing the mouse *Rd3* complementary DNA (cDNA) under the control of the human rhodopsin kinase promoter was delivered to photoreceptors of the *rd3* mice by subretinal injections. Optical coherence topography (OCT) and electroretinographic analyses showed that the gene therapy preserved the retinal structure in treated *rd3* mice and restored rod and cone function [7].

1.2.2 The Retinal Degeneration 12, rd12, a Mouse Model for LCA2

rd12 was discovered in a male B6.A-H2-T18a/BoyEg mouse at 10 months of age with small, discrete white dots present throughout the fundus. This male mouse was mated to a C57BL/6J female and the resulting F1 female progeny were mated back to the affected sire. The F1 mice had normal retinas, but some of the backcrossed mice showed a similar retinal phenotype as the founder with small, discrete dots in the fundus, suggesting a recessive mode of inheritance. These affected mice were intercrossed to produce the *rd12* mouse colony. Subsequently, the *rd12* stock was repeatedly backcrossed to C57BL/6J to make a congenic inbred strain on the B6 background, hereafter referred to as B6(A)-*Rpe65*^{rd12} (Stock No: 005379). The *rd12* retinal degeneration is caused by a nonsense mutation in exon 3 of the *Rpe65* gene. Functional and biochemical studies confirm that vitamin A metabolism and visual processing are disrupted in the *rd12* mouse. This naturally occurring mutation (*Rpe65*^{rd12}) provides another valuable mouse model for LCA [15]. About 10% of human LCA cases are caused by mutations in the gene encoding RPE65 [16, 17]. The *rd12* mouse displays a profoundly diminished rod electroretinogram (ERG), an absence of 11-*cis*-retinaldehyde and rhodopsin, an overaccumulation of retinyl esters in retinal pigmented epithelial (RPE) cells, and photoreceptor degeneration. *Rpe65* gene therapy was shown to restore normal vision-dependent behavior as well as retinal structure and function in the congenitally blind *rd12* mouse [18–20]. Because the human version of the *RPE65* gene was used to treat and restore vision in the *rd12* mouse, the *rd12* mouse is now routinely used to evaluate the efficacy of therapeutic interventions prior to use in human patient as an in vivo bioassay system [21, 22]. A number of gene and stem cell therapy trials addressing to correct *RPE65* are ongoing [23].

1.2.3 The Retinal Degeneration 16, rd16, a Mouse Model for LCA10

The *rd16* mouse was discovered in strain BXD-24/Ty (Stock No: 000031) at ~F140 generation of inbreeding. While all mice from this strain were found to be affected, BXD-24/Ty (Stock No: 005243) mice recovered from the embryo freezer, which were at ~F84 generation, had normal retinas, suggesting that it was a spontaneous mutation that had been fixed within the strain. Subsequently, the *rd16* mutation was backcrossed onto C57BL/6J for approximately five generations then intercrossed to make a homozygous line B6.Cg-*Cep290^{rd16}*/Boc (Stock No: 012283). The *rd16* mouse shows an early-onset retinal degeneration with autosomal recessive inheritance that maps to mouse Chromosome 10, and the homologous region in human Chromosome 12q21. Using recombinational mapping and a positional candidate strategy, an in-frame deletion in a novel centrosomal protein, CEP290 (also called NPHP6) was identified in the *rd16* mouse [24].

Homozygous *Cep290^{rd16}* mutants show a rapid retinal degeneration in the rod-rich mouse retina. In humans, mutations in the centrosomal-ciliary gene *CEP290/NPHP6* are associated with Joubert syndrome and are the most common cause of Leber's congenital amaurosis [25, 26]. The rapid progression of the disease in humans makes it difficult to pinpoint the stage at which gene augmentation therapy may be best implemented. But retinal degeneration in the *Cep290^{rd16}* mouse can be delayed by downregulating the expression of the protein Raf-1 kinase inhibitory protein (RKIP), for example by intercrossing rd16 mice with Rkip knockout mice resulting in double knockout mice, and such potentially making it amenable to additional therapeutic paradigms [27]. Targeting of such intermediates can be used in combination with other strategies, such as antisense oligonucleotide therapy for CEP290 mutations [28], to improve the outcome of the disease.

2 Materials

2.1 Mice

The mice are bred and maintained in standardized conditions in the Research Animal Facility at JAX. They are maintained on NIH31 6% fat chow and acidified water, with a 14-h light/10-h dark cycle in conventional facilities and are monitored regularly to maintain a specified pathogen-free environment. All experiments were approved by the Institutional Animal Care and Use Committee and conducted in accordance with the ARVO Statement for the Use of Animals in Ophthalmic and Vision Research. Mice from The Jackson Laboratory (JAX) (https://www.jax.org/jax-mice-and-services) are: C57BL/6J

(Stock No: 000664), B6.Cg-*Rd3rd3*/Boc (Stock No: 008627), B6(A)-*Rpe65rd12* (Stock No: 005379) and B6.Cg-*Cep290rd16*/ Boc (Stock No: 012283).

2.2 Drugs and Chemicals

1. 1% Atropine Sulfate Ophthalmic Solution (sterile), Alcon Laboratories, INC. Fort Worth, Texas 76134 USA.

2. 1% Cyclopentolate Hydrochloride Ophthalmic Solution USP (sterile), Bausch & Lomb Incorporated Tampa, FL 33637 USA.

3. Cyclomydril® (0.2% cyclopentolate hydrochloride, 1% phenylephrine hydrochloride ophthalmic solution, sterile), Alcon Laboratories, INC. Fort Worth, Texas 76134 USA.

4. 2.5% Gonioscopic Prism Solution (Hypromellose Ophthalmic Demulcent Solution, sterile), Wilson Ophthalmic, Mustang, OK 73064 USA.

5. Ketathesia (ketamine HCL injection USP, 100 mg/mL), Dublin, OH 43017 USA.

6. AnaSed® Injection (Xylazine sterile solution, 20 mg/mL), Shenandoah, Iowa 51601 USA.

7. 0.9% Sodium Chloride, INJ., USP (for use as sterile diluents), Hospira, INC., Lake Forest, IL 60045 USA.

2.3 Ophthalmic Instruments and Equipment

2.3.1 Micron III Image Guided OCT System for Rodents (Phoenix Research Labs, Pleasanton, CA 94566, USA)

The Phoenix Image-Guided OCT (optical coherence tomography) system is optimized for eye research using laboratory animals. The system allows for the visualization of the location of the OCT scan using the real-time Micron III bright-field image. A superimposed line placed directly on the image over the retinal feature being examined delivers precise cross-sectional information about the sample. Because the OCT is carried out on live animals, longitudinal studies that capture histologic detail to examine disease progression are possible. Documenting morphological changes over time using the Micron Image-Guided OCT has become an essential tool for total studying a variety of ocular diseases in the mouse.

2.3.2 Espion E³ Electro-retinography System

The Espion E³ (Diagnosys, Lowell, MA 01854 USA) electrophysiology system comprises hardware, software and stimulators to perform full-field dark-adapted and light-adapted electroretinograms (ERGs) and other electrophysiological tests. Custom protocols include single-flash rod response, single-flash cone response, flicker response, ON-OFF response, oscillatory potentials, scotopic threshold response, and photopic negative response.

3 Methods

Retinal vessel attenuation and retinal pigment epithelial disturbance are easily detected signs that are often associated with retinal disorders and retinal degeneration can be readily detected by OCT (*see* **Notes 1** and **2**). However, even if the retinal appearance (fundus) and retinal structure (OCT) are both normal, it is still possible that retinal functional abnormalities can exist. An ERG test is needed to detect any retinal functional defects, such as retinal cone photoreceptor function loss (achromatopsia) [29, 30], mouse model of LCA at younger ages [15], and no b-wave (*nob*) mutations [31, 32] (*see* **Note 1**). Heritability is subsequently established by breeding and genetic characterization [33, 34].

3.1 Mouse Fundus Examination and OCT Scan

The fundus of a mouse eye is the interior surface of the eye and includes the retina and optic disk. The color of the mouse fundus varies between pigmented (black) and albino (red). Fundus examination is a diagnostic procedure that employs the use of mydriatic eye drops (such as 1% atropine) to dilate or enlarge the pupil in order to obtain a better view of the fundus of the eye. Once the pupil is dilated, examiners can use specialized equipment such as the Micron III in vivo bright field retinal imaging microscope and the image-guided Optical Coherence Tomography (OCT) to view the inner surface of the eye. Abnormal signs that can be detected from observation of mouse fundus include hemorrhages, exudates, cotton wool spots, blood vessel abnormalities (tortuosity, pulsation, and new vessels), and changes in pigmentation.

Optical coherence tomography (OCT) is a noninvasive imaging test that uses light waves to take cross-sectional images of the mouse retina, the light-sensitive tissue lining the back of the mouse eye. With OCT, each of the distinctive layers within the retina can be seen, allowing examiners to assess changes in layer thickness and morphological alterations. These measurements can help with early detection, diagnosis, and as a guide to identify the best timing of therapeutic interventions. The results of the treatment for retinal diseases and conditions, including LCA, retinal degeneration, and other retinal disorders can also be seen and measured (*see* **Note 3**).

3.1.1 Preparation for Mouse Fundus Examination and OCT Scan

1. Pupil dilation: Remove the screw top from the vial containing the mydriatic (1% atropine). Restrain and hold the mouse firmly in one hand, pick up the vial containing the mydriatic and squeeze directly above an eye of the mouse allowing a drop to cover the surface of the eye. Repeat the procedure for the second eye. Return the mouse to its cage and allow at least 5 min for the effect of the mydriatic to take place.

2. Set up Micron III camera and OCT: Turn on the power and adjust Micron III camera to vertical position. Start Micron III camera application (App)-called Phoenix Streampix, adjust this App window size to about half of the computer screen and drag this App window to the right side of the computer screen as the fundus image panel, verify the desired camera settings (read user's manual for details). Start Micron OCT App called Phoenix

OCT and adjust this App window size to about half of the computer screen and place this App window to the left side of the computer screen as the OCT image panel, select and load the desired preferences (see user's manual for detail). Adjust OCT focus control for the best image and perform the beam alignment procedure if necessary (*see* **Note 4**). Click update background on the OCT panel before testing each mouse.

3. Set up computer file for documentation of fundus and OCT image files: Create a folder (for example: OCT files) and select this folder by clicking "File" on the OCT panel. Type in the mouse identifying information by clicking "File" on the OCT panel and selecting "Image prefix". This information must be changed/updated for each mouse tested.

4. Set up the microphone/headset for the voice control to save the fundus and OCT images: Open "Speech Recognition App" (Speech Recognition allows you to control your computer by voice, you have to set up the computer to recognize your voice for the first time—this need to be done only once) and engage the microphone to start commands (note: the mouse cursor must be in the OCT panel to operate the OCT control).

3.1.2 Examination of the Mouse Fundus and Use of the OCT Scan

1. Start mouse eye examination: Hold and restrain the mouse firmly in one hand to prevent mouse head movement and struggling (do not hold mouse too tightly!) and place the mouse eye under the light beneath the OCT lens to see on the computer monitor to be sure the pupil has fully dilated. Retract the mouse eyelids with two fingers from your other hand to hold the mouse eye fully open, orientate the field of view by visualizing the optic disk and then move the mouse eye around until you get the best fundus and OCT views displayed on the both Panels (the most focused). Save the fundus and OCT images using the Speech Recognition by clearly speaking "save" or "click save" (this step will save the fundus and OCT images in the folder selected with the file name typed in **step 3** above).

2. Check the images saved from above step: Explore to the folder selected in **step 3** above and open two image files just saved from above step to be sure the images quality is good. Repeat above **step 1** if the images saved are not good. Otherwise start the next mouse eye examination.

3. Example images from three LCA models: The retinal degeneration fundus (retinal vessels attenuation and retinal pigmentary changes) and OCT (photoreceptor cell layer thinning) can be seen on the *rd16* mouse at 3 weeks of age comparing with the *rd3*, *rd12*, and the wild type control C57BL/6J (B6) and the retinal degeneration phenotype gets worse with age in *rd16*, *rd3*, and *rd12* with age comparing with the B6 control mice at 3 and 8 months of age (Fig. 1).

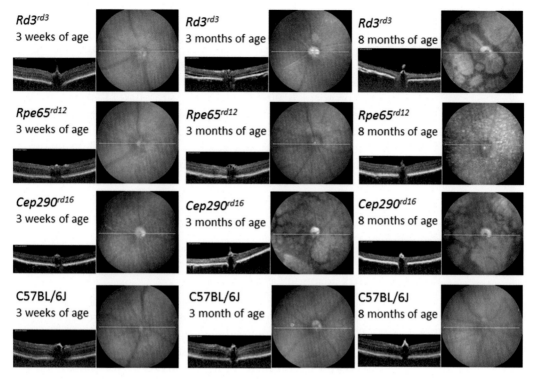

Fig. 1 Example images were taken from three LCA models and wild type C57BL/6J control mice

3.2 Electro-retinograpy

The basic method of recording the electrical response, known as the global or full-field Electroretinogram (ERG), is to stimulate the eye with a bright light source such as a flash produced by LEDs or a strobe lamp. The flash of light elicits a biphasic waveform recordable at the cornea. The two components that are most often measured are the a- and b-waves. The a-wave is the first large negative component, followed by the b-wave which is corneal positive and usually larger in amplitude. Two principal measures of the ERG waveform are taken: (1) the amplitude (a) from the baseline to the negative trough of the a-wave, and the amplitude of the b-wave measured from the trough of the a-wave to the following peak of the b-wave; and (2) the time (t) from flash onset to the trough of the a-wave and the time (t) from flash onset to the peak of the b-wave. These times, reflecting peak latency, are referred to as "implicit times" in the jargon of electroretinography. Scotopic ERGs (also called dark-adapted) are used to evaluate responses starting from rod photoreceptors exposed to flush light in darkness, and photopic ERGs (also called light-adapted) are used to evaluate responses starting from cone photoreceptors exposed to flush light under constant light exposure. For dark-adapted ERG, mice have to be placed in a dark room for at least 2 h or in a dark room overnight (this is called dark adaptation) before conducting the ERG test. For light-adapted ERG, mice need to be on a rod-saturating background (1.46 log cd/m²) for about 10 min of exposure to the background light to allow for complete light adaptation [35, 36]. With the Espion E³ electrophysiology system, it is simple to set up an automated protocol to run the dark-adapted and the light-adapted ERG on mice (*see* **Note 3**). Our ERG automatic protocol (called Rod-Cone-Auto) has 11 steps for dark-adapted ERG testing (light intensities are from low to high: 0.001, 0.0025, 0.006, 0.016, 0.04, 0.1, 0.25, 0.63, 4, 10, and 25 cd s/m²) and five steps for light-adapted ERG testing (light intensities are from high to low: 16, 8, 4, 2, and 1 cd s/m²) under constant background light (34 cd s/m²). There is a 10 min light adaptation under constant background light (34 cd s/m²) before the starting of light-adapted ERG testing.

3.2.1 Preparation for Mouse ERG Testing

1. Place the mice in the dark for at least 2 h (for screening purposes) or overnight (for characterization of the disease) for dark adaption (*see* **Note 5**).

2. Pupil dilation: Restrain and hold the mouse firmly in one hand, pick up the vial containing 1 % cyclopentolate (1 % Cyclopentolate Hydrochloride Ophthalmic Solution) and squeeze directly above the right eye of the mouse allowing a drop to cover the surface of the eye. Repeat the procedure for the second eye. Return the mouse to its cage and allow at least 5 min for the effect of the mydriatic to take place.

3. Anesthetize the mouse with an intraperitoneal injection: Weigh the mouse and record the weight. Restrain and hold the mouse firmly in one hand, pick up the second vial containing Cyclomydril (0.2% cyclopentolate hydrochloride, 1% phenylephrine hydrochloride ophthalmic solution, sterile) and squeeze directly above the right eye a drop to cover the surface of the eye, repeat the procedure for the second eye. Then inject the mouse with the anesthetic mixture solution (5 mL mixture containing 0.8 mL ketamine, 0.8 mL xylazine, and 3.4 mL 0.9% sodium chloride) at a dosage of 0.1 mL per 20 g of body weight.

4. Turn on the power for ERG unit and start the ERG App called Multifocal (Espion V6 by Diagnosys). Create a new database or select an exist database to store the ERG testing data by clicking "Database Center". Select a protocol to run (for example, Rod-Cone-Auto for dark-adapted and light-adapted ERG testing).

3.2.2 Testing Mouse ERG

1. Place the sufficiently sedated mouse on the heated bed. Insert a needle probe just under the skin at the base of the tail serving as ground, place the gold loop electrode between the gum and cheek, place the mouse head on the bar between the two active gold loop electrodes and contact the electrode on the cornea slightly below the middle of the eye, one at a time for both eyes. Pick up the third vial containing 2.5% Gonioscopic Prism

Solution and squeeze directly above the cornea and electrode a drop to assure a good contact, one at a time for both eyes.

2. Close the colordome and click Run to start testing. It takes about 20 min to run protocol "Rod-Cone-Auto" on one testing.

3. Save the test data by clicking "Exit" from ERG App menu, Type in the mouse identifying information, Press the "Save" button to save the ERG data to the database created or selected from above **step 4**.

4. Example ERGs from three LCA models: ERGs recorded from the *rd3* mice show a much reduced rod and cone response at 3 weeks of age compared to wild type B6 control mice and there is no rod and cone ERG response from the *rd12* and *rd16* mice at 3 weeks of age (Fig. 2).

3.3 Heritability Test   Heritability is established by outcrossing a retinal mutant mouse to a normal retina wild type (i.e., normal retina) mouse to generate F1 progeny, with subsequent intercrossing of the resultant F1 mice to generate F2 progeny. F1 and F2 mice are examined by fundus examination and OCT and/or ERG testing depending on which phenotype occurs first. If F1 mice are affected, the pedigree is designated as a dominant mutation. If F1 mice are not affected but ~25 % of F2 mice are affected, the pedigree is designated as a recessive mutation. Once heritability of the observed retinal phenotype is established, retinal mutants are bred and maintained for further characterization leading to gene identification [36, 37].

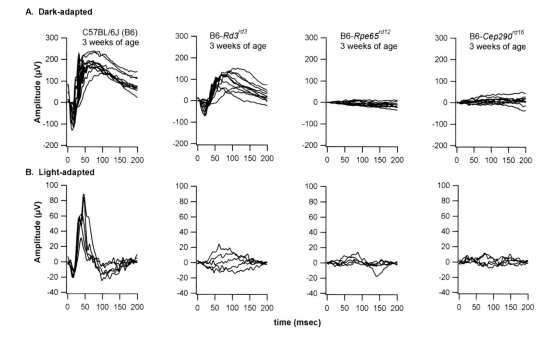

Fig. 2 Example ERGs are from three LCA models and wild type C57BL/6J control mouse

4   Notes

1. The status of the retinal health can be directly examined by non-invasive techniques such as fundus examination, optical coherence tomography (OCT), and electroretinography (ERG). These noninvasive techniques can also be used to monitor the retinal appearances (fundus), retinal structures (OCT), and retinal functions (ERGs) in real time to study the retinal disease progression or to observe the therapeutic efficacy such as mouse retinal degeneration 3 ($Rd3^{rd3}$) [7, 10], mouse retinal degeneration 12 ($Rpe65^{rd12}$) [15, 18], and mouse retinal degeneration 16 ($Cep290^{rd16}$) [24, 25].

2. It is important for investigators evaluating eyes to be aware of $Pde6brd1$ and its associated morphological findings, as it is a frequent strain background disease. Since the $Pde6b^{rd1}$ mutation is common in mice and the retinal phenotypes are very similar to some LCA models, it is important to avoid mouse strains or stocks carrying the $Pde6b^{rd1}$ allele, or to exclude the $Pde6brd1$ allele contamination in studying new LCA.

3. The Phoenix Micron III (or IV) Retinal Imaging Microscope and Phoenix Image-Guided OCT from Phoenix Research Labs (www.phoenixreslabs.com) as well as Espion E^3 Electroretinography System from Diagnosys (www.diagnosysllc.com) are designed for laboratory animals and are easy to use, but the users need to read user's manuals careful and are proficient in their Apps because a lot setups need to be done before using in the real testing.

4. The camera focus should be pre-adjusted to focus the fundus image and the OCT image at the same point. When properly set up, the focus of the two images can be adjusted by moving the mouse eye closer or further from the OCT lens.

5. The work is performed in the dark using red light.

Acknowledgements

This work has been supported by the National Eye Institute Grant EY019943 and EY011996. I am grateful to Dr. Patsy Nishina for her critical reading and editing of the manuscript.

References

1. Chung DC, Traboulsi EI (2009) Leber congenital amaurosis: clinical correlations with genotypes, gene therapy trials update, and future directions. J AAPOS 13:587–592

2. Wang H, Wang X, Zou X, Xu S, Li H, Soens ZT et al (2015) Comprehensive molecular diagnosis of a large Chinese Leber congenital amaurosis cohort. Invest Ophthalmol Vis Sci 56:3642–3655

3. Franceschetti A, Dieterle P (1954) Diagnostic and prognostic importance of the electroretinogram in tapetoretinal degeneration with reduction of the visual field and hemeralopia. Confin Neurol 14:184–186

4. den Hollander AI, Roepman R, Koenekoop RK, Cremers FP (2008) Leber congenital amaurosis: genes, proteins and disease mechanisms. Prog Retin Eye Res 27:391–419

5. Molday LL, Djajadi H, Yan P, Szczygiel L, Boye SL, Chiodo VA et al (2013) RD3 gene delivery restores guanylate cyclase localization and rescues photoreceptors in the Rd3 mouse model of Leber congenital amaurosis 12. Hum Mol Genet 22:3894–3905

6. Molday LL, Jefferies T, Molday RS (2014) Insights into the role of RD3 in guanylate cyclase trafficking, photoreceptor degeneration, and Leber congenital amaurosis. Front Mol Neurosci 7:44

7. Azadi S (2013) RD3: a challenge and a promise. JSM Biotechnol Biomed Eng 1:1016

8. Zheng Q, Ren Y, Tzekov R, Zhang Y, Chen B, Hou J et al (2012) Differential proteomics and functional research following gene therapy in a mousemodel of Leber congenital amaurosis. PLoS One 7:e44855

9. Burnight ER, Wiley LA, Drack AV, Braun TA, Anfinson KR, Kaalberg EE et al (2014) CEP290 gene transfer rescues Leber congenital amaurosis cellular phenotype. Gene Ther 21:662–672

10. Chang B, Heckenlively JR, Hawes NL, Roderick TH (1993) New mouse primary retinal degeneration (rd-3). Genomics 16:45–49

11. Danciger JS, Danciger M, Nusinowitz S, Rickabaugh T, Farber DB (1999) Genetic and physical maps of the mouse rd3 locus; exclusion of the ortholog of USH2A. Mamm Genome 10:657–661

12. Friedman JS, Chang B, Kannabiran C, Chakarova C, Singh HP, Jalali S et al (2006) Premature truncation of a novel protein, RD3, exhibiting subnuclear localization is associated with retinal degeneration. Am J Hum Genet 79:1059–1070

13. Azadi S, Molday LL, Molday RS (2010) RD3, the protein associated with Leber congenital amaurosis type 12, is required for guanylate cyclase trafficking in photoreceptor cells. Proc Natl Acad Sci U S A 107:21158–21163

14. Zulliger R, Naash MI, Rajala RV, Molday RS, Azadi S (2015) Impaired association of retinal degeneration-3 with guanylate cyclase-1 and guanylate cyclase-activating protein-1 leads to Leber congenital amaurosis-1. J Biol Chem 290:3488–3499

15. Pang J, Chang B, Hawes NL, Hurd RE, Davisson MT, Li J, Noorwez SM et al (2005) Retinal degeneration 12 (rd12): a new, spontaneously arising mouse model for human Leber congenital amaurosis (LCA). Mol Vis 11:152–162

16. Marlhens F, Bareil C, Griffoin JM, Zrenner E, Amalric P, Eliaou C et al (1997) Mutations in RPE65 cause Leber's congenital amaurosis. Nat Genet 17:139–141

17. Morimura H, Fishman GA, Grover SA, Fulton AB, Berson EL, Dryja TP (1998) Mutations in the RPE65 gene in patients with autosomal recessive retinitis pigmentosa or Leber congenital amaurosis. Proc Natl Acad Sci U S A 95:3088–3093

18. Pang J, Chang B, Kumar A, Nusinowitz S, Noorwez SM, Li J, Rani A et al (2006) Gene therapy restores vision-dependent behavior as well as retinal structure and function in a mouse model of RPE65 Leber congenital amaurosis. Mol Ther 13:565–572

19. Pang J, Boye SE, Lei B, Boye SL, Everhart D, Ryals R et al (2010) Self-complementary AAV-mediated gene therapy restores cone function and prevents cone degeneration in two models of Rpe65 deficiency. Gene Ther 17:815–826

20. Li X, Li W, Dai X, Kong F, Zeng Q, Zhou X, Lü F, Chang B et al (2011) Gene therapy rescues cone structure and function in the three-month-old rd12 mouse: a model for mid-course RPE65 Leber congenital amaurosis. Invest Ophthalmol Vis Sci 52:7–15

21. Cideciyan AV, Aleman TS, Boye SL, Schwartz SB, Kaushal S, Roman AJ et al (2008) Human gene therapy for RPE65 isomerase deficiency activates the retinoid cycle of vision but with slow rod kinetics. Proc Natl Acad Sci U S A 105:15112–15117

22. Cideciyan AV (2010) Leber congenital amaurosis due to RPE65 mutations and its treatment with gene therapy. Prog Retin Eye Res 29:398–427

23. Jacobson SG, Cideciyan AV, Ratnakaram R et al (2012) Gene therapy for Leber congenital amaurosis caused by RPE65 mutations: safety and efficacy in 15 children and adults followed up to 3 years. Arch Ophthalmol 130(1):9–24

24. Chang B, Khanna H, Hawes N, Jimeno D, He S, Lillo C, Parapuram SK et al (2006) In-frame deletion in a novel centrosomal/ciliary protein CEP290/NPHP6 perturbs its interaction with RPGR and results in early-onset retinal degeneration in the rd16 mouse. Hum Mol Genet 15:1847–1857

25. Cideciyan AV, Aleman TS, Jacobson SG, Khanna H, Sumaroka A, Aguirre GK et al (2007) Centrosomal-ciliary gene CEP290/NPHP6 mutations result in blindness with unexpected sparing of photoreceptors and visual brain: implications for therapy of Leber congenital amaurosis. Hum Mutat 28:1074–1083

26. McEwen DP, Koenekoop RK, Khanna H, Jenkins PM, Lopez I, Swaroop A, Martens JR (2007) Hypomorphic CEP290/NPHP6 mutations result in anosmia caused by the selective loss of G proteins in cilia of olfac-

tory sensory neurons. Proc Natl Acad Sci U S A 104:15917–15922

27. Subramanian B, Anand M, Khan NW, Khanna H (2014) Loss of Raf-1 kinase inhibitory protein delays early-onset severe retinal ciliopathy in Cep290rd16 mouse. Invest Ophthalmol Vis Sci 55:5788–5794

28. Collin RW, den Hollander AI, van der Velde-Visser SD, Bennicelli J, Bennett J, Cremers FP (2012) Antisense oligonucleotide (AON)-based therapy for Leber congenital amaurosis caused by a frequent mutation in CEP290. Mol Ther Nucleic Acids 1:e14

29. Chang B, Dacey MS, Hawes NL, Hitchcock PF, Milam AH, Atmaca-Sonmez P, Nusinowitz S, Heckenlively JR (2006) Cone photoreceptor function loss-3, a novel mouse model of achromatopsia due to a mutation in Gnat2. Invest Ophthalmol Vis Sci 47:5017–5021

30. Chang B, Grau T, Dangel S, Hurd R, Jurklies B, Sener EC et al (2009) A homologous genetic basis of the murine cpfl1 mutant and human achromatopsia linked to mutations in the PDE6C gene. Proc Natl Acad Sci U S A 106:19581–19586

31. Chang B, Heckenlively JR, Bayley PR, Brecha NC, Davisson MT, Hawes NL et al (2006) The nob2 mouse, a null mutation in Cacna1f: anatomical and functional abnormalities in the outer retina and their consequences on ganglion cell visual responses. Vis Neurosci 23:11–24

32. Maddox DM, Vessey KA, Yarbrough GL, Invergo BM, Cantrell DR, Inayat S et al (2008) Allelic variance between GRM6 mutants, Grm6nob3 and Grm6nob4 results in differences in retinal ganglion cell visual responses. J Physiol 586:4409–4424

33. Chang B, Hawes NL, Hurd RE, Wang J, Howell D, Davisson MT et al (2005) Mouse models of ocular diseases. Vis Neurosci 22:587–593

34. Won J, Shi LY, Hicks W, Wang J, Hurd R, Naggert JK, Chang B, Nishina PM (2011) Mouse model resources for vision research. J Ophthalmol 2011:391384

35. Hawes NL, Chang B, Hageman GS, Nusinowitz S, Nishina PM, Schneider BS et al (2000) Retinal degeneration 6(rd 6): a new mouse model for human retinitis punctata albescens. Invest Ophthalmol Vis Sci 41:3149–3157

36. Chang B, Hawes NL, Pardue MT, German AM, Hurd RE, Davisson MT et al (2007) Two mouse retinal degenerations caused by missense mutations in the "beta"-subunit of rod cGMP phosphodiesterase gene. Vis Res 47:624–633

37. Friedman JS, Chang B, Krauth DS, Lopez I, Waseem NH, Hurd RE et al (2010) Loss of lysophosphatidylcholine acyltransferase 1 leads to photoreceptor degeneration in rd11 mice. Proc Natl Acad Sci U S A 107:15523–15528

38. Acland GM, Aguirre GD, Ray J et al (2001) Gene therapy restores vision in a canine model of childhood blindness. Nat Genet 28:92–95

39. Cideciyan AV, Hauswirth WW, Aleman TS et al (2009) Human RPE65 gene therapy for Leber congenital amaurosis: persistence of early visual improvements and safety at 1 year. Hum Gene Ther 20:999–1004

40. Maguire AM, High KA, Auricchio A et al (2009) Age-dependent effects of RPE65 gene therapy for Leber's congenital amaurosis: a phase 1 dose-escalation trial. Lancet 374:1597–1605

41. Pawlyk BS, Bulgakov OV, Liu X et al (2010) Replacement gene therapy with a human RPGRIP1 sequence slows photoreceptor degeneration in a murine model of Leber congenital amaurosis. Hum Gene Ther 21:993–1004

INDEX

Gabriele Proetzel and Michael V. Wiles (eds.), *Mouse Models for Drug Discovery: Methods and Protocols*,
Methods in Molecular Biology, vol. 1438, DOI 10.1007/978-1-4939-3661-8, © Springer Science+Business Media New York 2016

Printed in the United States
By Bookmasters